American Institute of Chemical Engineers

7th Topical Conference on Natural Gas Utilization 2007

Held at the 2007 AIChE Spring National Meeting

April 22-27, 2007
Houston, Texas, USA

Printed from e-media with permission by:

Curran Associates, Inc.
57 Morehouse Lane
Red Hook, NY 12571
www.proceedings.com

ISBN: 978-1-60423-354-4

ISBN: 978-1-60423-354-4

American Institute of Chemical Engineers
Proceedings
Three Park Avenue
New York, NY 10016-5991
Phone: 212-591-8100

www.aiche.org

American Institute of Chemical Engineers

7th Topical Conference on Natural Gas Utilization
2007

TABLE OF CONTENTS

Session 5 – Dr. Harry H. West Memorial: LNG Technology

Session 30 – Coal and Gas to Liquids I

Session 31 – Dr. Harry H. West Memorial: LNG Regasification

Improving Design, Training and Start-up using a Virtual Plant

Vibhor Mehrotra
Bechtel Oil, Gas and Chemicals, Inc.
Houston, Texas

AIChE Spring Meeting, April 2007
7th Topical Conference on Natural Gas Utilization
Houston, Texas, 22-26, 2007

ABSTRACT

Bechtel has constructed grass-root facilities at various remote locations worldwide. As part of the technology initiative, Bechtel Oil, Gas and Chemicals (Bechtel) has developed a Virtual Plant Simulator (VPS) for testing design changes, improving operator training and conducting de-bottlenecking studies.

Typically the model development for the VPS is started in the Front End Design Stage where the basic design package is being established. Typical applications include driver selection and or process studies that require validating the basic design and control strategies. For example, one grass roots facility required a single fixed speed driver for two refrigeration services. The licensor and the owner/operator wanted to ensure safety and reliability. A dynamic model was developed to validate the turbine and helper motor configuration. In addition, a new control strategy was implemented to increase reliability.

Following the Front End Design, the dynamic simulation model is further developed with engineering detail. This model is then utilized to confirm anti-surge valve sizes and testing start-up strategies. One example was in evaluating start-up of a propane refrigeration system under high pressure conditions. Typical motor size recommended for the project was confirmed and additional helper motor was installed due to the high pressure observed during start-up conditions.

Once the distributed control system (DCS) of the plant is configured, the dynamic model is linked to the DCS to uncover any errors in the configuration during Factory Acceptance Test (FAT). Typical configuration errors that are not easy to uncover with traditional testing are rapidly identified during the integration phase. This produces superior DCS configuration and flawless start-up. The Dynamic Model integrated with the post FAT DCS System is now called a Virtual Plant Simulator (VPS).

The VPS is installed at the site 3-6 months prior to start-up to conduct operator training. An extensive training program is conducted by the Bechtel operations team to train all operators by simulating plant behavior such as start-up, shutdown, normal and emergency conditions. Many practice sessions are conducted to focus on key areas of the Plant.

THE ADVENT OF DYNAMIC SIMULATION

Computer simulation technology has progressed with time from the late 80's into a proven tool for performing design, control and plant operation. Steady-state process simulation tools started in the 80's and 90's are being employed by all engineering and technology licensing companies to develop new and improved design. Plant Distributed Control Systems are an important component of all modern facilities for performing plant operations. At the turn of the 21st century a transformation is happening in the process and control simulation area. *"Dynamic Simulation"* which marries the benefits of process simulation with DCS controls is coming into mainstream design and operations. As owner/operators demand larger plant capacities and/or installation of greener technologies, it becomes imperative that dynamic simulation be used to verify the design and operational viability prior to start-up. The resulting simulation tool can be transformed into a *"Virtual Plant"* by integrating the distributed controls and safety system and validating the model with plant data. Thus a model that started in the design phase has become a lifecycle cost reduction tool for EPC, Owner and Operator of the facility. The following Diagram lists the evolution of simulation models at the various stages of a project and plant lifecycle.

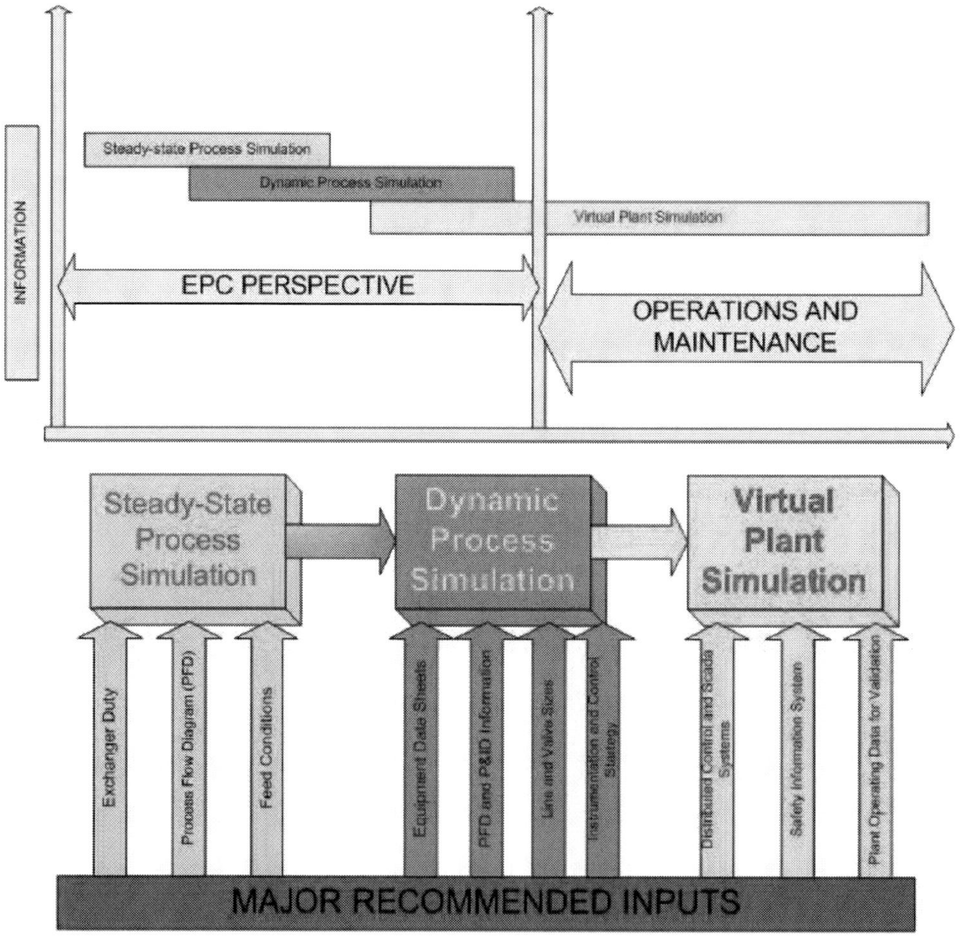

Figure 1 Evolution of Dynamic Model in Plant Lifecycle

APPLICATION IN LNG INDUSTRY

The LNG industry has experienced rapid expansion in remote areas in recent years and several new trains are being built around the world. This is coinciding with the biggest boom in the E&C industry in 20 years. This has put an unprecedented strain on resources and has resulted in shortage of skilled designers and operators for LNG Plants. The resulting shortage has raised concerns over quality, reliability and plant performance.

The size of LNG trains has been increasing over the years. During 2005-2006, ten new LNG Trains were brought on-line worldwide.

Train Name	First Cargo	Size (MTPA)	E&C Lead
Damietta 1	Jan '05	5	JGC/KBR/TR
Idku 1	May '05	4.1	Bechtel
Idku 2	Sep '05	4.1	Bechtel
RasGas 4	Dec '05	4.7	Chiyoda/MS
Oman 3	Dec '05	3.7	Chiyoda/FW
Atlantic 4	Jan '06	5.2	Bechtel
NLNG 4	Jan '06	4.2	TSKJ
Darwin 1	Feb '06	3.7	Bechtel
RasGas 5	Dec '06	4.7	Chiyoda/MS
NLNG 5	Mar '06	4.2	TSKJ

Figure 2 - New LNG Trains went on-line in 2005-06 (Avidan, 2006)

Currently there are twenty five new trains proposed by various owner/operators and are at various stages of development by the E&C Companies. These projects include *expansion* trains in: Equatorial Guinea, Egypt (Idku and Damietta), Algeria (Arzew and Skikda), Nigeria (NLNG), Qatar, Australia (Darwin), and Russia (Sakhalin). In addition, many *grass-roots* projects are being proposed in Nigeria (Brass and OK), Australia (Pluto, Gorgon), Angola, Peru, Venezuela, Russia (Shtokman, Baltic, and Yamal) Iran, and others (Avidan, 2006).

This growth rate cannot be sustained without the use of global engineering resources and automation tools that enable high productivity throughout the supply chain. The current engineering community faces some serious challenges due to the shrinking schedule of some EPC projects. Oil and Gas owners are aggressively looking at monetizing their reserves due to the high prices of oil and gas. Shaving a few months off the schedule fetches top dollars to the owner/operators. Sometimes a share of the profit is offered as incentives to the EPC contractor to improve schedule. Due to a combination of resource crunch and aggressive schedules, the EPC contractors have to innovate and improve their work process without sacrificing quality and performance. Some contractors like Bechtel are actively adopting Six Sigma methodology to reduce defects in the Engineering and Construction design and start-up. This Virtual Plant Simulation is currently being used by Bechtel throughout the LNG project lifecycle to improve design, training, start-up and operation.

REFERENCES CITED

1. "Designing for Efficient Execution of Large-Scale Natural Gas Projects in an Era of Rapid Growth", Amos Avidan, GASTECH 2006, Abu Dhabi, December 4-7, 2006

BIOGRAPHY

Vibhor Mehrotra is the Manager of the Advanced Simulation and Analysis group at Bechtel. His primary responsibility includes the deployment of CFD, Dynamic Simulation and Operator Training Simulation. Vibhor is also responsible for managing a group of 16 specialists and in hiring and training new experts/engineers in advanced simulation. Vibhor is an expert in advanced computational technologies such as computational fluid dynamics (CFD), dynamic simulation and process simulation and has applied modeling on a wide range of Bechtel projects and outside consulting projects involving petroleum plants, chemical processes, metallurgical operations, and environmental problems.

Vibhor holds a Bachelor in Science from Indian Institute of Technology and a Doctor of Philosophy in Chemical Engineering from the University of Utah. He is also a member of the American Institute of Chemical Engineers.

Vibhor Mehrotra, Ph.D.
Chief Engineer and Manager
Advance Simulation Group
Bechtel Oil, Gas and Chemicals Inc.
Houston, TX
Email: vmehrotr@bechtel.com
Ph: 713-235-4291

Non-Ideal Carnot Model provides Insights on LNG Liquefaction Processes

ChoLee Poh, Kaman (Ida) Chan, Stanley Huang
LNG Process Engineer
Chevron Energy Technology Company
Houston, Texas

AIChE Spring Meeting, April 2007
7th Topical Conference on Natural Gas Utilization
Houston, Texas, 22-26, 2007

INTRODUCTION

The provision of refrigeration compression represents a major cost item in cryogenic applications, such as an LNG plant. Therefore, a simple and reliable method for estimating the compressor duty would greatly facilitate process evaluations during initial project stages, where data of high precision are typically not required.

The global demands for natural gas have generated strong interests in LNG recently. As a result, planning activities for large-scale LNG plants have escalated. However, due to the complexities of LNG plants, each rigorous simulation of the entire process may take several hours, if not days. This is particularly true for systems using mixed refrigerants because the compositions are project-specific. As comprise between speed and accuracy, the estimation of total compression duty may be performed by scaling existing plants of similar designs. The shortcoming of this approach is the tedious corrections for key project parameters, such as heat sink (ambient air or seawater) temperatures, gas inlet pressures, and feed gas compositions.

It is recognized that a major factor contributing to the complexity of LNG plant simulations resides in the refrigeration circuits. Therefore, this work presents an approach to simplify the simulation tasks by focusing on the process streams only. The conversion from the process refrigeration requirement to mechanical compression duty is performed by a simple mechanism based on non-ideal Carnot model.

In addition to estimating the total compression duty of an LNG plant, the non-ideal Carnot model is also extended to represent crossover temperatures, or CT's, between different refrigeration circuits inside the plant. Except a few early installations, all LNG plants include two or three refrigeration circuits. The CT is determined by many factors, including process requirements, refrigerant physical properties, and constraints in mechanical provision of compressors and drivers. As a result, the determination of CT remains more of an art than science nowadays. This work will shed light on the relations between CT and distribution of compression duties between the refrigeration circuits.

TECHNICAL BACKGROUND

Refrigeration technology has been widely practiced in both domestic and industrial applications after World War II. This can be attributed to the advances in compressor machinery and cryogenic exchangers. Several refrigerants, such as ammonia and propane, were popular in the early days. Since the introduction of synthetic compounds, the Freon family dominates the domestic applications due to its superior properties in boiling temperatures and pressures, JT chilling effects, and safety. Modified formulae have been developed in recent years to mitigate the damages to the ozone layer. Although the Freon family is ideal for refrigeration applications, the relatively high costs are prohibitive for large-scale applications in industry. Consequently, the traditional HC-based refrigerants remain common up to this date.

The first attempt to estimate the compression duty of a refrigeration system can be traced back to the second law of thermodynamics [Elliot and Lira, 1998]. The famous Carnot engine defines the thermodynamic minimum work required to move a unit of energy from T2 to T1 as (T1/T2 – 1). Also, the Carnot efficiency is independent of the working medium.

On the practical side, large-scale refrigeration systems used in LNG industry can be classified into three categories: cascade, MR, and gas refrigerant. Figure 1 shows typical cooling-curves for the first two categories. The gas-loop refrigeration is a relatively new development. Figure 2 shows an example of using nitrogen as the working medium [Dunbar, 2002]. The art of estimating refrigeration compression duty received some attention before the 1980s when resources in mainframe computers were not always accessible. Since the arrival of inexpensive personal computers in the early 1980s, plant simulations have become common practices. Consequently, the interest in this art subsided. There were several reports on efficiencies of industrial refrigeration systems. They were summarized by Bogart [1982]. Some engineering design books also provide good information [GPSA, 2000].

Figure 3 plots miscellaneous simulation data for rejecting heat from different source temperatures to a heat sink of 40 °C. The simulations include both cascade and gas loops. Here, the cascade process refers to the traditional nine-stage, closed-loop process as shown in Figure 1. No attempt was made to perform open-loop cascade configurations in this work. Rigorous HYSYS simulations were performed. Unless stated otherwise, the simulation parameters are listed in Table 1.

As a general observation, the results show significant scattering resulted from different processes, process parameters, and refrigerants. The refrigerant-dependency is contrary to the Carnot assertion that the efficiency is independent of working media. Also, cascade processes are generally more efficient than gas-loops. This is anticipated because the phase changes in cascade loops eliminate the excessive pressure boosting in gas loops. Also, the results of gas-loops show wider scattering, indicating the sensitive impacts by the process parameters.

Also presented in the figure are curves corresponding to 100%, 50%, and 30% of Carnot efficiencies. As an example of calculation, to move 1 MW of heat from -60 °C

(213 K) to 40 °C (313 K), a Carnot engine requires 0.47 (= 313/213-1) MW of compression duty. In case of 50% efficiency, twice of the amount, or 0.94 MW, is required due to the reduced efficiency.

The curve by Bogart is also plotted for comparison. The curve agrees well with the 50% Carnot efficiency down to about -120 °C, then it turns up sharply. This up-turning cannot be substantiated by the simulation results.

As another point of reference, a value listed in the GPSA Engineering Data Book with a heat sink temperature around 40 C is also indicated [2000, Table 14-10]. This point is consistent with the prediction by 50% Carnot efficiency.

The curve with 50% Carnot efficiency defines the minimum compression duty reasonably well. Further fine-tuning of this number is possible. In the following discussion, the reference of this 50% efficiency is nominal. The fine-tuned value of the non-ideal Carnot efficiency will not change significantly the conclusions of this presentation.

SCOPE AND APPROACHES

This work intends to develop a simple and universal methodology for estimating the compressor duty for any specified LNG application. It is recognized that the compression duty is strongly influenced by process parameters, such as the temperature approaches used in exchangers, condenser designs, and the compressor efficiency. This recognition is a direct contradiction to the stated intention of keeping the methodology simple and universal. The intention can only be achieved by avoiding detailed process information. Hence, the following implicit assumptions must be made. The verification of these assumptions will be through the comparison of estimations and actual plant data.

- All practical process designs should maintain the temperature approaches in exchangers within a reasonable range, which is commonly accepted by most players in the field.
- Likewise, the compressor efficiencies should be consistent among players in this field. This appears plausible in current industry because all manufacturers are striving to achieve the best efficiencies of available technology.

Furthermore, the measure described below is adopted in the establishment of the simple thermodynamic models.

The minimum temperature approaches of all vaporizers, air (or water) coolers, and air (or water) condensers are neglected. There is to simplify the procedure and eliminate any ambiguities in references to temperatures of heat sinks or sources. The impact of neglecting finite temperature approaches will be absorbed in the model parameters.

MODEL DEVELOPMENT

The development of LNG plant models is divided into two levels, according to the complexities involved.

Level 1 Model: total compression duty

The Level 1 model focuses on estimating the total compression duty related to process refrigeration after essential project parameters are fixed. The essential parameters include natural gas composition, flow rate, heat sink temperature, and approximate product heating values. The heating values could be used to determine the separation level of intermediate components.

Figure 4 shows a typical cooling curve of an LNG plant. The leveling-off in the middle section is caused by the release of latent heat during liquefaction of the feed gas. The entire curve is divided into a finite number of segments as shown. The cooling duty of each segment is determined by HYSYS simulations. The cooling duty and its corresponding temperature are then fed to the Eq. (1) to calculate the required compression duty of the segment:

$$W_i = F(Q_i, T_i, non-ideal\ Carnot\ Efficiency) \quad (1)$$

Here the function F represents the non-ideal Carnot efficiency with nominal 50% of efficiency. The summation of all segment heat duties is the total heat in the process stream to be rejected to the heat sink. The summation of all the segment works is the total refrigeration work required. Eqs. (2) and (3) show the relations. The combined Q_{tot} + W_{tot} is the total plant heat rejected to the heat sink. There is no cross-segment heat transfer in this model.

$$W_{tot} = \sum W_i \quad (2)$$

$$Q_{tot} = \sum Q_i \quad (3)$$

The Level 1 model significantly simplifies the simulation work. As shown in Figure 5, a typical simulation for cascade process can be reduced to finite steps of simple chilling. The design of the removal of intermediate components is optional. Experiences indicate that it is non-critical in the initial phase of a project.

Level 2 Model: distribution of compression duties between refrigeration circuits

Except a few installations in the early years, all LNG plants include two or three refrigeration circuits. The most popular process currently in operation is the C3MR process by Air Products. The process uses a propane circuit to provide the first level of chilling. Similar propane chilling is used in the Optimized Cascade process by ConocoPhillips. Recently, there have been developments using dedicated MR circuits in the first stage chilling.

Figure 6 shows a typical cooling curve and generalized provision of refrigeration circuits in an LNG plant. Based on its shape, the curve can be divided into three stages: pre-chilling, liquefaction, and sub-cooling. During the initial chilling, the feed remains mostly in a gaseous phase. The relatively steep cooling curve is due to the small heat capacity of the gas. In the liquefaction stage, the latent heat is released as the gas undergoes a phase change. The slope of the cooling curve somewhat levels off. After the feed is totally liquefied, the cooling curve slope becomes sharp again because further cooling (subcooling) removes the sensible heat of the natural gas liquid.

The physical provision of refrigeration circuits may not coincide with the phase boundaries. In most existing plants, the process provides two or three refrigerant loops; one covers the pre-chilling and the other(s) take(s) care of liquefaction and sub-cooling stages. The crossover temperatures among the refrigeration loops are marked as CT1 and CT2 in Figure 6.

In the Level 2 Model, the focus is on the CT1. All other lower-temperature loops are lumped into a single pseudo refrigeration loop. For the sake of generalization, the high-temperature loop will be designated as MR1 and the low-temperature loop, MR2. It is understood that MR1 and MR2 could be single-component refrigerant systems. Since there is only one crossover temperature in Level 2 model, there is no distinction between CT and CT1 in the following discussions. The impact of CT on the split of compression duties between the two loops will be illustrated.

Level 1 model shown in Figure 4 is a simplified picture that each segment rejects its own heat to the heat sink. Figure 7 shows a sketch of Level 2 model, which is a refinement of Level 1. Specifically, the heat in MR2 range will be partially transferred to MR1 for final rejection to the heat sink. As a result, the MR1 compressor duty would increase to shoulder the additional load.

Figure 7 also shows symbols used in the development of the model. Definitions and explanations are also listed in Section 8: Symbols and Abbreviations. For segments above CT1, Q_1 is the heat from process stream to be rejected to T_{sink}. W_1 is the total work required to reject Q_1 to T_{sink}.

$$Q_1 = \sum q_i, \quad T \geq CT \tag{4}$$

$$W_1 = \sum w_i, \quad T \geq CT \tag{5}$$

Similarly, Q_2 is the heat from process stream to be rejected to T_{sink}. W_2 is the total work required to reject Q_2 to T_{sink}.

$$Q_2 = \sum q_i, \quad T < CT \tag{6}$$

$$W_2 = \sum w_i, \quad T < CT,\ final\ T = T\text{sink} \tag{7}$$

The rejection of Q_2 follows the steps as described below. First, MR2 brings the heat to CT. Then Q_2 is split into Q_{21} and Q_{22}. Q_{21} is transferred to MR1 loop where it is ultimately rejected to T_{sink}. Q_{22} stays within MR2 loop and is ultimately rejected to T_{sink}. Physically, Q_{22} represents the heat being rejected to the environment through the MR2 compressor discharge cooler.

In the following, W_{20} is defined as the work required to bring Q_2 to CT, and W_{21} and W_{22} are defined as the works required to handle Q_{21} and Q_{22} per their definitions, respectively. The following equations can be written:

$$W_{20} = \sum w_i, \quad T < CT, \; final\ T = CT \tag{8}$$

$$W_2 = W_{20} + W_{21} + W_{22} \tag{9}$$

$$W_{tot} = W_1 + W_2 = W_1 + (W_{20} + W_{21} + W_{22}) \tag{10}$$

Since Level 1 and Level 2 models are addressing the same physical phenomenon, the total compression duty should remain unchanged. Eq. (11) rearranges the relation to show the shift of compression duties from MR2 loop to MR1 compressor:

$$W_{tot} = W_1 + W_2 = (W_1 + W_{21}) + (W_{20} + W_{22}) = W_{MR1} + W_{MR2} \tag{11}$$

In summary, there are four unknowns: Q_{21}, Q_{22}, W_{21} and W_{22}. Correspondingly, there are four constraints to solve them uniquely:

$$Q_{21} + Q_{22} = Q_2 \tag{6}$$

$$W_{21} + W_{22} = W_{tot} - (W_1 + W_{20}) \tag{12}$$

$$W_i = F(Q_i, T_i, non-ideal\ Carnot\ Efficiency) \tag{1}$$

$$\frac{Q_{21}}{(Q_{21} + Q_{22})} \propto \frac{W_{21}}{(W_{21} + W_{22})} \tag{13}$$

The last equation (13) is the only additional assumption made in the Level 2 Model. A spreadsheet has been developed to solve the above set of simultaneous equations. The results are presented in the next section.

RESULTS

There are about 20 LNG production complexes in the world; the total number of production trains ranges around 80. These LNG trains are located in different climate regions, use varieties of processes, and have different utility provisions. Collectively, they form an excellent database for testing the models described above. However, it is not a general industrial practice to publish their plant design and operation data.

Nonetheless, some data are scattered in the open domain. A couple of examples are listed in the references [Kikkawa and Nakamura, 1998; Suprapto, 2001]. The results presented below are mainly based on collections from various sources. The purpose is to verify the models, not to compare different process designs and efficiencies.

Figure 8 shows the comparison between Level 1 model predictions and rigorously simulated duties for various LNG trains. Individual data points are blurred into overall contours to hide their identities. Two observations can be made: (1) the data fall within a narrow region along the diagonal regardless of the process selected, and (2) there is no systematic bias by using the nominal 50% efficiency in the Level 1 model. There are individual cases which show unusually large deviations. These cases are further investigated for verification. In this sense, the Level 1 model serves as a useful tool for screening.

Although not presented, it is worth mentioning that the micro structures of different processes within the data contour can be used for further improving the predictive accuracy of the Level 1 model. For example, suppose data of Process A indicate a consistent over-prediction by the model, a separate correlation can be established specifically for this Process A. Generally, a prediction accuracy of 1 or 2% can be achieved.

Figure 9 shows the predictions of Level 2 Model. The ordinate is the split ratio of compression duties between MR1 and MR2 circuits. The abscissa is the CT between MR1 and MR2 loops. Note that MR1 and MR2 may represent pure-component refrigerant. As the CT decreases, more duty is shifted to MR1. Consequently, the ratio should rise. Also, MR1 shoulders the major impacts of ambient temperature variations. As T_{sink} increases, the load on MR1 is expected to increase. The model predictions are in agreement with these expected trends.

Similar to Figure 8, the actual data in Figure 9 are grouped into two contours based on heat sink temperatures. The data are consistent with model predictions. Of special interest is the CT temperature range. The majority of existing LNG plants use propane in the MR1 loop. As a result, the CT ranges from -30 to -40 °C, limited by the propane boiling temperature at the ambient pressure. Accordingly, the ratio of compression duties ranges between 0.3 and 0.5. Since it is more economical to provide two identical main drivers in an LNG plant, various methods have been resorted to equalize the loads between the two drivers. For example, the compressor casings are arranged on two drivers in a way that the total load on one drive is equal to the other.

A true MR loop could break the aforementioned CT limitation and push the ratio of compression duties toward unity. As shown Figure 9, for an even split of the compressor duty at a heat sink temperature of 26 °C, the CT should be about -60 °C. A data point is located in a distinct range, between -40 and -70 °C [Paradowski and Rojey, 2000], away from the bulk data. The model correctly predicts the trend.

CONCLUSIONS

The conclusions of this work are listed as follows:

- For refrigeration systems, the minimum compression duty required to move a specified heat from low to high temperatures can be estimated by using a nominal 50% Carnot efficiency.
- The Level 1 model can be used to estimate the total refrigeration compression of an LNG plant. The process cooling curve is divided into finite segments and the heat content in each segment is independently rejected to the heat sink. The compression duty can be estimated by the aforementioned method.
- The Level 2 model can be used to estimate the split of compression duty between MR1 and MR2 refrigeration circuits when the CT between them is given. The model partially transfers the total heat content below CT to MR1 circuit for final rejection to the heat sink.
- From process domain, it is possible to split evenly the total compression duty between MR1 and MR2 circuits. The Level 2 model correctly predicts the CT of an industrial case. The CT resides in a low temperature region between -40 and -70 °C.

REFERENCES

- Bogart, M. J. P., "Ammonia Absorption Refrigeration in Cryogenic Services", AIChE Symposium Series, Cryogenic Processes and Equipment, No.224, Vol 79, pp. 57-64 (1982).
- Dunbar, C. A. T., "Liquefaction Process and Apparatus", U. S. Patent 6,446,465 (2002).
- Elliott, J.R., Lira, C.T.; "Introductory Chemical Engineering Thermodynamics", Prentice Hall, NJ (1998).
- GPSA, "Engineering Data Book", 11th Edition, Gas Processors Suppliers Association, Tulsa OK (2000).
- Kikkawa, Y., Nakamura, M., "New Liquefaction Process of Natural Gas", paper presented at AIChE Spring Meeting, New Orleans (1998).
- Paradowski, H., Rojey, A., "Method and Device for Liquefying a Natural Gas without Phase Separation of the Coolant Mixtures", U. S. Patent 6,105,389 (2000).
- Suprapto, Y. P., "Concept and Techniques for Grassroots LNG Plant cost Optimization", paper presented at LNG 13, Seoul, Korea (2001).

SYMBOLS & ABBREVIATION

AD	Average Deviation
AAD	Average Absolute Deviation
CT	Crossover Temperature between refrigeration circuits within a plant. In this work, CT = CT1 since only one CT is discussed.
CT1, CT2	Refer to Figure 6, crossover temperatures of different levels
HC	Hydrocarbon
JT	Joule-Thompson
MR	Mixed Refrigerant

	Refer to Figure 7 for symbols below:
MR1	Mixed refrigerant circuit for highest-temperature services
MR2	Mixed refrigerant circuit for lower temperature services
Q_{tot}	= $Q_1 + Q_2$; total heat in process stream to be removed;
Q_1	Total heat in process stream (> CT1) to be removed
Q_2	= $Q_{21} + Q_{22}$; total heat in process stream (< CT1) to be removed;
Q_{21}	The portion of Q_2 to be rejected from CT1 to T_{sink} through MR1 loop
Q_{22}	The portion of Q_2 to be rejected from CT1 to T_{sink} through MR2 loop
T_{sink}	Heat sink Temperature
W_{tot}	= $W_{MR1} + W_{MR2} = W_1 + W_2$; Total work required to reject Q_{tot} to T_{sink}
W_1	Work required to reject heat in process stream (> CT1) to T_{sink}
W_2	= $W_{20} + W_{21} + W_{22}$; Total work required to reject Q_2 to T_{sink}
W_{20}	Work required to reject heat in process stream (< CT1) to CT1
W_{21}	Work required to reject Q_{21} from CT1 to T_{sink}, performed by MR1 loop
W_{22}	Work required to reject Q_{22} from CT1 to T_{sink}, performed by MR2 loop
W_{MR1}	= $W_1 + W_{21}$; Work corresponding to MR1 circuit;
W_{MR2}	= $W_{20} + W_{22}$; Work corresponding to MR2 circuit;

LIST OF TABLES

Table 1 Simulation parameters used in this work

Natural Gas inlet	
Inlet temperature	40 °C
Inlet pressure	45 bar
Components	Mole Fraction
Nitrogen	0.0012
CO_2	0.0000
Methane	0.8605
Ethane	0.0640
Propane	0.0425
i-Butane	0.0181
n-Butane	0.0117
Pentane plus	0.0020
TOTAL	1.0000
HHV, MJ/Sm^3	40.3
Btu/SCF	1,082
Heat sink temperature	40 °C
Pressure drop through plate-fin exchanger	0.2-0.3 bar
Pressure drop through air coolers	0.6 bar
Minimum delta T in plate-fin exchanger	2.5 ± 0.2 °C
Compressor/Expander polytropic efficiency	80%

LIST OF FIGURES

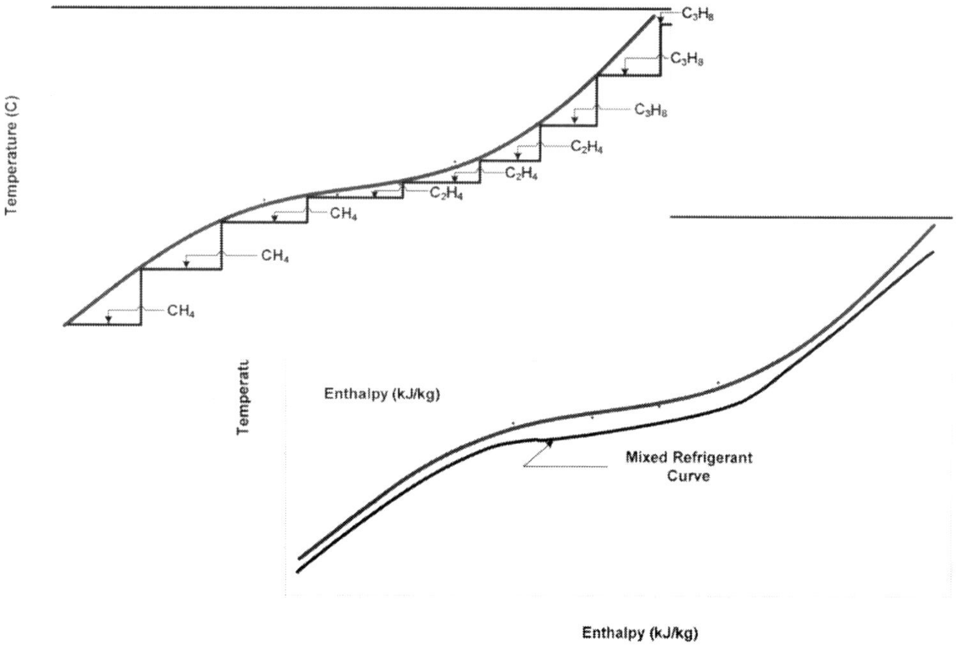

Figure 1 Typical cooling curves for cascade (top) and MR systems

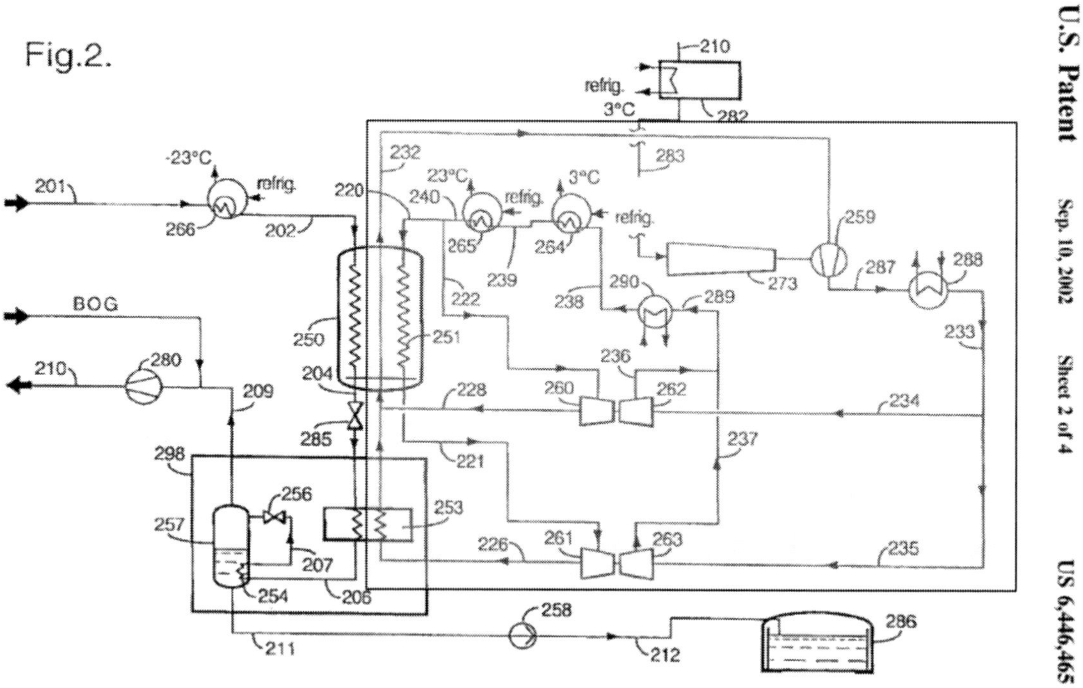

Figure 2 Example of gas-loop refrigeration

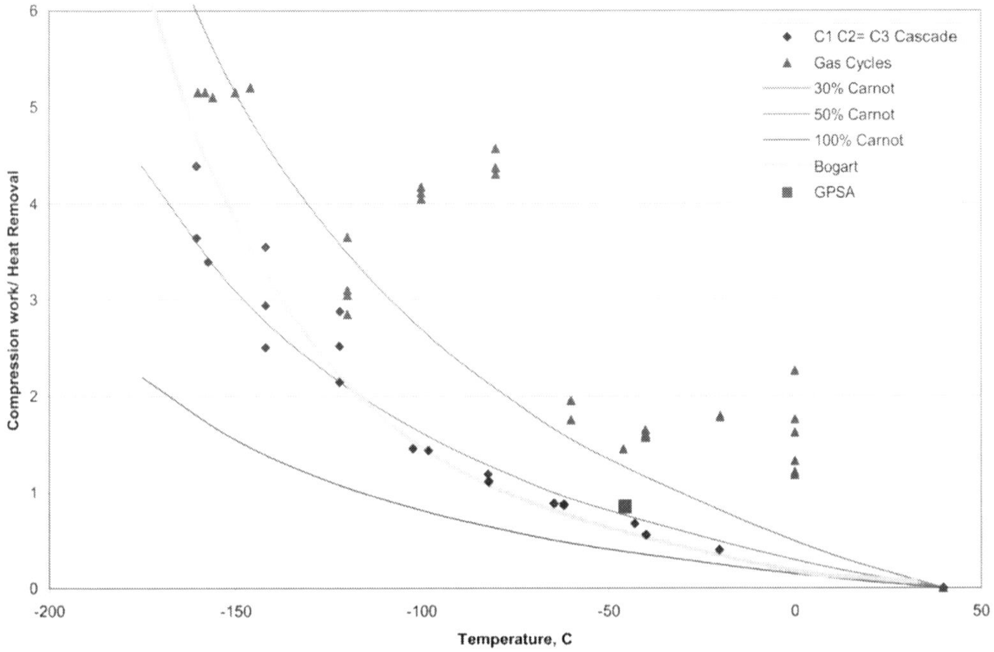

Figure 3 Converting refrigeration duty to compression

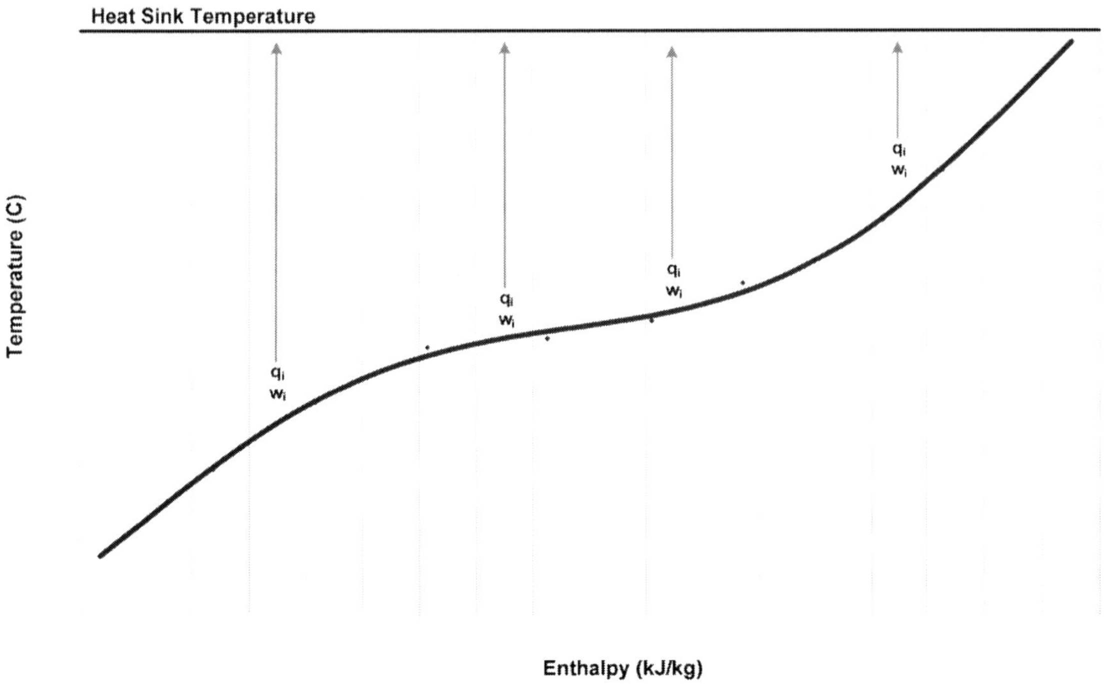

Figure 4 Level 1 model schematic

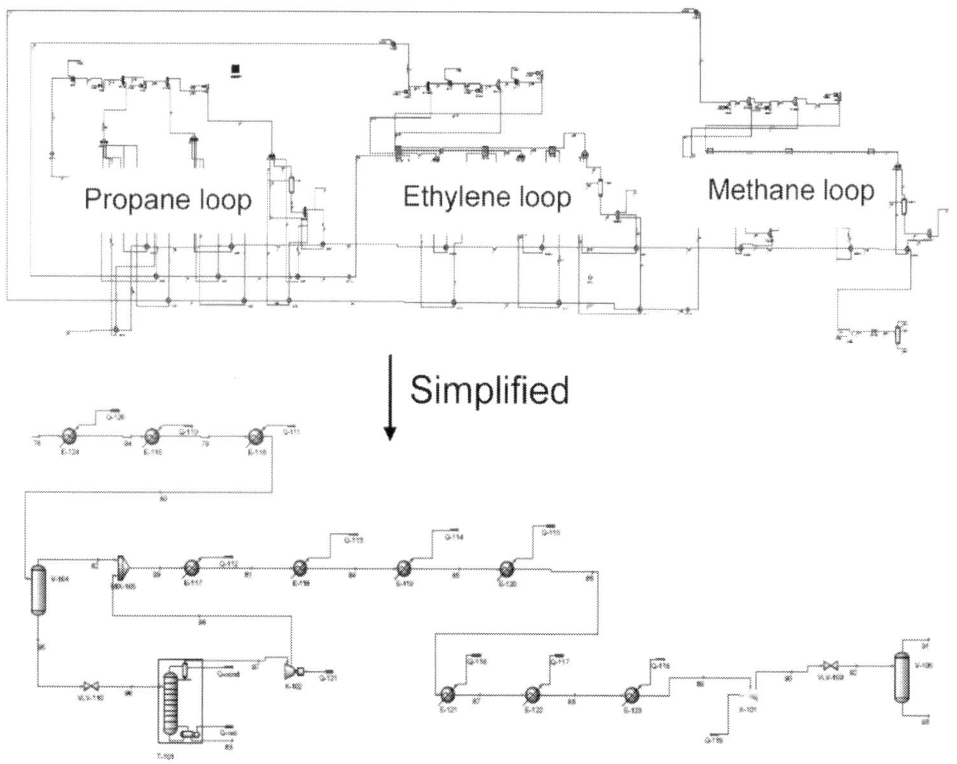

Figure 5 Simplified HYSYS set-up for LNG processes

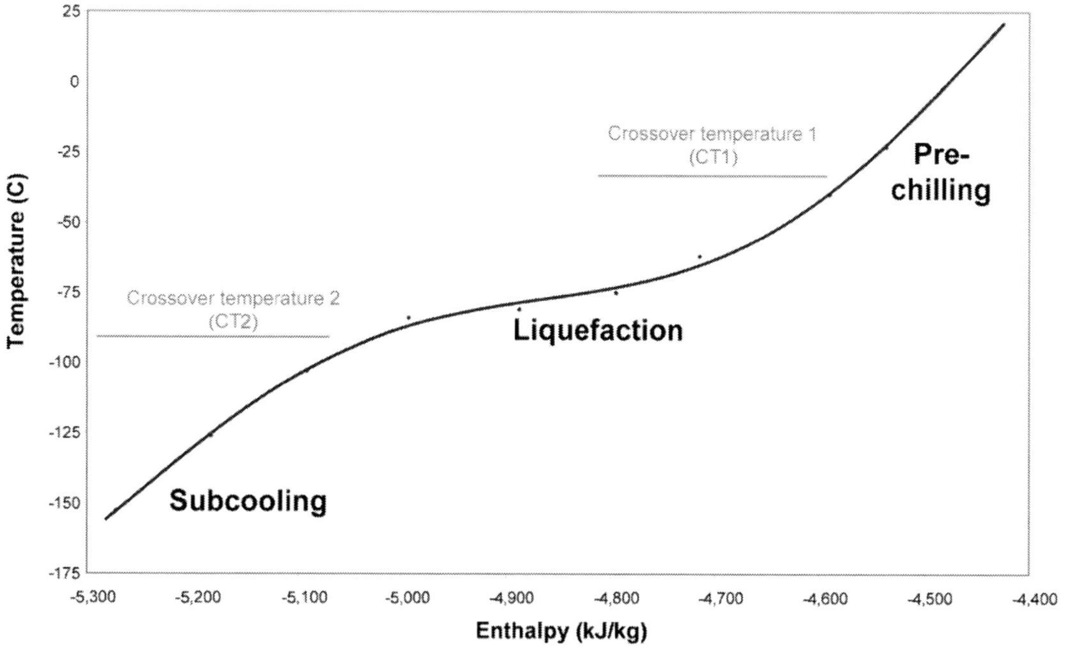

Figure 6 Chilling curve in a typical LNG plant

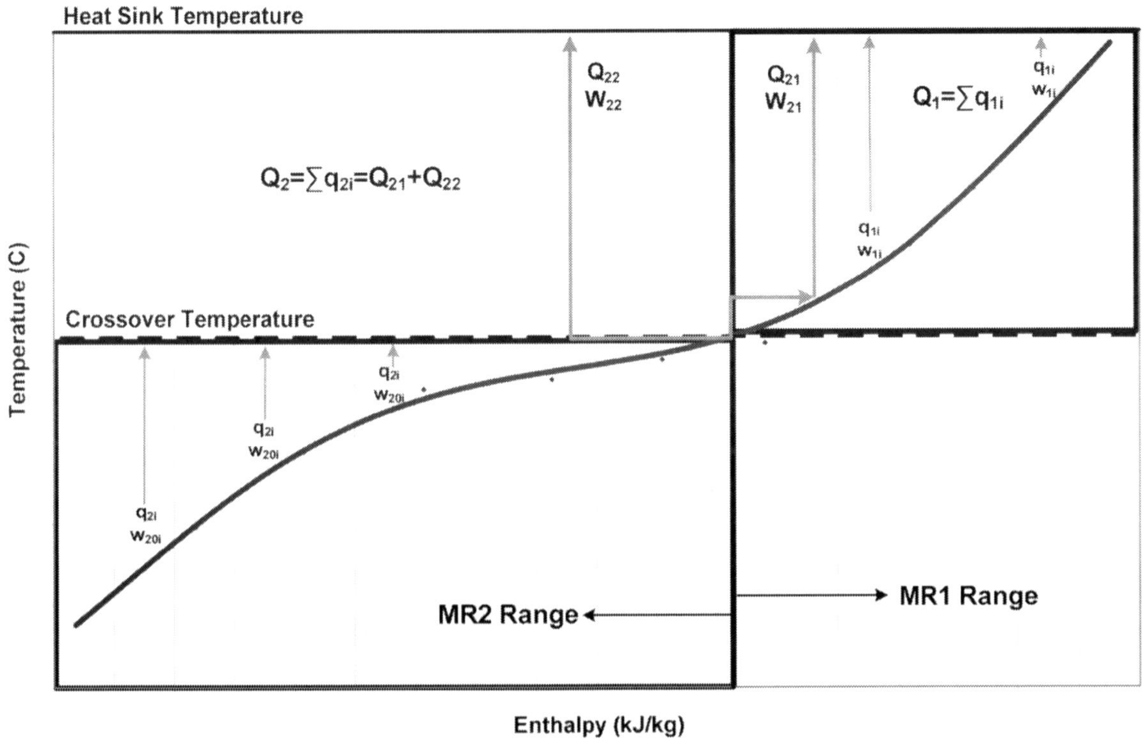

Figure 7 Level 2 model schematic

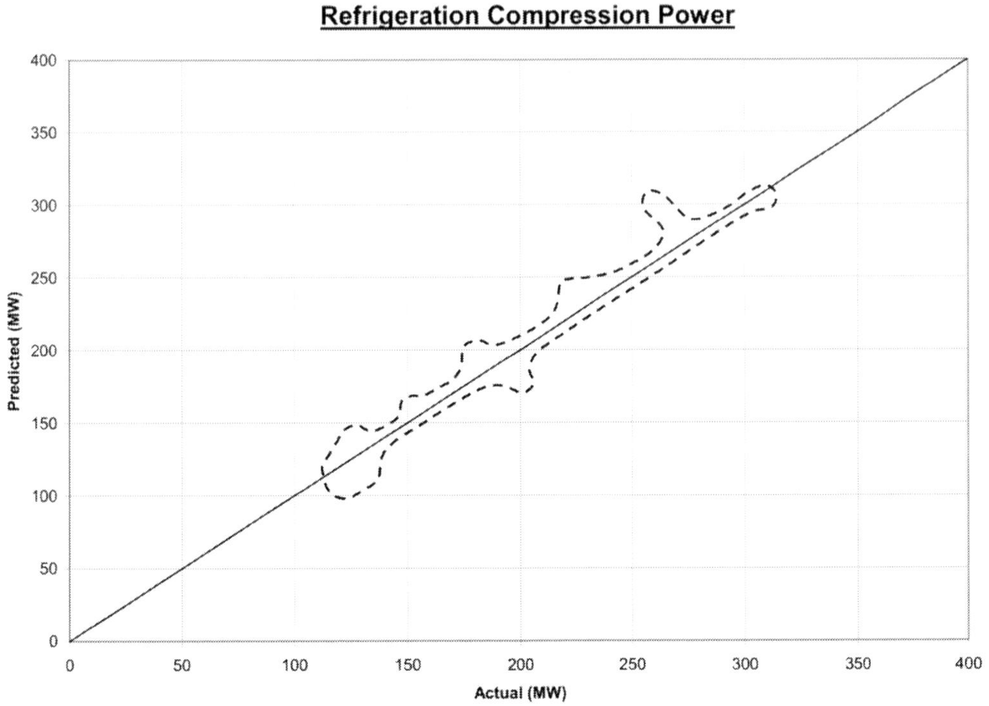

Figure 8 Verification of Level 1 model

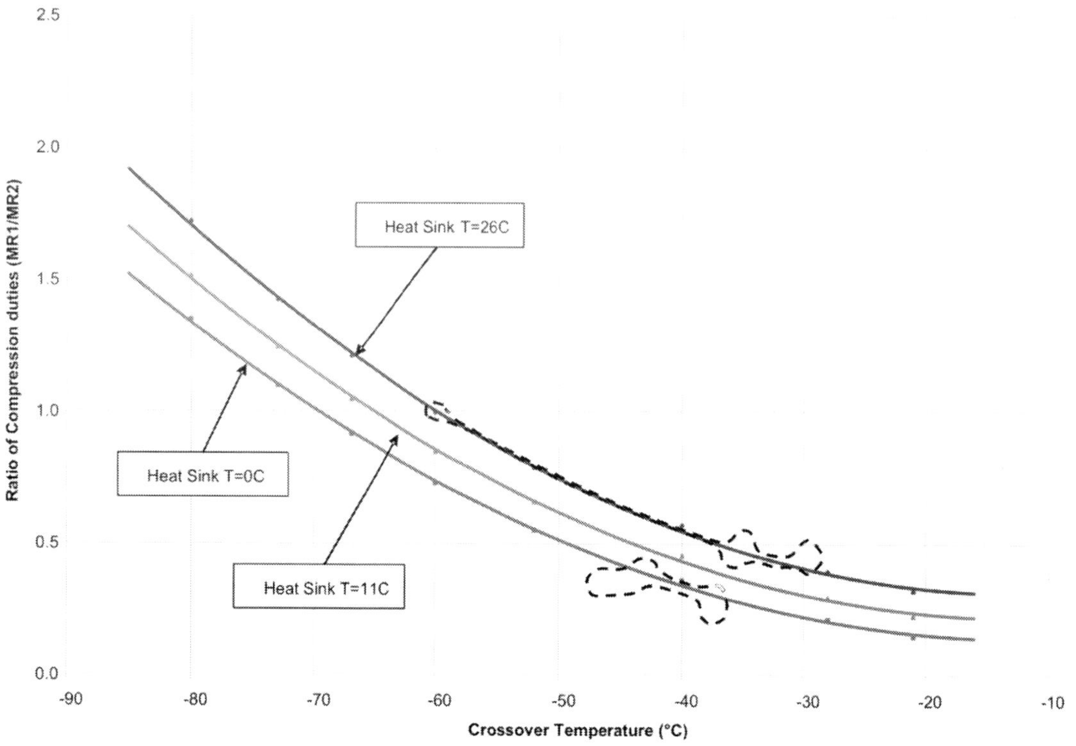

Figure 9 Verification of Level 2 model

Biographies

ChoLee Poh, graduated with a Bachelor (Honors) degree in Chemical & Process Engineering in 1986 from University of Canterbury, New Zealand. She is a registered professional engineer in Texas and has been in the oil & gas industry for 15 years. Her current responsibilities in Chevron include technical support to gas interchangeability/quality, LNG receiving terminals and LNG projects.
E-mail: ChoLee.Poh@Chevron.com

Kaman Ida Chan is a process engineer in Chevron Energy Technology Company. She graduated in 2001 from the University of Texas at Austin with a chemical engineering degree. Ida has been in the industry for five years focusing on LNG and Gas Processing.
E-mail: Kaman.Chan@Chevron.com

Dr. Stanley Huang has a specialty area in cryogenic applications, particularly in LNG and gas processing. He graduated from National Taiwan University with a B.S. degree and attended Purdue University in 1981. He earned his Master and Ph.D. degrees there, all in Chemical Engineering. Additionally, he also acquired a Master of Science in physics. Stanley has 20 years of industrial experiences and is a Registered Professional Engineer in Texas. Currently he is a Staff LNG Process Engineer of Chevron.
E-mail: Shhuang@Chevron.com

Design Considerations for Offshore Liquefaction Process Facility

Chen-Hwa Chiu
Senior Technology Advisor
Chevron Energy Technology Company
Houston, Texas

AIChE Spring Meeting, April 2007
7th Topical Conference on Natural Gas Utilization
Houston, Texas, April 22-27, 2007

INTRODUCTION

The objective of this paper is to address design considerations for offshore liquefaction facilities. These considerations are related to the development of technologies for both the liquefaction process and equipment. The effects of motion, transfer of LNG from offshore facilities to LNG ships, operational flexibility, availability of the plants and safety concerns due to the congestion of areas in the offshore facilities are important items. Offshore LNG transfer needs to meet both marine and environmental challenges. Safety considerations on the plant layouts, spacing and level of risks are also important. This paper discusses the industrial development, design considerations, further improvement and safety aspects.

INDUSTRIAL DEVELOPMENT

Most offshore LNG schemes use subsea well completions to eliminate costly production platforms and large transmission pipelines to transport the gas to shore. They also eliminate costly site preparation or harbor dredging that is generally required for a land-based plant. In the 1990s, the Gas Utilization Research Forum (GURF) addressed the marinization of liquefaction processes, including an assessment of safety issues. Then the Azure project demonstrated the integrity of the membrane containment system in partially filled mode and developed the design for a novel concrete hull.

The Azure project targeted two offshore scenarios: a dual mixed refrigerant process liquefaction cycle for a stand-alone gas field in Asia and a nitrogen expander cycle for the liquefaction of the associated gas from a deep sea oil field in West Africa. It also developed steel hull design for an LNG FPSO and developed the topside layouts to meet safety and operability requirements. It proved aspects of the control of an LNG transfer system. The transfer of LNG in open seas used the Boom-To-Tanker design of FMC. The project included both steel and concrete hull designs for the LNG FPSO. Chantiers de l'Atlantique developed the steel hull options, while Bouygues Offshore designed the concrete hull alternative.

There were other offshore floating LNG studies performed by Mobil with a moon-pool layout using the single mixed refrigerant process of PRICO, Shell on Greater Sunrise offshore Australia and Kudu, offshore Namibia, Western Africa, Statoil and other companies on Nnwa-Doro, offshore Nigeria and several Japanese and French outfits on various offshore LNG production schemes.

Figure-1 shows an offshore LNG plant with the LNG carrier alongside. Floating offshore LNG production can avoid the flaring or re-injection associated with crude production if sufficient gas quantities are available. Shell has developed concepts such as the floating LNG (FLNG) and floating oil and natural gas (FONG) for processing natural gas and associated gas respectively.

Figure-1 Floating LNG Plants

Courtesy of Moss Maritime

OFFSHORE LIQUEFACTION PROCESSES

In this section the major LNG processes considered for potential offshore application are briefly described. Because of the ranges of production capacities can be considered, no single process is suitable for all applications. Some processes are suitable for smaller capacity, while others may be more suitable for medium and higher capacities.

- **Black & Veatch PRICO Process**

The PRICO process is an all mixed refrigerant process using the plate-fin heat exchanger (PFHE). The process was first used in the Skikda LNG plant in Algeria. This process is simple and operationally well suited for offshore application. It was considered favorably by Mobil for offshore liquefaction of natural gases. The PRICO process is attractive as it has few equipment components and freedom of arrangement. Thermal efficiency of the process is lower than the dual mixed refrigerant process.

- **ABB Dual Expander Process**

The ABB Dual Cycle, called the NicheLNG process is for small and mid-scale application for less than 3.5 MTPA. It is a turbo-expander process with the use of two independent cycles, one uses methane, while the other uses nitrogen as refrigerant. The refrigerants are always in gas phase. The process is relatively simple and uses one brazed aluminum heat exchanger. Thermal efficiency of the process is lower than the single or dual mixed refrigerant process.

- **BHP Dual Nitrogen Expander Process**

BHP has developed a variation of the commonly adopted Dual Nitrogen Expander Cycle which they coined the cLNG process. It utilizes a brazed aluminum heat exchanger in a compact modular design. It is a relatively safer process with nitrogen refrigerant being an inert, but its thermal efficiency is lower.

- **Statoil - Linde MFC Process**

The Statoil - Linde mixed fluid cascade, MFC, process uses a three mixed refrigerant cascade cycles. This process is being used for Statoil's Snøhvit LNG project. It uses an all electric drive design, but can be designed for direct turbine drive as well. The refrigeration process and a cold Arctic climate help to boost the liquefaction plant's energy efficiency. This in turn cuts polluting emissions substantially. For pre-cooling the brazed aluminum plate-fin exchangers are used. The liquefaction cycle uses one spiral wound heat exchanger and the sub-cooling cycle uses a second spiral wound heat exchanger. The liquefaction cycles use the exit liquid expanders to control the process. This process reportedly can be used for a capacity up to 8 MTPA.

- **Shell Single Mixed Refrigerant Process**

In this process, Shell considers the use of a single mixed refrigerant process for plants producing LNG up to 2 MTPA. The main parts of this process are formed by two electrically driven parallel mixed refrigerant compressors. The mixed refrigerant is compressed, cooled and condensed. Part of the MR flow is to precool the feed gas and the mixed refrigerant with the remaining refrigerant used to cool the LNG to cryogenic temperatures. A single spiral wound heat exchanger is used. End flash vapor and boil off gas from LNG tanks are compressed to fuel gas system.

Air Products Dual Mixed Refrigerant Process

Air Products has developed a dual mixed refrigerant process. This process is characterized by twin wound coil heat exchangers for precooling and liquefaction respectively. Each exchanger will be approximately half the height of a single unit. This process eliminates the propane precooling and propane inventory, maintains high thermal efficiency, and is suitable for the compact arrangement required in floating LNG applications.

- **Shell Dual Mixed Refrigerant Process**

The Shell dual mixed refrigerant process consists of two process strings that share certain components to provide a high overall availability. Splitting process functions such as refrigerant compression and acid gas removal limits the size of equipment. It allows one string to shutdown while the other string remains in service. Thus, it avoids the difficulties associated with warming up and cooling down such cryogenic equipment and saves time for total facilities shutdown, if only one string of cryogenic heat exchanger were used for the whole plant. Both refrigerant circuits are designed with higher pressure equipment to avoid flaring of refrigerant during upsets.

CONSDERATIONS FOR LIQUEFACTION PROCESS

The following are various factors to consider in the selection of liquefaction process for offshore application.

- **Compactness of Equipment**

Floating or offshore LNG plants usually have to withstand greater wind load and motion than the land-based plants, therefore, it is better to have less height and weight of the equipment. For instance, the aero-derivative gas turbines such as the LM-6000 or LM-2500 would be able to provide power output equivalent to comparable frame size 2-shaft mechanical drive heavy duty turbines, but at one-third the weight and one-quarter of the footprint area.

Also, the height of heat exchangers such as that of the Air Products coil-wound main cryogenic heat exchanger can be reduced by having separate bundle in separate units rather than having two or three bundles in one unit, which could have a height of more than 150 feet. The requirements of compact footprint and lightweight equipment will be obvious. This is not only good for land-based LNG plants but especially good for offshore locations, where limited space is available.

- **Reduction of the Effect of Motion**

The floating facility inevitably needs to face the effect of motion on equipment. Motion effects on separator vessels are generally understood from oil FPSO applications. The effects of motion on heat exchangers or distillation columns are manifested in the mal-distribution of fluid phases and decrease heat and mass transfer efficiencies. Special packing or redistribution devices are used in towers. Also special designs need to be considered in heat transfer equipment by either forced flow or redistributors for fluid phases. These are also potential impacts of motion-induced acceleration forces on equipment.

For LNG storage tanks the effects of sloshing and the impacts on the tank structures are being addressed. Also the effect of relative motion between the LNG FPSO and LNG carrier ship needs to be considered in the loading or unloading of LNG through common LNG loading arms or specially designed loading devices.

- **Flexibility**

The process selected should have the flexibility to handle varying feed composition and ambient temperatures. It should have simplicity and operability to maximize production and to avoid downtime. This is important as the offshore location is quite isolated and any kind of upset and resulting in shutdown is costly. While thermodynamic efficiency is important, emphasis should be placed on the simplicity of design and ease of operations.

- **Increase of Safety**

LNG plants located on offshore locations tend to be more crowded and compact in equipment arrangement. More rigorous safety codes applicable to ships and offshore facilities need to be adhered to. The effects of waves, wind, relief flares radiation and leaks and spills especially need to be considered. Special HAZOP and safety reviews need to be conducted to document those measures and precautions taken in any conceivable risky scenario. The docking of ships for loading or unloading of LNG needs to be modeled either in the real scaled down model or by computer simulation.

- **Reduction of the Refrigerant Storage**

Liquefaction processes, except the nitrogen expander process, invariably require hydrocarbon refrigerants to provide the precooling and liquefaction of the feed gas. However, to the extent possible the amount of hydrocarbon refrigerant storage should be minimized. This will reduce the hazard of the plant and also the footprint required. Processes that utilize pure hydrocarbon refrigerants such as propane require multiple vessels and piping to prevent leakage, and thus, are considered disadvantageous for offshore application.

- **Modularization**

The offshore location is especially suitable for equipment modularization and fabrications under more controlled conditions. This will reduce the footprint of equipment and also the ease of construction and installation. This method will considerably reduce the amount of piping and thus reduce the risks of leaks and spills.

- **Synergy with Other Facilities**

The gas liquefaction plant on the FPSO can benefit from synergy with other facilities on the same floating units. Oil and gas production facilities offshore can be handled with condensate recovery or the NGL plant, which are physically close to or on the same

FPSO of the LNG facilities. Thus some of the utilities, such as electric generation, could be shared or integrated with the LNG plant and achieve significant savings both in the capital and the operating costs of the projects. It may be better also to consolidate the power drivers with the refrigeration drivers by producing electricity for all the refrigeration compressors. This will reduce the footprints of the equipment, and increase the overall system availability. For the smaller LNG plants the compressors will be smaller and more suitable for the electrical driven options.

IMPROVEMENT AND SAFETY

There are many aspects of the offshore LNG production plant that can be improved. In general one can consider the different areas of liquefaction process, utilities, storage, and LNG transfer. In each area, one can further divide the improvement into technology, equipment and safety. Table 1 shows the results of such efforts.

Table 1 Improvement Concepts for Offshore LNG Plants

	Technology	Equipment	Safety
Liquefaction	Better compressor fit with aero-derivative turbines	Improve heat exchanger performance under effects of motion	Reduce relief loads with better control and design. Use expander cycle for inherently safer design.
Utilities	Use gas turbine to generate electricity for all electric driven refrigerant compressors	Synergy and sharing of utilities with other units such as LPG and condensate recovery and oil production	Reduce flare loads and sizes.
Storage	Improve other storage concepts for effects of motion	Improve membrane and other type tanks for offshore large storage	Reduce risks of spills, collisions and proximity to cryogenic processing equipment
LNG Transfer	Transfer of LNG of side by side or in tandem between FPSO and ship	Loading arms: testing of tandem loading and the cryogenic flexible hose loading	Reduce risks of LNG spill and fire during transfer of LNG

One can pursue similar efforts for the future improvements of the offshore LNG production. The offshore LNG plant considered here are more suitable for the smaller to medium sizes such as from 1 to 3 MTPA. Larger size plants, to around 5 MTPA and above, will possibly be the limit, due mainly to the plot size.

The safety record of land-based LNG plants will need to be extended to offshore LNG plants. Besides adapting process option and equipment design for a marine and moving environment, Quantitave Risk Analysis (QRA) has been used to analyze the safety risks of equipment congestion, effect of refrigerant and to understand mitigation measures that can minimize overpressure and incident escalation. This will lead to a better equipment layout and a safer offshore LNG plant.

CONCLUSION

This paper reviews industrial development, offshore liquefaction processes and design considerations on offshore liquefaction of natural gases. Current thinking on various factors such as compactness of process, reduction of the effect of motion, flexibility, increase of safety, reduction of the refrigerant storage, modularization, and synergy with other facilities is presented. Further, ideas of improvement and importance of safety are emphasized.

REFERENCES

1. Chiu, Chen-Hwa and Quillen, D., Commercial and Technical Considerations in the Developments of Offshore Liquefaction Plant. 23rd World Gas Conference, Amsterdam, the Netherlands, June 2006.
2. McDonald, D., Chiu, Chen-Hwa and Adkins, D., Comprehensive Evaluations of LNG Transfer Technology for Offshore LNG Development. LNG 14 Conference, Doha, Qatar, March 2004.
3. Quillen, D. and Chiu, Chen-Hwa, New Developments in Floating LNG Technology. North American LNG Conference, Houston, November 2002.
4. Chiu, Chen-Hwa, Challenges of Small and Medium Scale LNG Projects. International Forum on Liquefied Natural Gas '99, London, October 1999.
5. Chiu, Chen-Hwa, Dimitroff, R. and, Shah, P., Methodology for LNG Technology Selection. Gastech 2006, Abu Dhabi, December 2006.
6. Barclay, M. and Denton, N., Offshore Natural Gas Liquefaction Process Selection. AIChE Spring National Meeting, Atlanta, April 2005.
7. Roberts, M. J., Liquefaction Cycles for FPSO Service. LNG Journal, July/August 2001.
8. Faber, F., Resweber, P. S., Jones, A. E. and Bliault, A. E. J., Floating LNG Solutions from the Drawing Board to Reality. Offshore Technology Conference, Houston, May 2002.
9. Sheffield, J. A., Offshore LNG Production – How to Make it Happen. Business Briefing: LNG Review 2005.

10. Persaud, M., Chamberlain, G. and Kaufman, S., Safety Drivers in the Lay-out of Floating LNG Plants. AIChE Spring National Meeting, New Orleans, March 2003.
11. Johnsen, K. R. and Christiansen, P., LNG Production on Floating Platforms. Gastech, Dubai, November-December 1988.
12. Naklie, M., Mobil's Floating LNG Plant. LNG 12 Conference, Perth. May 1998.
13. Eriksen, R., Brandstop, J. M. and Cramer, E., Evaluating the Viability of Offshore LNG Production and Storage. Gastech, Qatar, October 2002.
14. Miyake, H., Kishimoto, N. and Kakutani, Y., Small Scale LNG FPSO for Marginal Gas Fields. LNG12 Conference, Perth, May 1998.
15. Bliault, A., Reijnen, D., Runbalk, D. and Nagelvoort, R. K., Shell's Floating LNG Plant for the 21st Century. 21st World Gas Conference, Nice, June 2000.
16. Miyata, K., Nakamura, A., and Uchina, K. Case Study on LNG FPSO (in Japanese). Journal of the Japan Institute of Energy, October 2006.

BIOGRAPHY

Dr. Chen-Hwa Chiu is a Senior Technology Advisor of the Chevron Energy Technology Company. Dr. Chiu has served as a company representative to the Joint Industry Projects of Azure Floating LNG Chain, LNG FPSO Safety, and Offshore Cryogenic Loading for LNG transfer. Dr. Chiu is a Fellow of American Institute of Chemical Engineers and was a Chair of AIChE Fuels and Petrochemicals Division. He is the organizer and LNG Session chair of the AIChE Topical Conference on Natural Gas Utilization. Dr. Chiu graduated from National Taiwan University with a B.S. and earned his Master of Engineering and Ph.D. from the University of Oklahoma, all in Chemical Engineering. He is a Registered Professional Engineer in Texas and Pennsylvania.
Dr. Chiu can be contacted at 1500 Louisiana Street, Houston, TX 77002, USA. Phone: 832-854-5305, Cell: 713-397-4097, and E-mail: Chen-Hwa.Chiu@Chevron.com.

Medium-Scale LNG Plants have started making Economic Sense

Tim Cornitius
Editor, Syngas Refiner
Zeus Development Corporation
Houston, Texas

Gerry Ohlhaver
Executive Director
Zeus Development Corporation
Houston, Texas

Pat LaStrapes
Managing Director
Zeus Development Corporation
Houston, Texas

AIChE Spring Meeting, April 2007
7th Topical Conference on Natural Gas Utilization
Houston, Texas, April 22-27, 2007

Introduction

Medium-scale liquefaction projects have captured the imagination of a number of serious players over the last few years. This newfound interest has been fueled partly by (1) the need to eliminate flaring and venting while cutting back on the reinjection of associated gas and (2) the focus of mid-level players on monetizing gas resources and serving markets basically inaccessible by pipeline. Until recently, the high CAPEX for liquefaction, transport and regasification has resulted in unfavorable economics for monetizing reserves of less than 4.0-5.0 trillion cubic feet.

Four years ago, installation of medium-scale liquefiers in the 0.1- to 1.0-million metric-ton-per-year (mty) range would have cost considerably more than $400 per mty of capacity. However, advances in process technology, standardized designs and manufacturing efficiencies are beginning to favorably impact the costs of small- and medium-scale plants – despite the obvious rapid escalation of material and labor costs associated with larger (world-class) baseload facilities.

Honing a Definition of "Medium-Scale"

The numerical definition of "medium-scale" tends to vary considerably, depending on the user's role and vantage point in the industry. Numerous operating utility peakshaving plants can be found in the 5-20 MMcfd range. Recent flare-gas reclamation applications can also be found in this range, while LNG refueling stations and small liquefiers are typically found in 5-10 MMcfd range.

Certain industry vendors consider 100 MMcfd to be medium-scale, while other firms see the medium-scale market as starting around 50 MMcfd. Still, other, smaller companies might consider medium-scale to fall within the 10-100 MMcfd range.

Notwithstanding, for our purposes here, applications of less than 50 MMcfd of liquefaction will be referred to as small-scale. The term medium-scale will be reserved for applications between 50 and 250 MMcfd, a zone that has remained largely unpopulated to date.

Medium-Scale Liquefaction Technology

Flare-gas reclamation projects in Asia and particularly China, as well as Chinese projects that liquefy pipeline and tanker-truck feed for distribution, have recently generated a market for cost effective medium-scale liquefaction plants.

Figure 1: At its western Guanghui LNG Plant in Xinjiang, which uses the Linde MR Process to liquefy 55 MMcfd, Chinese technologists have devised a way to loadout a convoy of 100 trucks in 16 hours (See footnote 1.).

A 55 MMcfd medium-scale LNG plant has come on line at Shan Shan in the remote northwestern part of China. It is the first facility to encroach on the 50-to-250-MMcfd capacity gap.[1] Its upstream systems are sized for a small baseload facility, while the cryogenic systems are of a size more in keeping with a large peakshaving or vehicle-fueling plant.

Traditionally, export plants have not been built with capacities below 250 MMcfd because of the disproportionate cost of related infrastructure. The requisite storage systems, pipelines and marine facilities would not downsize cost effectively enough to extend profitability into this region.

Xin Jiang Guanghui LNG Development Co. Ltd.'s new Shan Shan liquefaction plant in Xin Jang Province, China, the first facility to fall inside our medium-scale range, has found a novel approach that appears to have solved this conundrum. The plant, landlocked, does not utilize a pipeline or loading trestle or jetty. Rather it employs a large fleet of Mercedes Benz cryogenic tanker-trucks organized in convoy systems to deliver LNG over 4000 km of road to widely scattered distribution terminals in the more populous East, including Shanghai and Fujia (See Figure 1.).

---

[1] "Baseload Plant in Xinjiang, China – Commercialization of Remote Gas Resources for an Eco-Responsible Future," Eginhard Berger, Linde AG – Linde Engineering Division; Xiang Dong, Xinjiang Guanghui Industry and Commerce Group Co. Ltd.; Jin Guo Qiang, Shanghai Pharmaceutical Industry Design Institute of SINOPEC (SPIDI); Albert Meffert, Tractebel Gas Engineering; Lothar Atzinger, Linde AG – Linde Engineering Division.

Receiving terminals such as the Dehua satellite station, for example, have eight vertical cylinder storage tanks with an overall capacity of 450 cu. m., with regassing done in natural convection fin-tube air exchangers. The terminals serve coastal areas that are becoming increasingly dependant on natural gas for their burgeoning industrial growth and might otherwise be considering marine delivery from international sources.

Shan Shan's liquefaction process is based on an efficient Linde mixed-refrigerant (MR) process design tailored to a remotely located medium-scale environment. The natural gas is cooled, condensed and subcooled in a three-section Linde spiral wound heat exchanger (SWHE) with 3 JT valve letdown systems on cascading pressure levels. The three-stage compressor system is gas turbine-driven, and the turbine exhaust hot-oil heat-recovery system fulfills process-heating requirements. The SWHE's rugged characteristics are well suited to remote operations with a less mechanically skilled workforce.

Downstream of the plant, while one may question the cost effectiveness of the above-mentioned rubber tire pipeline from the standpoint of standard Western practice, it undoubtedly gets the job done. It also provides the ultimate in flexibility for catering to market shifts – and, perhaps, even future redeployment to other gas fields.

One might also look at this transportation mode as a scale-up of the system of road distribution utilized by Clean Energy, located in Willis, Texas, for its 10 MMcfd of LNG destined for road fuel markets as far away as California.

Remote liquefaction plants, such as Shan Shan, as well as larger peakshaving and vehicle fuel plants will also likely be built, allowing this capacity gap to close over time and creating new opportunities for developers, technology providers, vendors and local distribution companies.

In the near future, export plants that load large LNG tankers will not necessarily require today's lengthy jetties and pipeline trestles, but rather rely, possibly, on vacuum-insulated subsea pipelines that could reduce plant CAPEX significantly.

Meeting Higher Demand

Shan Shan represents an important step in monetizing indigenous gas in markets not economically reachable by pipeline, enabling penetration of smaller scale regional markets and increasing the importance of indigenous gas in the country's overall energy mix.

More than three times the capacity of the typical utility peakshaving plant, the Shan Shan plant is still considerably smaller and less expensive than other medium-scale liquefaction plants for baseload export of stranded or remote gas. This provides an important bridge to monetizing heretofore uneconomical sources.

Linde is not alone in the ability to provide up-to-date and cost effective medium-scale liquefaction processes. The following competitors offer a range of choices in this area:

- Kryopac is the leader in terms of the number of plants onstream in the combined small- and medium-scale arena. Its EXP single-cycle turbo-expander refrigeration process uses inlet process gas as the sole refrigerant, a plus in remote locations. Kryopac also offers single-cycle MR and precooled MR systems in the SCMR and PSCMR processes, respectively. The claim is that Kryopac can, thereby, more closely conform to requisite conditions over a broad range of capacities.

- Black & Veatch's PRICO process has been extensively modified and improved since its introduction in the sixties in large-scale processes. The single MR loop, with a single refrigeration compression system, is quite comfortably downsized into the medium-scale capacities we are discussing. The MR is compressed and partially condensed upstream of the cold box housing core-type plate-fin heat exchangers. Black & Veatch has a total of 25 plants on line or under development.
- Chart Energy & Chemicals specializes in cold box design and undertakes EP&C services on medium-scale and other units. It is a leading manufacturer of core-type plate-fin exchangers.
- Mustang Engineering's LNG Smart utilizes inlet process gas as the sole refrigerant, a plus in remote areas. It also offers a medium- to large-scale ambient air heated regassing system that has generated interest.
- Hamworthy offers a small-scale plant utilizing a closed-system N2 loop for liquefaction.

Proliferation in China

In China, the new center of activity for the small- and medium-scale LNG plant (with half the world's current and planned units), there are plans to provide a number of other plants of somewhat lesser capacity for similar use in similar situations. The following are currently in the planning stage:

- Shanxi LNG Ltd. is planning a 50-MMcfd plant.
- Wuxi Yongda Gas Ltd. is planning a 37-MMcfd plant.
- Pacific Asia CBM's has announced a 30-MMscfd plant using Kryopak technology.
- Also in China, Black & Veatch technology is being used for plants that will be built with 12-MMcfd, 20-MMcfd, 34-MMcfd and 36-MMcfd capacities.

Rising Interest in Medium-Scale

The elimination of flaring, venting and reinjecting of associated gas in countries that signed the Kyoto Treaty has particularly helped to raise interest in medium-scale liquefaction projects in recent years. Second-tier players are also entering the market to monetize natural gas resources and tap markets that are inaccessible by pipeline. While more than 5.0 tcf per year is currently flared worldwide, initiatives to eliminate gas flaring are well under way in Russia, Nigeria and Indonesia.

As already mentioned, historically, high infrastructure costs associated with liquefaction, transport and regasification have rendered the economics unfavorable for monetizing gas reserves of less than 4.0-5.0 tcf.

However, advances in process technology, standardized designs and manufacturing efficiencies are reducing the costs of medium-scale plants. Also, flare-gas reclamation projects in China and elsewhere in Asia, as well as those Chinese projects that liquefy pipeline feed and coalbed methane for distribution, have stimulated demand for engineering and construction services to build smaller-scale liquefaction plants.

TABLE 1: Small-Scale and Medium-Scale Applications

PRODUCTION FACILITY	LOCATION		STATUS	PRODUCT RATE (MMcfd)
Clean Energy	Willis, TX,	USA	On Line	10
Alt USA	Topock, AZ,	USA	On Line	8.6
ExxonMobil	Shute Creek, WY,	USA	On Line	7
Williams	Ignacio, CO,	USA	On Line	3
BP	Evanston, WY,	USA	On Line	3
Pioneer Natural Resources	Satanta, KS,	USA	On Line	2
Guanghui LNG Plant	Xinjiang (Shan Shan),	CHINA	On Line	55 (Linde)
Hainan LNG Plant	Hainan Island,	CHINA	On Line	10
Naturgass Vest	Bergen	NORWAY	On Line	120mtd (Linde)
Xinao LNG Plant	Weizhou Island,	CHINA	On Line	6 (Kryopak; flare gas)
Bayerwerk AG	Gablingen,	GERMANY	On Line	45 kmol/h
KeySpan Energy	Brooklyn, NY,	USA	In Planning	8.5 (Expander)
Washington Gas	Chillum, MD,	USA	In Planning	5 (PRICO)
Peoples Gas Light/Coke	Fisher, IL,	USA	In Planning	15 (PRICO)
Erdos-Xingxing Gas Co.	Xingxing,	CHINA	In Planning	36 (BV; remote gas; industry/vehicles
Zhuhai,	Guandong,	CHINA	In Planning	20 (BV; 2007)
Dazhou	Dazhou,	CHINA	In Planning	34 (BV; clone of Erdos)
Yongda	Yongda,	CHINA	In Planning	36 (BV; clone of Erdos)
Lanzhou	Lanzhou,	CHINA	In Planning	12 (BV; municipal gas dist.)
Hong Kong Towngas CBM	Shanxi,	CHINA	In Planning	6 (Kryopak, 2007)
Shanxi LNG Ltd.	Shanxi,	CHINA	In Planning	50 (Kryopak, 2007)
Wuxi Yongda Gas LNG	Congqing, Sichan,	CHINA	In Planning	37 (2007)
Pacific Asia CBM	Guizhou,	CHINA	In Planning	30 (Kryopak, 2008)
Transco National Energy	Scotland,	UK	In Planning	12 (EP)

Until 2003, medium-scale liquefiers in the 0.1-to-1.0-million-mty range could be built for something between $300 to $400 per mty of capacity. The substantial costs associated with marine facilities such as long jetties, breakwaters and channel dredging are more easily borne by world-scale baseload liquefaction plants with capacities in excess of 5.0 million mty. However, small-to-medium-scale LNG appears to offer a viable alternative for monetizing associated gas from remote locations.

Producing Alternate Fuel

CNOOC's Zhuha LNG facility in China will have a feed-gas capacity of some 20 MMcfd and will produce an estimated 390 metric tons per day (mtd) of LNG. The plant will include turbine-driven refrigeration compression and will be fuel self-sufficient. Black & Veatch and partner Chemtex will provide design, equipment procurement and startup services for this plant and the projects that follow.

Erdos-Xingxing Gas Co.'s LNG plant has a liquefaction capacity of 37 MMcfd, and production will be used as an alternate fuel. The liquefaction unit will be a motor-driven refrigeration system. Before liquefaction, water and CO2 present in the pipeline are removed in two steps. Amine treating is used to remove CO2, and a dehydration section using a two-bed molecular sieve system removes the water. Boil-off gas and regeneration gas are recycled internally to eliminate waste-gas production. What follows is a list of additional alternate-fuel projects along with some high points:

- Erdos Gas Co.'s Huixin LNG production facility at Dazhou, China, has a capacity of 37 MMcfd. The LNG will be used as an alternate fuel, and the plant is an exact duplicate of the Erdos-Xingxing plant described immediately above.
- The Yongda LNG plant, with 37 MMcfd of capacity, is a duplicate of the highly successful Erdos and Dazhou projects.
- The Lanzhou Gas Co.'s 12-MMcfd Lanzhou LNG facility will be used for peakshaving. The LNG will also be used as an alternative fuel.

Cooling, Liquefying, and Subcooling in the Linde Plant

The SWHE accommodates the gas pre-cooling, liquefaction and sub-cooling processes, where the refrigerant cycle and product streams reach temperatures of -160°C. The outer dimensions (length x diameter) of the three SWHE sections of the 55-MMcfd unit are: pre-cooler: 15 m x 3.0 m; liquefier: 17 m x 3.0 m; and sub-cooler: 11 m x 2.0 m.

After concentric stacking and welding in the steel structure, the combined SWHEs have an overall height of 43 m. The heat exchanger is relatively insensitive to thermal shocks that may occur during startup, shutdown or process perturbations.

Figure 2: Linde's SWHE is employed in MR cycles for gas cooling and liquefaction. Linde claims that its mixed-fluid cascade (MFC) LNG process offers relatively high energy efficiency See Footnote 1.).

The SWHE is central to plant geometry as well as process flow. The plant pictured in Fig. 1 exemplifies the small footprint of the vertical, open structure installation, of Central European design.

Using Single-Cycle Turbo EXP

In remote locations, where properly proportioned MR is difficult to maintain, Kryopak uses the single-cycle turbo EXP process with a turboexpander cycle providing the full refrigeration effect.

In the Weizhou Island plant, located offshore from mainland China, three stages of gas compression are employed. The compressed gas passes through a cold box and is then expanded to low pressure, becoming the cryogenic refrigerant stream.

This stream remains in the gas phase throughout, meaning that no liquid refrigerant is involved anywhere in the process. Liquefaction of natural gas into LNG does, however, require a proportionately large refrigerant gas flowstream. While this process may appear inefficient, the use of available stranded gas in remote locations eliminates the need for MR and minimizes equipment requirements.

Lessening Horsepower Requirement

Kryopac's pre-cooled MR (PCMR) process design incorporates an ammonia pre-cooling step. This step knocks water out of the process gas, resulting in size reduction of the dehydration bed system. The ammonia refrigerant is also used to pre-cool the process stream prior to heat exchange with MR. Downsizing the dehydration, compression and MR heat exchange systems at the expense of installing a conventional ammonia refrigeration package can result in interesting first-cost economies. This process is more efficient than the expander cycle above.

A single-cycle MR (SCMR) process dispenses with the pre-cooling ammonia refrigeration circuit in the PCMR process above and typically consumes 2-3% less energy than that process. Furthermore, it addresses all Kryopac design requirements quite directly.

Processing Associated Gas

XinAo's 6.0-MMcfd LNG plant on Weizhou Island is the first LNG plant to be built in China in recent years. It is operating successfully. CNOOC had originally processed the island's associated gas by extracting the LPG and flaring the methane-plus. Now, this residual gas is piped across the island to the LNG plant. All the equipment on the three-quarter-acre site is skid-mounted, eliminating most of the "stick building" that would have been required. The plant is driven by dual 2,400-hp Caterpillar diesel engines, which power six-cylinder reciprocating compressors. On each compressor, two cylinders compress process gas, and four compress refrigeration to varying pressure levels.

Kryopac Equipment Sourcing

A major equipment component of any Kryopac unit is the insulated brazed-aluminum heat exchanger. The Weizhou Island plant, for example, runs four passes through these exchangers. Kryopak works closely with Chart Industries on heat exchanger and headering design (and cold box design, where applicable). An MDEA amine plant is normally used to remove CO2 and H2S from the gas feed on entering the plant. Dehydration beds are sized to optimize plant fuel-gas requirements. The China plant uses a three-bed dehydration process scheme followed by a UOP mercury-polishing sieve.

Producing Two LNG Grades

The company's Karratha project in Australia is called a *virtual pipeline*. The project sponsor, Energy Developments Ltd., supplies LNG to five remote power stations in northwestern Australia. A 200-mtd LNG plant produces two grades of LNG. The first grade, a bi-fuel, is for trucks and is comprised of 80% LNG with a 93-methane-number and 20% diesel. The second, a 75-methane-number LNG, supplies the power plants. The LNG plant, expected to start up in the first quarter 2007, employs a PCMR cycle, a Chart cold box running six passes and a two-bed dehydration process.

Offering Cost Advantages

Kryopak is currently working on a 4.5-MMcfd plant to be located on the topsides of a jackup rig that employs a small-process design using motor-driven centrifugal compressors. In the current market, Kryopak still requires only 36 weeks from order to delivery of a compressor. Equipment costs have escalated an average of 62% over the past two years. With cost escalation running at unprecedented rates, modular ("cookie-cutter"), skid-mounted designs offer significant cost advantages. Except for larger companies such as Linde and Black & Veatch, Kryopak has relatively few serious competitors.

Author Biographies

Tim Cornitius – Editor, *Syngas Refiner*, a publication of Zeus Development Corporation

Tim Cornitius, editor of *Syngas Refiner*, brings more than 20 years of experience in covering such areas as natural gas markets, petrochemicals, lubricants and coal. His experience includes analyzing supply and demand for Exxon Coal USA, compiling data on fuel use and mileage for the U.S. Marine Corps and conducting and maintaining energy statistical surveys and databases for major energy research firms. Cornitius holds a BA in Journalism from the University of Houston.

Gerhard H. Ohlhaver – Executive Director, Zeus Development Corporation

Gerry Ohlhaver has worked in 52 countries and brings experienced senior leadership to the Zeus consulting team. He has been in supervisory charge of (or has directly led) projects in synthetic fertilizer, refining and allied fields with a total investment cost of some $12 billion.

With more than 35 years of project execution and program management experience, he brings to Zeus expertise in nitrogenous fertilizer production, FT synthesis, gas separation, cryogenic systems, sulfur removal and other gas processing and monetization technologies, both proprietary and open art.

He started up and served as VP and general manager of Black & Veatch's Houston business unit. Prior to that, as VP of international projects at B&V's process division, he held overall responsibility for projects in the Middle East, South America and Europe. Before joining B&V, as VP of project management for ABB Lummus Global, Mr. Ohlhaver held principal responsibility for successful execution of domestic and international projects. Prior to that, as VP with The Badger Co., he directed ~ 930 home office engineering and design professionals and ~ 530 field workers concurrently. Before joining Badger, he worked for M.W. Kellogg on five continents, engineering and building fertilizer units and plants in the US, Canada, Ireland, Iran and Pakistan.

Mr. Ohlhaver served as business advisor to the Emirate of Sharjah, UAE, working in Sharjah, the UK and Taiwan on fertilizer industry developments in the UAE. He has (1) led multinational teams in Syria, the UK and the US in the development of associated natural gas beneficiation for the Suedi and Karatchok fields in Syria; (2)

developed business plans for the French firm Coflexip's penetration of the US market for large-diameter armored hose; (3) led an integrated client/contractor team in Uruguay and the US to assess a radical expansion of the Montevideo refinery operations for the Uruguayan National Petroleum Co.; and (4) provided process plant design advisory services to Design Power Inc. (a subsidiary of Nokia) on their artificial intelligence 3D software plant design optimization program.

Mr. Ohlhaver holds a BS degree in Mechanical Engineering from the University of Michigan, Ann Arbor. He has also completed coursework at New York University in the fields of Law and Business Administration.

J. Patrick LaStrapes – Managing Director, Zeus Development Corporation

A senior manager and global energy consultant, Mr. LaStrapes brings the leadership necessary to take complex energy deals to fruition. He has developed world-class integrated energy projects with investments/revenues in the $billions and has established new international business platforms in rapidly evolving energy markets. He brings experience in leading multi-disciplinary international development and asset acquisition teams. Mr. LaStrapes brings a deep background in project development, project/structured finance, due-diligence, risk mitigation, contracts, operations, energy policy and strategic planning.

Prior to joining Zeus, Mr. LaStrapes served in various management roles, including President and member of the board of directors of Texas-Ohio Power, VP/Principal of EnerTran Technology Company, VP of Heartland Development Corporation and VP/General Manager of Entra Technologies.

As a global energy consultant, he has testified before state (provincial) and federal authorities and has prepared expert testimony in contested proceedings. A partial list of client services includes: (1) global market research and broad-based value-chain analysis of the markets for natural gas via LNG, pipeline, CNG, GTL, gas-by-wire, etc.; (2) economic studies of super-critical pulverized coal and integrated gasification combined-cycle power generation; (3) comparative economic studies of competing major gas pipeline projects; (4) linear programming model construction for gas-utility planning purposes; (5) electric generation expansion planning; (6) reservoir simulation, estimation of proved hydrocarbon deposits and forecasting of field deliverabilities; (7) economic studies of liquefaction, LNG peak-shaving and regasification; and (8) due-diligence on behalf of lenders and developers in connection with non-recourse project financing.

A registered professional engineer, Mr. LaStrapes holds an MBA degree from the University of Houston and a BS degree in Chemical Engineering from Texas A&M University.

Small Scale LNG Facility Developments

Brian C. Price, VP- LNG Technology
Black & Veatch Corp.
Overland Park, KS

Shawn D. Hoffart, LNG Section Lead
Black & Veatch Corp.
Overland Park, KS

AIChE Spring Meeting, April 2007
7th Topical Conference on Natural Gas Utilization
Houston, Texas, 22-26, 2007

ABSTRACT

While base load LNG plants provide a large quantity of the LNG market, there is an increasing need for small scale plants to provide LNG for distributed markets. The small scale LNG industry in the US was initially developed in the 1960-70 time period and generally consisted of peak shaving facilities for winter gas supply. The numerous facilities developed during that time period are now reaching the end of their useful life and are being replaced with new, modern designs which are more efficient and easier to operate. In addition, new uses for LNG from small scale facilities are filling needs outside the US for vehicle fuel and distributed gas supply. These facilities are based on the same process as the peak shavers with radically different LNG storage and sendout facilities.

This paper will cover the process options for new facilities and the unique facilities being developed around the world to fill the need for clean fuel from LNG.

US PEAK SHAVERS

Small scale LNG facilities have been used for over 30 years in the US for gas supply "peak shaving" applications. In these facilities the LNG produced from the liquefier is stored in a large LNG tank for use in peak demand periods. When additional gas supply is needed, such as in the winter, the LNG is pumped from storage, vaporized and sent to the customer. These plants are set up to send out large volumes of gas at a moments notice. The tank inventory is replenished during the off peak periods. Typical plants have liquefaction capacities of 5-20MMSCFD and are intended to fill a large LNG tank in 150-200 days. Thus, these facilities will have one or more LNG tank(s) to provide 1-3 BCF storage.

There are currently 58 liquefaction facilities in the US. More than 80% of these are well over 20 years old. Most of these facilities are in need of process replacement due to aging equipment. Also, modern technology provides more efficient and lower cost liquefaction capacity. Operators are also striving to reduce operating cost by minimizing staff and optimizing running times to fill storage. Many of the older plants struggle with disposal of regeneration gas from gas treating system and tank vapor recovery. As plants age, the parameters of pipeline supply and take away can change dramatically.

In addition to plant replacements and upgrades, new peaking facilities are needed to support the increased use of natural gas and the increased peaking nature of gas demand. The primary issue with new facilities is the siting and cost of the LNG storage tanks. While all LNG tanks in the US are currently single containment, new facilities are often faced with installing full containment LNG tanks to meet siting requirements. These tanks can cost 40% more than single containment designs.

TECHNOLOGY SELECTION

Existing LNG production facilities in the small scale area employ a variety of LNG production technologies. These older plants are either 1) Cascade refrigeration, 2) Expander, 3) Single Mixed Refrigerant (SMR), 4) Nitrogen Refrigeration, or 5) Other multiple refrigerant schemes. After the initial developments in the 1960-70's, the predominant technology has been the mixed refrigerant system with over 60% of the installations. Modern LNG production facilities are either Expander, N2 Refrigeration or SMR designs. The cascade and multiple loop systems have proven to be too complex and costly to operate.

The Expander process (Figure 1) uses the feed gas pressure expanding to a lower pressure to drive the liquefaction process. In many instances, plants are located where the natural gas is taken from a high pressure main line and is to be delivered to a local distribution system. In these cases, the liquefaction of a portion of the feed gas can be accomplished with little or no compression or mechanical refrigeration. A typical facility would liquefy about 15-20% of the feed gas with the balance going to the downstream system. These expander plants have been used frequently and are interesting where the pressure drop is available. If the tail gas could not be dumped to a low pressure system, this type of process would not be considered. Recompression of the tail gas would result in a process with the highest cost and worst efficiency of all processes.

The Nitrogen Refrigeration process (Figure 2) has been used on a limited basis for liquefaction, especially on smaller systems. The process is similar to the expander process in using a vapor refrigeration stream with an additional large compressor for nitrogen circulation. Since the process relies on nitrogen vapor for condensing, the refrigeration flow is quite large and the system will be significantly larger than a mixed refrigerant system. Also, the power to run this system is much larger than other technologies.

The SMR process (Figure 3) is normally the lowest cost design for small scale liquefaction systems. Also, unless free pressure drop is available for the expander process, the mixed refrigerant system will have the lowest power requirement. The

PRICO process offered by Black & Veatch is just such a mixed refrigerant process that has now been applied in 20 facilities around the world. This technology, in contrast the others discussed, has only a single compression system for the refrigeration. Also, the main exchanger is a very simple plate-fin unit with a minimal number of connections. This results in a simple, easy to operate liquefaction process. The system is designed such that during a shutdown the refrigerant inventory is maintained in the system. No venting or pressure relieving is needed. This type of process has thus become the workhorse of the industry.

Improvements in the SMR process by B&V over the years have resulted in a 25-35% reduction in power versus older facilities. Generally, these processes require about 350-500 hp per MMSCFD of LNG capacity. The exact value depends on the system design parameters such as feed pressure, ambient conditions and process specifications. Besides the refrigerant compressor, the main exchanger is the key piece of equipment in the liquefaction system. The main exchanger (Figure 4) is an aluminum plate-fin core or cores in a carbon steel box. The box is filled with perlite for insulation. All connections are external to the box which eliminates any leak potential inside the box.

Modern SMR plants have much more efficient main exchanger designs. Figure 5 shows a comparison of the modern plate fin box design versus a 1970 vintage unit. Another good example of a modern installation is the Alagasco LNG plant in Figure 6. This is a new installation with 12 MMSCFD liquefaction capacity.

FOREIGN DEVELOPMENTS

The small scale LNG business has traditionally been focused in the US with a modest number of plants around the world. In the past 4 years, this landscape has changed with multiple plants in China, Australia and South America. As a liquid fuel, LNG can be transported long distances for gas supply to a variety of facilities and industries. In Brazil, White Martins has recently put a PRICO plant (Figure 7) into production for vehicle fuels. This facility is capable of producing over 160,000 gal/day of fuel.

The portability of LNG makes it an ideal fuel from supplied from a single point to multiple customers. Also, users can access supply from a variety of sources as with other liquid fuels. The model for the China developments is a central plant for distribution of LNG to multiple gas users in industrial and residential markets (Figure 8).

Process facilities for these new markets are quite similar to peak shaving facilities but differ widely in the storage and sendout areas of the plant.

LNG PRODUCT ISSUES

Many new and existing facilities are facing issues of changing feed gas streams while maintaining LNG product quality. Also, LNG product specifications in new facilities require process modifications to the basic liquefaction process. One of the areas of impact in the US is the processing of gas from imported LNG in addition to traditional pipeline sources. The imported streams have no CO2 and no heavies which are key design issues in small scale plants. In contrast, these steams have large quantities of ethane and propane which must be handled. In some cases the ethane content can be over 12%. Two facilities recently designed by Black & Veatch have demethanizer systems for BTU control. Figure 9 shows this type of system for tailoring the LNG product from varying feed streams.

PROJECT COSTS

The authors published cost curves for small scale facilities in 2003. However, recent dramatic cost increases in all phases of LNG facilities have made these curves obsolete. Updated curves for liquefiers and tanks are shown in Figures 10 and 11. These curves are for planning purposes and facility studies. The true cost of these facilities is site specific and must be developed and adjusted for local conditions.

CONCLUSION

The small scale LNG market is a robust market with many new developments and technical challenges. As the industry grows and pushes into new areas, proven designs and technology will be relied on to continue the industry's proud tradition of innovation and safety. While costs are escalating, it is important to focus on the scope of the project. While liquefaction systems are similar for most facilities, new developments have driven the industry to provide increased flexibility in feed gas and LNG product streams.

Another important item in the project development is the selection and specification of the LNG storage tank(s). The storage capacity selected has a great influence on the project capital cost. Over-conservatism in this area can severely impact project economics.

BIOGRAPHY

Brian C. Price is Vice President for LNG Technology for Black & Veatch Corp. in Overland Park, KS, USA. He is in charge of technology development and process design for LNG production facilities, LNG import terminals and related gas processing and NGL recovery facilities. Mr. Price has over 33 years experience in gas processing and related technology areas. Prior to joining BVPI, he worked for ARCO Oil & Gas Co. in various positions including Manager of Process Engineering and Projects Manager. Mr. Price is a member of AIChE and is active in the Gas Processor Suppliers Association. He currently serves as Chairman of Editorial Review Board for the GPSA Engineering Data Book and is past Chairman of the Technical Committee for GPA. Mr. Price has BS and MS degrees in Chemical Engineering from Oklahoma State University.

Shawn D. Hoffart is a LNG Section Lead for Black & Veatch Corp. in Overland Park, KS, USA. He is an expert in LNG process design and has been design lead on numerous liquefaction facilities in the US and internationally. Since joining BVPI, Mr. Hoffart has worked for 15 years in a broad range of technologies in addition to LNG including: sulfur recovery, cogeneration, NGL recovery, NGL fractionation, and cryogenic CO2 separation. Prior to joining BVPI, he worked in product development for the Goodyear Tire and Rubber Company Industrial Products Division. Mr. Hoffart is a member of AIChE and a registered Professional Engineer in the State of Nebraska. He holds a BS degree in Chemical Engineering from the University of Nebraska and an MS degree in Engineering Management from the University of Kansas.

Figure 1
Expander Process

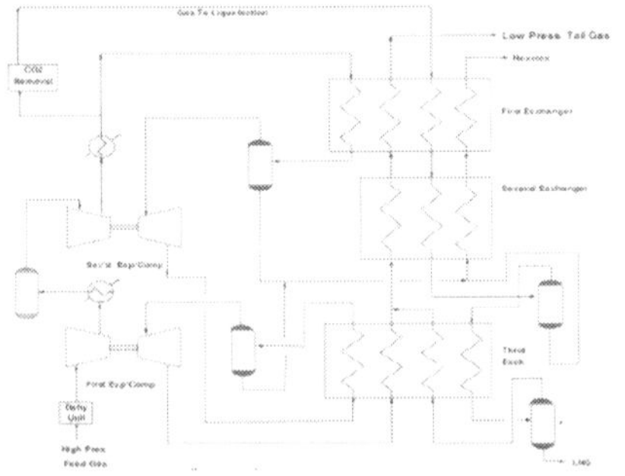

Figure 2
Nitrogen Refrigeration / Expander Process

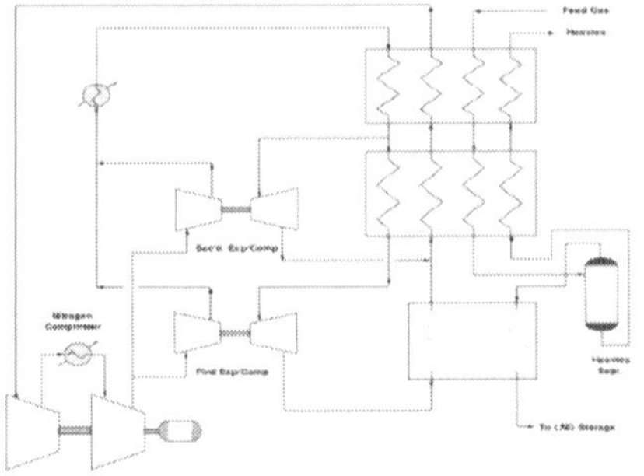

Figure 3
Single Mixed Refrigerant PRICO® Process

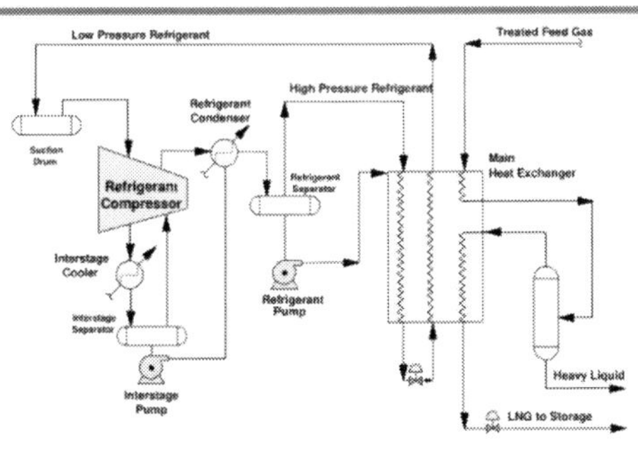

Figure 4
Typical Main Heat Exchanger

Figure 5 Wrenshall, MN
Process Replacement

Figure 6
12 MMSCFD Alagasco LNG Plant

Figure 7
White Martins LNG Plant

Figure 8
DISTRIBUTED LNG - China Model

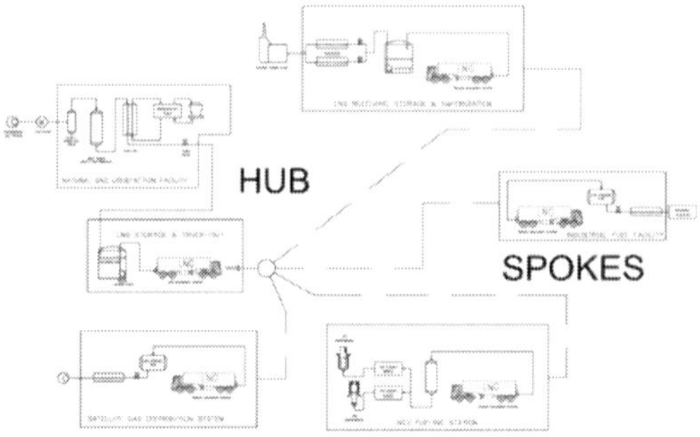

Figure 9
NGL Recovery for BTU Control

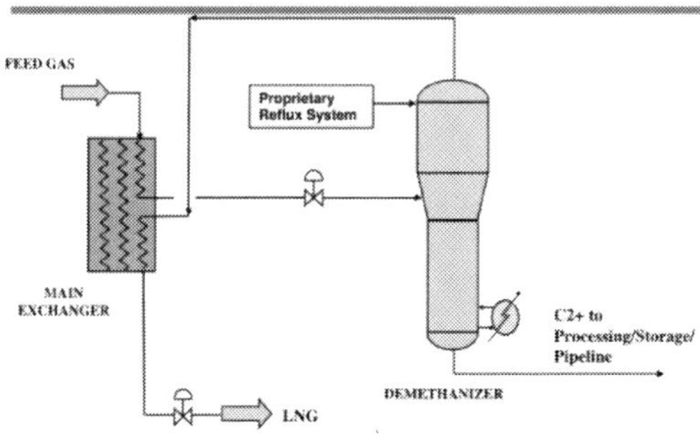

Figure 10
Indicative LNG Liquefaction Plants Cost

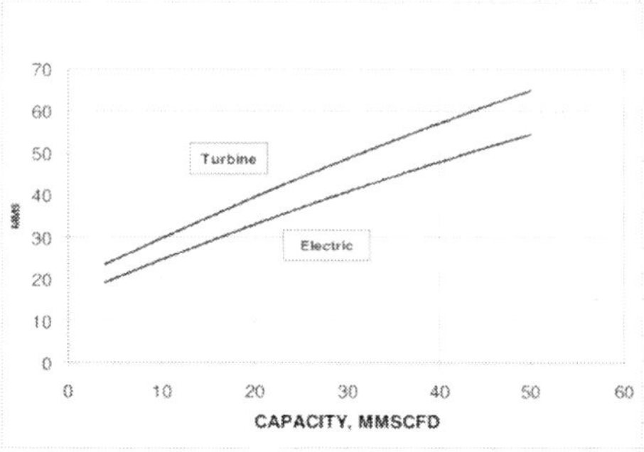

Figure 11
Indicative LNG Storage Cost

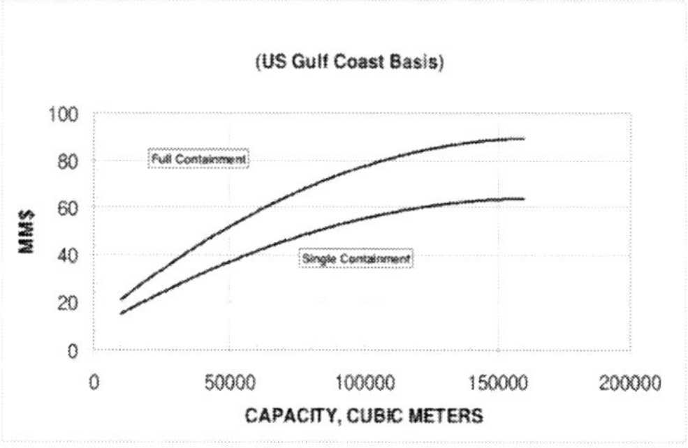

Figure 1
Expander Process

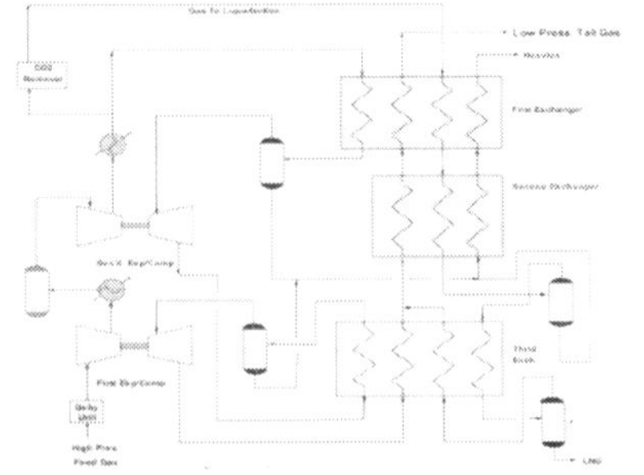

Figure 2
Nitrogen Refrigeration / Expander Process

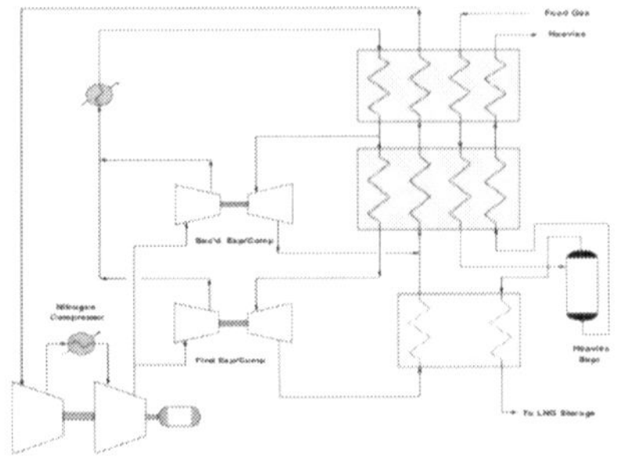

Figure 3
Single Mixed Refrigerant PRICO® Process

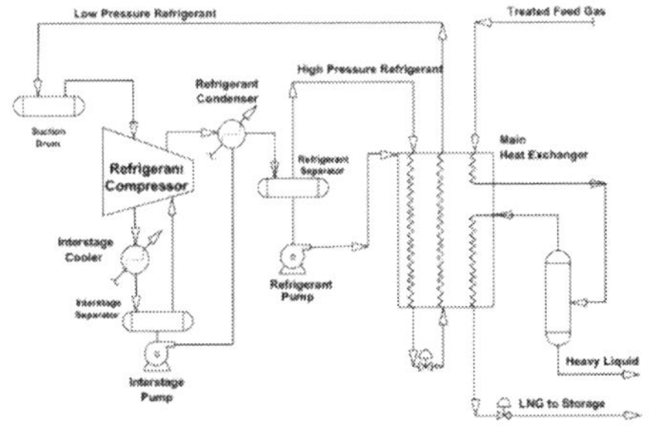

Figure 4
Typical Main Heat Exchanger

Figure 5 Wrenshall, MN
Process Replacement

Figure 6
12 MMSCFD Alagasco LNG Plant

Figure 7
White Martins LNG Plant

Figure 8
DISTRIBUTED LNG - China Model

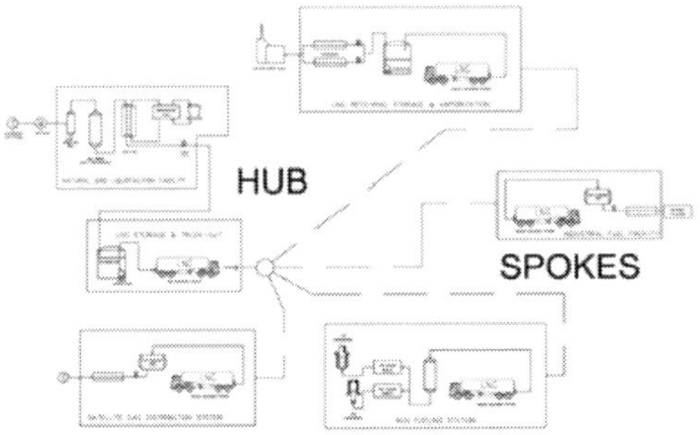

Figure 9
NGL Recovery for BTU Control

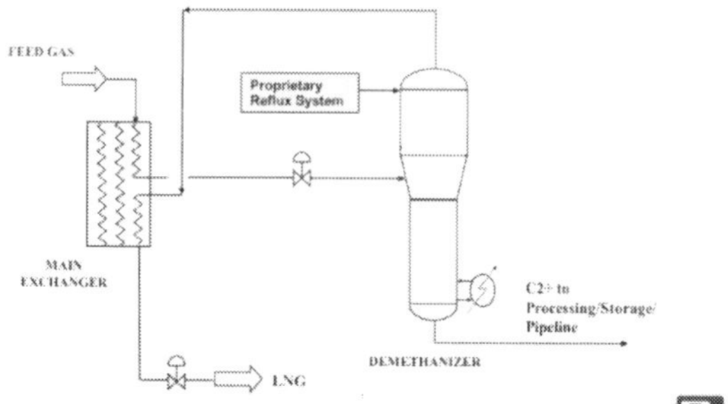

Figure 10
Indicative LNG Liquefaction Plants Cost

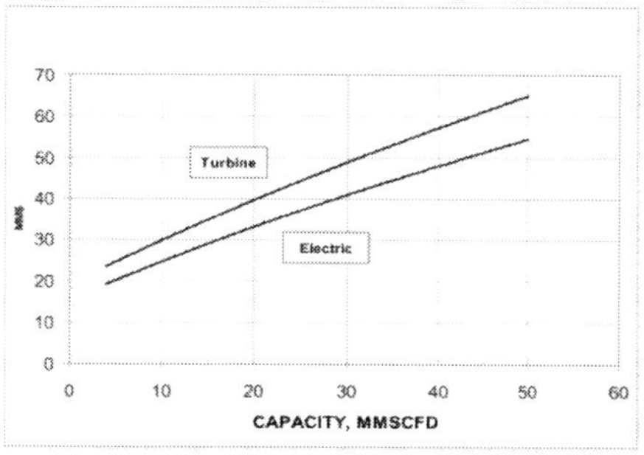

Figure 11
Indicative LNG Storage Cost

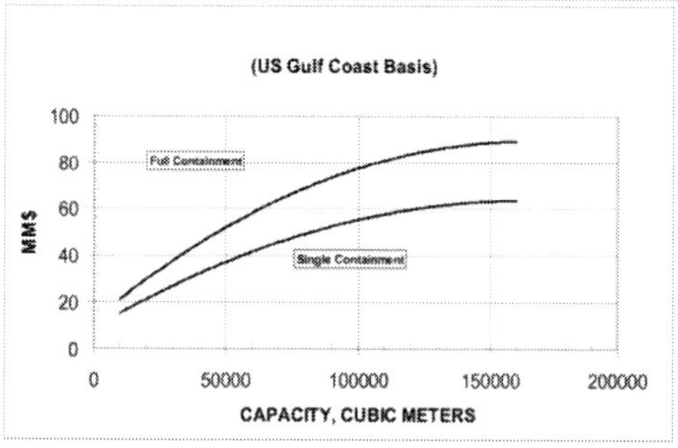

Viability of Combined Cycle Drivers for LNG Plants

Ram Tekumalla, Maheep Jain and Vibhor Mehrotra
Bechtel Corporation
Houston, TX, USA
rptekuma@bechtel.com

Karl Masani
ConocoPhillips Company
Houston, TX, USA
Karl.Masani@conocophillips.com

AIChE Spring Meeting, April 2007
7[th]Topical Conference on Natural Gas Utilization
Houston, Texas, 22-26, 2007

Introduction

Liquefied natural gas plants have increased in size to take advantage of economies of scale and rapidly growing markets. Designers of LNG facilities must consider the impact of numerous variables such as capital and operating costs, thermal efficiency, availability, reliability, environmental regulations, and project schedule. For many LNG projects, driver configuration remains one of the largest areas of study. This is because driver and compressor configuration sets the plant capacity, represents a large share of capital costs for the liquefaction facility, and determines the fuel gas consumption and emissions.

This paper presents the viability of combined cycle drivers for LNG compressors, based on the ConocoPhillips Optimized Cascade process. The basis for the paper is a dynamic simulation study sponsored by the ConocoPhillips-Bechtel Global LNG Collaboration Product Development Center (PDC). The paper also presents preliminary economic analysis supporting the combined cycle driver configuration.

Throughout this paper, combined cycle refers to a combination of gas turbines, waste heat recovery, and steam turbines.

Combined cycle drivers provide an attractive design alternative for LNG plants. Qualls and Hunter, 2003, described how a combined cycle plant successfully reduces capital costs and increases thermal efficiency for Optimized Cascade plants. The main advantage of the combined cycle drivers is the reduction in fuel

[*]Optimized Cascade services are provided by ConocoPhillips Company, Phillips LNG Technology Services Company and Bechtel Corporation via a collaborative relationship with ConocoPhillips Company. Optimized Cascade, the Optimized Cascade logo, ConocoPhillips and its logo are trademarks of ConocoPhillips Company. Bechtel and its logos are trademarks of Bechtel Group Inc.

consumption by increasing cycle efficiency. Figure 1 demonstrates the value of converting fuel gas savings into LNG for a 5.0 MTPA LNG plant.

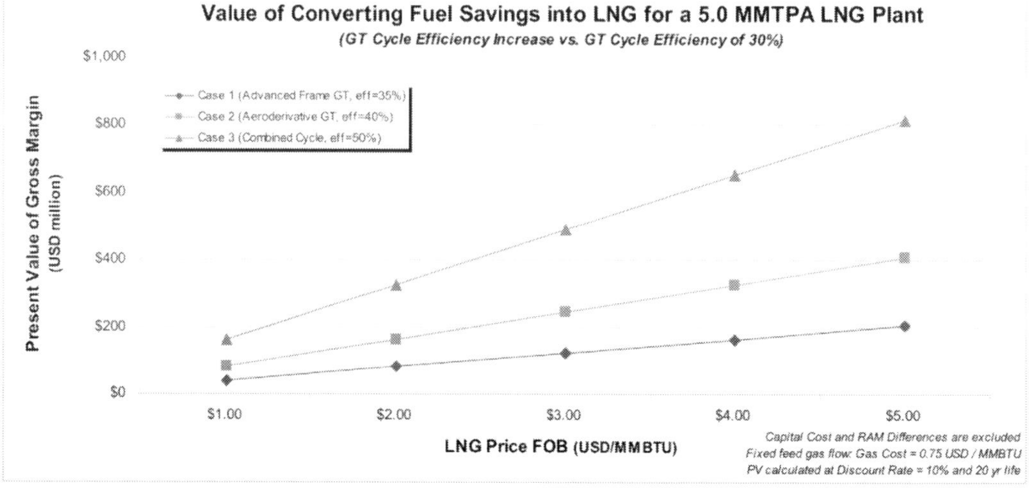

Figure 1: Value of converting fuel gas savings into LNG

In combination with key aspects of ConocoPhillips LNG process, described in a later section, combined cycle driver configuration could have a significant impact over the life-cycle of the facility.

ConocoPhillips Optimized Cascade Process

The ConocoPhillips Optimized Cascade process has been used in nine plants with capacities of 1.5 to 5.2 MTPA. Figure 2 provides a simplified over-view of the process. Plant availability, reliability, and overall production efficiency are the hallmarks of the technology. The standard design for Optimized Cascade process incorporates a "two trains in one" concept. Historically, each compressor is driven by a gas turbine of appropriate size. Each refrigerant cycle (propane, ethylene, and methane) includes a minimum of two compressors operating in parallel. This parallel configuration allows the plant to operate at production rates in excess of 50% when any single gas turbine compressor is off-line. Furthermore, Avidan et al (2003) and Redding et al (2005) have demonstrated that this operating flexibility, equipment reliability, and overall design lead to production efficiencies greater than 95%.

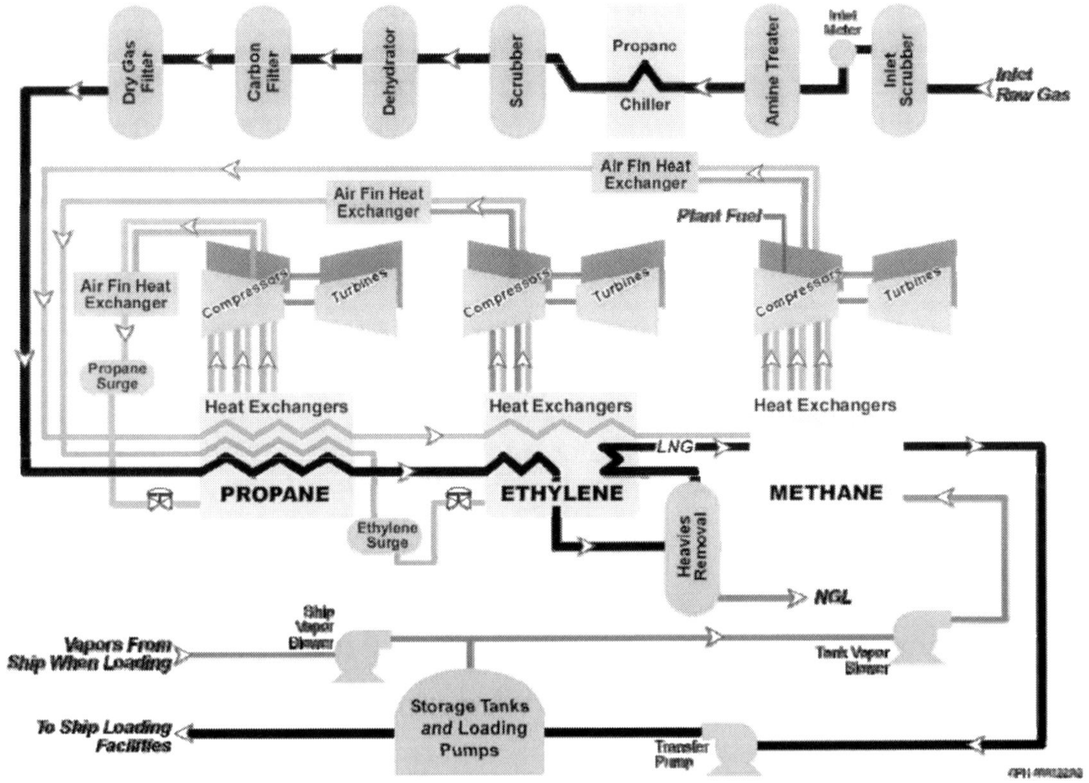

Figure 2: ConocoPhillips LNG Process - Schematic

Simulation Studies

An alternative to the configuration with all gas fired turbines is a design that uses steam turbines to drive the methane compressors. The steam is generated from waste heat off the propane and ethylene turbine exhaust.

A dynamic simulation study was undertaken to determine the operational viability of the combined cycle driver design. The dynamic model for an existing LNG plant was modified to incorporate the waste-heat system and turbines for the methane compressor. For simplicity, it was assumed that each steam turbine had an independent steam header. The heat recovery steam generator (HRSG) was a single pressure HRSG. The HRSG was equipped with duct-firing to provide 25% more heat. The dynamic model was built in Aspen Hysys 2002, utilizing a proven high-fidelity approach (Valappil et al, 2004). The combined cycle driver configuration was designed utilizing GT Pro and modeled on Invensys Dynsim platform.

The combined cycle plant operation was compared with that or the base LNG plant under normal and abnormal conditions. In order to establish equivalency of the two configurations, feed gas volume to the combined cycle plant was reduced by the difference in fuel gas consumption. This resulted in same amount of LNG as the base plant.

During normal operation, the combined cycle plant produces an equivalent amount of LNG as the base plant without requiring supplemental firing. Efficiency of the turbine heat cycle increased from 28% (lower heating value) for the base plant to 43% for the combined cycle driver configuration. The two most critical abnormal scenarios in an LNG plant are compressor trips and turn down to half-rate. The study shows that the combined cycle driver configuration had a minimal impact on the operational flexibility.

Operational Analysis

Half-rate Turn-down Production

A dynamic analysis was performed for each configuration by changing the set-point of the main feed gas rate controller to 50% of the design value. As shown in Figure 3, the LNG production rate is very similar to the base configuration.

A decrease in feed rate reduces the compressor power demand and gas turbine fuel consumption. This leads to a decrease in steam flow and hence a reduction in the power delivered by steam turbines. However, supplemental firing is not required because the reduced steam production is sufficient to satisfy the methane compressor power demand.

Figure 4 shows a comparison between the propane suction pressure for the base and the combined cycle plant configurations. Although, the dynamic response of the suction pressure differs slightly, both trends are very similar. This transient effect is mainly attributed to the additional time lags in the steam system for the combined cycle drivers. Similar behavior is demonstrated by the ethylene suction pressure (Figure 5) and methane suction pressure (Figure 6).

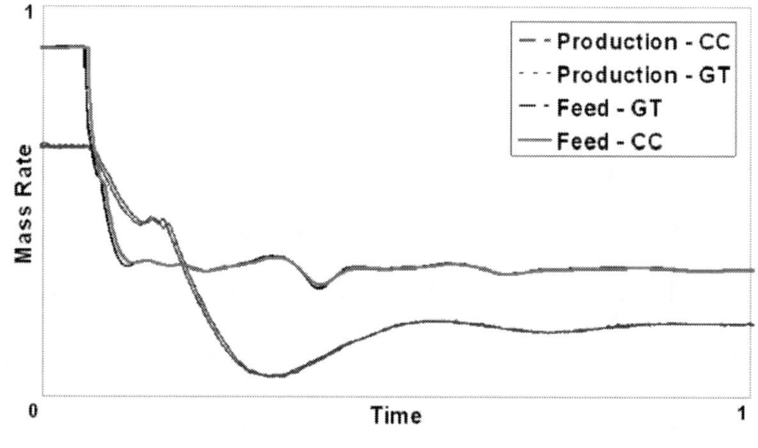

Figure 3: Comparison of Feed Gas Rate and LNG Production rate during half-rate turn-down production

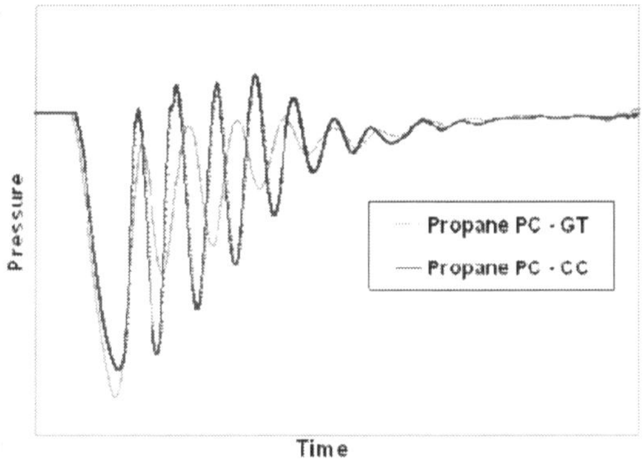

Figure 4: Propane Suction pressure during half-rate turn-down production

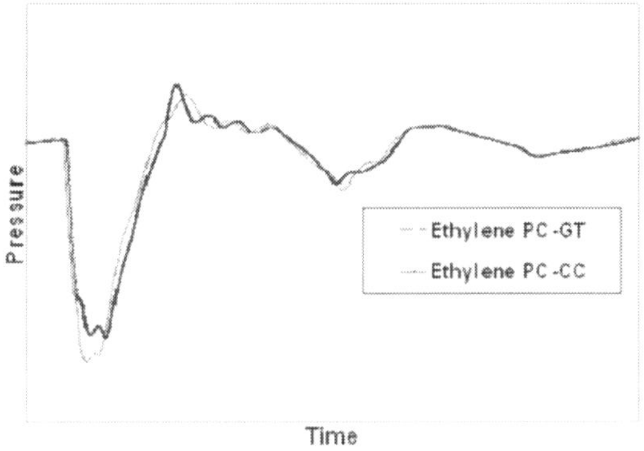

Figure 5: Ethylene Suction pressure during half-rate turn-down production

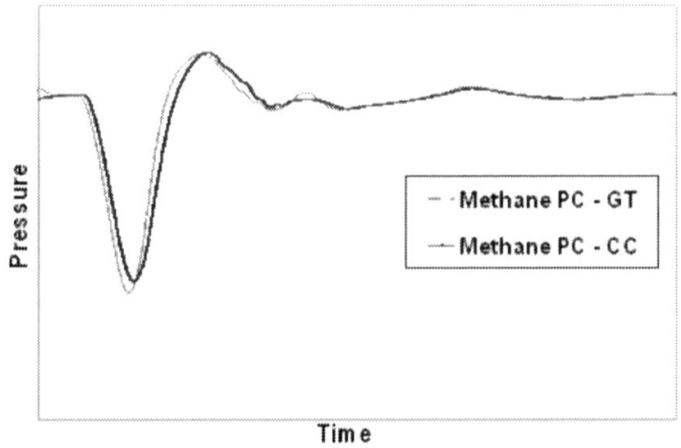

Figure 6: Methane Suction Pressure during half-rate turn-down production

Propane Compressor Trip

A propane compressor trip was initiated in the process model and subsequent behavior was studied for each configuration. In the event of a compressor trip, a safety shut-down signal is sent to the propane gas turbine driver and the inlet feed control reduces the plant feed to half-rate. Following the compressor trip, LNG production rate (Figure 7) reduces to half for both configurations.

The propane gas turbine driver trip drastically decreases the amount of waste heat available for steam generation, thereby reducing the power delivered by the steam turbine. Consequently, the additional demand from the methane compressor is satisfied by duct firing. As shown in Figure 8, the supplemental boiler output peaks shortly after the compressor trip.

A comparison of the propane suction pressures for each configuration (Figure 9) shows a brief increase in load on the operating propane compressor. Both plant configurations return to the normal operating value once the feed flow is reduced to half-rate. In contrast, the ethylene (Figure. 10) and methane (Fig. 11) suction pressure trends display a substantial difference in the dynamic behavior. This is due to the inherent inertia and time lag associated with the steam system and supplemental boiler, respectively. However, the suction pressures settle out to the same steady-state value for each configuration.

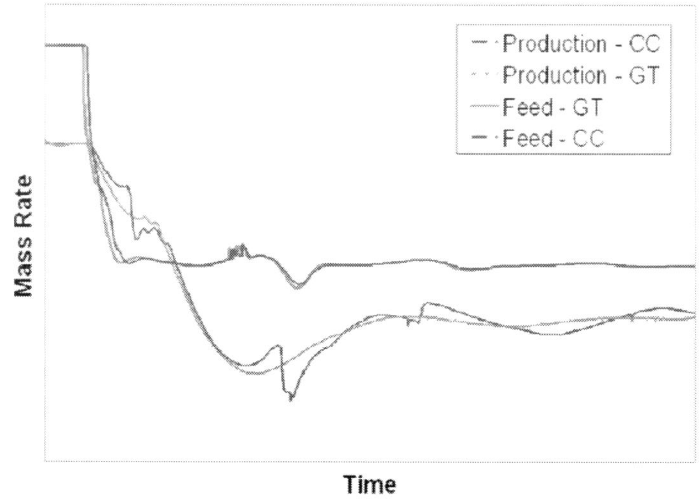

Figure 7: Comparison of Feed Gas Rate and LNG Production rate for propane compressor trip

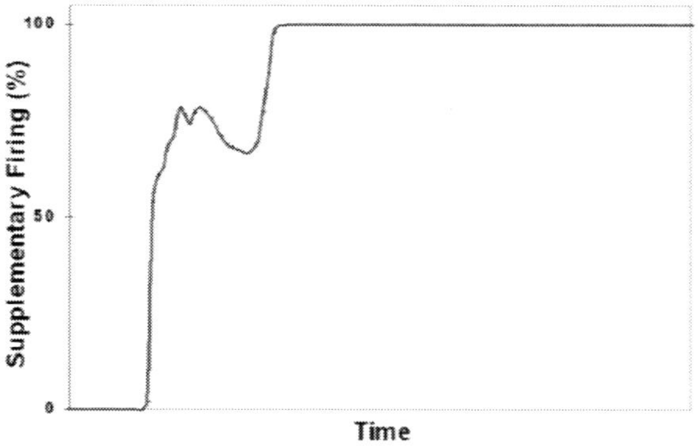

Figure 8: Supplemental firing during propane compressor trip

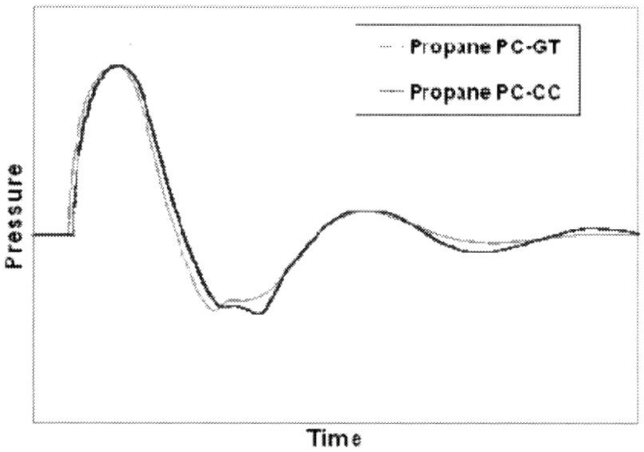

Figure 9: Propane Suction pressure during propane compressor trip

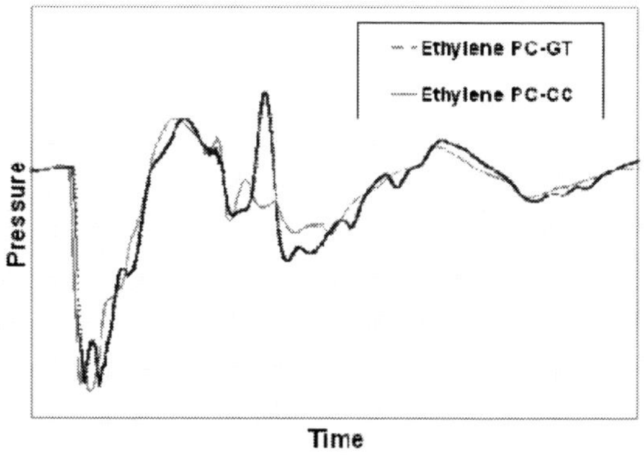

Figure 10: Ethylene Suction pressure during propane compressor trip

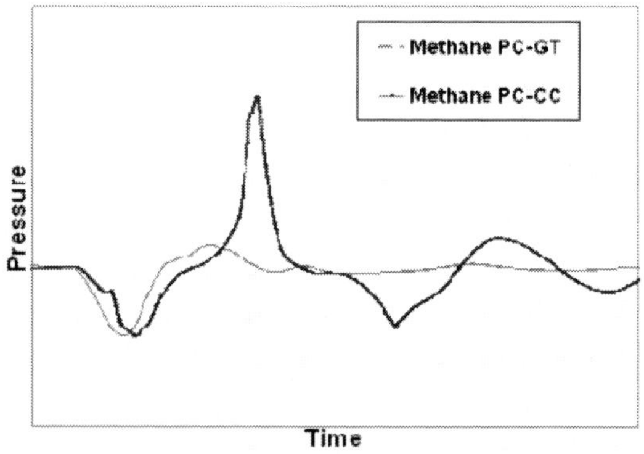

Figure 11: Methane Suction pressure during propane compressor trip

Economic Analysis

Reliability, availability, and maintainability (RAM) are three primary components that influence LNG production at a liquefaction facility. As described earlier, reliability is a key feature for the Optimized Cascade process using the "two-in-one" train concept.

The largest single impact on production is the availability of the turbine drivers. Reliability of most proven gas turbines is in excess of 99% and availability is over of 96%. The availability is highly dependent on the owner's scheduled maintenance strategy and the turbine selected (i.e. frame or aeroderivative).

A detailed economic evaluation is needed to justify a combined cycle liquefaction plant. The value provided by the increase in thermal efficiency of the combined cycle plant will have to be evaluated against the increase in capital and maintenance costs and the decrease in plant availability. The cost of operations and maintenance does not have a major impact on the economic analysis and can be ignored in a preliminary evaluation.

The major factors affecting the economics are the cost of gas supply and the LNG sales price. Gas supply to a liquefaction facility is also an issue, as it is usually constrained in some manner. A combined cycle design conserves some feed gas that can be valued in at least two ways.

The first method assumes that the feed gas rate decreases while keeping the LNG production constant. The savings would be the value of the feed gas saved using the cost of the gas. The second technique assumes that the fuel gas saved would be converted into LNG. The extra LNG could be produced by two different schemes. In the first, the feed gas rate remains constant and the LNG production rate increases. This increase will likely require some additional investment to increase capacity. The second way would be to produce the LNG when the gas supply is no longer at plateau, but has declined. The actual gas production would keep the plant full longer. Both methods should be evaluated in pre-FEED in order to maximize the value to the owners.

Conclusion and Discussion

This paper demonstrated the operational viability of a combined cycle driver for ConocoPhillips Optimized Cascade process. For the chosen configuration, the hallmark flexibility of the Optimized Cascade process was retained. A HRSG with duct firing may be necessary to maintain performance over various operating upsets.

The decision on driver selection should take into account a rigorous economic analysis. Steam turbines lead to decreased availability and increased start-up times. The impact of these variables should be balanced against potential gains from efficiency improvements. Further, it should be recognized that a combined cycle configuration need not be limited to the liquefaction. There is increasing interest for LNG facilities driven by electric motors alone (Martinez et al, 2005). In such cases, the combined cycle concept could be utilized for on-site power generation to monetize efficiencies.

References

- "A Focus on Balance – A Novel Approach Taking the Phillips Optimized Cascade LNG Process Into The Future", Qualls, W. and Hunter, P., AIChE Spring National Meeting, 2003.
- "All Electric Motor Drives for LNG Plants", Martinez, B., Meher-Homji, C., Paschal, J. and Eaton, A., Gastech 2005.
- "Natural Gas Liquefaction Process Designers Look for Larger More Efficient Liquefaction Designs", Avidan, A., Varnell and Martinez, B., Oil & Gas Journal, August 2003.
- "Egyptian LNG, the Value of Standardization", Redding, P., Hernandez, R., Qualls, W. and Avidan, A., Gastech 2005.
- "Virtual Simulation of LNG Plant", Valappil, J., Mehrotra, V., Messersmith, D. and Bruner, P., LNG Journal, Jan/Feb 2004

Biography

3000 Post Oak Blvd., Houston, TX 77056
Phone: 713-235-4340
E-mail: rptekuma@bechtel.com.

Ram Tekumalla is a Senior Simulation Specialist with the Advanced Simulation & Analysis Group, Bechtel Corporation, Houston. He is involved in development of dynamic models for a variety of projects. Over the past 4 years, he has used his expertise to develop Operator Training Simulators for 3 LNG plants. Ram is actively involved in various studies sponsored by the Bechtel – ConocoPhillips LNG Project Development Center.

Ram Tekumalla started his career in 1998. He worked at Cape Software and Invensys prior to joining Bechtel. Ram earned his M.S in Chemical Engineering from the University of Massachusetts, Amherst; and B.E. (Hons) in Chemical Engineering from Birla Institute of Technology & Science, Pilani.

A Graphical Approach to Assess Gasification: Novel Routes for F T Synthesis

Lwazi Ngubevana
University of the Witwatersrand
Johannesburg, South Africa
lwazi@prme.wits.ac.za

Diane Hildebrandt
University of the Witwatersrand
Johannesburg, South Africa
diane.hildebrandt@comps.wits.ac.za

David Glasser
University of the Witwatersrand
Johannesburg, South Africa
david.glasser@wits.ac.za

Abstract:

The gasification of coal as a process is one of the most important in supporting our daily modern life, as it is used for electricity generation for our cities and in countries like South Africa; it is used for the production of CO and H2 (syngas) that is used as feedstock for the production of fuels through the Fischer Tropsch (FT) process.

It is of the utmost importance that we have a proper understanding of the gasification process; as optimising this process can help in the production of cheaper electricity, optimising the FT process and therefore lessens the dependence on natural gas and oil. There are a number of ways that one can interpret and optimise gasification and the purpose of this paper is to use graphical techniques to do this without the use of experiments. We are currently working with the Golden Nest Company to build a demonstration FT plant in China and we are using this approach in its design.

It is envisaged that we can use the fundamental principles of mass balance, energy balance, reaction equilibria and thermodynamics, to predict and therefore to optimise gasification under specific conditions. Employing these principles together with defined possible chemical reactions in the gasifier and arbitrarily chosen reaction extents, one is then able to achieve the objective and that is to optimise the feed composition to the process and the process conditions (temperature and pressure), so as to achieve a specified composition of syngas. There were eight possible reactions that were chosen and two of those were assigned reaction extents □1 and □2 and these were the syngas production from carbon and steam reaction (gasification reaction), and the other was the water gas shift reaction (WGS), respectively.

In solving the problem, the assumptions made were, the amount of carbon fed into the reactor was just enough for the production of syngas and for combustion with oxygen and that the gases in the system exhibit ideal gas behaviour. The feasible region of gasification was determined in three ways; firstly a mass balance was performed on the reaction components (hydrogen, steam, carbon monoxide, and carbon dioxide; and this can be viewed as an atom balance on hydrogen, carbon and oxygen), with the use of the defined reaction extents. Secondly, the reaction equilibrium constants were defined making use of the defined reaction extents and the ideal gas behaviour assumption and finally the energy balance was set up; also making use of the defined reaction extents and the resulting equations were solved numerically and their respective curves were plotted. The area within the abovementioned curves then gives us an idea where gasification is feasible. The feed composition and the process conditions were varied so as to investigate the effects this would have on the feasible gasification region. Also investigated were the H2:CO and the H2:CO2 ratios and the purpose of this was to determine the 'best' feed composition for the optimum production of hydrogen that is needed as a reactant in the FT process.

The results of varying the feed composition can be summarised as follows, increasing the amount of steam and increasing the amount of oxygen have the effect of increasing the feasible region but these have the consequence of high energy losses from the gasifier and this energy is lost as either temperature or large amounts of unreacted steam. It is also found that at relatively low temperatures, carbon deposition becomes a problem and therefore high temperatures are necessary. There are however in real world operations, limits to how high the temperatures can go depending on the nature of the coal used. An overall analysis of the reaction equilibria shows that high pressures are also necessary as this minimises carbon deposition and methane production, but this also has limitations such as downstream pressure requirements. As an overall result of the analysis it was concluded that graphical techniques can be employed to understand gasification.

Characterization of Active Species in Rb Promoted Iron Catalyst during Fischer-Tropsch Synthesis: an Exafs, Xanes and Mossbauer Study

Amitava Sarkar
University of Kentucky Center for Applied Energy Research
Lexington, KY 40511
sarkar@caer.uky.edu

Gary Jacobs
University of Kentucky Center for Applied Energy Research
Lexington, KY 40511
jacobs@caer.uky.edu

Mingsheng Luo
University of Kentucky Center for Applied Energy Research
Lexington, KY 40511
luo@caer.uky.edu

Burtron H. Davis
University of Kentucky Center for Applied Energy Research
Lexington, KY 40511
davis@caer.uky.edu

Abstract:

Group I alkali metal promoted iron-based catalysts for Fischer-Tropsch (FT) synthesis offer a number of benefits regarding product selectivity. Rubidium promoted iron catalysts are prepared with two Rb/Fe atomic ratios (1.44 and 5) using rubidium nitrate and rubidium carbonate as rubidium precursors. Results of catalytic activity and deactivation studies in a CSTR revealed that Rb promoted catalyst results in a steady conversion with a lower deactivation rate than the corresponding unpromoted catalyst although the initial activity of the promoted catalyst is almost half that of the unpromoted catalyst. Mössbauer spectroscopic measurements of CO activated and working catalyst samples indicate that the composition of the iron carbide phase formed after carbidation is χ-Fe_5C_2 for both promoted and unpromoted catalysts. However, in the case of the rubidium promoted catalyst, ε'-$Fe_{2.2}C$ becomes the predominant carbide phase as FTS continues and the overall catalyst composition remains carbidic. The carbide content of the unpromoted catalyst is found to decline very quickly with synthesis time. EXAFS and XANES synchrotron techniques are used to determine the chemical compound that most closely represents the state of the Rb promoter in the working catalyst under FTS environment. Results of XANES measurements suggest that Rb is present in the oxidized state and that the compound most prevalent in the active catalyst samples closely resemble rubidium carbonate. Relative intensity of the XANES white line indicate that the catalyst with a Rb/Fe atomic ratio of 5 is more oxidized than the catalyst with a Rb/Fe atomic of 1.44. Fourier transformed magnitude spectra for the Fe K-edge from EXAFS measurement indicate that the Fe-Fe coordination peaks are in good agreement with those observed for the iron carbide reference sample at approximately 2.2 and 2.5 Å. Furthermore, there are low R peaks (e.g., ~1.5 Å) that exhibit good agreement with the iron carbide reference, and may correspond to Fe-C coordination. However, it is not possible to rule out the Fe-O coordination from the contribution of iron oxide in the samples.

Evaluation of a Long-Term Fischer-Tropsch Test and the Resulting Spent Iron Catalyst

Santosh K. Gangwal
RTI International
Research Triangle Park, NC 27709
skg@rti.org

Velu Subramani
RTI International
Research Triangle Park, NC 27709

David A Green
RTI International
Research Triangle Park, NC 27709

Abstract:

A precipitated iron Fischer-Tropsch catalyst (100 Fe: 5Cu: 4.2 K: 17 SiO2) was run continuously for 500 hours in a 1.0 liter CSTR under highly carburizing conditions. The slurry initially consisted of 10 g of the catalyst dispersed in Oronite synfluid. The simulated syngas feed for the test consisted of H2 and CO with a small amount of argon as an internal standard. The H2 to CO mole ratio of the syngas feed was maintained at 0.7 to simulate a syngas obtained from a biomass or coal gasifier. Following catalyst activation in syngas, steady state conditions were established at 375 psig and 240oC at a space velocity of 2,200 per hour. Under these conditions, the reactor outlet H2 to CO mole ratio was approximately 0.6. The CO conversion and C5+ hydrocarbon yield over the run were constant at about 40 % and 2 g/h, respectively, for about 425 hours when the activity began to decline rapidly. Over-carburizing of the catalyst is thought to be the primary reason for this decline, as such a decline has not been observed in previous long-term tests of 500 hours under less carburizing conditions. The catalyst sample was recovered from the product wax and analyzed by XRD, XPS, and Mossbauer spectroscopy to understand the phase changes associated with the activity decline.

Fischer-Tropsch Synthesis: Investigation of the Metal-Support Interaction of Cobalt-Based Catalysts by Standard Tpr and Synchrotron-Based Tpr-Xanes/Tpr-Exafs Techniques

Gary Jacobs
University of Kentucky Center for Applied Energy Research
Lexington, KY 40511
jacobs@caer.uky.edu

Yaying Ji
University of Kentucky Center for Applied Energy Research
Lexington, KY 40511
ji@caer.uky.edu

Burtron H. Davis
University of Kentucky Center for Applied Energy Research
Lexington, KY 40511
davis@caer.uky.edu

Donald Cronauer
Argonne National Laboratory
Argonne, IL 60439
cronauer@cmt.anl.gov

Christopher L. Marshall
Argonne National Laboratory
Argonne, IL 60439
marshall@cmt.anl.gov

A. Jeremy Kropf
Argonne National Laboratory
Argonne, IL 60439
kropf@cmt.anl.gov

Abstract:

The application of a multi-sample holder to carry out TPR-XANES/EXAFS [1] provided key information for verifying the nature of the chemical transformations occurring during the activation of cobalt-based Fischer-Tropsch catalysts activation in H2, as well as providing insight into the resulting crystallite size as a function of the catalyst support interaction with the Co- oxide species. A 2-step reduction process involving Co3O4 to CoO to Co0 transformations over standard calcined catalysts was quantified over catalysts exhibiting both weak (e.g., Co/SiO2) and strong interactions (e.g., Co/Al2O3) with the support. The impacts of Co loading and reduction promoter addition (e.g., Pt) on extent of reduction and Co crystallite size were also investigated. Results were in good agreement with, and assisted in the interpretation of, standard H2-TPR and H-chemisorption / pulse reoxidation data [2]. The results are in contrast with reduction models that assume direct reduction of Co oxide species to Co0 in both TPR peaks [3,4]. Support type, and not surface area, was found to be a key factor in determining the strength of the interaction and the rate at which the cobalt oxides underwent reduction. Yet not only that, the more weakly interacting support (in this case, SiO2) yielded a much larger cobalt crystallite size which is detrimental to the resulting active Co metallic surface area (a linear correlation exists between Co0 surface and CO conversion rate [5]). In contrast, after a standard reduction treatment, despite a much lower extent of reduction, the strongly interacting support (Al2O3) yielded much smaller Co crystallites, which in fact provided a higher cobalt metallic surface area. Further gains in extent of reduction were managed by either (a) increasing the loading to provide a larger particle size that weakened the interaction with the support or (b) utilizing a noble metal promoter (i.e., Pt) to facilitate reduction, most likely by a hydrogen dissociation and spillover mechanism [6].

Acknowledgements: Argonne's research supported by U.S. Department of Energy (DOE), Office of Fossil Energy. Use of the Advanced Photon Source was supported by the U. S. DOE, Office of Basic Energy Sciences, under Contract No. DE-AC02-06CH11357.MRCAT operations are supported by the Department of Energy and the MRCAT member institutions

REFERENCES 1. M.K. Neylon, C.L. Marshall, A.J. Kropf, JACS 124 (2002) 5457. 2. G. Jacobs, T. Das, Y. Zhang, J. Li, G. Racoillet, B.H. Davis, Appl. Catal. 233 (2002) 263. 3. P. Arnoldy, J.A. Moulijn, J. Catal. 93 (1985) 38. 4. A. Lapidus, A. Krylova, V. Kazanskii, V. Borovkov, A. Zaitsev, J. Rathousky, A. Zukal, M. Jancalkova, Appl. Catal. 73 (1991) 65. 5. E. Iglesia, S.L. Soled, R.A. Fiato, J. Catal. 137 (1992) 212. 6. G. Jacobs, J.A. Chaney, P.M. Patterson, T.K. Das, J.C. Maillot, B.H. Davis, J. Synch. Rad. 11 (2004) 414.

Minimizing Carbon Dioxide Emissions from Fuel Plants by Integrating Processes

Bilal Patel
University of the Witwatersrand, Johannesburg
Johannesburg, South Africa
Bilal.Patel@wits.ac.za

David Glasser
University of the Witwatersrand, Johannesburg
Johannesburg, South Africa
david.glasser@wits.ac.za

Diane Hildebrandt
University of the Witwatersrand, Johannesburg
Johannesburg, South Africa
diane.hildebrandt@wits.ac.za

Brendon Hausberger
University of the Witwatersrand, Johannesburg
Johannesburg, South Africa
brendon.hausberger@comps.wits.ac.za

Abstract:

Increased environmental awareness and efficient energy utilization requires the development of improved methodologies for the synthesis and design of chemical processes. A new approach of synthesizing process from a mass, energy and entropy point of view has been developed (Patel et. al., 2006). This design approach can be utilized in the very early design stages and requires only simple thermodynamic data to effectively synthesize a process. This tool effectively sets targets for the overall process, determines the limiting variable (mass, energy or work) and allows one to scan all the various options easily.

It can be shown that the mass, energy and entropy balance for an overall process are all linear and by employing a linear programming model, with the various constraints and a suitable objective function one can determine the feasible region for the overall chemical process i.e. the overall mass balance.

We employ an example of co-generating hydrocarbon liquids and methanol (and heat) from a methane and/or coal feed. We will show that it is possible to determine an overall process which requires no energy or work input and does not produce carbon dioxide. Thus, in principle it is possible to design a zero-emissions overall process.

These ideas are to be tested in a pilot plant currently being built in China by the Golden Nest Company.

References:

Patel B, Hildebrandt D, Glasser D, Hausberger B; The synthesis and integration of chemical processes from a mass, energy and entropy perspective. Submitted, 2006.

Synchrotron Computed Microtomography (Cmt) and Micro X-Ray Fluorescence (ìXrf) as Tools to Map Fischer-Tropsch Catalysts

Devinder Mahajan
Brookhaven National Laboratory and Stony Brook University
Upton, NY 11973
dmahajan@bnl.gov

Mouzhgun Anjom
Stony Brook University
Stony Brook, NY 11794
manjom@gmail.com

Keith Jones
Brookhaven National Laboratory
Upton, NY 11973
kwj@bnl.gov

Abstract:

Gas-to-liquids (GTL) remains an attractive pathway to produce clean fuels from carbon feedstocks via the synthesis gas route. Technological advances in the Fischer-Tropsch (F-T) option to synthesize clean hydrocarbon fuels continue but a highly efficient process remains elusive. Heterogeneous catalyst assemblies undergo changes during the GTL conversion reaction and an in-depth characterization of these catalytic materials would help design the next-generation more efficient catalysts. Computed Microtomography (CMT) and micro X-ray fluorescence (ìXRF), techniques at the National Synchrotron Light Source (NSLS), Brookhaven National Laboratory (BNL), are complementary to allow mapping of catalyst constituents. The experimental set-up, collected data, and the potential of CMT and ìXRF techniques for monitoring changes in the catalyst composition during F-T synthesis will be presented.

Transient Flow Analysis in LNG Regasification Terminal Design

Brett Wink
Principal Process Engineer
Aker Kvaerner
Houston, Texas

Stephen Marcum
Process Manager
Aker Kvaerner
Houston, Texas

AIChE Spring Meeting, April 2007
7th Topical Conference on Natural Gas Utilization
Houston, Texas, April 22-27, 2007

As the US energy demand rises and domestic energy production remains relatively flat, liquefied natural gas (LNG) is gaining importance as an economically favorable energy source. As a result, many new LNG regasification terminals will be built in the US. The design of the regasification terminal must provide reliable and safe operation, both to ensure a consistent supply of energy to consumers and to minimize the Terminal impact on the surroundings. The Terminal's safety systems typically include fire and gas release monitoring. The final element of the safety system includes emergency shutdown valves (ESDVs) installed throughout the Terminal, to isolate inventory, thereby preventing intensification and escalation of a possible event. To achieve the goal of safe operation, the Terminal's safety system must be analyzed to balance rapid detection and isolation against the effects of transient flow (i.e. waterhammer) which can produce significant pressure rises in process piping. This paper discusses the areas of the Terminal where transient flow may be expected, outlines transient flow analysis methodology and provides recommendations for minimizing the safety and risks associated with transient flow.

INTRODUCTION

Regulatory requirements and good engineering practice necessitate the consideration of transient flow (i.e. waterhammer) in the design of piping systems. Transients may be caused by starting or stopping of pumps or by intentional or unintentional changes in valve position. The effect of these events is the sometimes rapid conversion of kinetic energy, in the form of fluid momentum, to pressure energy. This conversion can cause the pressure in the piping system to exceed the design pressure and can exert high forces on piping supports.

THEORY

Neglecting frictional losses, an instantaneous change in the velocity, v, of a fluid with density ρ, results in a head increase of:

$$\Delta H = -\frac{a\Delta v}{\rho} \quad (1)$$

Where a is the wavespeed. For complete flow stoppage, the head increase is:

$$\Delta H = -\frac{a v_0}{\rho} \quad (2)$$

where v_0 is the initial velocity of the fluid. This value represents the maximum head rise as it neglects the conversion of a portion of the kinetic energy to heat in the form of frictional losses in the flow of the fluid.

The wavespeed is calculated from the continuity equation, considering the accumulation of mass in a control element due to radial and axial extension of the piping segment and an increase in the fluid density (compression) within the control element. As a result:

$$a = \sqrt{\frac{\dfrac{\Delta p}{\rho}}{\dfrac{\Delta A}{A} + \dfrac{\Delta \rho}{\rho}}} \quad (3)$$

where A is the cross sectional area of the pipe. The bulk modulus of a fluid, K, is defined as the change in pressure required to cause a fractional change in density:

$$K = \frac{\Delta p}{\Delta \rho / \rho} \quad (4)$$

The wavespeed then becomes:

$$a = \sqrt{\frac{K/\rho}{1 + K\left(\dfrac{\Delta A}{A\Delta p}\right)}} \quad (5)$$

The change in pipe dimension under stress may be approximated as:

$$\frac{\Delta A}{A\Delta p} = \frac{D}{Et} \quad (6)$$

Where D is the pipe diameter, t is the pipe wall thickness and E is the bulk modulus of the pipe material. As a result:

$$a = \sqrt{\frac{K/\rho}{1 + K\left(\frac{D}{Et}\right)}} \tag{7}$$

Table 1. Wavespeed for various line classes

Pipe size, in	**8**	**24**	**30**	**36**
Wall thickness, in	0.594	0.250	0.312	0.375
Diameter, in	7.437	23.500	29.376	35.250
Design pressure, psig	2160	210	210	210
Material	304SS (bulk modulus, E = 4.28 Glb/ft2)			
Fluid	LNG (density = 28.7 lb/ft3) (bulk modulus, K = 0.0198 Glb/ft2)			
Wavespeed, ft/s	4538	3942	3937	3934

RESULTS

Typical results of a transient flow analysis for an LNG regasification terminal are presented below. The following portions of the Terminal were modeled (see Figure 1 for reference:)

- carrier unloading system
- LNG storage and BOG recovery
- high pressure sendout pumps

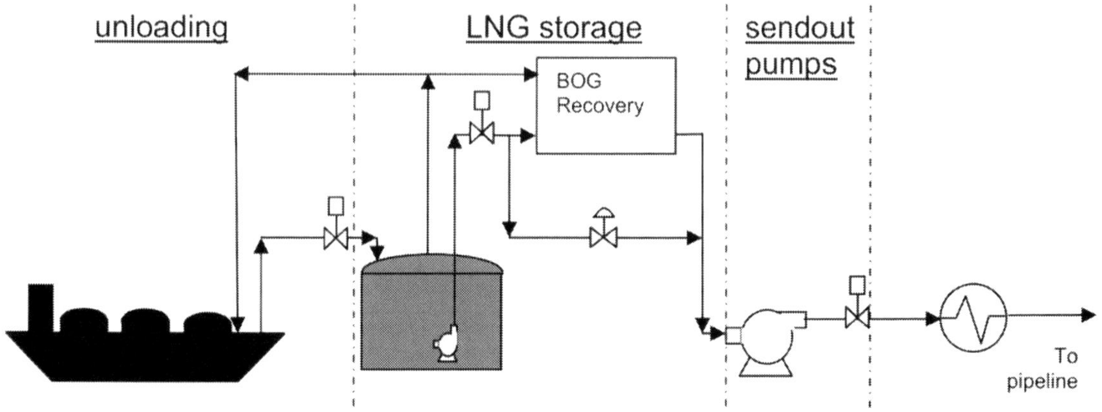

Figure 1. Typical LNG Regasification Terminal

The unloading and storage areas of the Terminal are characterized by relatively low operating pressures, generally consisting of piping with an ANSI flange rating of 150#.

The piping systems in the storage area of the Terminal are in continuous service. Operating fluid velocities are on the order of 10 ft/s. Operation in the unloading area of the Terminal is intermittent. Normal fluid velocities in this portion of the Terminal can approach 20 ft/s, depending upon required carrier unloading rate and line size. As indicated by Equation (2), the initial fluid velocity can be used to predict that maximum transient pressure rise that a given fluid system can experience.

Carrier Unloading System

LNG is typically delivered to the Terminal by large oceangoing carriers. Cargo pumps installed on the carrier are used to offload the LNG cargo to the Terminal's LNG Storage Tanks. Due to the motion of the LNG carrier even while moored, a flexible connection between the Terminal and carrier is required, the most common of which are marine unloading arms.

Due to the size and draft of the LNG carrier, the Jetty is usually a large distance from the Terminal. Piping between the Jetty and Terminal is typically large bore (>30") in order to accommodate required unloading rates of 12,000 to 17,500 m3/hr (52,800 to 77,000 gpm).

The results presented below in Figure 2 and Table 2 are from a carrier unloading model. Pipe size is 36" with a length of 3100 ft. Portions of the 12" jetty keep cool line are utilized during carrier unloading, resulting in a parallel flow path from the carrier to the Terminal. Initial carrier unloading rate is as listed in Table 2, via four marine unloading arms. Closure of a 36" butterfly valve (equal percentage performance characteristic) is initiated at t = 5 s. The pressure profile at the valve inlet, for various closure times, is presented in Figure 2.

Upon completion of valve closure, the transient pressure rise reaches its maximum. The pressure wave eventually dissipates due to the elastic and friction forces imposed on the fluid by the piping system. The piping system then attains a steady state pressure. For simplicity and clarity, the pressure data calculated after completion of valve closure are omitted from Figure 2, Figure 3, Figure 4 and Figure 5.

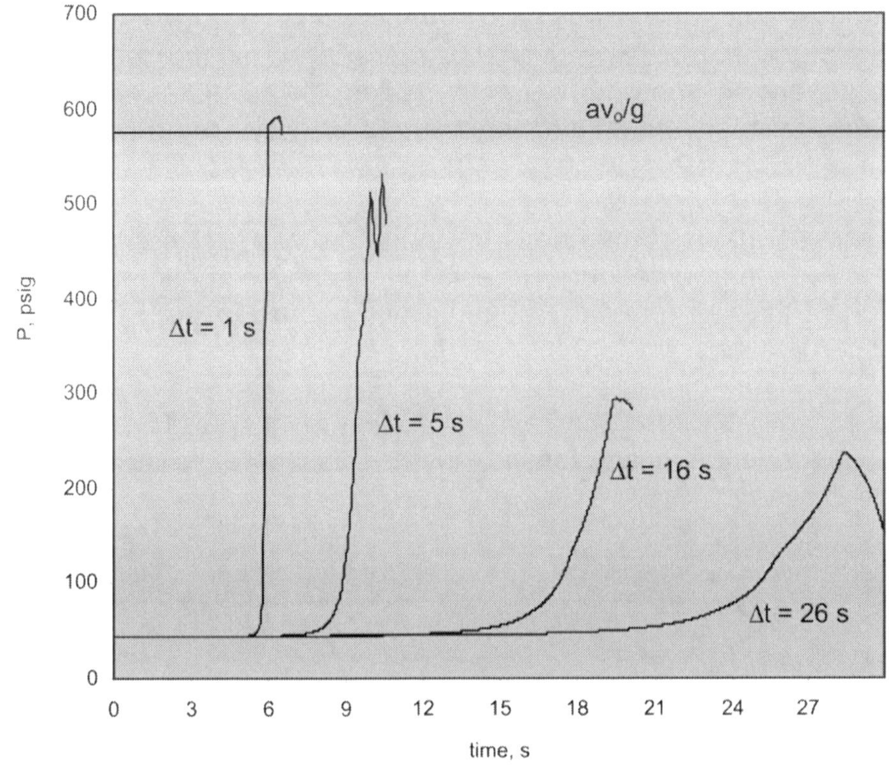

Figure 2. Pressure profile for valve closure during carrier unloading.

As shown in Figure 2, rapid valve closure can result in a peak transient overpressure of many times operating pressure. This is caused by the rapid conversion of momentum (kinetic energy) to pressure (static energy.) Relatively long valve closure times are required in order to minimize the transient pressure rise. The maximum pressure rise in the unloading circuit for various valve closure times is listed in Table 2. Note that these values compare well to the maximum head rise predicted by Equation (2).

Table 2. Pressure rise for carrier unloading system.

Valve closure time, s	**26**	**16**	**5**	**1**
Flow rate, gpm	66332	66332	66332	66332
Line size, in	36 (0.375" wall thickness)			
Initial velocity, ft/s	21.8	21.8	21.8	21.8
Initial pressure, psig	44	44	44	44
Maximum pressure, psig	238	296	529	592
pressure rise, psi	194	252	485	548
Head rise, ft	973	1264	2433	2749
av0/g, ft	2663			

BOG Recovery System

LNG is extracted from the Terminal's storage tanks by the intank pumps. These submerged motor pumps deliver LNG to the suction of the high pressure sendout pumps and, depending upon the Terminal design, to the BOG recovery system. LNG may be sourced from several tanks simultaneously by operating combinations of intank pumps.

In this example, LNG flows from each tank to the discharge header through 300 ft of 24" piping. The 30" discharge header connects the discharge of each tank and carries LNG to the suction of the high pressure sendout pumps and to the BOG recovery system. The 30" discharge header is 1000 ft in length.

Figure 3 and Table 3 depict the results of closure of the 30" butterfly valve (equal percentage performance characteristic) in the discharge header. Valve closure is initiated at t = 5 s.

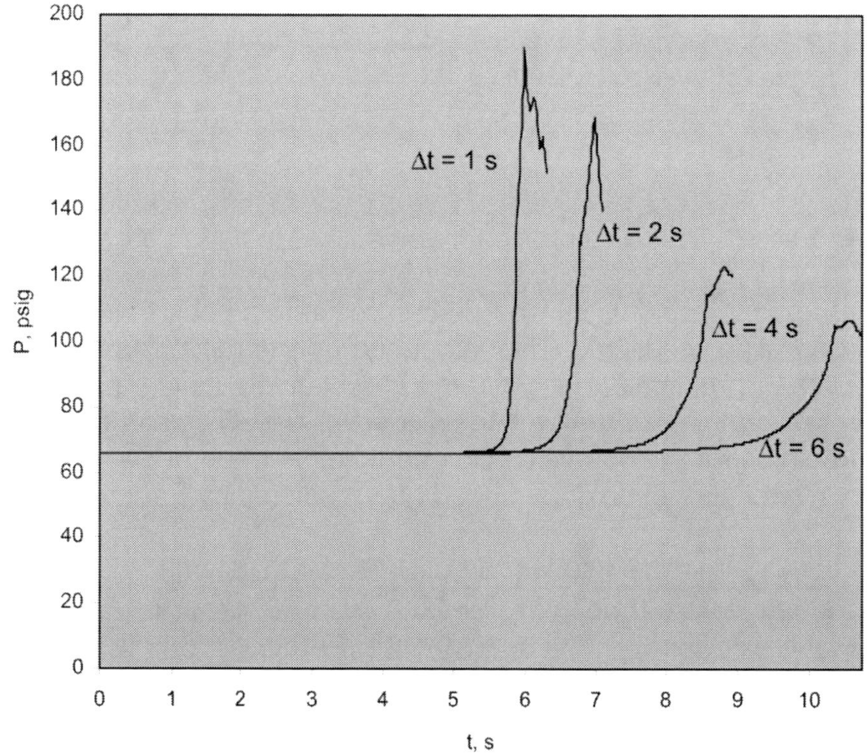

Figure 3. Pressure profile for valve closure upstream of BOG recovery.

As shown in Table 3, rapid valve closure leads to the largest transient pressure rise. Note that the transient pressure rise is relatively small compared to the carrier unloading system, even for the 1 s valve closure duration case. This is a result of the low initial fluid velocity.

Table 3. Pressure rise at BOG recovery.

Valve closure time, s	6	4	2	1
Flow rate, gpm	23448	23448	23448	23448
Line size, in	30 (0.3125" wall thickness)			
Initial velocity, ft/s	11.1	11.1	11.1	11.1
Initial pressure, psig	66	66	66	66
Maximum pressure, psig	109	123	170	208
pressure rise, psi	43	57	104	142
Head rise, ft	216	286	522	712
av0/g, ft	1357			

LNG Storage System

The results given below are from a model of an individual tank discharge line. A 24" shutdown valve is installed at the end of the discharge line. The line size is 24". The length is 300 ft. The results for this case are shown in Figure 4 and Table 4. Closure of a 24" butterfly valve (equal percentage performance characteristic) is initiated at t = 5 s.

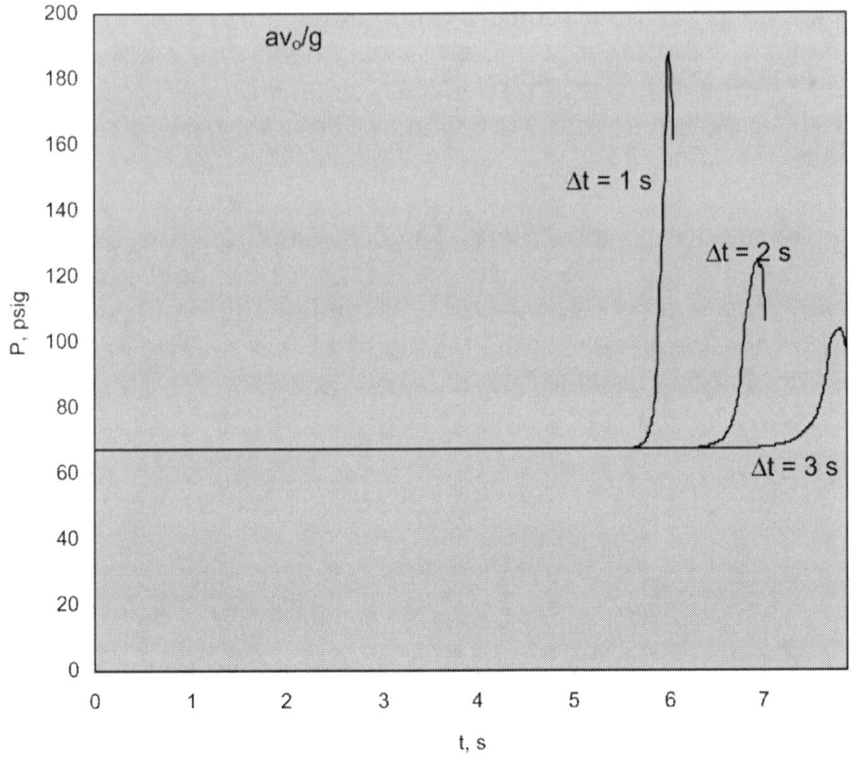

Figure 4. Pressure profile for valve closure at discharge of storage tank.

This model does not consist of parallel flow paths or multiple sources of flow or pressure. As a result, the pressure profiles for this case a fairly smooth compared to those depicted in Figure 2 and Figure 3.

Table 4. Pressure rise at individual storage tank discharge.

Valve closure time, s	**3**	**2**	**1**
Flow rate, gpm	7474	7474	7474
Line size, in	24 (0.250" wall thickness)		
Initial velocity, ft/s	5.5	5.5	5.5
Initial pressure, psig	67	67	67
Maximum pressure, psig	104	125	186
pressure rise, psi	37	58	119
Head rise, ft	186	291	597
av0/g, ft		673	

High Pressure Sendout Pumps

The high pressure sendout pumps take suction from the intank pumps. These pumps increase the LNG pressure to above the critical point and deliver the stream to the Terminal's vaporizers. Operating pressures in this portion of the Terminal are on the order of 1500 psig. Piping components in this area of the Terminal are typically 600# or 900# ANSI flange rating in order to withstand the high operating pressures.

In this model, multiple sendout pumps are connected in parallel to a 24" discharge manifold. This piping manifold routes the supercritical LNG to the vaporizer header. In this model, multiple submerged combustion vaporizers are connected in parallel to the vaporizer feed header. The vaporized LNG is collected in a discharge header, metered and sent to the pipeline for consumption. The vaporizer discharge header is not considered here.

For the case depicted in Figure 5 and Table 5, simultaneous closure of all vaporizer inlet valves is considered. The model consists of a sendout pump discharge header 300 ft in length. The feed to each vaporizer includes 100 ft of 8" line. Closure of the 8" isolation valve at the vaporizer inlet is initiated at t = 3 s. The inlet valve has an equal percentage performance characteristic.

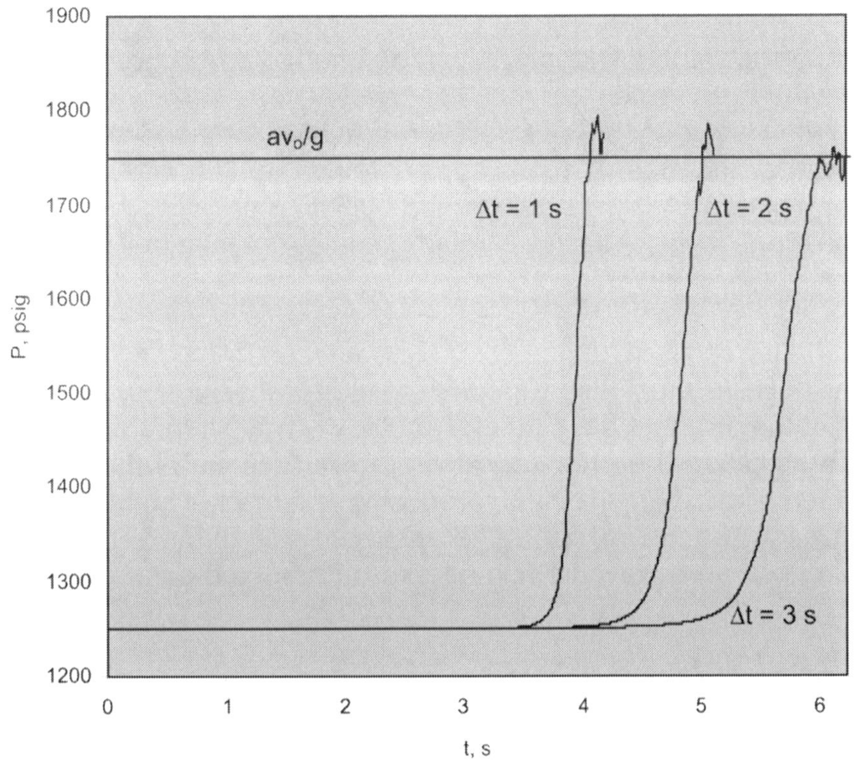

Figure 5. Pressure profile for valve closure at sendout pump discharge

As predicted by theory, values depicted in Figure 5 are listed in Table 5 indicate that shorter valve closure times result in a higher transient pressure rise.

Table 5. Pressure rise at sendout pump discharge for various valve closure times.

Valve closure time, s	**3**	**2**	**1**
Flow rate, gpm	2305	2305	2305
Line size, in	8 (0.594" wall thickness)		
Initial velocity, ft/s	17	17	17
Initial pressure, psig	1250	1250	1250
Maximum pressure, psig	1759	1785	1795
pressure rise, psi	509	535	545
Head rise, ft	2554	2684	2734
av0/g, ft		2396	

CONCLUSION

Transient flow analysis can be utilized to determine the forces exerted on piping systems during an initiating event, such as valve closure or pump start-up. These forces are caused by the rapid propagation of a pressure pulse through the fluid within the piping. A thorough piping support design will consider the forces that can be generated from the impingement of a pressure pulse on an elbow, weldcap, valve or other piping component.

The results of a transient flow analysis can be used to prescribe minimum closure times for isolation valves, in order to limit the transient pressure rise to within acceptable limits. Since the valves under consideration are usually equipped with spring return (fail closed) actuators, the design must consider limiting the rate at which the actuator depressures. This is typically accomplished by fitting the vent port of the actuator with a restriction orifice. Orifice sizing should consider minimum Instrument Air supply pressure.

Although not discussed here, liquid column separation, a phenomenon in which pockets of fluid vapor develop must also be considered in transient flow analysis. This can be experienced at the downstream side of a shutdown valve if the momentum of the fluid is large enough to cause the pressure at the valve face to fall below the bubble point. Eventually, the liquid column rejoins, producing a large transient pressure rise. Liquid column separation can also occur at the upstream side of the valve if the momentum of the rebounding pressure wave is high enough to reduce the fluid pressure to below the bubble point.

REFERENCE

Wylie, E. Benjamin and Streeter, Victor L. Fluid Transients in Systems; Prentice-Hall, 1993.

BIOGRAPHY

PICTURE

3600 Briarpark Drive, Houston, Texas 77042
Phone: 713.414.3384
Email: brett.wink@akerkvaerner.com

Brett Wink is a Principal Process Engineer at Aker Kvaerner. He is involved in the design and construction of LNG regasification terminals. Mr. Wink also has experience in the offshore oil and gas, energy and petrochemical sectors. Mr. Wink has experience in fire protection engineering and risk assessment studies.

Mr. Wink started his career with Kvaerner John Brown in 1998. He has worked on onshore and offshore LNG regasification terminal design projects in the USA and Europe. He has worked on siting studies for US regasification Terminals.

Mr. Wink graduated from the University of Missouri-Rolla in 1997 with a B.S. in Chemical Engineering. He is a Registered Professional Engineer in Texas and Louisiana.

PICTURE

3600 Briarpark Drive, Houston, Texas 77042
Phone: 713.270.2568
Email: stephen.marcum@akerkvaerner.com

Stephen Marcum is a Process Manager at Aker Kvaerner. He is involved in the design and construction of LNG regasification terminals.

Mr. Marcum began working at Kvaerner John Brown in 1989.

Mr. Marcum graduated from Texas Tech University in 1976 with a B.S. in Chemical Engineering. He is a Registered Professional Engineer in Texas.

Gas Interchangeability – Quality Issues and Solutions for US LNG Terminals

Kamal Shah
Technical Vice President
Aker Kvaerner, Houston, Texas

AIChE Spring Meeting, April 2007
7th Topical Conference on Natural Gas Utilization
Houston Hilton Hotel, April 22-26, 2006

Introduction

LNG continues to be increasingly important in meeting United States (U.S.) growing energy demand. US Energy Information Administration (EIA) [1] estimates natural gas consumption in the U.S. at approximately 22 trillion cubic feet for 2004. EIA also forecast the U.S. consumption for natural gas to increase to 26 trillion cubic feet in 2010 and further increase to approximately 31 trillion cubic feet by 2025. At present, imported Liquefied Natural Gas (LNG) makes up part of this natural gas demand. The supply of imported LNG will make up from 0.65 trillion cubic feet in 2004 to approximately 6.4 trillion cubic feet by 2025.

The LNG supplies originating from different natural gas fields around the world are unique and different from U.S. natural gas supplies. One common differential is a higher hydrocarbon content; also the natural gas derived from LNG is bone dry with very little inert content. The overall effect will be a higher gas calorific value (also referred in this paper as High Heating Value or HHV) in the gas delivered to the pipelines. The changing supplies and quality characteristics associated with the higher hydrocarbon content impacts market (economic) and operational practices of the interstate natural gas pipelines.

U.S. Pipeline Gas Quality

The natural gas pipeline quality specification in the U.S. has evolved over the last 150 years. The quality specification evolved as technical solutions developed for the problems faced in transporting the gas such as, liquid drop out, hydrate formation, corrosion to improving mechanical design for higher pressure pipelines. Along with dew point control requirements to avoid liquid drop out in the pipeline, the economic value for extraction of gasoline (C5 plus components) was recognized. Later on the economic value for LPG (propane and butane or C3's and C4's) extraction and finally extraction of ethane or natural gas liquids (NGL) for petrochemical industry feedstock gained importance in the U.S. Gulf Coast regions.

The result was a gas specification for interstate pipelines, which was very lean with a High Heating Value (HHV) not exceeding 1075-1100 Btu/scf. This lean gas is delivered to industries and residential customers in various regions of the U.S. According to U.S. Department of energy [2], most of the natural gas flowing to the U.S. pipelines is pre processed, to strip out ethane and heavier components leaving mostly methane in the natural gas. The existing infrastructure for gas processing and petrochemical industries in the Southern region supports this type of processing, but similar Natural Gas Liquids (NGL) pipeline and petrochemical infrastructure does not exist in the East coast and Northeast regions.

Gas Quality – Current Trend

Due to the tight natural gas supply, the pressure has mounted to allow natural gas with some of the ethane and heavier components to be imported. This so called "richer" gas with higher HHV is not desirable due to a flame that may be too large or too hot to burn for certain applications and appliances.

Over the last three years there has been a concerted effort to address this issue and Federal Energy Regulatory Commission (FERC) hosted public meetings to address it. The current responsibility lies with FERC to establish a formal set of gas quality standards to address the issue. At the request of FERC, the Department of energy (DOE) has undertaken a study researching gas quality and interchangeability. DOE has commissioned the National Energy Technology Laboratory (NETL) to conduct a detailed study to address the issue of gas interchangeability. FERC is likely to resist forcing the national standards due to geographic variation in the gas compositions. They are more likely to issue broad set of guidelines based on SoCal and NGC white paper[3], NETL study and findings based on cases such as AES v. FGT[4]. It is likely that the final rule making will address some sort of Wobbe Index (WI) characteristics, similar to the one utilized in Europe, to establish the gas pipeline quality standards for certain regions. The WI is also referred to as an interchangeability factor and is defined as follows:

$$WI = HHV / (SG)^{1/2}$$

where HHV is the Gross Heating Value for the gas in Btu/scf and SG is the gas specific gravity (Air=1.0). The current trend is to have the WI on a gas which results in a value of 1350 to 1400; this avoids various operational and environmental issues. The maximum and minimum value of WI may vary slightly depending on the regional agreement. The Figure 1 below is a graphical representation of gas quality. The pipeline quality specifications are also likely to establish some maximum inert specifications, similar to various pipelines requiring maximum 2-3 % nitrogen. In some cases such as AES v. FGT, there can also be limit on maximum CO_2 at 1.0 mole % and CO_2 injection to adjust the WI may not be allowed.

FIGURE 1

GAS QUALITY – CURRENT TREND

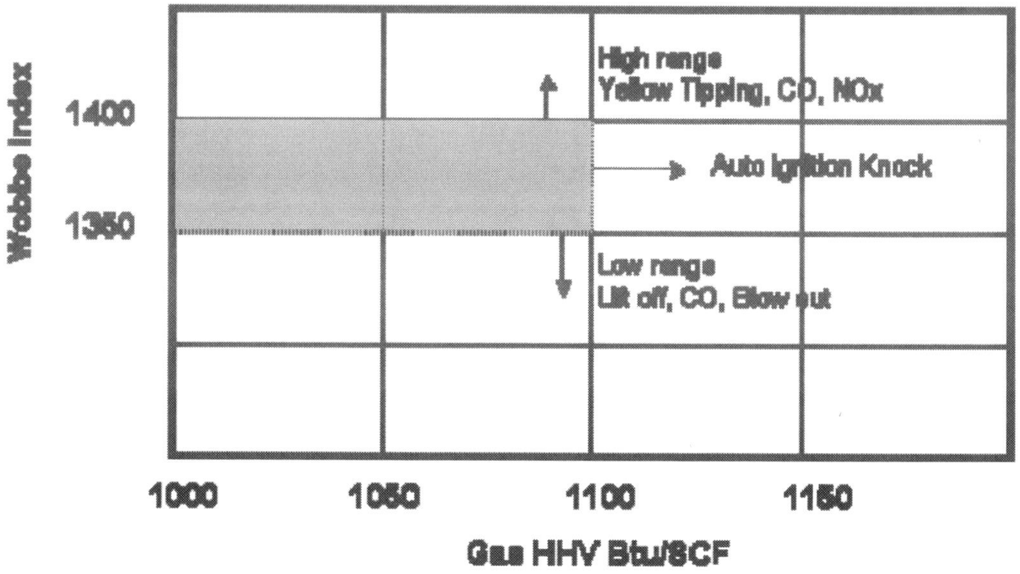

This paper presents various technical approaches that LNG supplier or LNG terminal operator may choose for different regions of the U.S. to meet the export pipelines gas quality requirement. It is desirable to provide flexibility in LNG supply sources to meet the ever increasing future energy demand. Depending on the source of LNG supply, the terminal operators can choose appropriate option to meet export gas quality requirements. The paper also discusses the advantages and disadvantages for each option.

Quality of Typical Imported LNG

At present there are five LNG import terminals with the supply of LNG coming from various countries [5]. The composition of imported LNG and the country of origin are listed in Table 1. It lists HHV for imported LNG, and the value varies between 1037 Btu/scf and 1146 Btu/scf. In most cases, the imported LNG shows a HHV value above 1100 Btu/scf, exceeding most of the U.S. pipeline standards. The pipeline gas quality standards at present vary from region to region. These regions are divided into three basic areas of the U.S. They are Pacific, Atlantic and Southern region. In the Atlantic region, especially the Northeast part, the pipeline companies desire to limit their gas below 1075 Btu/scf and would prefer to have it below 1050 Btu/scf.

The following discussion provides various technical approaches that LNG terminal operators can take in reducing HHV to meet export gas pipeline quality requirements.

TABLE 1

TYPICAL COMPOSITION OFIMPORTED LNG BY COUNTRY

Imported LNG Origin	Methane C_1 %	Ethane C_2 %	Propane C_3 %	Butane[+] $C_4{}^+$ %	Nitrogen N_2 %	HHV Btu/scf
Algeria	87.6	9.0	2.2	0.6	0.6	1119
Australia	89.3	7.1	2.5	1.0	0.1	1123
Malaysia	89.8	5.2	3.3	1.4	0.3	1128
Nigeria	91.6	4.6	2.4	1.3	0.1	1109
Oman	87.7	7.5	3.0	1.6	0.2	1146
Qatar	89.9	6.0	2.2	1.5	0.4	1118
Trinidad and Tobago	96.9	2.7	0.3	0.1	0.0	1037

Pipeline Gas Quality Control - Nitrogen Injection

Nitrogen injection is one of the most common methods utilized for gas HHV control. There are two ways to inject the nitrogen in the LNG terminal; low pressure and high pressure, the method depending on the amount of nitrogen injected. Figure 2 below is a block diagram to show the two alternatives.

FIGURE 2

EXPORT GAS BTU CONTROL WITH NITROGEN INJECTION

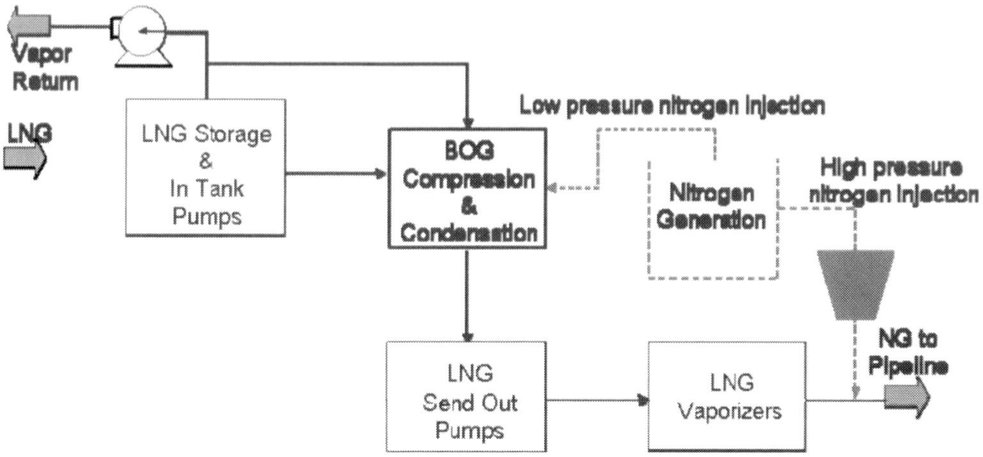

Obviously low pressure nitrogen injection is preferred due to savings in utility requirement, but there is a limit to this option. The low pressure nitrogen is generally injected in BOG condenser which operates around 75 -100 psig. The amount of low pressure nitrogen injection is dependent on the sub-cooling effect available in the LNG from the in-tank pumps, the amount of boil-off gas generated in the terminal from the LNG storage tanks / LNG carrier unloading operations etc., and the amount of export gas delivered from the terminal. The high pressure nitrogen is injected into the export gas directly downstream of the LNG vaporizers. The design of the system should maximize the low pressure nitrogen injection and the additional amount of nitrogen required at high pressure can be injected up to a maximum limit allowable by the pipeline.

Nitrogen is an inert having no heating value and reduces the HHV. It also helps reduce the WI due to its higher specific gravity compared to the natural gas. However, there are limitations in the amount of nitrogen that can be injected into the gas. The maximum amount of nitrogen in the gas varies between 2-3 mole percent depending on the pipeline. Figure 3 provides a chart to quickly calculate the effect of nitrogen injection on HHV for the product gas (export gas).

FIGURE 3

BTU CONTROL WITH NITROGEN INJECTION

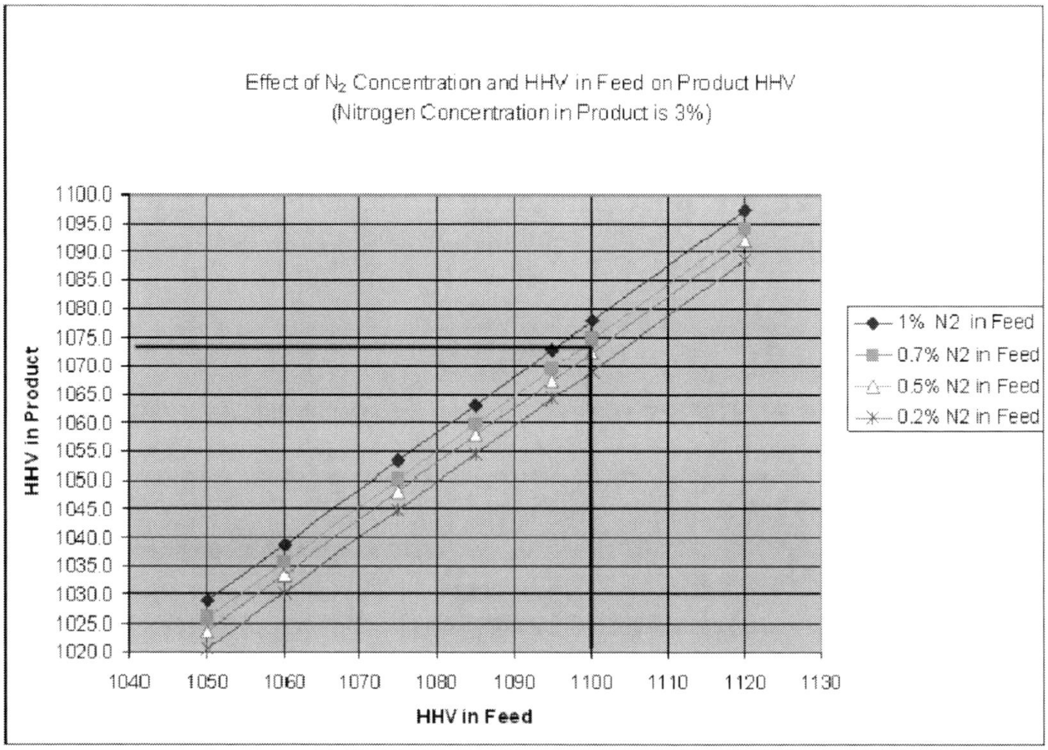

Some of the advantages for nitrogen injection include lower CAPEX, readily available as it can be separated from the air, fewer external factors to consider and easy to implement. The disadvantages for nitrogen injection are higher utility consumption and a limitation in the maximum nitrogen concentration in the export gas pipeline that may restrict the maximum reduction of HHV.

Pipeline Gas Quality Control – NGL Extraction

Aker Kvaerner [6] and other suppliers have developed NGL extraction technology to strip out ethane and heavier components from the imported LNG. Figure 4 below describes a typical LNG import terminal configuration with integrated NGL extraction. The NGL extraction process is usually integrated within the LNG terminal to strip out ethane and heavier components from the richer LNG. The liquid NGL product containing propane, butanes and most of the ethane has high vapor pressure and is transported to petrochemical complex in the NGL pipeline operating at high pressure.

FIGURE 4

TYPICAL LNG TERMINAL WITH NGL EXTRACTION

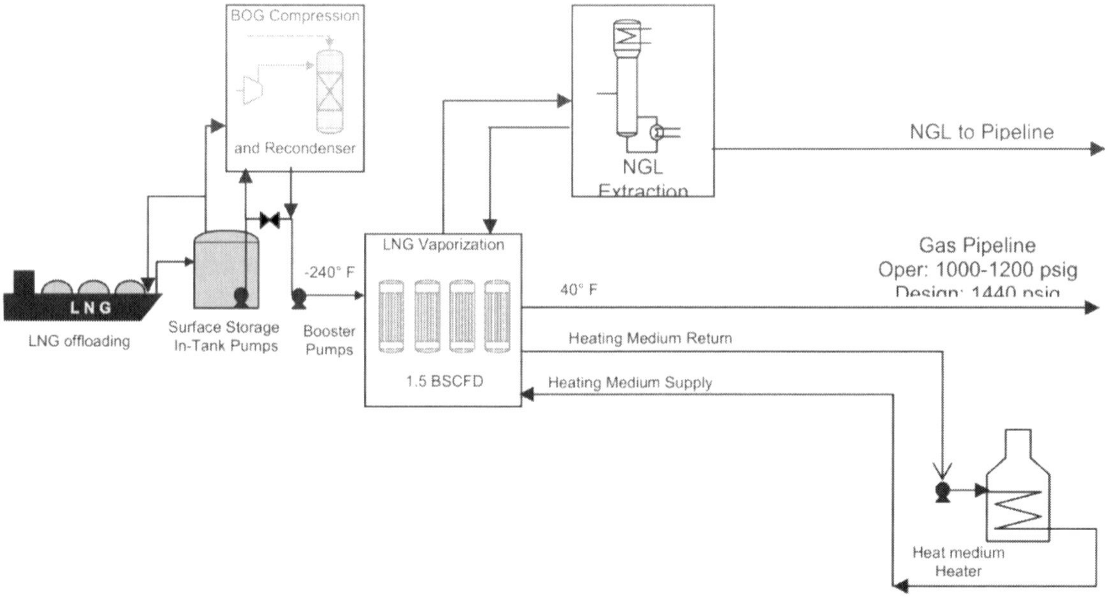

TABLE 2

REDUCTION IN HHV FROM LNG WITH NGL EXTRACTION

Imported LNG Origin	Import LNG HHV Btu/scf	Export Gas HHV Btu/scf	Total NGL Production Barrels /Day [2]	Export Gas Volume MMSCFD[1]
Algeria	1119	1052	57825	1412
Australia	1123	1043	65258	1401
Malaysia	1128	1026	77472	1384
Nigeria	1109	1023	67724	1398
Oman	1146	1046	76884	1385
Qatar	1118	1031	68092	1398
Trinidad and Tobago	1037	1012	30683 [3]	1500 [3]

Notes: 1. Gas volume in the table is based on inlet LNG equivalent of 1500 MMSCFD
2. NGL production is based on most LPG and approximately 30,000 BPD of ethane.
3. NGL extraction is not required for Trinidad and Tobago; Volume for NGL shown is available if extracted.

NGL extraction from the LNG is one of the most economically attractive methods for BTU control. This method is very effective in controlling pipeline quality gas BTU content at a very low operating cost. The Gulf Coast region's pipeline and petrochemical infrastructure supports this solution to control pipeline gas quality. The NGL can be sold as a feedstock to petrochemical industry in the Southern region. The main disadvantage of this process is the higher CAPEX, and in the Pacific and East Coast /Northeast regions the lack of infrastructure. Table 2 provides the level of reduction in HHV with NGL extraction from various imported LNG sources.

Pipeline Quality Control - Power Generation and LNG terminal Integration

The East Coast and the Northeastern regions have a market for LPG (propane and butanes). Also LPG can be separated from the NGL and easily transported locally or to other regions by truck or barges due to its relatively lower vapor pressure. The issue is ethane content in the imported LNG. In many cases, the ethane content is very high and it will result in much higher BTU value compared to methane, and sometime exceeds pipeline quality standards. Where the imported LNG contains large amounts of ethane, overall HHV can not be reduced to a desired level.

This following concept described in Figure 5 utilizes a combination of proven processes to reduce HHV. The NGL extraction methods from LNG described in the previous section can be extended further to separate ethane from the NGL using a de-ethanizer column. The LNG terminal has adequate refrigeration available from cold LNG to provide the necessary cooling for the column reflux system. The ethane separated from the NGL can be utilized to fuel the power generators. The removal of ethane from the imported rich LNG further reduces the BTU for the pipeline gas.

The power plant facilities are usually combined cycle with multiple power generation turbine units with heat recovery steam generation (HRSG). Figure 6 provides a concept of power generation system with combined cycle.

The waste heat from the steam condensing at lower temperature is normally removed by ambient air cooling or a cooling tower system. The integration of the power plant will allow recovery of the heat normally rejected. This heat is recovered using a circulating heating medium to supply the heat for LNG re-gasification in the terminal. The cold heating medium, returned from the LNG terminal vaporizers, is utilized for condensing steam saving the energy required to run the ambient air cooler fans or cooling tower fans.

The method described above is illustrated using a typical 1.5 BSCFD re-gasification terminal and a nominal 500 MW combined cycle power plant. Table 3 describes how much BTU reduction can be achieved. The Table 3 also provides quantities of LPG and ethane extracted from each imported LNG.

FIGURE 5

INTEGRATED LNG TERMINAL WITH POWER PLANT

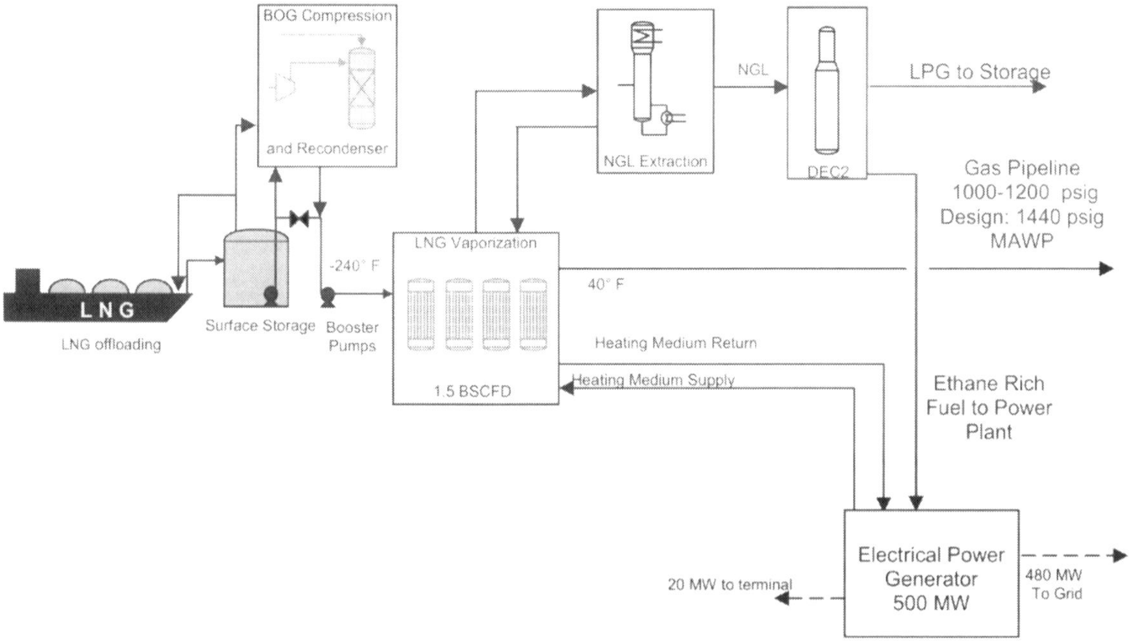

FIGURE 6

COMBINED CYCLE POWER PLANT WITH HEATING MEDIUM SYSTEM

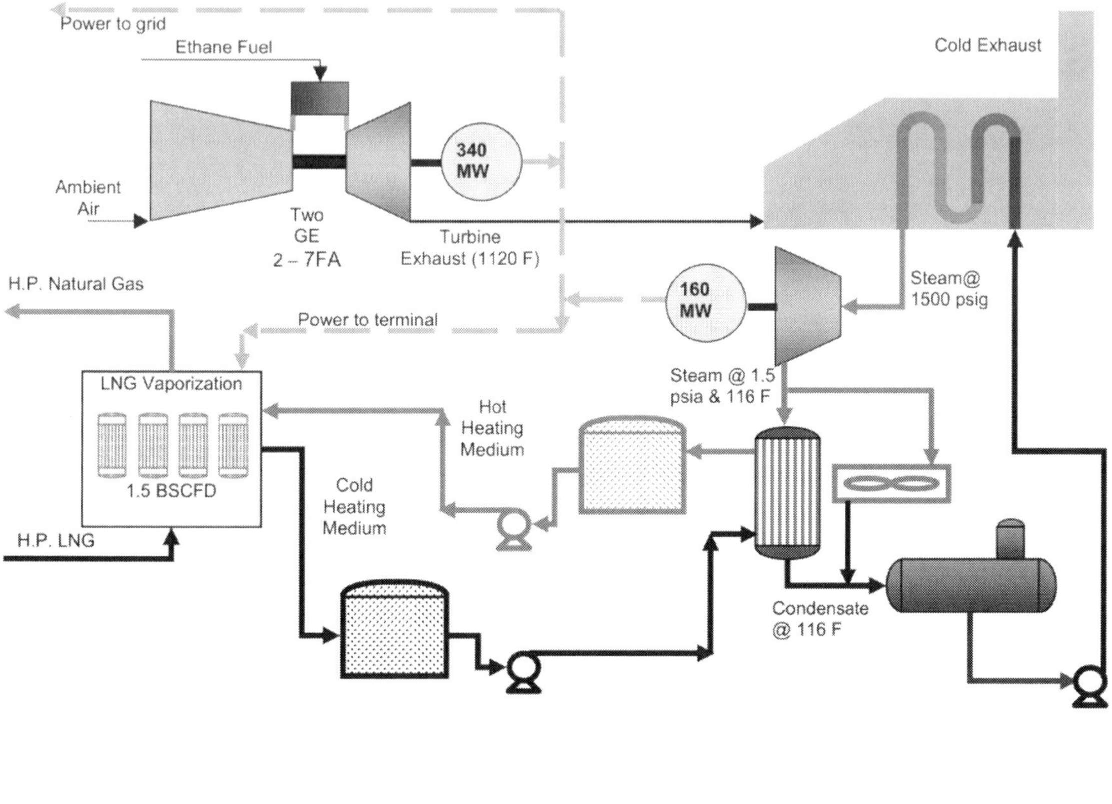

TABLE 3

REDUCTION IN HHV OF IMPORTED LNG WITH POWER PLANT INTEGRATION

Imported LNG Origin	Import LNG HHV Btu/scf	Export Gas HHV Btu/scf [5]	Ethane To Power Plant MMSCFD[3]	Total LPG Production[4] Barrels /Day	Export Gas Volume MMSCFD[1]
Algeria	1119	1052	47.62	27571	1412
Australia	1123	1043	47.62	35004	1401
Malaysia	1128	1026	47.62	47218	1384
Nigeria	1109	1023	47.62	37470	1398
Oman	1146	1046	47.62	46630	1385
Qatar	1118	1031	47.62	37838	1398
Trinidad and Tobago	1037	1012[6]	47.62	429	1452

Notes: 1. Total equivalent inlet LNG is 1500 MMSCFD. Export gas volume after shrinkage & fuel.
2. Power plant: Total 500 MW net output using 2-GE 7FA machines and 1- steam turbine.
3. Ethane rich fuel (LHV: 1618.5 Btu/scf, GHV: 1769.4 Btu/scf utilized for generators.
4. Estimated LPG production (propane and butanes) from 1.5 BSCFD inlet
5. Expected send out gas quality to pipeline.
6. The inlet LNG meets the pipeline quality. The data shown are if LNG is processed.

The main advantage for the concept described above is that it provides a solution to control the natural gas quality for Northeast and other regions where NGL infrastructure is not available. It also enhances the overall energy efficiency and emission reductions. Generally the fuel utilized by the LNG re-gasification terminal is approximately 1.5% of the inlet LNG. The integration of the power plant will realize this fuel savings as well as a reduction in the import of LNG by the same amount. In addition the reduction in greenhouse gas emissions are also a key for Northeastern "Signatory States" in meeting their future goals.

TABLE 4

FUEL AND EMISSION (GREENHOUSE GAS) COMPARISON

Type of Unit	Fuel Gas Consumption MM Btu/hr		Annual CO_2 Emissions 1000 TPA	
	Non Integrated	Integrated	Non Integrated	Integrated
Power Plant	3212	3212	1835	2017
LNG Terminal	931	0	527	0
TOTAL	4119	3212	2362	2017

Overall fuel and environmental emission reductions are provided in Table 4 for the integrated facility using 2 GE-7FA machines and 1.5 BSCFD LNG re-gasification capacities. The comparison in the table indicates approximately a

28% reduction in overall fuel gas consumption. The integrated facility will also reduce the green house gas (CO_2) emissions by 17% compared to the non-integrated facility.

Figure 7 below provides a similar concept with GE LMS100 machine, developed recently showing additional integration opportunity for future heat recovery from their compressed air intercooler system.

FIGURE 7

LNG TERMINAL INTEGRATION WITH POWER PLANT USING GE LMS100

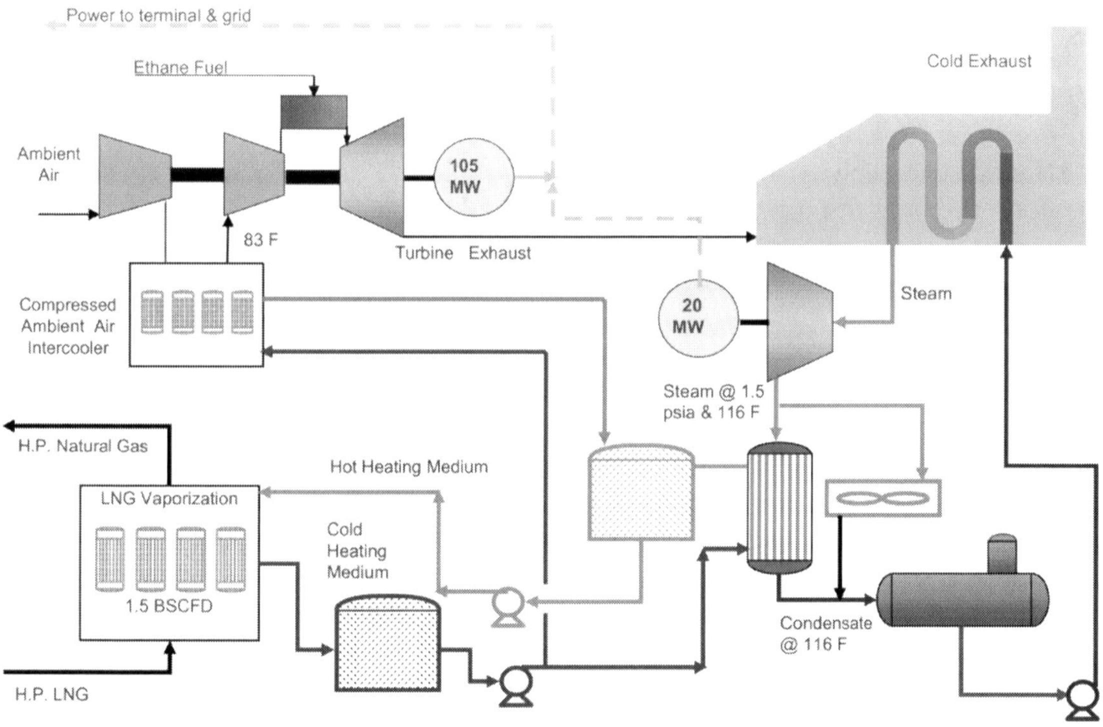

GE design currently utilizes cooling water with intermediate fluid to intercool the interstage compressed air. The LNG terminal integration with LMS100 can provide the cooling required for compressed air while recovering heat for the LNG re-gasification. This concept if implemented using ethane as a fuel for the turbine, can provide necessary Btu control for the export gas quality as well as additional enhancement in fuel efficiency and environmental emission reduction. Expected reduction in overall energy consumption is about 360 Btu/KW.

There are some limitations and disadvantages for this LNG integration with power plant concept. The expected CAPEX will be higher. The concept may require working across various industry sectors for terminal developers. Also the LPG produced must be marketed locally, and additional infrastructure may be required for shipping LPG elsewhere due to the seasonal local market.

Conclusions

The paper discusses issues associated with gas interchangeability when importing LNG in the U.S. The paper presents several methods to control the gas quality to meet export gas pipeline specifications. The solutions presented here have their advantages and disadvantages depending on the regional pipeline requirements and other infrastructure to support the methods utilized. The solutions to meeting gas quality standards in each region will be different depending on the quality requirements by the pipelines, future marketing and FERC regulations. Industry and regulators must continue to advance in developing gas quality standards to support the future energy demand and flexibility in importing LNG from different parts of the world.

It is also important to realize that LNG terminal developers may benefit by working across various industry sectors for integration opportunities. The integration of LNG terminal provides opportunities for energy efficiency and emission reduction as well as providing a solution to export gas quality solution in regions such as the Northeast.

References Cited

1 Country analysis briefs: United States; source of information is US EIA website at http://www.eia.doe.gov/emeu/cabs/usa.html

2 U.S. Department of energy, Office of Fossil Energy, document DOE/FE-0489, Ensuring Consistent Quality for End Use, Source of information website at http://www.fossil.energy.gov

3 Report on Joint Workshop on Natural Gas Quality standards (CPUC R.04-01-025, CEC 04-IEP-01), California Public Utility Commission, California Energy Commission, April 4, 2005

4 AES Ocean Express v. Florida Gas Transmission, Docket No. RP04-249-002

5 U.S. Department of energy, Office of Fossil Energy, document DOE/FE-0489, Typical Composition of LNG Imports by Country, Source of information website at http://www.fossil.energy.gov

6 Kulish, Shah and Wanvik, "Terminal considerations", LNG Industry, A supplement to Hydrocarbon Engineering, Autumn 2005.

Biography

Kamal Shah is a Technical Vice President, Aker Kvaerner Inc., Part of the Aker Group, Houston, Texas. He has been working with Aker Kvaerner, Inc., Houston office for 20 years and has been involved in the process design, management and technology development in many oil and gas and energy projects. Prior to joining Aker Kvaerner, he was chief process engineer at Key Engineering, Inc. in Houston, Texas. He attended Michigan State University and received his BS and MS degree in chemical engineering. He is a registered professional engineer in Texas.

E-mail: kamal.shah@akerkvaerner.com

Floating LNG Regasification with Smart® Air Vaporization

Brad Hubbard
Director Technology Development
Mustang Engineering, L.P.
Houston, Texas

AIChE Spring Meeting, April 2007
7th Topical Conference on Natural Gas Utilization
Houston, Texas

Introduction

Increasing global demand for natural gas and the resulting growth in gas value continues to fuel the rapid development of the Liquefied Natural Gas (LNG) industry. Planning and permitting for new LNG liquefaction and regasification facilities are at high levels since the LNG industry slumped in the late 1970s and numerous innovative design concepts are emerging. This paper presents concepts for utilizing fin-fan air exchangers on floating LNG terminals for the capture of heat from the air to vaporize LNG.

Existing LNG regasification terminals have traditionally been operating near the shoreline in protected harbors in the United States for decades. Much of the current LNG regas permitting activity has been centered on the expansion of these facilities or the construction of new onshore regas plants. While expanding existing LNG receiving terminals or building new ones is attractive economically, there is increased concern about the potential environmental impact and safety of onshore or near shore LNG facilities. In the US, permitting of onshore LNG regas terminals is often difficult due to movements by local residents and regulatory jurisdictions to block construction efforts. As a potential alternative, there has been an interest in the development of LNG regas terminals for offshore environments using gravity based fixed concrete structures (GBS) and floating LNG storage regas barges and ships (FSRUs). While one GBS is being built for LNG regas service in the Adriatic, it appears the application in the US may be prohibitively expensive. Investigation of the viability of offshore floating LNG regas has been particularly active for use in the Gulf of Mexico, East Coast and West Coast, but is applicable to any global locale where there is sufficient demand and a desire for gas importation. In fact, Excelerate Energy has built and is operating the first new LNG regas terminal built in the US in 20 years, Gulf Gateway, offshore Louisiana. It uses a specially built LNG carrier with regas equipment that can connect to a sub-sea buoy and discharge gas to a subsea pipeline. Excelerate and Suez are close to having permits for two similar Deep Water LNG terminals offshore Massachusetts, the Northeast Gateway and Neptune, respectively. BHP Billiton is seeking a permit decision for a FSRU and Woodside Natural Gas has started the permit application process for a LNGRV, each offshore California near Los Angeles. The TORP Corporation has applied for offshore permits in the Gulf of Mexico, Bienville and is developing a similar project offshore California. All of these offshore LNG terminals either use sea water in an open loop or burn gas to generate the heat of vaporization for regasifying the imported LNG.

Mustang Engineering, L.P. has developed, with others, four offshore floating LNG regas concepts utilizing its patented LNG Smart® Air Vaporization (SAV) technology for the economic and environmentally compatible regasification of LNG at a safe distance offshore and beyond visibility from shore, in waters as deep as 300 meters. Each of these facilities would be developed independently as a newly constructed fit-for-purpose vessel for receiving, storing and vaporizing LNG and delivering high pressure pipeline gas. The floating LNG regas concepts are novel and innovative and are actually just new arrangements and applications of components and technologies proven in the offshore oil and gas production, gas processing, LNG and marine operations industries. Floating offshore oil production, storage and offloading facilities (FPSOs) similar to the LNG FSRUs have been used safely, reliably and in adverse marine conditions for more than 20 years.

The Mustang concepts utilize innovative floating hull, LNG storage and LNG regas equipment designs that are, for the most part, extensions of successfully tested and commercially used designs. Each floating unit would attach to a single point mooring system, allowing it to weathervane for easier berthing with LNG carriers and to better position itself in relation to changing wave, current and wind forces. The single point mooring includes the mooring lines for station keeping, a gas swivel and flexible subsea gas risers that connect to a subsea pipeline to shore. One of the concept vessels would also have a quick disconnect capability from its mooring buoy in order to escape severe weather situations and to allow the LNG FSRU to meet a LNG carrier or travel to a LNG supply terminal for a new load of LNG. The disconnectable concepts are much like the Excelerate Energy's LNG Regas Vessel but mounted with air vaporizers for heating the LNG. Each concept has advantages and disadvantages, but all have viability with the Mustang SAV process for the purpose of regasifying LNG offshore.

Disconnectable Floating Storage Regas Unit (DFSRU)

The Disconnectable Floating Storage Regas Unit (DFSRU) is designed to receive LNG from either a liquefaction plant or a conventional LNG carrier by ship to ship transfer of LNG which transfers the cryogenic liquid between ships at a safe location in calm waters away from the gas discharge buoy. The Mustang SAV process, with send-out capacities of 800 mmscfd and greater is installed over the deck of the DFSRU and integrated with the ship systems. Utilizing the heat in the air and water vapor is the most efficient technology for supplying heat of vaporization to the LNG and results in lowest operating cost and lowest air and water emissions, since the least amount of gas possible is burned and no sea water is used. The DFSRU propulsion is driven by electric motors; the power for which can be redirected to power the Mustang SAV process and utilities during LNG vaporization and gas send-out. The self-propelled DFSRU docks without tug assistance to a normally submerged buoy that provides a single point mooring, gas turret and flexible gas risers tied into a submarine pipeline to shore. By eliminating the need for dedicated tugs, operating costs and air emissions for berthing are effectively lowered. In the event of severe weather, the DFSRU disconnects from the buoy and propels itself into safer waters. To achieve continuous send-out and greater reliability, the Deep Water Port LNG terminal can utilize two or more DFSRUs and two submerged offloading buoys. LNG storage in capacities of 150,000 m^3 to 240,000 m^3 or more is provided by Semi-Prismatic Type B (SPB) tanks, reinforced membrane tanks or spherical Moss type tanks. The ship to ship LNG transfer system from the LNG carrier to the DFSRU is accomplished either with purpose-specific flexible cryogenic hoses or specially-designed mechanical loading arms modified for operation in open seas.

As a precedent, a similar concept is currently in use in the U.S. Gulf of Mexico, offshore Louisiana by Excelerate Energy. Their Gulf Gateway Deepwater Port utilizes the Energy Bridge concept and has operated since March 2005 in 300 feet of water. It was able to continue gas send-out while Hurricane Katrina approached, producing seas of Hs 6 m height. Excelerate, with its partner EXMAR, has also successfully transferred LNG cargo from ship to ship using cryogenic hoses. Below is a drawing of an concept DSFRU ship with 150,000 m^3 storage in reinforced membrane tanks, the patented Mustang SAV regas process on deck with 800 mmscfd send-out capacity and the disconnectable Submerged Turret Loading buoy receiving apparatus which is a proprietary product of APL, Inc.

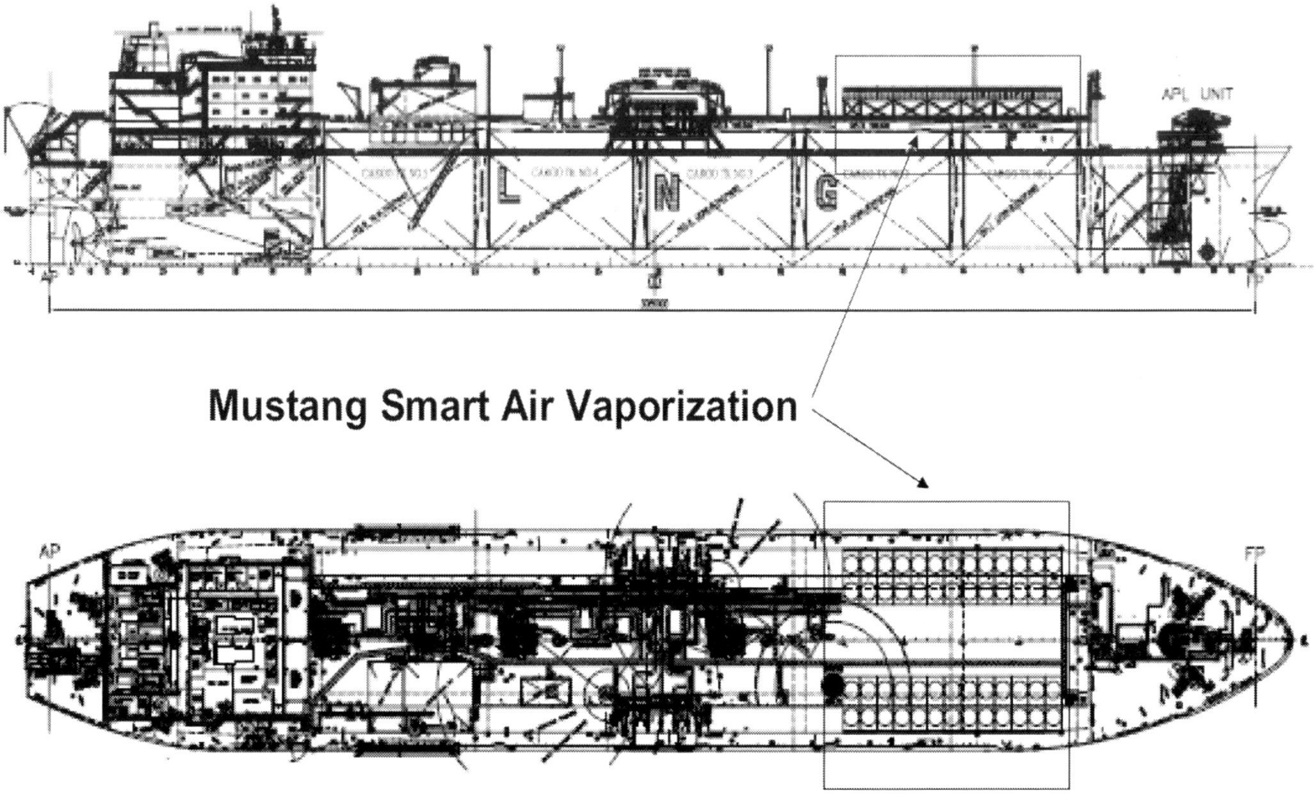

Figure 1. Disconnectable FSRU with Mustang Smart® Air Vaporization

Column Stabilized Floating Platform (CSFP)

The Column Stabilized Floating Platform concept shown in the rendering below was conceived by FMC Technologies as a LNG regas terminal operating with a LNG carrier moored alongside to provide the LNG supply. FMC designed the hull and mooring and Mustang designed the LNG regas facilities for 800 mmscfd and more send-out using the Mustang SAV process and Mustang LNG horizontal tanks for 36,000 m³ of interim storage during swap out of LNG carriers. This facility provides the advantages of minimal LNG storage resulting in lower cost, coupled with the capability to berth conventional LNG carriers with 200,000 m^3 capacity and mid-ship manifolds.

The option incorporates a hull concept not unlike that currently used in semi-submersible drilling vessels and other offshore vessels where stability is paramount. The structure is characterized by a rectangular deck frame supported by a series of buoyant columns attached to two submerged pontoons. Its design provides exceptional stability for offloading in inclement weather and high sea conditions. It also offers an open deck space with flexibility for the arrangement of regas equipment and living quarters. Stern thrusters allow the vessel to alter its mooring path in the event of high sea states and to maneuver to the most desirable heading during LNG carrier berthing operations.

The CSFP facility provides a continuous pipeline gas send-out capability by including small (36,000 m^3) LNG storage capability onboard. This minimal capability allows time for one LNG carrier to disconnect and depart the facility, followed by the berthing and offloading of an awaiting carrier. The vaporization rate could vary to accommodate delayed LNG carrier arrival and departure schedules.

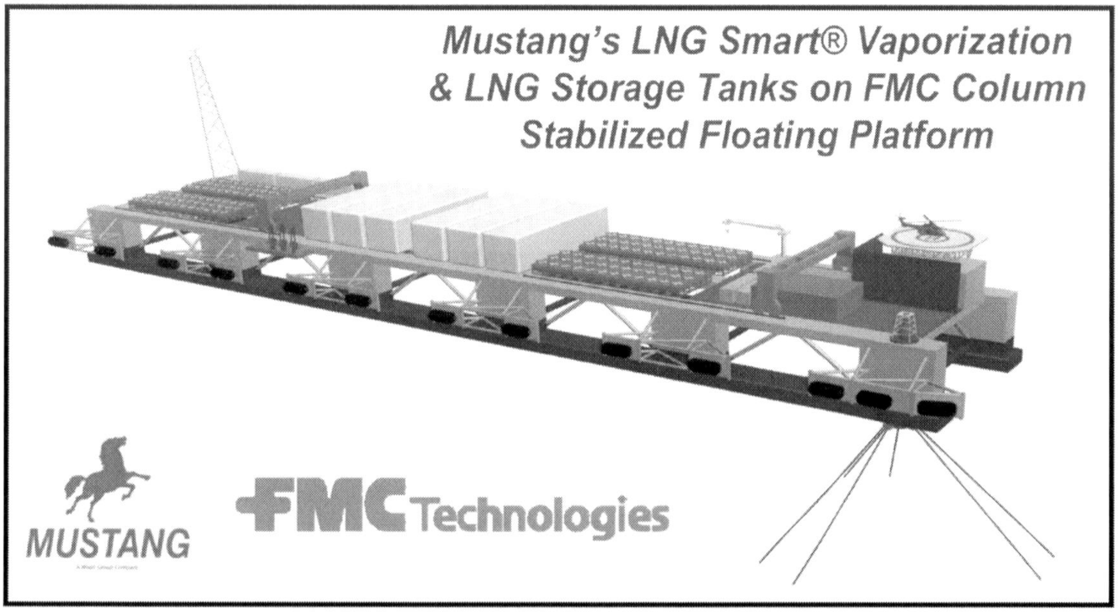

Figure 2. Column Stabilized Floating Platform with Mustang Smart® Air Vaporization & LNG Storage Tanks

Concrete Floating Regas Storage Unit (Concrete FSRU)

This concept, as shown in the rendering below, is a floating concrete barge capable of storing 230,000 m^3 of LNG in cryogenic steel horizontal tanks each fully-contained within a concrete compartment. The concrete construction offers an alternative to steel construction which will result in cost and schedule savings. The LNG vaporization equipment and utilities for send-out rates from 800 to 2,000 mmscfd fit on the deck of the barge. LNG is delivered to the FSRU by LNG carriers that moor alongside and offload through enhanced mechanical loading arms. This FSRU incorporates two stern thrusters for maneuvering to a preferred heading to assist in the LNG carrier berthing and to minimize motions from adverse swells and changes in the wind direction. Two dedicated sea going tugs for berthing the LNG carrier would also be required.

The FSRU is moored with an external turret and nine-leg mooring system that provides single point mooring and weather vaning capability. The gas swivel in the turret connects to a submarine pipeline to shore via flexible gas risers.

The concrete hull and horizontal tank storage configurations for the FSRU are modeled after the Ardjuna Sakti, a liquefied petroleum gas (LPG) liquefaction barge which has been successfully operated for the past 30 years, during which time the hull has remained water tight. The use of concrete provides for an advantageous construction schedule and maximum use of host country resources when compared with steel hulled vessels, in that the construction can be accomplished in a graving dock rather than a shipyard dry dock. Using concrete for the hull material should also cost less than a steel hull in view of the recent rises in steel prices. Horizontal tanks can be fabricated in controlled shops and shipped to the construction site to be installed in the hull as it is being built.

Figure 3. Concrete Hull FSRU Barge with Mustang Smart® Air Vaporization and Horizontal LNG Storage Tanks

Steel Hull Floating Storage Regas Unit (Steel FSRU)

This steel hull LNG FSRU barge would be similar in function to the concrete FSRU, except that it has an internal mooring table and gas turret for delivery of vaporized gas through flexible risers to a submarine pipeline. It would have LNG storage capacity of 240,000 m^3 in SPB Type B tanks and a send-out capacity of 800 mmscfd or more. This new-build steel hull design, with comparable capacities and features as the concrete FSRU, would be the most expensive of all options because of the dramatically escalated steel prices. It would have the longest delivery schedule due to the capacity constraints at global shipyards for undertaking this work. Lengthy fabrication delays with industry demand on shipyards are currently extending booking dates by 3 ½ - 5 years. The drawing below is a representative layout of the steel hull FSRU with the Mustang SAV process for 800 mmscfd send-out.

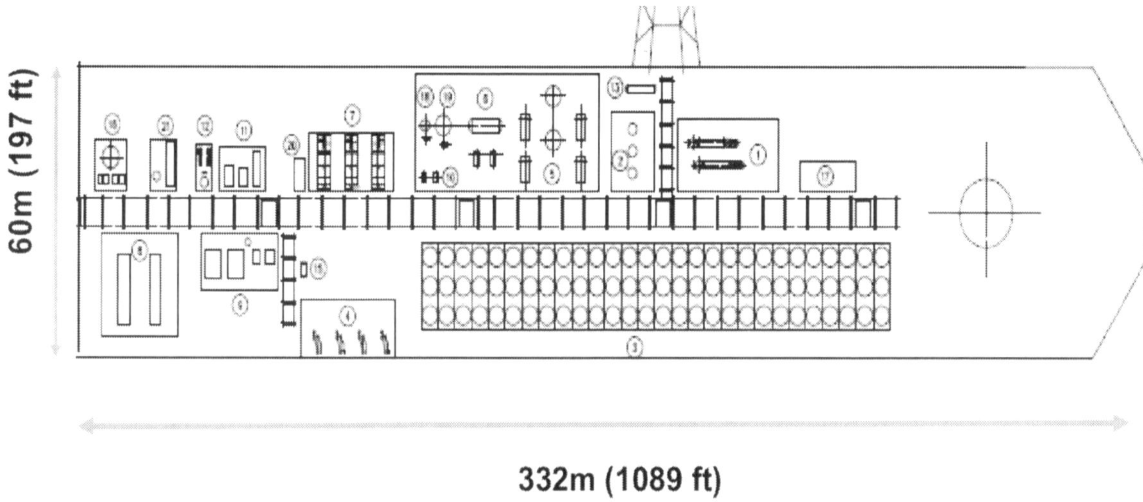

Figure 4. Steel Hull FSRU Barge with Mustang Smart® Air Vaporization and SPB LNG Tanks

Mustang LNG Smart® Air Vaporization Technology

The common denominator in all four floating concepts presented is the patented Mustang LNG Smart® Air Vaporization (SAV) technology for the regasification of LNG which is shown in the diagram below. The technology is an alternative to the submerged combustion vaporizer (SCV) or other gas fired system and to open loop sea water vaporizer techniques currently used or proposed. The SAV regas process transfers heat from the air to the LNG which reduces gas usage and air emissions dramatically, uses no sea water and creates no water pollution.

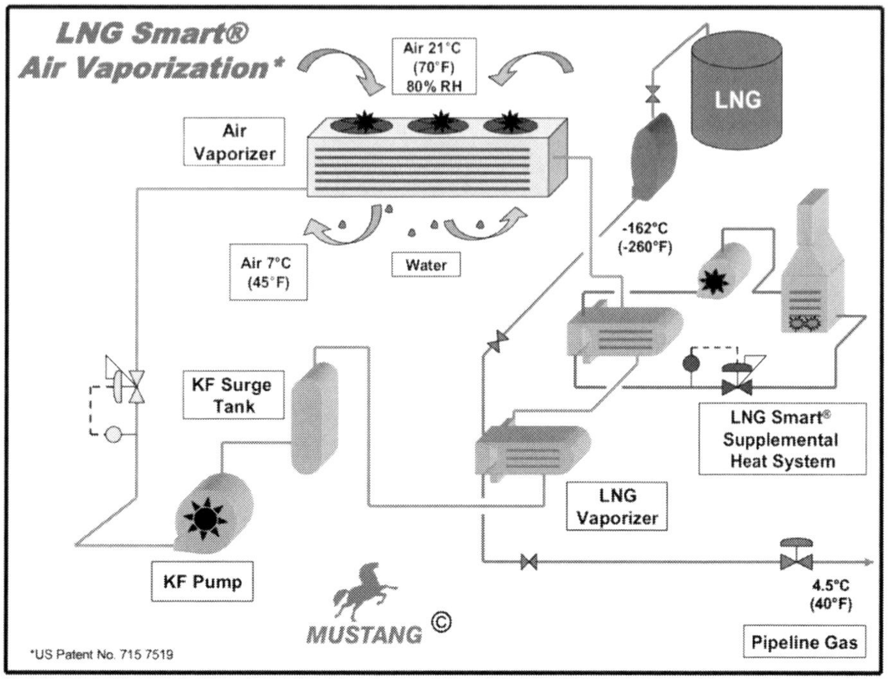

Figure 5. Mustang Smart® Air Vaporization Technology with Supplemental Heat

The Mustang SAV technology uses specially designed fin-fan heat exchangers as shown in the photo below to extract the sensible heat from the air and the latent heat from condensing water vapor and to inject the heat into an intermediate fluid such as potassium formate solution or propane, which in turn transfers the heat to the LNG for vaporization. The SAV process, at design ambient conditions of 15.6° C (60° F) and 80 % relative humidity provides all the heat required to vaporize the LNG, dramatically reduces air emissions, especially NOx, CO and CO_2 (84, 62 and 275,000 tonnes per year, respectively), and lowers fuel consumption by as much as 100%, resulting in an operating cost savings of $25 mm or more per year for a 1,000 mmscfd send-out.

While the SAV air heaters are designed to transfer sufficient heat from ambient air for 100% of LNG vaporization, on cold days they require a supplemental heat source. For those instances, a closed-loop, gas fired backup heater is used. The equipment used in this air exchange system is commonly found in the industry in similar offshore applications. The process has been thoroughly tested using a scalable pilot unit and is now being installed in an onshore LNG regas terminal with 2,100 mmscfd send-out capacity.

Figure 6. Mustang Smart® Air Vaporization 3-Fan Air Heater (2-fan heater used for floating applications)

Using the Submerged Combustion Vaporizer (SCV) method, the LNG is heated from its cryogenic state to required pipeline gas temperatures using heat exchange tubes running through a water bath heated by submerged gas burning. The technique demands a high usage of natural gas, requiring as much as 2.0% shrinkage of the total LNG imported. The fuel cost for a LNG regas terminal burning 1.3% of the LNG to vaporize 1,000 mmscfd send-out would be $28 mm per year assuming $6 per mmbtu value for the gas. The terminal would also produce air emissions that come from combustion such as NOx, CO, PM10, and CO_2. The Mustang SAV system can save 60 to 100% of this fuel cost and air emissions depending on average ambient conditions at the terminal location.

The open loop sea water technique requires a huge amount of seawater be pumped through shell and tube heat exchangers, heating the LNG to a gaseous state. While the technology has relatively low operating cost and low air emissions, it requires that the continual flow of seawater be treated with biocides which are harmful to marine life.

Conclusion

As long as the demand for natural gas remains high, new LNG regasification technologies and developments will continue. Numerous engineering, procurement and construction options need to be explored to circumvent lengthy delivery schedules, increasing costs, environmental emissions, permitting hurdles and safety concerns. Each of the four floating LNG regasification concepts presented in this paper are viable for offshore vaporization of LNG into gas that is, in many ways, more acceptable than an onshore receiving terminal. Utilization of the Mustang LNG Smart® Air Vaporization technology and equipment can be done safely, with significant operational cost savings and environmental benefits.

References:

- "Four Offshore Floating LNG Regas Concepts Offer Air Vaporization", Brad S. Hubbard, LNG Journal, November/December, 2006.
- "An Overview of Mustang Engineering's LNG Smart® Vaporization Technology", Randy M. Sharp, AIChE Spring Meeting, Orlando, Florida, April 2006.
- "A Concrete Solution", Brad S. Hubbard, LNG Industry, Autumn 2006.
- "LNG Debuts for Smaller Fields", Don Lyle, E&P Floating Production Systems, September 2004.

Author

Brad Hubbard
Mustang Engineering, L.P.
16001 Park Ten Place
Houston, Texas 77084
713 215 8384
Brad.hubbard@mustangeng.com

Brad S. Hubbard, P.E., has 30 years experience in the development, management, engineering, fabrication and installation of large oil and gas field projects and midstream conversion projects. He is the Director of Technology Development for Mustang Midstream and has been the project developer and manager of gas monetization project studies including LNG Regasification, LNG Liquefaction and Gas to Liquids. Mr. Hubbard holds BS degree in General Engineering and a MS degree in Ocean Engineering from Texas A&M University. He is a Registered Professional Engineer in Texas and a PMI certified Project Management Professional.

SIMAR Applications 1: Evaluating Expander-Based C_2+ Recovery in Gas Processing

Justin Trinter and Stanley Huang
Chevron Energy Technology Company
Houston, Texas

AIChE Spring Meeting, April 2007
7th Topical Conference on Natural Gas Utilization
Houston, Texas, 22-26, 2007

Introduction

System Intrinsic Maximum Recovery (SIMAR) is a hypothetical operating scenario in which an expander-based recovery system can draw refrigeration freely from the environment. SIMAR operation is approximated when the feed stream to the system is liquefied natural gas (LNG) instead of natural gas. LNG contains abundant refrigeration, so no additional external refrigeration is required. [Huang et al., 2006 a, b]. Given a C_2+ recovery process configuration, SIMAR curves can be generated following a few simple steps. SIMAR curves are shown to characterize the given C_2+ recovery process configurations very well.

The expander-based process for C_2+ recovery has been the mainstay technology since the 1970s [Lee et al., 1999]. There are more than one hundred filed patents in the U. S. alone with individually claimed advantages. Despite the great technical and commercial success, a methodology for objectively evaluating different process configurations remains missing [Pennybaker et al., 2000]. Two major hurdles prevent the development in this aspect:

1. Concerns over possible impact on commercial interests

2. Many process parameters obscure the underlying reasons of claimed advantages.

The second hurdle is particularly noteworthy as some proposed configurations actually have adverse impacts on the recovery. However, the adverse impacts can be compensated by artfully performed refrigeration integrations over the entire system.

Figure 1 shows a generalized scheme for C_2 recovery based on expander-technology. The process is intended to strip the inlet natural gas of its heavier components. The residue gas is recompressed and returned to the pipeline. Sweet, dry natural gas goes through an inlet chilling section, where the gas is chilled to a suitable level before entering the heart of the plant, an expander-based C_2 recovery scheme. The refrigeration of the inlet chilling section is mainly provided by the returning residue gas and supplemented by side-draws (not shown) from the demethanizer column (DeC1). Depending on actual requirements, external refrigeration may be required (not shown).

The main components of the expander section include a separator, expander, and DeC1. A sub-cooler is usually provided for improved refrigeration integration in the low temperature regions. The exact design of this section is an art [Pitman et al., 1998]. Over the past 30 years, quantum progresses and incremental improvements have pushed the art to an unprecedented level of efficiency.

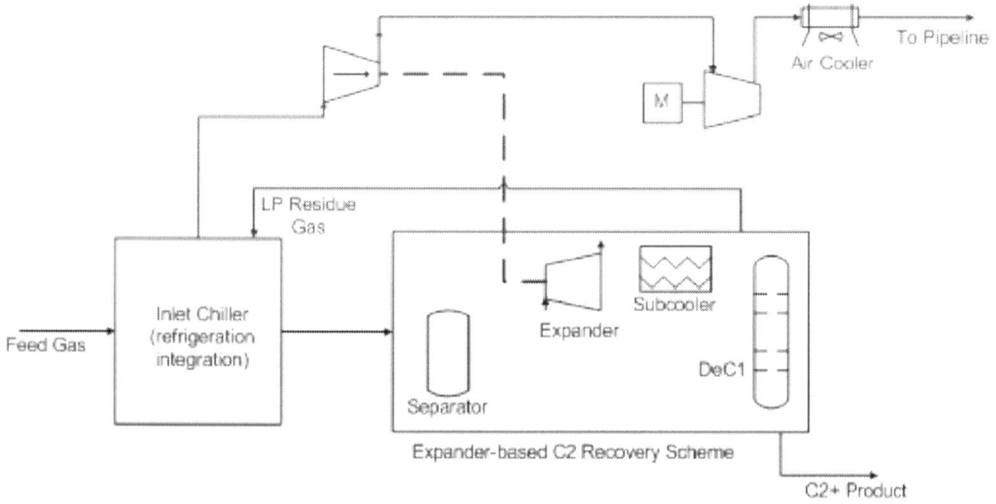

Figure 1: Generalized Gas Processing Scheme for C_2 Recovery

This work will generate SIMAR curves for a wide spectrum of processes configurations described in the patent literature. The curves will reveal the underlying efficiencies of the C_2+ recovery process configurations. The actual process selection may be affected by additional factors like equipment cost versus improved efficiency. More detailed performance, operating, and economic calculations are required before making the final process selection. The results of this work are presented only to shed light on current practices of the art. The conclusions and recommendations will be useful to guide future developments in C_2+ recovery systems.

Technical Background

As previously mentioned, the main components of the expander section in Figure 1 include a separator, expander, DeC1. Usually, a sub-cooler is provided for improved refrigeration integration in the low temperature regions. Following the previous works [Huang et al., 2006 a, b], the expander schemes are grouped into four categories. The first three categories, shown in Figure 2, are:

- Separator top not rectified (XPDR 1)
- Separator top fully rectified (XPDR 2)
- Separator top self-rectified (XPDR 3)

This specific XPDR 3 configuration will be used as a benchmark comparison for all categories. The fourth category, which includes heat pumps in the process scheme, is not shown and will be explained later.

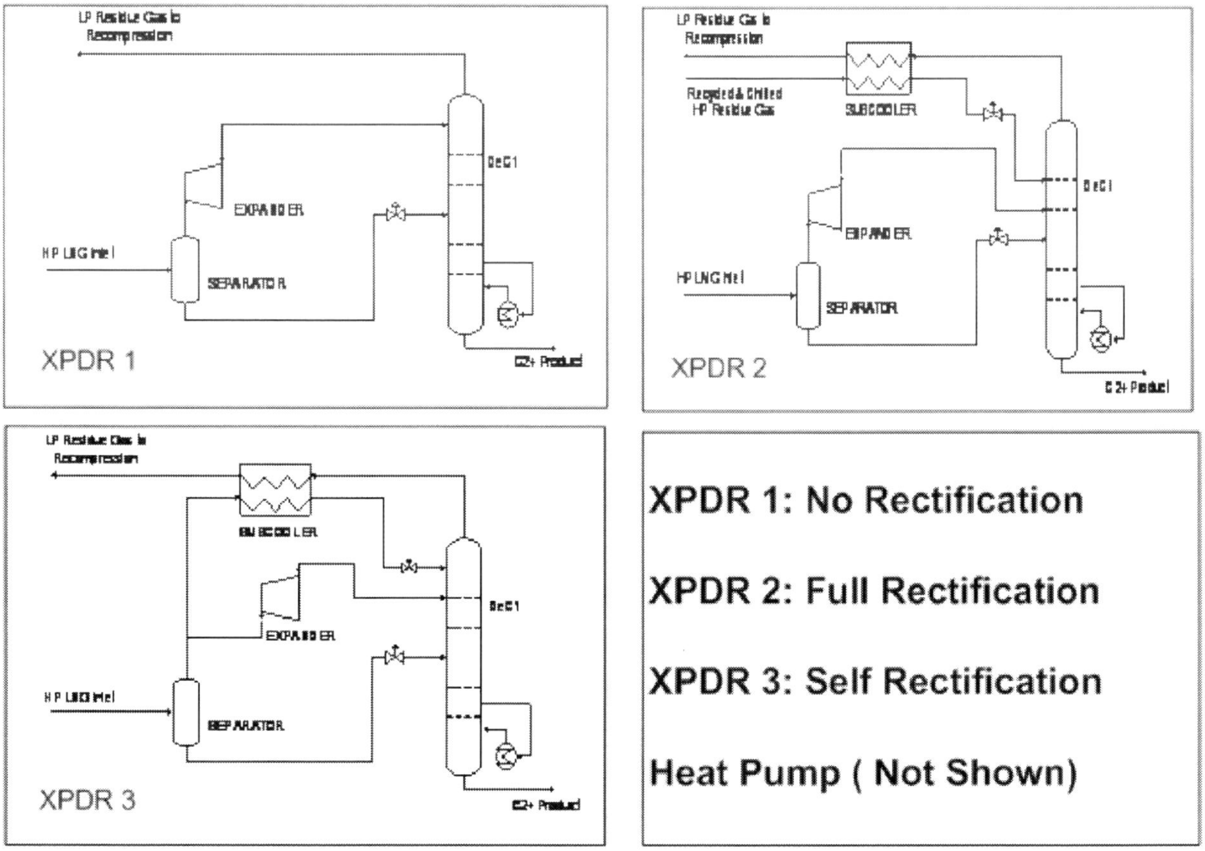

Figure 2: Categorizing Expander-Based Schemes

If the feed stream into the system is changed from natural gas to LNG, the entire pre-chilling section in Figure 1 is not required because LNG contains excess refrigeration. Figure 3 depicts this change in feed stream. Although this physical configuration helps visualizing the process, it is not required to define SIMAR.

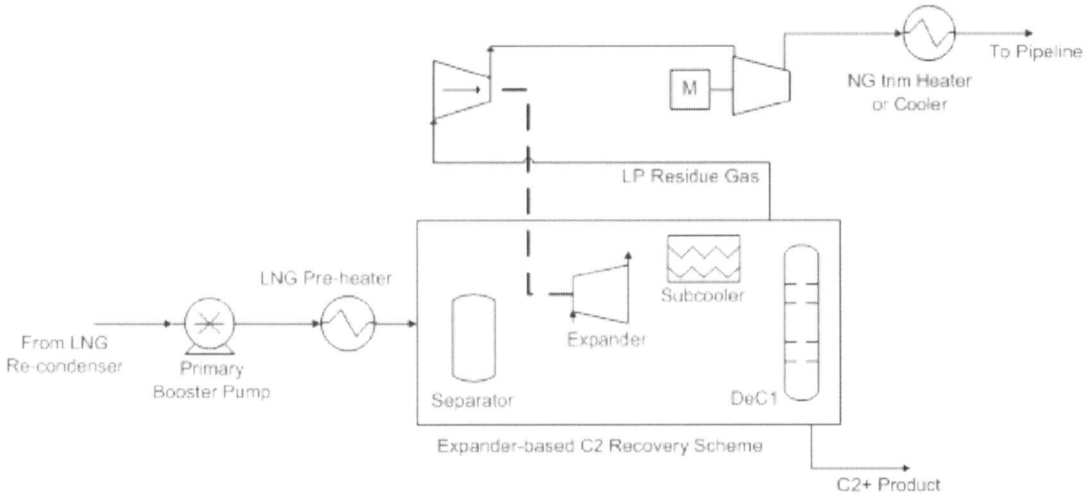

Figure 3: Generalized Processing Scheme for C_2 Recovery with LNG as Feed

The following paragraphs describe procedures to construct SIMAR curves.

Figure 4 shows the phase envelope of a medium LNG composition. Lean LNG can be very high in methane composition and is not of interest in this work. The procedure starts with a relatively high temperature at a reasonable pressure level, which is maintained at a level equivalent to 85% in reduced pressure (ratio of actual pressure to the critical pressure of sample) in this work. The track of testing starts with temperature penetrating through the two-phase region, and ends at an arbitrarily chosen level of -100°C. The fluid would remain liquid at and below this temperature level.

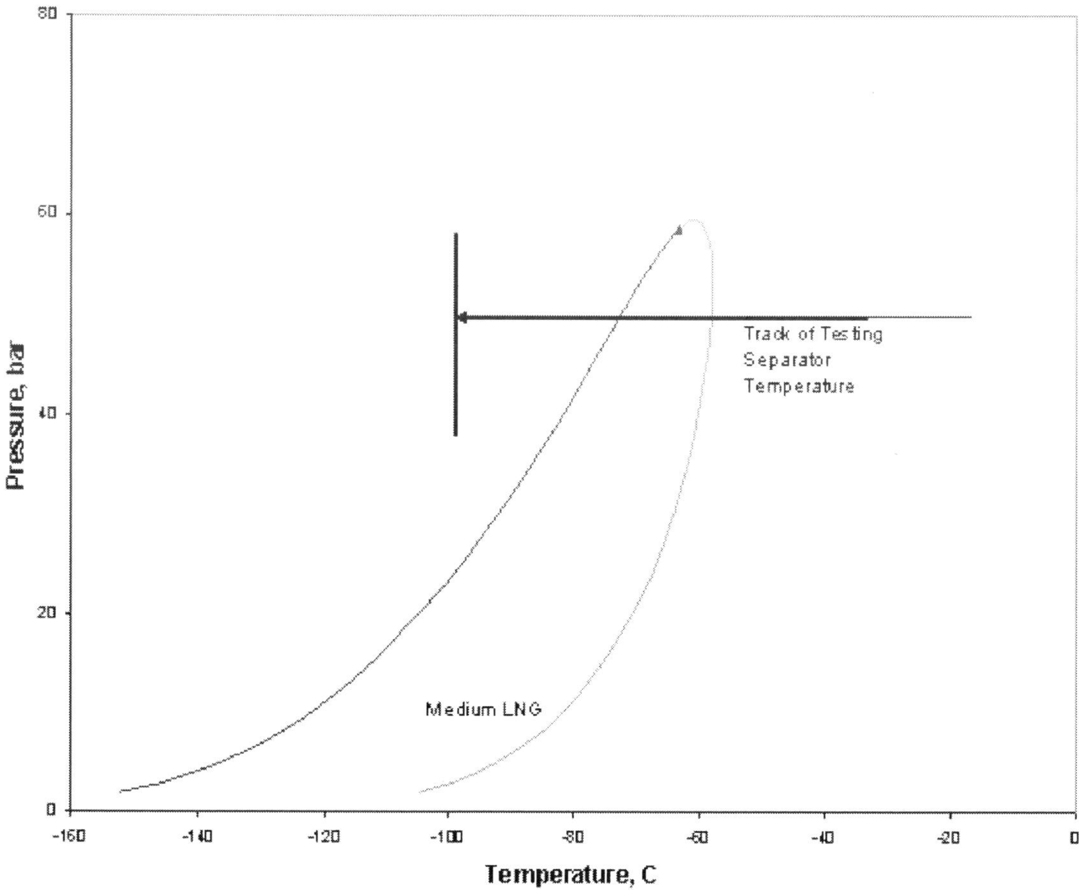

Figure 4: Phase Envelope of Inlet LNG

Figure 5 plots the C_2 recovery level against the separator temperature for the aforementioned three types of process configurations. Both XPDR 1 and XPDR 2 show monotonic trends of improvement as the temperature decreases. The trend continues even after the inlet gas is totally liquefied, when the expander is replaced by a Joule-Thomson (JT) valve and no expansion work is recovered.

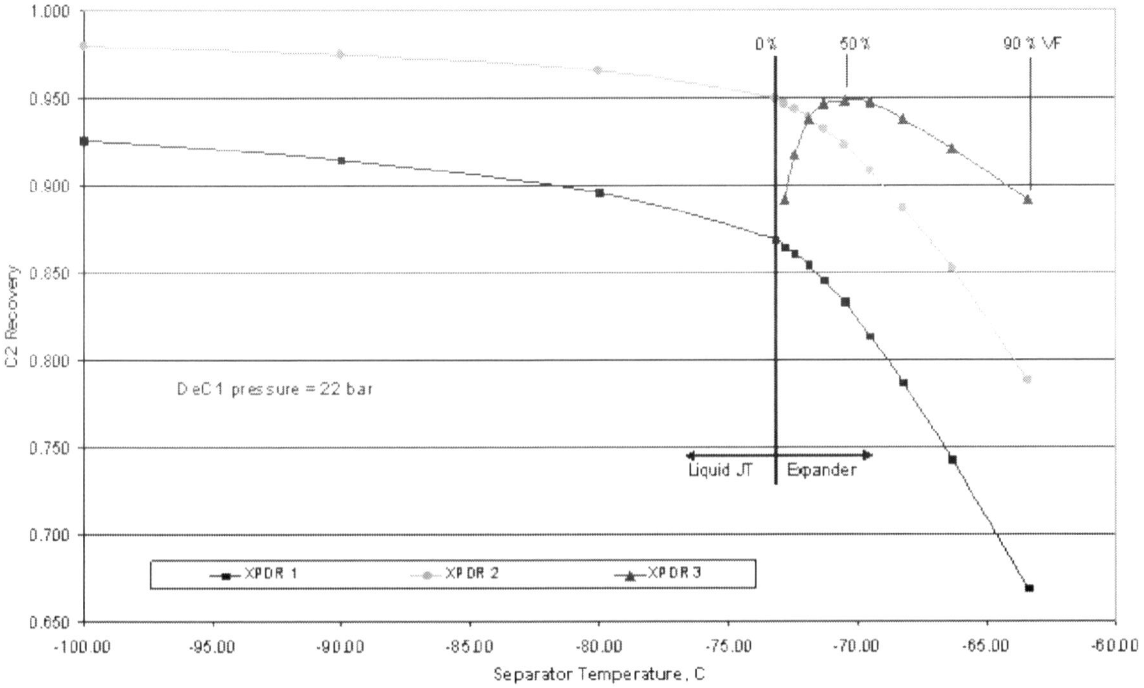

Figure 5: Two Types of Behavior for Different Categories (VF = Vapor Fraction)

XPDR 3 shows a different trend, though. As the separator temperature decreases, the C_2 recovery goes through a maximum. In other words, additional refrigeration at the separator hurts the C_2 recovery. For XPDR 3, the SIMAR condition is defined as the maximum of the curve. For XPDR 1 and XPDR 2, SIMAR is defined at the temperature level where the separator fluid has 30% of vapor fraction (VF). The choice of 30% is based on a practical consideration that no expanders would be installed if the gas flow is below this level.

Figure 6 depicts typical SIMAR curves for the three types of process schemes. A SIMAR curve is a collection of SIMAR conditions plotted against the recompression duty, which is a strong function of DeC1 operating pressure. Because refrigeration in the residue stream is not recovered to support pre-chiller operations, SIMAR curves define the minimum compression duty required by the specified process scheme. The plot shows that the efficiencies of XPDR 2 and XPDR 3 are superior to XPDR 1, while each has its advantages over a certain range.

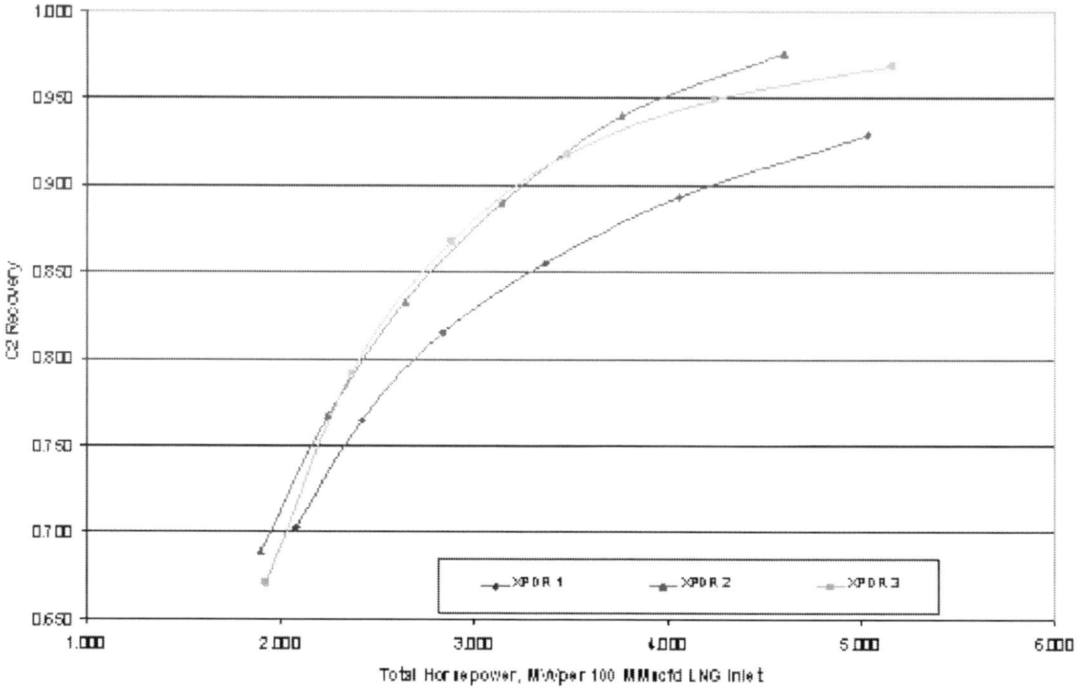

Figure 6: Comparing SIMAR Curves for Three XPDR Categories

Study Basis and Approach

Table 1 lists the LNG composition used in this work [McCartney, 2003]. In order to have a leveled comparison basis, the feed composition is the same for all configurations. The purpose is to highlight the underlying working principals of the art. The phase envelope is shown in Figure 4. Table 1 also lists pertinent simulation parameters used when generating process models in HYSYS.

For each selected process, the DeC1 operating pressure is fixed. A series of C_2 recovery levels at the separator temperatures can then be uniquely determined according to the SIMAR procedures previously defined. The DeC1 operating pressure is then changed and the procedure is repeated to cover the entire pressure (or C_2 recovery) range of interest.

Heat pumps change the temperature distribution of a system and should be handled separately. A heat pump is a mechanism capable of moving heat from low to high temperatures. A heat pump includes a compressor, a cooler for rejecting heat to a high-temperature sink, an expander or JT valve, and an exchanger to transfer heat from the low-temperature source. The most famous example of a thermodynamic heat pump is the Carnot engine [Elliott and Lira, 1998]. Figure 7 shows two examples of heat pumps in gas processing.

The heat pumps in this work are included into the models after determining the SIMAR separator temperature. The discharge pressure of the internal compressor is limited to the inlet pressure. In some cases, further improvement in C_2 recovery could be obtained by relaxing this limit.

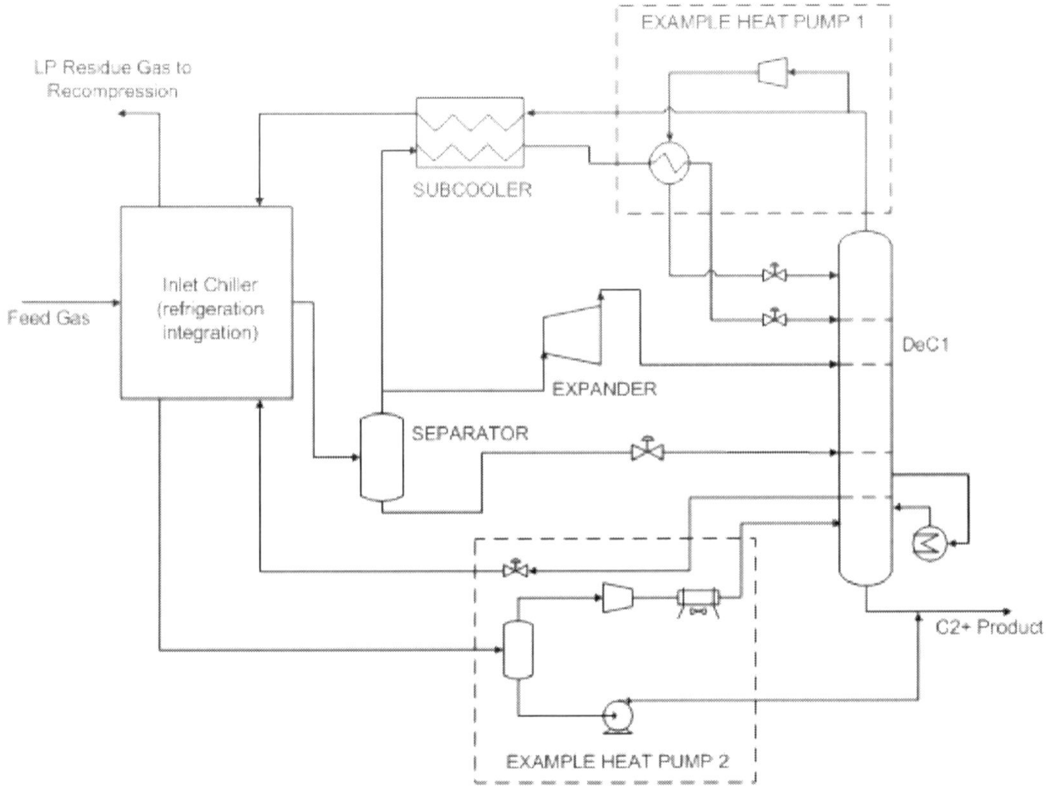

Figure 7: Example – Heat Pumps Embedded in XPDR 3

This work will first collect the patents filed to the U. S. Patent and Trademark Office and other open-domain sources. Henceforth, these sources will be referred to as literature. The processes are then categorized through visual inspections. Afterwards, HYSYS® simulations will be performed on selected configurations to facilitate the comparison among different categories. Characteristics of configurations and categories revealed by simulations will be reported with conclusions of this work.

Results

The results are broken down into three sub-sections. The first sub-section presents the results of sensitivity analysis, which provides optimal consistent process variables that could also serve as a reference for future optimizations. The second sub-section presents the results of simulating processes. The third sub-section demonstrates the improvements of adding heat pumps to process schemes.

•

Sensitivity Analysis

The areas of sensitivity analysis include:

- The split ratio of separator vapor to the expander and the sub-cooler in XPDR 3
- Side inlet locations on the DeC1 column using XPDR 3 as an example
- Recycle ratios in XPDR 2 for fully rectified processes

Split Ratio of Separator Vapor to the Subcooler

The separator vapor is split into two branches, one goes to the sub-cooler to recover the refrigeration from the DeC1 top and the other goes to the expander to generate refrigeration. The top branch also generates refrigeration through a JT valve.

Figure 8 shows that, as more separator vapor is directed toward the sub-cooler the process, efficiency improves. The trend continues until it reaches 80% through the sub-cooler, after which the efficiency retreats. In this work, the split is set to a fixed value between 40% and 60% due to practical considerations that a larger proportion of the vapor facilitates the expander operations with minimal efficiency losses.

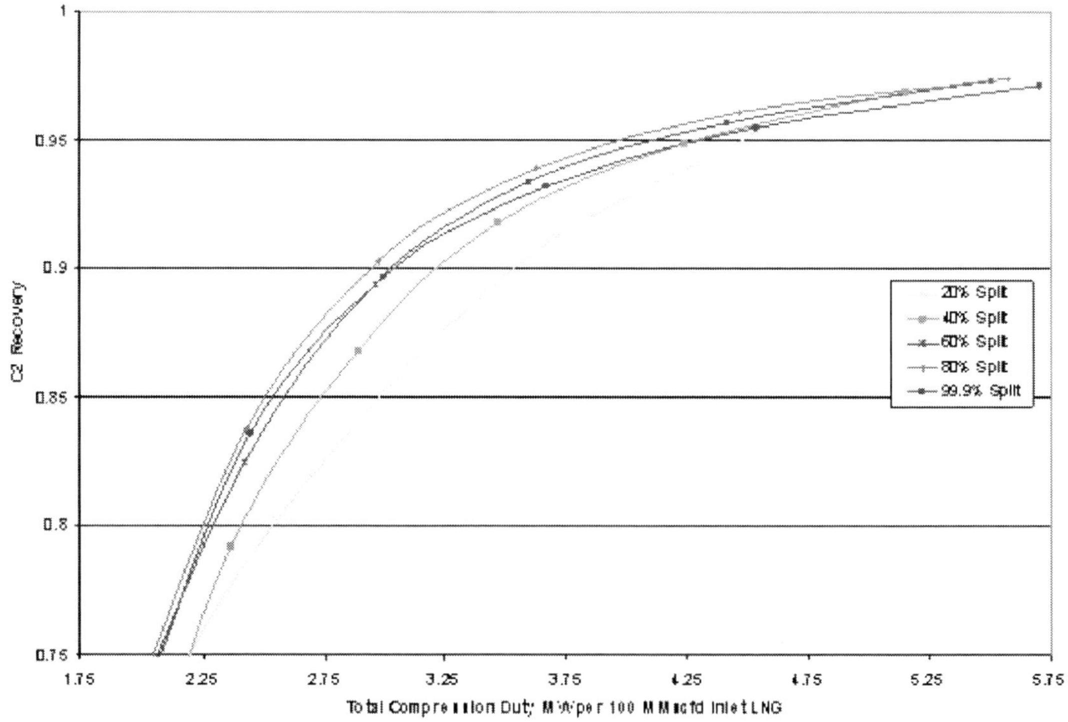

Figure 8: Impact of the Separator Value Split (to Sub-Cooler) in XPDR 3 Schemes

Location of Inlet Trays in DeC1

Figure 9 shows the results of varied inlet stream locations to the DeC1. The different locations are represented by two numbers: the first is the inlet tray of the gas from the expander and the second is the liquid inlet from the bottom of the separator. There is one tray between the two inlet streams. The column tray numbering scheme is from the top down.

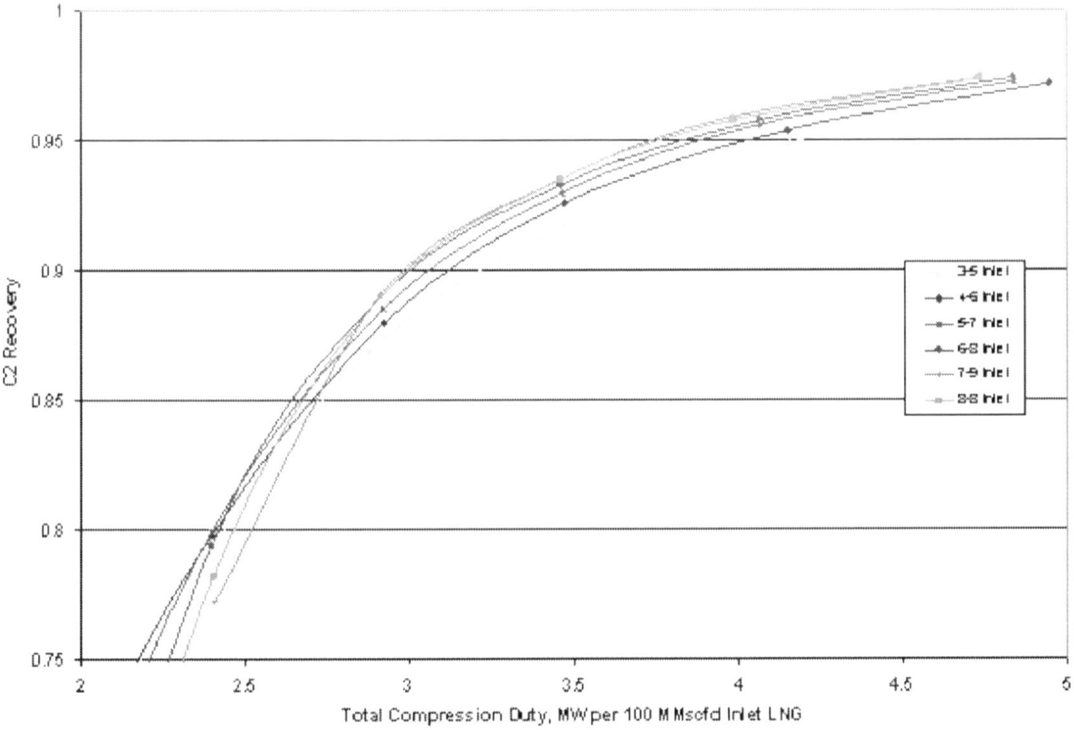

Figure 9: Impact of the Location of the Inlet Tray in DeC1

The most efficient location is the 7-9 configuration. One drawback to this configuration is the difficulty in converging the calculations around the DeC1 at lower column pressures. Another consideration with the lower side inlets is that in real applications, side-draws are required for refrigeration recovery. Side-draws from the bottom half of the column are used to pre-chill the inlet natural gas. The sideboilers can raise the temperature of trays above the return point, so additional trays between the sideboilers and the side inlets are needed. Based on these considerations, the 5-7 configuration was used since it leaves more trays below for the side re-boilers. This configuration is about 1% less efficient the 7-9 configurations.

Most of the configurations described in the literature can be simulated according to the descriptions above. However, slight modifications to the column design were made for some of the processes configurations in the literature. For example, some cases had more side inlets or defined the stream order differently than the common design above. The standard used during this study was the 5-7 configuration with one trays separating any additional streams, so in some cases there was an additional side inlet on the third tray. If the configuration was not well defined, then the side inlets were ordered based on their temperature, with the coldest on the top, and with an attempt to keep a 3-5-7 configuration.

Recycle Flow and Location

This subsection shows the impact of the flow ratio of the rectifying stream to the column for the XPDR 2 category, fully rectified processes. To optimize this variable, the XPDR 2 process from Figure 2 was modeled and the rectifying stream comes from the lean product. The split ratio on the recycle was varied as seen in Figure 10.

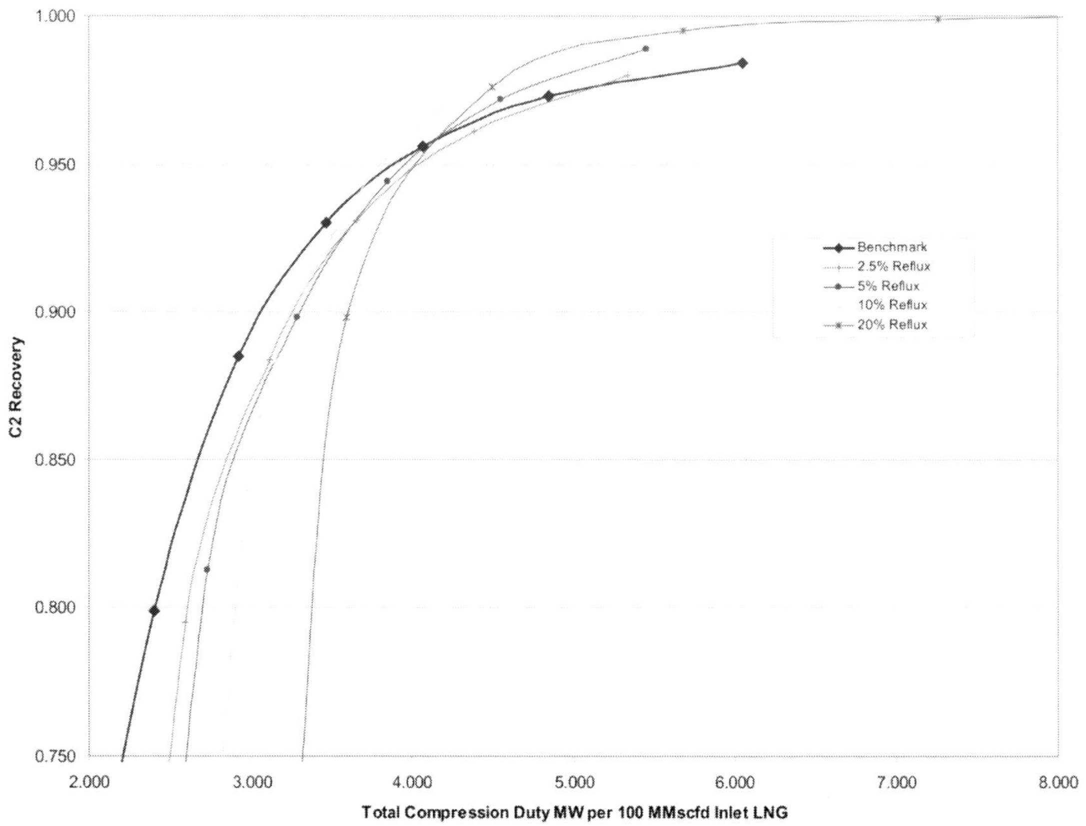

Figure 10: Impact of the Recycle Rate on XPDR 2 Schemes

In the case presented, the stream was taken from the final pipeline conditions (after compression and cooling). This figure shows the different rectifying splits compared to the benchmark process. As the rectifying split is decreased, the process performance becomes more similar to the benchmark curve. Larger rectifying splits show an increase in C_2 recovery at a given compression duty.

Values above 20% were not tested because larger rectifying splits would increase equipment size and are usually uneconomical. These processes could be capable of higher recoveries with larger rectifying splits, and can be optimized per project requirements. In the figures, all processes modeled with variable rectifying splits note the reflux percentage.

Modeling Various Processes

This sub-section presents simulation results based on various processes described in the literature. The presentation is arranged based on process categories presented earlier. Each category includes the benchmark curve as reference and comparison.

XPDR 1: Not Rectified

This category provides a baseline case for other rectified processes. Two different processes without rectification are presented in Figure 11.

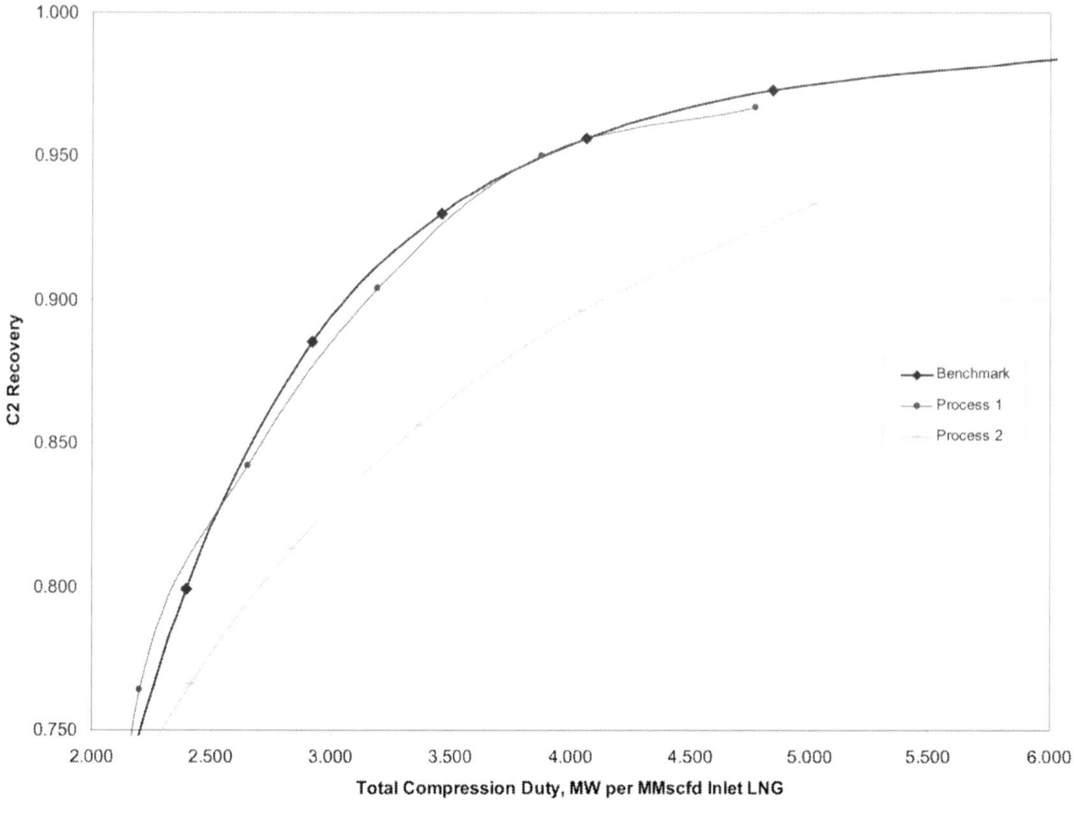

Figure 11: Comparing XPDR 1 Schemes

Generally, processes in this category require more compression duty than the benchmark process, which is consistent with the results shown Figure 6. However, through creative enhancements, the efficiency can be comparable.

XPDR 2: Fully Rectified

A variety of fully rectified processes are compared to the benchmark in Figure 12.

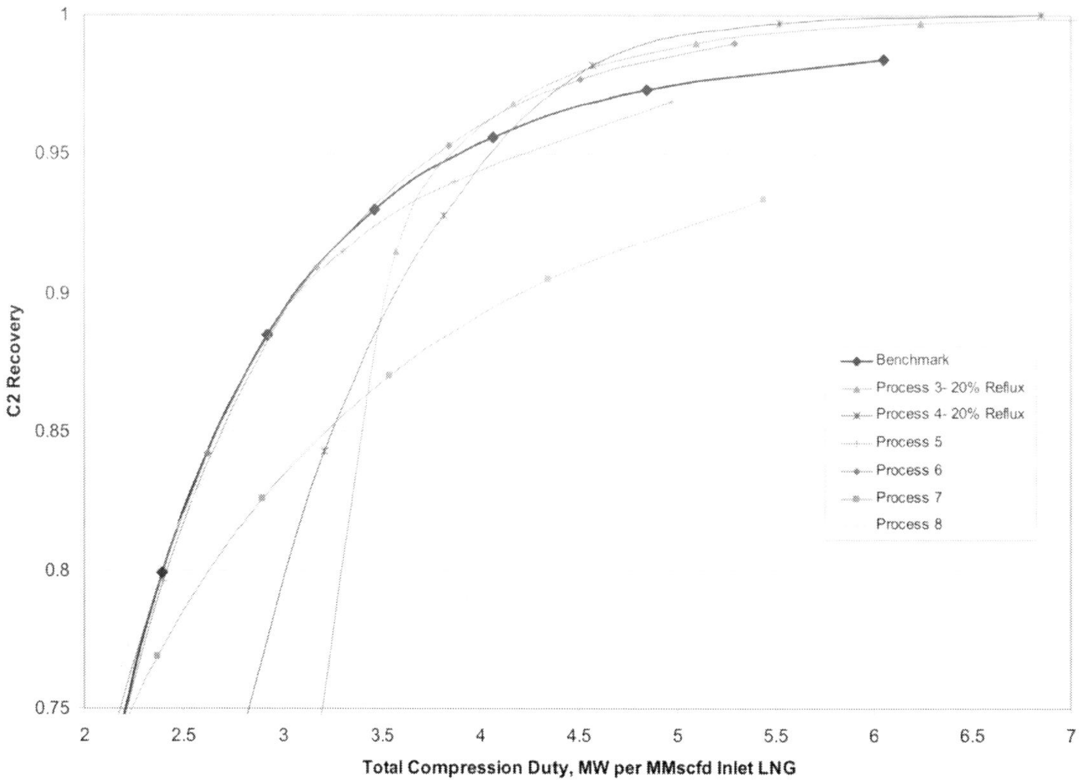

Figure 12: Comparing XPDR 2 Schemes

Some show better recovery and some have poor performance. Of those that show improvement above the benchmark, each has a certain C_2 recovery range where they require the least total compression duty. Different processes may be a better fit to a project, depending on the desired C_2 recovery.

XPDR 3: Self Rectified

There were four XPDR 3 processes modeled and the results are shown in Figure 13.

At high and low C_2 recoveries, the compression duties are similar; however, in the middle range there is variance between different processes. Above 95%, it is possible to achieve savings in compression duty. Although the benchmark appears to be superior in Figure 13, the conclusion should not be generalized because only one set of LNG composition and inlet pressure was used in this work. Also, SIMAR curves do not include pre-chiller sections. Practical experiences indicate improvements by various configurations for different feed conditions.

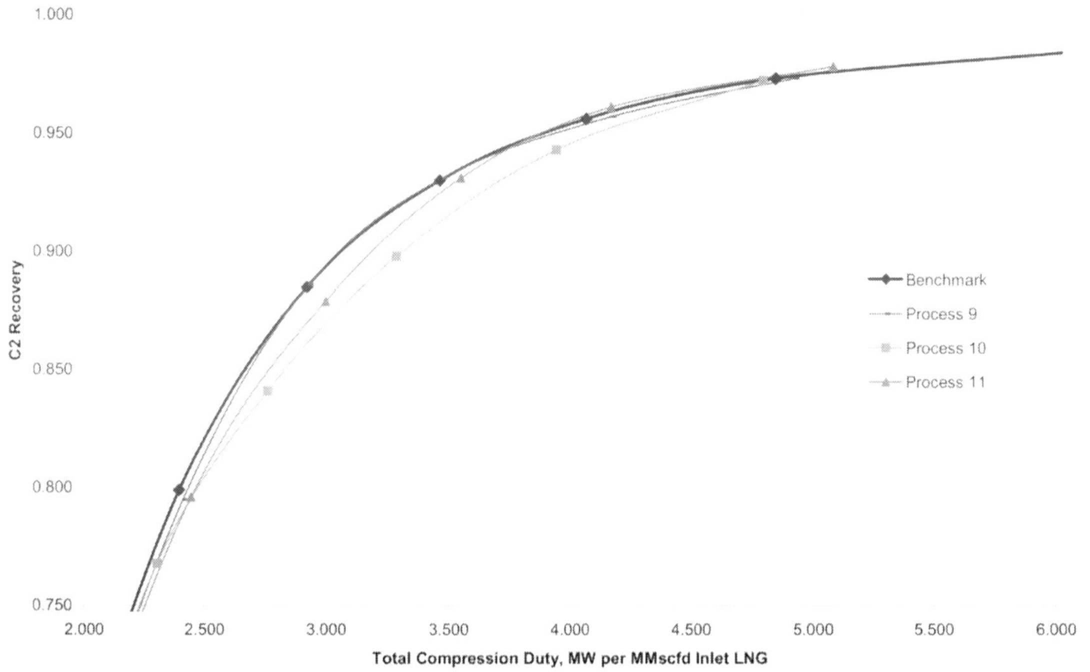

Figure 13: Comparing XPDR 3 Schemes

Impact of Heat Pumps

The performance of three processes with heat pumps compared in this study is shown in Figure 14.

All three show improvement over the benchmark at high C_2 recoveries. Heat pumps are very versatile tools that can achieve higher C_2 recoveries for the same compression duty than the other expander configurations. A heat pump can be added to just about any configuration and provide additional refrigeration that also increases the compression duty. The increase in compression duty is outweighed by the benefit of the additional refrigeration into the system and the C_2 recovery increases.

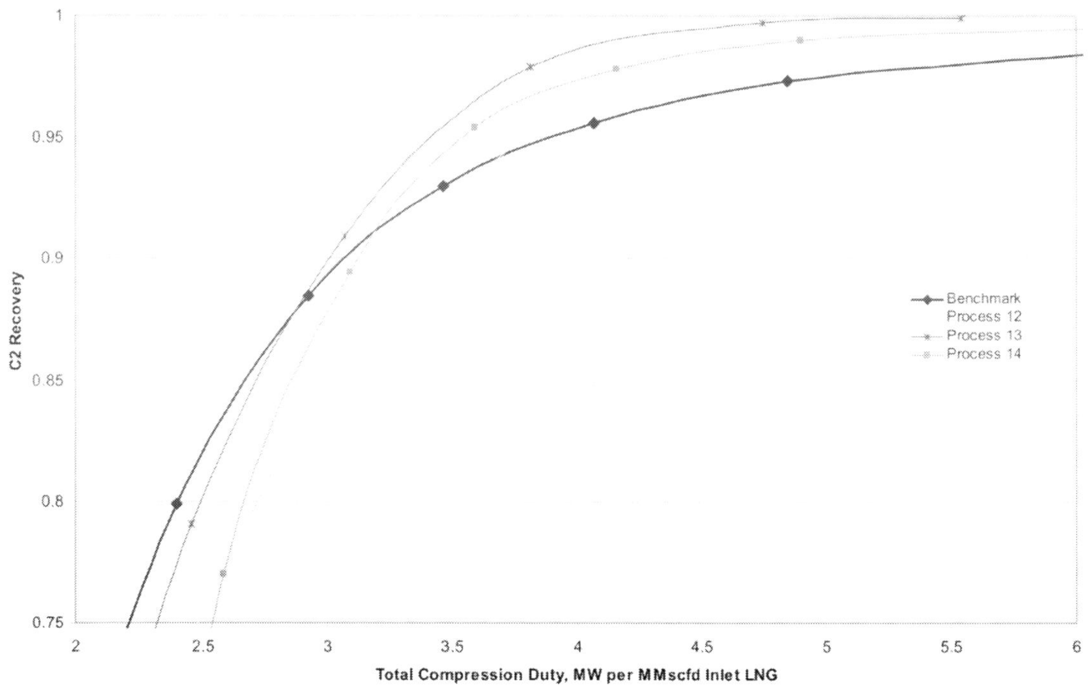

Figure 14: Impact of Heat Pumps on Process Schemes

Conclusions

The conclusions and recommendations of this work are summarized as follows:

- SIMAR curves can be generated following a few simple steps for different types of gas processing configurations that offer an opportunity to compare expander based processes based on total compression duty for certain C_2 recovery levels.

- Results of sensitivity analysis may reveal areas for optimization. For configurations containing a vapor subcooler, the split between the expander and subcooler should be set to the optimal operating level. The inlet tray locations of liquid and vapor stream to the DeC1 needs to be standardized with the consideration of side-draw reboilers in the lower trays of the column. In XPDR 2 processes, the rectifying stream to the DeC1 is not explicitly defined, and can be optimized for C_2 recovery.

- SIMAR curves for the three different expander based categories of processes have been generated and presented. Rectified column configurations have a higher C_2 recovery than not rectified designs. In the individual categories, there is some variance in process performances.

- Process designs that include heat pumps can achieve high C_2 recoveries with lower total compression duties. Heat pumps are powerful tools to increase refrigeration inside a process and the results show that the increased compression duty is offset by the increased C_2 recovery.

References

Elliott, J.R. and Lira, C.T. Introductory Chemical Engineering Thermodynamics. New Jersey: Prentice Hall, 1998.

Huang, S., Chen, R.J.J., Yao, J., and Elliot, D. Processes for High C_2 Recovery from LNG – Part II: Schemes Based on Expander Technology. AIChE Spring National Meeting, Orlando, FL. 23-27 April 2006.

Huang, S., Chen, R.J.J., Cook, D., Elliot, D.G. Processes for High C_2 Recovery from LNG – Part III: SIMAR Applied to Gas Processing. AIChE Spring National Meeting, Orlando, FL. 23-27 April 2006.

Lee, R., Yao, J., Elliot, D . *Flexibility, Efficiency to Characterize Gas-Processing Technologies. Oil & Gas Journal*. 13 Dec. 1999.

McCartney, D. Gas Conditioning for Imported LNG. 82nd GPA Annual Convention, San Antonio, TX. 9-12 Mar. 2003.

Pennybaker, K.A., Wolverton, S.E., Chafin, S.W., Ruddy, T.R., Pritchard, C.W. Comparative Study of Ethane Recovery Processes. 79th Annual GPA Convention, Atlanta, GA. 13-15 Mar. 2000.

Pitman, R.N., Hudson, H.M., Wilkinson, J.D. Next Generation Processes for NGL/LPG Recovery. 77th Annual GPA Convention, Dallas, TX. 116 Mar. 1998.

Table 1: Simulation Parameters Used in This Work

LNG Inlet	Medium LNG
Inlet temperature	-160°C
Inlet pressure to C_2+ facilities	10 bar
Primary Booster Pump discharge pressure (85% of critical pressure shown below)	50 bar
Plant capacity, inlet LNG	100 MM scf/D
Send-out gas pressure	100 bar
Send-out gas temperature	1°C
Components	**Mole Fraction**
Nitrogen	0.00127
CO_2	0.00008
Methane	0.90692
Ethane	0.08617
Propane	0.00552
i-Butane	0.00003
n-Butane	0.00001
i-Pentane	0.00000
n-Pentane	0.00000
TOTAL	1.00000
HHV, MJ/Sm^3	40.3
Btu/SCF	1,082
Critical pressure, bar	58.63
Expander polytropic efficiency	80%
Compressor polytropic efficiency	80%
Pressure drop-through plate-fin exchanger	0.2 to 0.3 bar
Pressure drop-through air coolers	0.6 bar
Minimum delta T in plate-fin exchanger	2.5 ± 0.2°C
DeC1 operating pressure	17 to 42 bar
Number of trays in DeC1 column	15
C1 composition in DeC1 bottom product	1%

Biography

1500 Louisiana Street, Houston, TX 77002, USA
Phone: 832-854-3332, Fax: 832-854-2039
Email: J.Trinter@Chevron.com

Justin Trinter is a process engineer with Chevron specializing in LNG and Gas Processing. He graduated with a Batchelor of Science in Chemical Engineering from the University of Idaho in 2006.

Dr. Stanley Huang has a specialty area in cryogenic applications, particularly in LNG and gas processing. He graduated from National Taiwan University with a B.S. degree and attended Purdue University in 1981. He earned his Master and Ph.D. degrees there, all in Chemical Engineering. Additionally, he also acquired a Master of Science in physics. Stanley has 20 years of industrial experiences and is a Registered Professional Engineer in Texas. Currently he is a Staff LNG Process Engineer of Chevron.

SIMAR Applications 2: Optimal Design of Expander-Based C2+ Recovery in Gas Processing

Christopher M[c]Mullen
LNG Process Engineer

Stanley Huang
Staff LNG Process Engineer

Chevron Energy Technology Company
Houston, Texas

AIChE Spring Meeting, April 2007
7thTopical Conference on Natural Gas Utilization
Houston, Texas, 22-26, 2007

1.0 INTRODUCTION

Since its acceptance by the industry in the 1970's, the expander-based process has become the mainstay technology in ethane recovery applications [Lee et al, 1999]. Despite the great technical and commercial success, a systematic methodology leading to the optimal design of an expander-based ethane recovery system has yet to be developed. Typically, a new process engineer could spend several years mastering this art as it has a steep learning curve which can be frustrating. This may significantly hamper the spreading of this art to other engineers.

This work intends to establish a systematic methodology to flatten the learning curve. The methodology is based on SIMAR, which stands for System Intrinsic Maximum Recovery [Huang et al, 2006a and b]. SIMAR was discovered when liquefied natural gas (LNG) was used as the feed to an expander-based C2 recovery process, in lieu of natural gas (NG). Since this scenario represents an extreme scenario where the refrigeration supply is unlimited, the only constraint for increasing C2 recovery resides in the system configurations.

The principles of SIMAR were extended to gas processing applications, where the supply of refrigeration is finite. In addition to the intrinsic constraints imposed by the physical configuration of the system, the limited refrigeration supply should be considered as well. For instance, the refrigeration supply to the Pre-Chiller may be drawn from the returning residue gas and the lower section of the Demethanizer (DeC1). If there is still a deficiency, then external refrigeration must be provided. The total compression work should include both re-compression of the residue gas and external refrigeration.

This effort is an extension of the previous SIMAR work. The main focus of this paper will be the presentation of a systematic methodology which leads to optimal operating conditions of a process configuration. This methodology can be followed by any process engineer with a suitable background in process simulations. Real industrial cases will be used as illustrations.

2.0 TECHNICAL BACKGROUND

Figure 1 shows two generalized schemes for C2 recovery based on expander-technology. The top scheme uses natural gas (NG) as the feed, whereas the bottom one uses LNG. Both schemes are intended to strip the feed of its heavier components. The residue gas is then recompressed and delivered to the pipeline.

In the top scheme, sweet and dried natural gas goes through the Pre-Chiller, where the gas is chilled to a suitable level before entering the heart of the plant: an expander-based C2 recovery scheme. The refrigeration of the pre-chiller is mainly provided by the returning residue gas and supplemented by side-draws (not shown) from the DeC1. Depending on actual requirements, external refrigeration may be required (not shown). The main components of the expander section include a separator, expander and DeC1. A subcooler is usually provided for improved refrigeration integration in the low temperature regions. The exact design of this section remains an art [Pitman et al, 1998].

The major difference between the two schemes in Figure 1 is the inlet pre-chilling section. When LNG is used as the feed, the entire pre-chiller can be eliminated. The refrigeration in the residue gas can be retained, thus dramatically reducing the recompression duty.

Figure 2 provides further details of the expander section. Following previous works, the expander schemes are grouped into three types: separator top not rectified (XPDR 1), separator top fully rectified (XPDR 2), and separator top self-rectified (XPDR 3). A fourth category of heat pumps is not discussed here, but can be easily recognized by the inclusion of recirculation compressors within the process configurations.

Figure 3 shows two phase envelopes corresponding to two natural gas compositions. The two LNG compositions are designated as Medium and Rich LNG. Lean LNG can be very high in methane composition and is not of interest here. The Rich LNG has an enlarged two-phase region toward higher temperatures and pressures. Unless otherwise specified, the medium LNG composition is the default set in all discussions.

A SIMAR curve can be constructed following a few simple steps. The procedure starts from a relatively high temperature at a reasonable pressure level. The track of testing temperatures penetrates through the two phase region and ends at an arbitrarily chosen level of -100°C. The fluid would remain liquid at and below this temperature level.

Figure 4 plots the C2 recovery level against the test separator temperature for the three categories of process configurations. The DeC1 operating pressure is 22 bar. Both XPDR 1 and XPDR 2 show monotonic trends of improvement as the temperature decreases. This trend continues even after the inlet gas is totally liquefied, when the expander is replaced by a Joule-Thompson (JT) valve and no expansion work is recovered. XPDR 3 shows a different trend though. As the separator temperature decreases, the C2 recovery goes through a maximum. In other words, too much refrigeration at the separator may hurt the C2 recovery.

For XPDR 3, the SIMAR is defined as the maximum of the curve. For XPDR 1 and XPDR 2, the SIMAR is defined at the temperature level where the separator fluid has 30% vapor fraction (VF). The choice of 30% is based on a practical consideration that no expanders would be installed if the gas flow is below this level.

A SIMAR curve is the collection of all the SIMAR conditions which cover the entire range of DeC1 operations of interest. Figure 5 depicts typical SIMAR curves corresponding to the aforementioned three categories. The characteristics of the three categories are observed. (1) The reflux stream makes XPDR 2 more efficient than XPDR 1 throughout the entire range, corresponding to the operating pressures of DeC1 from 17 to 42 bar. (2) The efficiencies of XPDR 2 and XPDR 3 are comparable, while each has its advantages over a certain span.

Figure 6 shows typical results when the feed is changed from LNG to NG, based on the XPDR 3 category. Since the refrigeration in the residue gas must be recovered to cool the inlet gas, the recompression duty increases significantly by this change in feed. As indicated by the two thin curves on the left, there is a big gap between them.

In addition to the recovered refrigeration in the residue gas, external refrigeration may also be needed. When this is true, the compression duty required in the external refrigeration duty should be added to the aforementioned recompression duty to form the total compression duty. The two thick curves on the right represent the recompression duties by using one or two side-boilers (SB). The gap between the curves of total compression and the recompression curves on the left in Figure 6 represents the external refrigeration. Using two SB reduces the external refrigeration because of improved refrigeration integration in the pre-chilling section.

The gap narrows when the DeC1 operating pressure decreases. This indicates the decreased demands for external refrigeration. Also observed in Figure 6 is that when the DeC1 operating pressure decreases, or C2 recovery level increases, the need for external refrigeration also decreases. The expander provides more refrigeration for process needs at lower DeC1 operating pressures.

3.0 STUDY BASIS

Table 1 lists the two LNG compositions used in this work [McCartney, 2003]. The phase envelopes of both LNG compositions are shown in Figure 3. Also listed are two industrial cases designated as Cases 1 and 2, respectively. Case 1 is taken from the open domain [Wilkinson and Hudson, 1992]. Industrial Case 2 is taken from an operating plant, built in the early 1980's, at an undisclosed location.

All simulations in this work are performed using HYSYS. Unless specified otherwise, the parameters listed in Table 2 are used in the simulations.

In order to keep discussions simple and conceptual, this work adopts thermodynamic heat pumps to account for external refrigeration supplies. Typically, external refrigeration implies closed-loop designs of propane or cascade designs of propane in combination with ethylene circuits. The adopted thermodynamic heat pumps are assumed to be 50% of Carnot efficiency. For example, to move 1 MW of heat from -60 °C (213 K) to 40 °C (313 K), a Carnot engine requires 0.47 (= 313/213-1) MW of compression duty. In this work, twice of the amount, or 0.94 MW, is required due to the reduced efficiency.

4.0 SCOPE

This work will first investigate the impacts of pre-chiller configurations. The default pre-chiller is represented by a multi-path brazed-aluminum heat exchanger (BAHE), as shown in Figure 7. However, individual thermal siphons as side-boilers are sometimes preferred in certain applications. The impact of such splits will be investigated. Figure 8 shows a typical arrangement with side draws from the DeC1 column used in this work.

Figure 9 presents a systematic methodology of searching for optimal operating parameters. For a specified process configuration, there are two independent variables determining the system performance: separator temperature and DeC1 pressure. The criteria of process optimization are: (1) the specified C2 recovery level should be met, and (2) the provision of external refrigeration should be within specified constraints.

The conventional optimization method follows a two dimensional (2-D) search pattern, as shown in the lower part of Figure 9 which helps to explain the high requirement for the process engineer's experiences. The role of SIMAR is presented in the top part of Figure 9. Basically, the constructed SIMAR curve serves as a guiding curve throughout the entire search procedure. With this guidance, the 2-D search is reduced to one-dimensional (1-D) which greatly simplifies the process.

Finally, two industrial cases will be used to demonstrate that the process conditions derived following the SIMAR approach are consistent with existing industrial practices.

5.0 RESULTS

This section is divided into three subsections according to the topics described above.

Impact of Pre-Chiller Configurations

In an integrated pre-chiller design, the utilization of available refrigeration is efficient because the flow of refrigeration among streams is unlimited. In a split design, as shown in Figure 8, this unlimited flow of refrigeration cannot be realized due to physical limitations in equipment provisions. As a result, the utilization of available refrigeration cannot match the integrated design. Certain penalties in the total compression power are anticipated.

Figure 10 compares the performances of two pre-chiller configurations using two side-boilers. The two curves on the left are identical to those in Figure 6. The split of pre-chiller results in an increased demand for external refrigeration, hence, total compression duty. Similar trends are observed in Figure 11 using one side-boiler.

Although the split design is less efficient from a process viewpoint, practical considerations may push the design toward this option. Examples include size limitation in BAHE manufacturing capabilities, preferences toward providing Shell & Tube exchangers for unfiltered fluids, and wide temperature swings in operations, among other considerations.

Sub-SIMAR Operations

Results of Sub-SIMAR operations have been reported previously [Huang et al., 2006b]. Figure 12 plots the two Model cases operating at SIMAR and sub-SIMAR conditions. For Model case Medium, two sub-SIMAR scenarios are presented with the separator temperature being progressively raised by about 10 °C each. The thin curves on the left side are recompression works. As anticipated, the C2 recovery suffers with each raise of the separator temperature. As a compensation, the external refrigeration duty decreases with each rise in the separator temperature. The combined effect is that total compression remains within a narrow range, as shown by the thick curves on the right side. Similar observations can also be made for the Model case Rich.

Based on compositions listed in Table 1, Industrial Case 2 is richer than both Model cases. SIMAR and sub-SIMAR operations for Industrial Case 2 were systematically investigated and . Figure 13 shows the results. Similar to the Model cases, the total compression duties for all cases fall within a relatively narrow range. However, there appears to be a systematic drift in the total compression duty with respect to the separator temperature.

It is important to note that the total compression work remains roughly unchanged regardless of deviations from SIMAR operations. In other words, once the process configuration is fixed, the total compression required to achieve a specified C2 recovery level is insensitive to pre-chilling levels. The choice of the separator operating conditions should be optimized by other criteria, such as economics. For example, if operating a system at SIMAR level requires a supply of external refrigeration at -50 °C the decision may shift toward sub-SIMAR operations in order to avoid cascaded external refrigeration circuits including ethylene. Ethylene refrigeration systems are relatively expensive to provide. Therefore, given the conditions of an inlet gas, SIMAR can serve as an expedient tool to establish the relations between C2 recovery and total compression work. Then, the exact pre-chilling level will be determined by other factors defined by the projects.

Industrial Cases

Industrial Case 1 has a high-pressure feed stream and is in the supercritical region. As a result, the expander generates sufficient refrigeration for the entire facility. The energy balances in the pre-chiller can be fulfilled without the need for any external refrigeration. Figure 14 shows both the SIMAR curve and the recompression curve. The recompression curve coincides with total compression because no external refrigeration is needed.

Industrial Case 2 has a low-pressure feed stream and requires external refrigeration. Furthermore, the plant specifies an external, relatively high-temperature propane refrigeration system. Accordingly, the designed facilities have to operate at sub-SIMAR conditions. Figure 15 shows the systematic approaches to match the specifications. All the simulations were based on an efficiency of 80% for compressors and expanders. As the separator temperature deviates further away from the SIMAR condition, the matching between simulations and plant design data seems to improve. At the level of Sub-SIMAR 2, the separator temperature agrees with the design data. However, the simulation results remain superior to the actual design data.

It was decided to lower the efficiency of compressors and expanders from the default value of 80% to 77%, considering the year of design for Industrial Case 2. In the past two decades or so, significant progresses in the machine's design have been well recognized. Figure 16 indicates that this practice would allow full replications of the design data.

As a summary, results following SIMAR methodology are in agreement with industrial data. Also, SIMAR methodology provides entire sets of feasible operating conditions for a given process configuration. The final optimization should be based on other criteria, such as plant economies and product desirability.

6.0 CONCLUSIONS

The conclusions of this work are summarized as follows:

- Instead of an integrated design, splitting the Pre-Chilling heat exchanger system would increase the total compression required for the same task. However, this option may be preferable due to other considerations, such as vendor availability and plant operability.
- The presented SIMAR methodology comprises straightforward steps to generate the SIMAR curve for any specified process configuration. Once available, the SIMAR curve can be used to guide other calculations in pre-chiller refrigeration balances or sub-SIMAR searches.
- Sub-SIMAR operating conditions are fairly common in industrial applications. For a given process configuration, SIMAR methodology provides entire sets of feasible operating conditions. The final optimization should be based on other criteria, such as plant economies and product desirability.

8.0 REFERENCES

Elliott, J.R., Lira, C.T., "Introductory Chemical Engineering Thermodynamics", Prentice Hall, NJ (1998).

Huang, S., Chen, R.J.J., Yao, J., Elliot, D., "Processes for high C2 recovery from LNG – Part II: Schemes based on Expander Technology", paper presented in AIChE 2006 Spring National Meeting, Orlando, FL (2006a).

Huang, S., Chen, R.J.J., Cook, D., Elliot, D.G., "Processes for High C2 Recovery from LNG – Part III: SIMAR Applied to Gas Processing", Paper presented in AIChE Spring Meeting, Orlando, Florida (2006b).

Trinter, J., Huang, S., "SIMAR Applications in Gas Processing: 1. Evaluating Expander-Based Ethane Recovery Process", paper to be presented in AIChE Spring Meeting, Houston, Texas (2007).

Lee, R., Yao, J., Elliot, D., "Flexibility, efficiency to characterize gas-processing technologies", Oil & Gas Journal, Issue of December 13th (1999).

McCartney, D., "Gas Conditioning for Imported LNG", paper presented in 82nd GPA Annual Convention in San Antonio, TX (2003).

Pennybaker, K.A., Wolverton, S.E., Chafin, S.W., Ruddy, T.R., Pritchard, C.W., "Comparative Study of Ethane Recovery Processes", paper presented at the 79th Annual GPA Convention, Atlanta, GA (2000).

Pitman, R.N., Hudson, H.M. & Wilkinson, J.D., "Next generation processes for NGL/LPG recovery", paper presented at the 77th Annual GPA Convention, Dallas, TX (1998).

Wilkinson, J. D. & Hudson, H. M., "Improved NGL Recovery Designs Maximize Operating Flexibility and Product Recoveries", paper presented at the 71st Annual GPA Convention, Anaheim, CA, March 16-18, 41-45 (1992).

9.0 List of Tables

Table 1: Gas compositions used in models for this study

	Model Cases		Industrial Cases	
Equivalent LNG	**Medium**	**Rich**	**Case 1**	**Case 2**
Inlet temperature, C	40	40	40	37.77
Inlet pressure, bar	50	65	73.40	50.67
Critical T, C	-63.45	-48.40	-63.69	-34.09
Critical P, bar	58.63	75.69	63.89	92.4
Components	Mole Fraction	Mole Fraction	Mole Fraction	Mole Fraction
Nitrogen	0.00127	0.0000	0.0027	[redacted]
Methane	0.90700	0.8695	.9247	
Ethane	0.08617	0.0792	.0423	
Propane	0.00552	0.0332	.0132	
i-Butane	0.00003	0.0074	.0039	
n-Butane	0.00001	0.0055	.0032	
i-Pentane	0.00000	0.0037	.0015	
n-Pentane	0.00000	0.0015	.0009	
n-Hexane	0.00000	0.0000	.0026	
n-Heptane	0.00000	0.0000	0.00000	
n-Octane	0.00000	0.0000	0.00000	
Carbon Dioxide	0.0	0.0	.0027	
TOTAL	1.00000	1.0000	1.00000	1.0000
Btu/SCF	1,082	1,165	1087	1227
C2+ content, GPM (Note 1)	2.45	3.63	1.43	3.97
C3+ content, GPM	0.15	1.52	0.58	1.88

Notes:
1. GPM stands for Gallons Per Thousand SCF.

Table 2: Simulation parameters used in models of this study

		Model Cases	Industrial Cases	
Process parameter	**Unit**	**Value**	**Case 1**	**Case 2**
Send-out residue gas pressure	Bar	100	73.4	50
Send-out residue gas temperature	°C	40	40	37.77
General				
Expander polytropic efficiency		80%	80% - 85%	80%
Compressor polytropic efficiency		80%	80% - 85%	80%
Pressure drop through plate-fin exchanger	Bar	0.2-0.3	Same as model	Same as model
Pressure drop through air coolers	Bar	0.6	"	"
Minimum delta T in plate-fin exchanger	°C	2.5 ± 0.2	"	"
DeC1 operating pressure	Bar	17 to 42	"	"
Number of trays in DeC1 column		15	"	"
C1 composition in DeC1 bottom product		1%	"	"
Minimum delta T in plate-fin exchanger	°C	2.5 ± 0.2	"	"
Max impact on C2 recovery by side-boilers for operations at SIMAR conditions		-0.005	"	"
High-temperature (HT) sink	°C	40	"	"

10.0 List of Figures

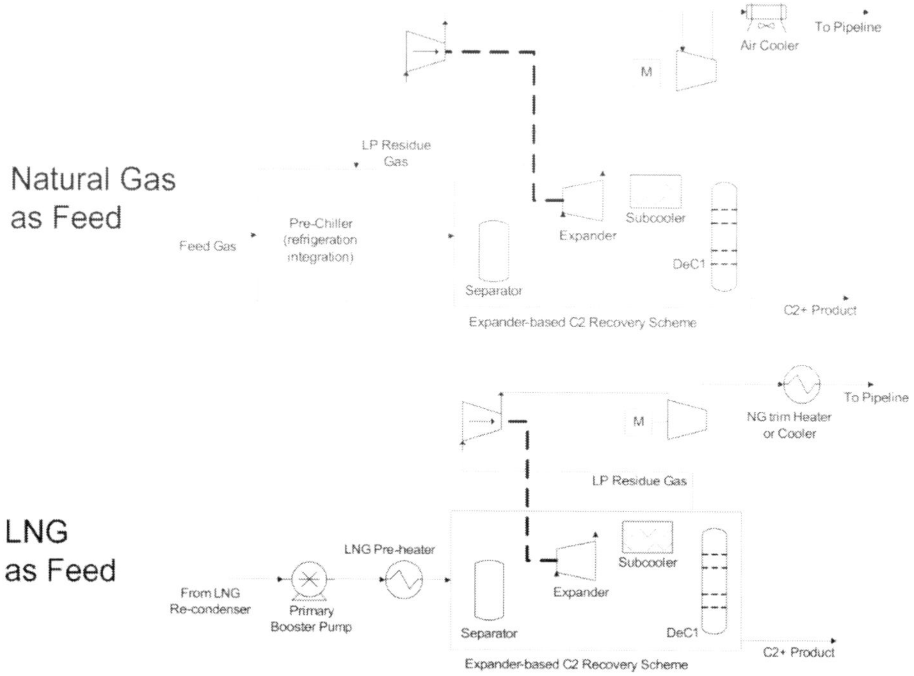

Figure 1 Generalized C2 recovery schemes with natural gas (top) and LNG (bottom) as feed streams

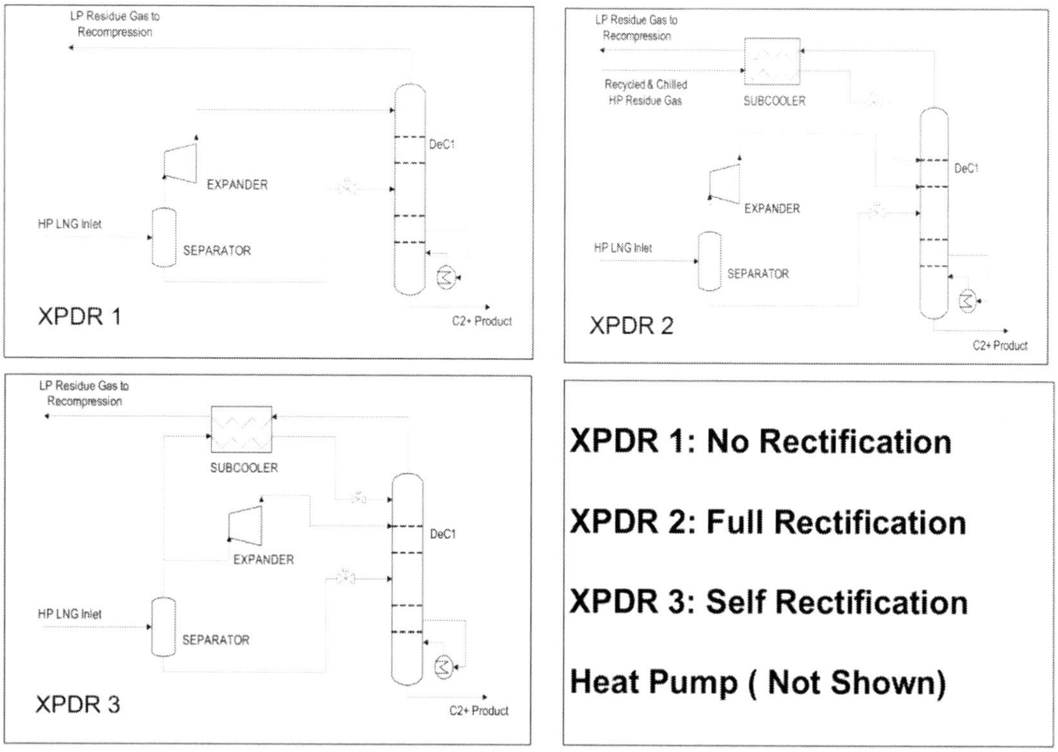

Figure 2 Categorizing expander-based schemes

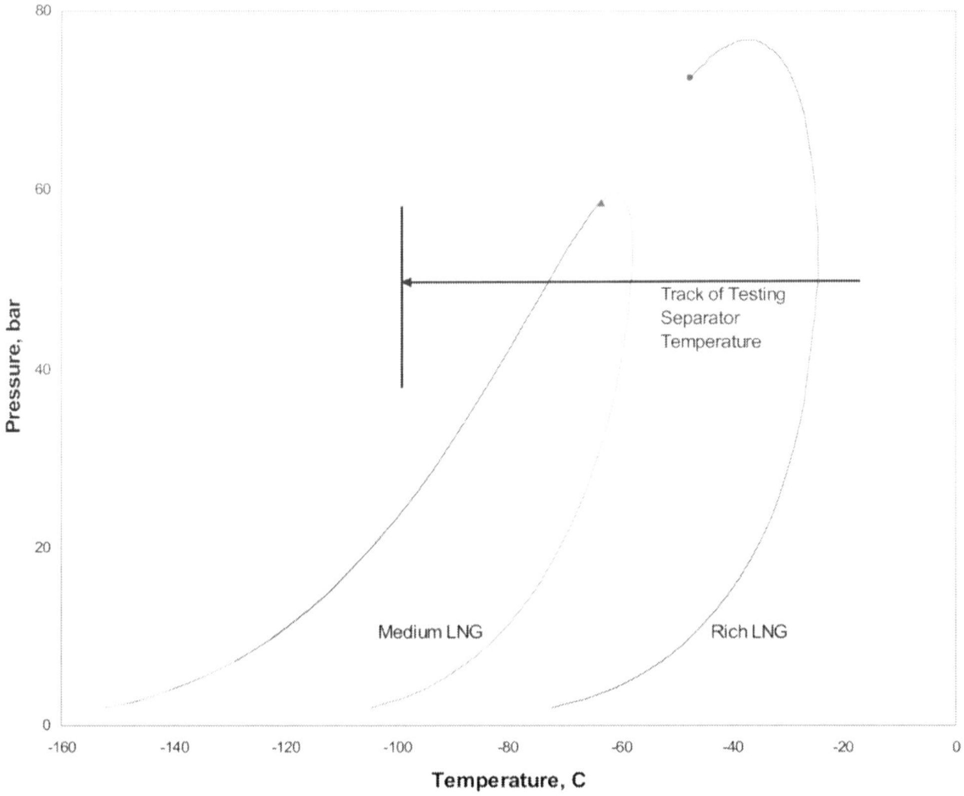

Figure 3 Phase envelope of inlet gases

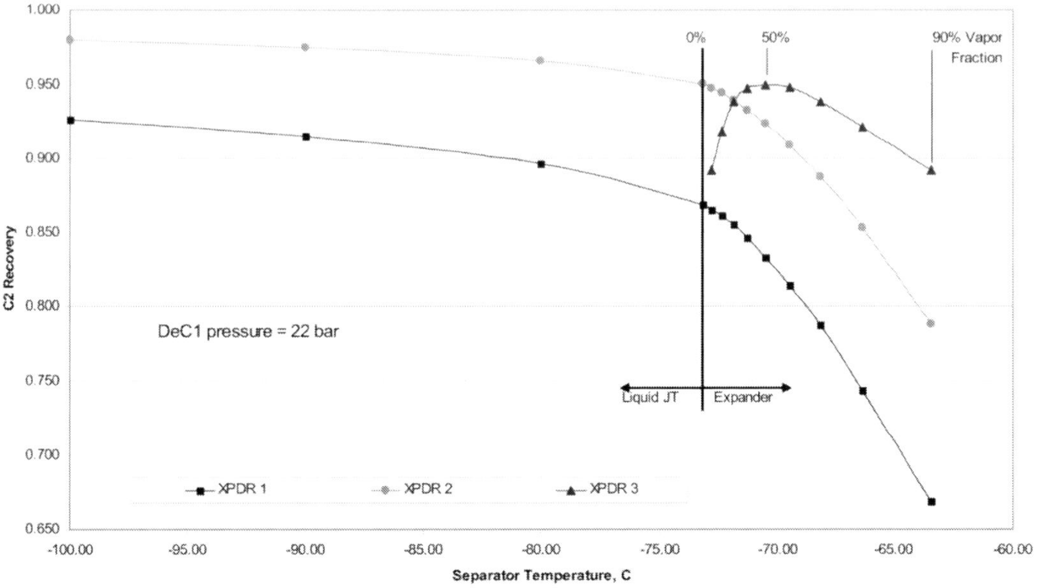

Figure 4 Two types of behavior for different categories. VF stands for Vapor Fraction.

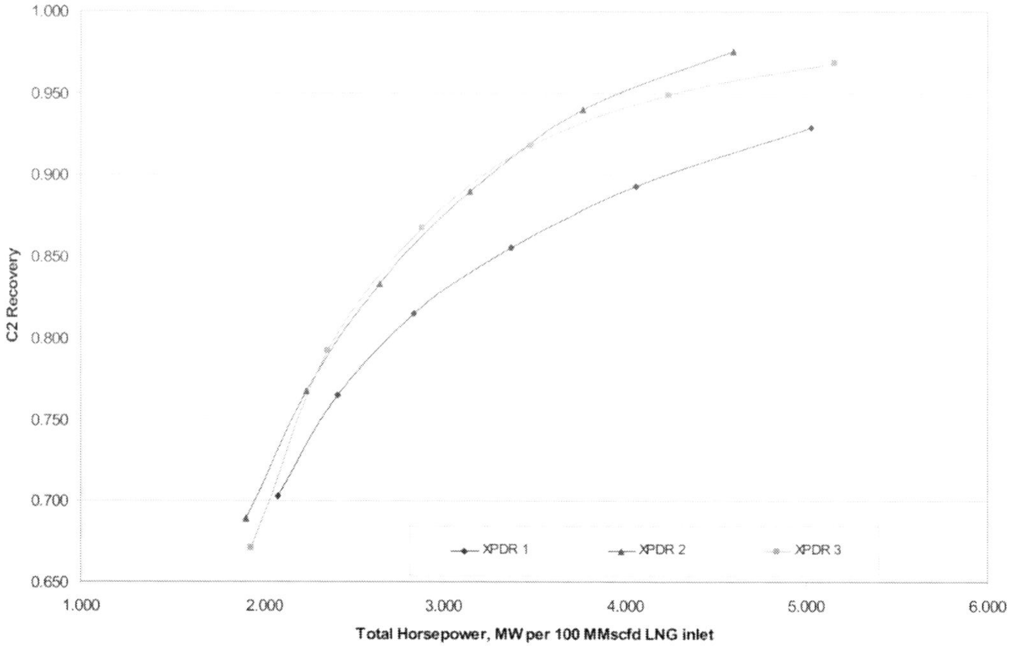

Figure 5 Comparing SIMAR curves for three XPDR categories

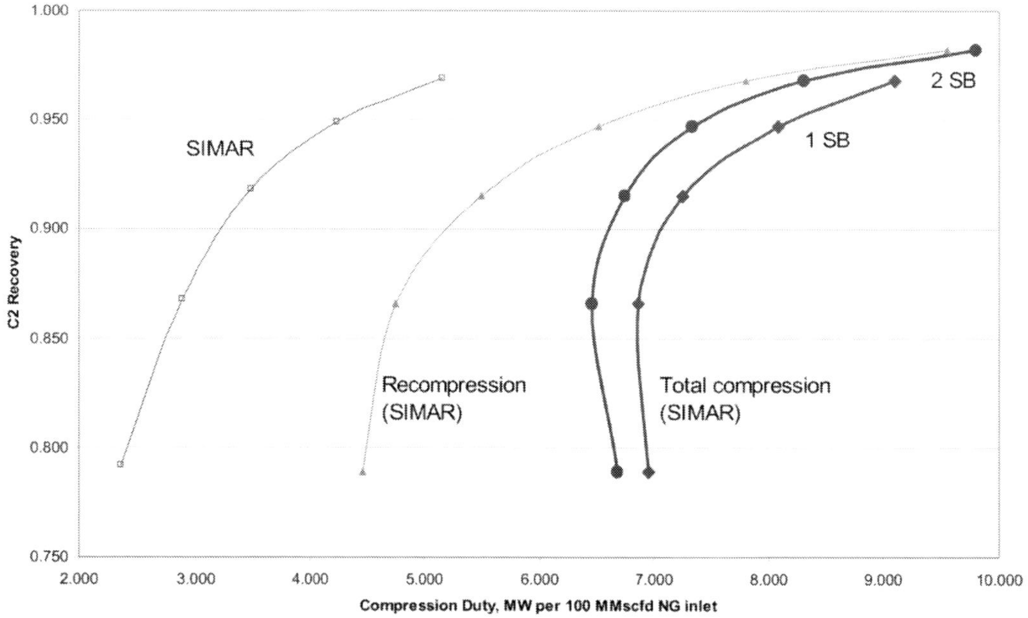

Figure 6 Comparing compression duties and Impact of side-boilers (SB)

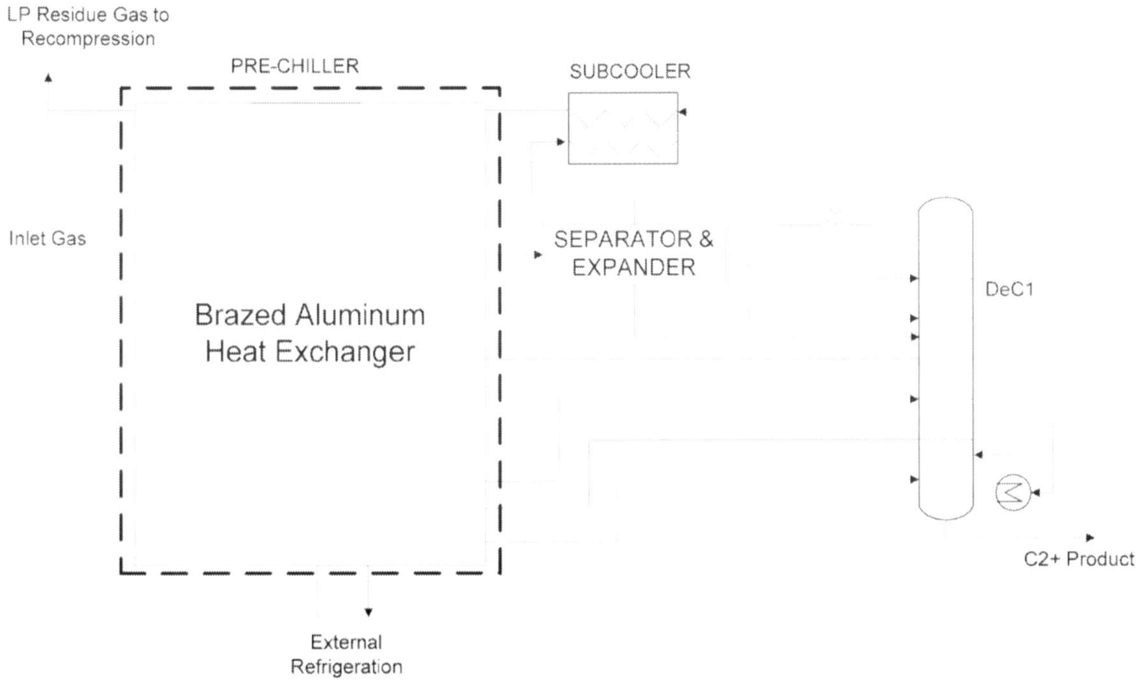

Figure 7 Integrated pre-chiller design

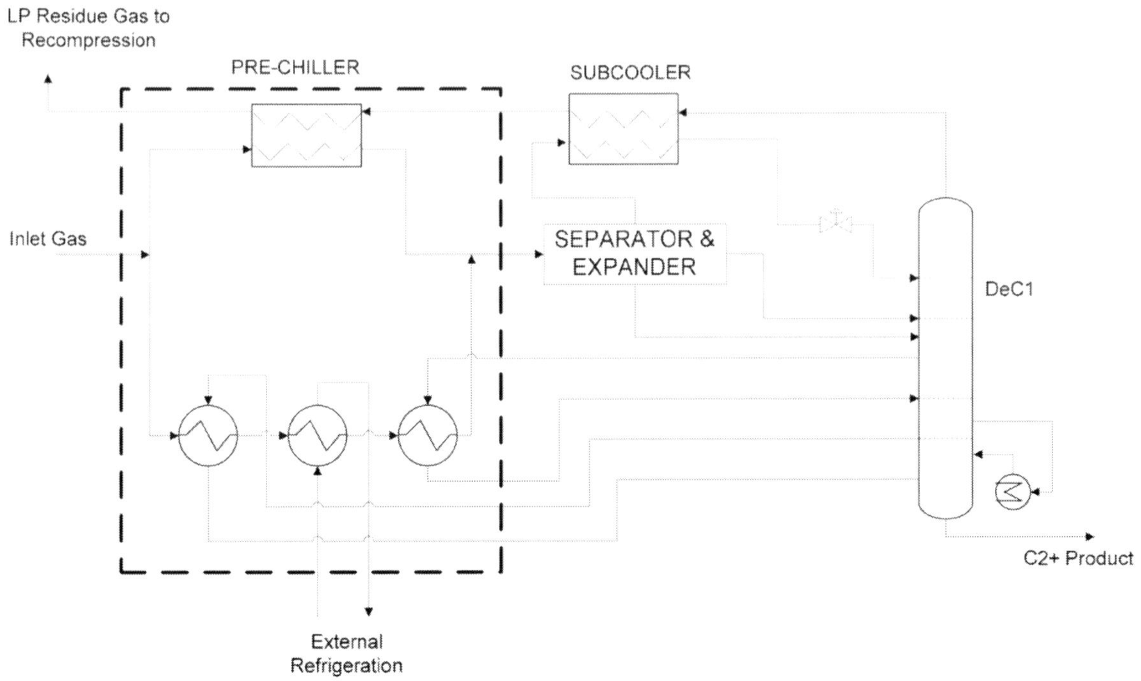

Figure 8 Using individual thermo-siphons in pre-chiller

Figure 9 SIMAR as guiding tool in search for the optimal design in gas processing

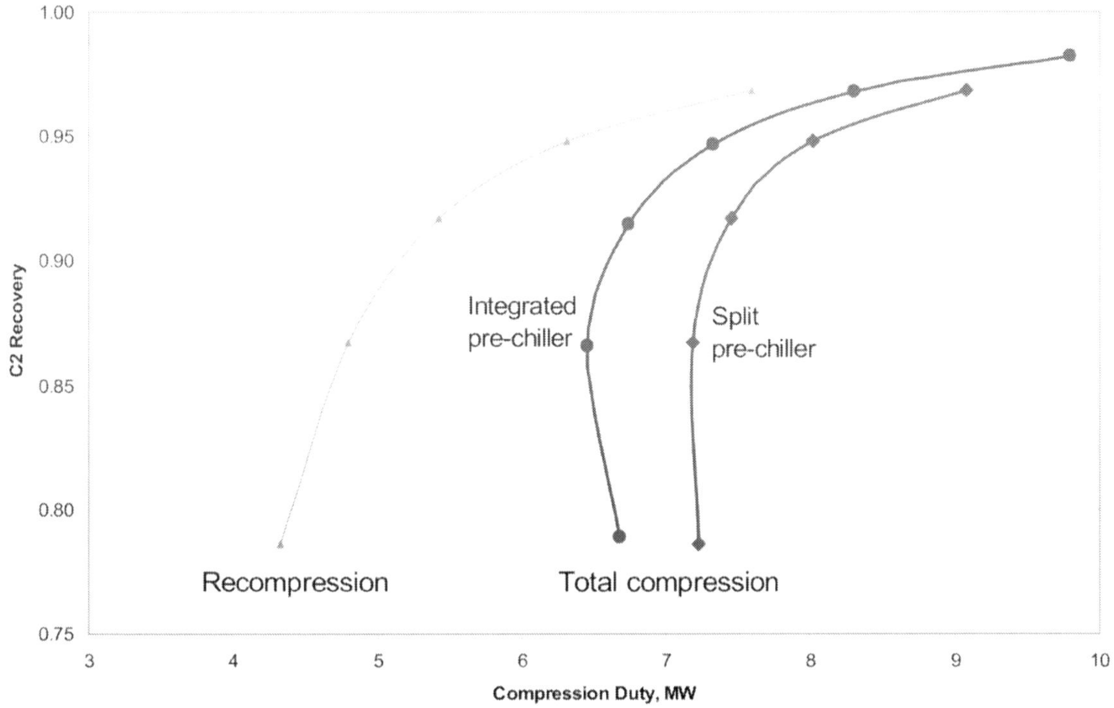

Figure 10 Impact of split pre-chiller with two side-boilers

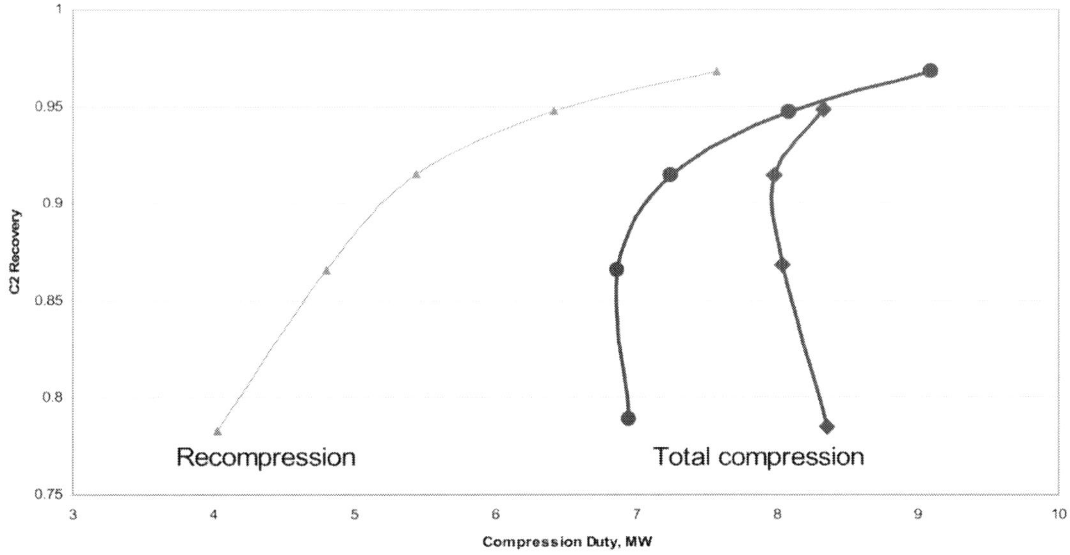

Figure 11 Impact of splitting Pre-Chiller design with one side-boiler

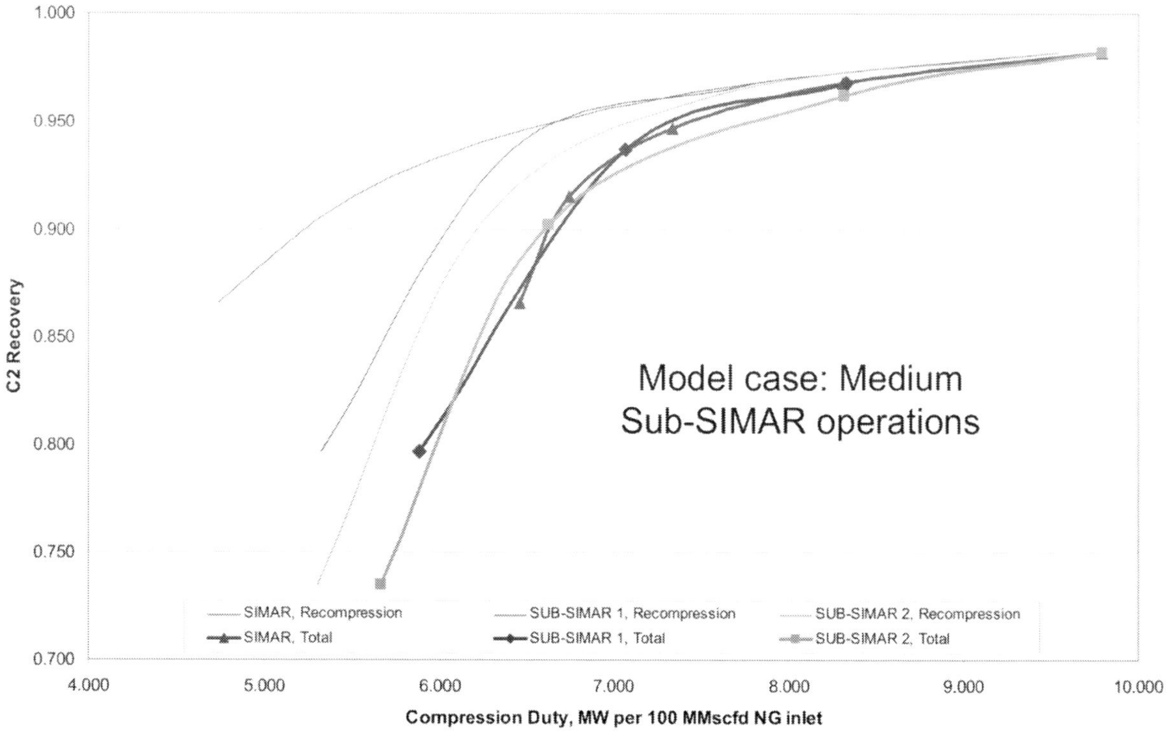

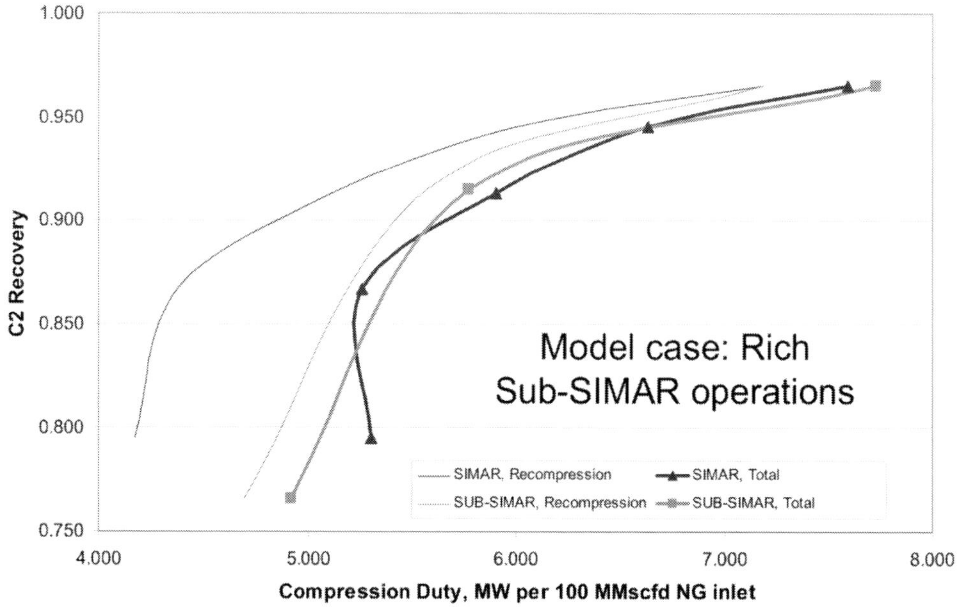

Figure 12 Sub-SIMAR operations for two model cases

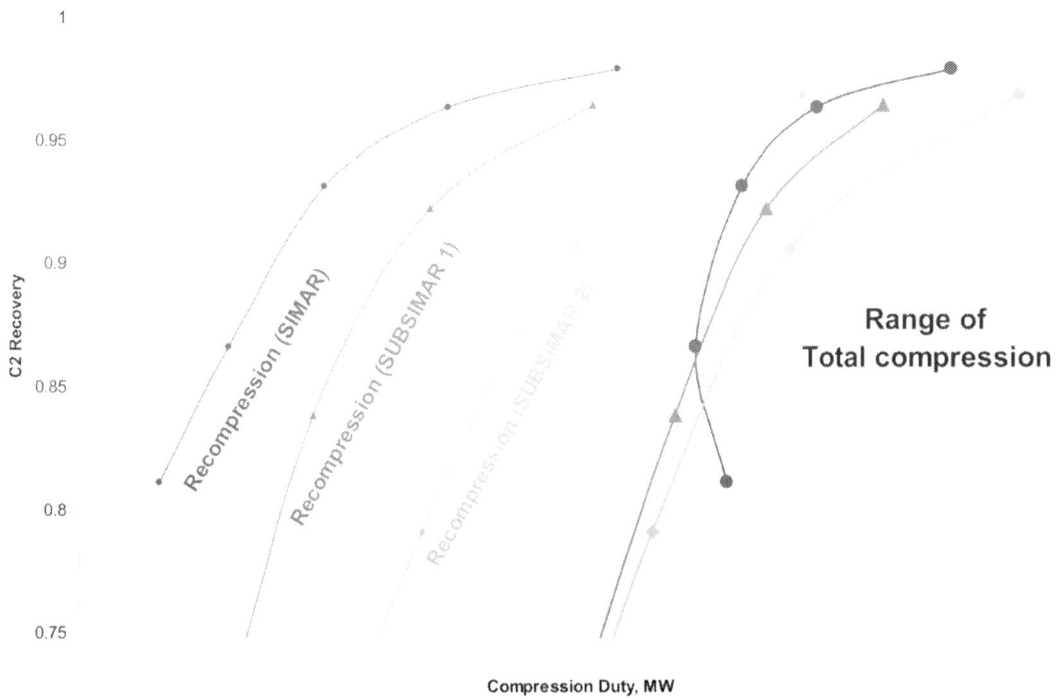

Figure 13 Sub-SIMAR operations for Industrial Case 2

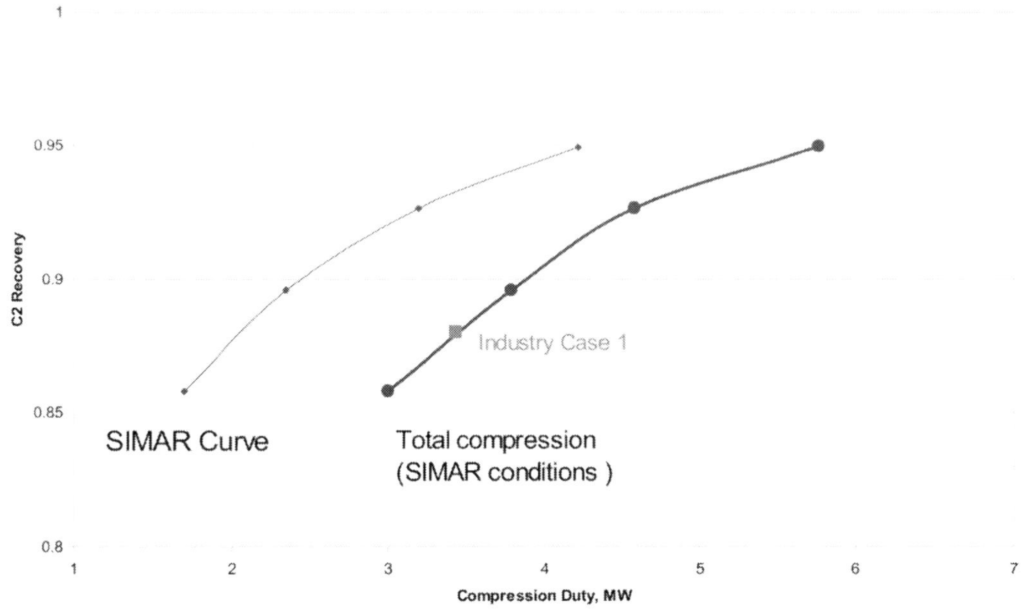

Figure 14 SIMAR operations for Industrial Case 1

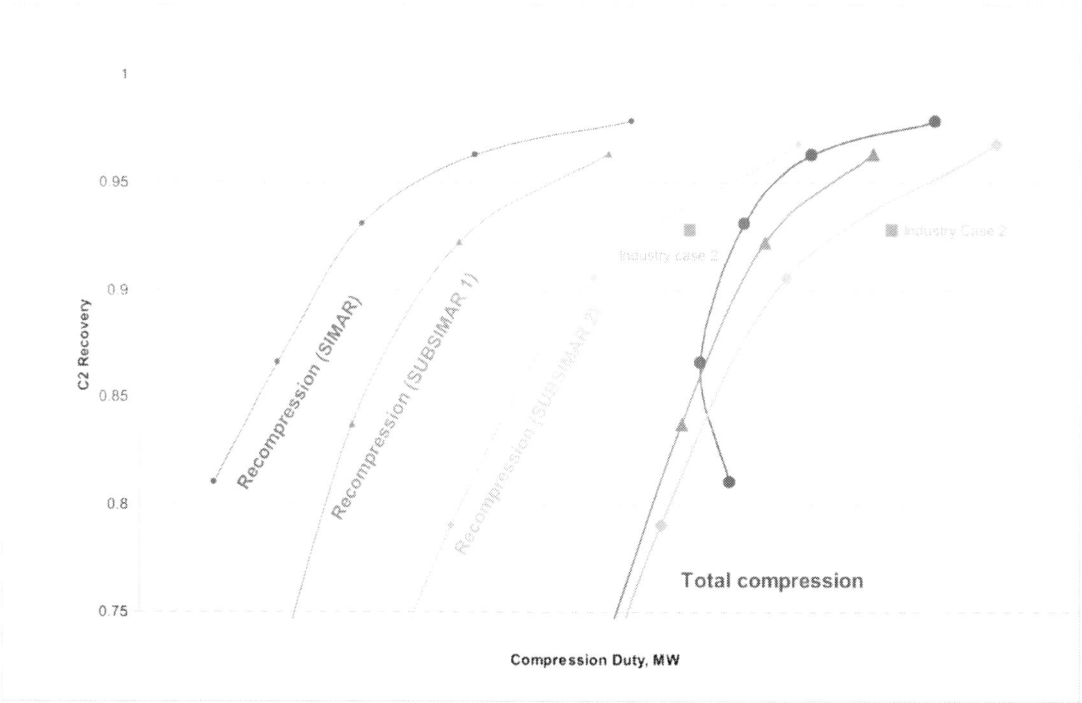

Figure 15 Sub-SIMAR operations for Industrial Case 2

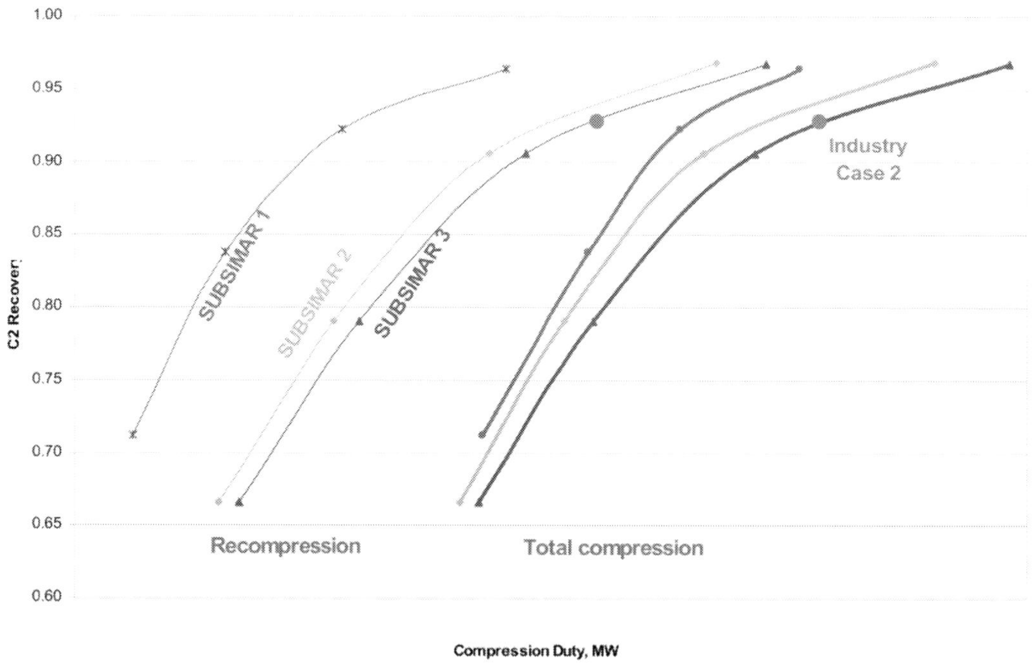

Figure 16 Sub-SIMAR operations for Industrial Case 2

Christopher M[c]Mullen is a Chemical Engineer in Chevron's Energy Technology Company. He graduated in 2004 from the Curtin University of Technology in Perth, Western Australia with a double degree in Chemical Engineering and Commerce (Management). Christopher has been in the industry for two years focusing on CO_2 injection wells and LNG/Gas Processing.
c.mcmullen@chevron.com

Dr. Stanley Huang has a specialty area in cryogenic applications, particularly in LNG and gas processing. He graduated from National Taiwan University with a B.S. degree and attended Purdue University in 1981. He earned his Master and Ph.D. degrees there, all in Chemical Engineering. Additionally, he also acquired a Master of Science in physics. Stanley has 20 years of industrial experiences and is a Registered Professional Engineer in Texas. Currently he is a Staff LNG Process Engineer of Chevron.
shhuagn@chevron.com

LNG Expert : the Latest Evolution in LNG Storage Tank Management

Dominik Uznanski
Senior Research Engineer
Research & Development Division
Gaz de France
Paris, France

Pieter Versluijs
Vice President/Director of Sales & Marketing
Whessoe S.A.
Calais, France

AIChE Spring Meeting, April 2007
7th Topical Conference on Natural Gas Utilization
Houston, Texas, April 22-26, 2007

Introduction

In the past, in order to avoid stratification and possible consequential roll-over, specific grades of Liquefied Natural Gas (LNG) were stored in specific storage tanks, each having a grade identical to the newly arriving cargo. Such a policy called for multiple tanks and made it difficult for the gas companies in entering into additional contracts covering a wider variety of LNG grades, all to be stored at their receiving terminals.

However, with an increasing diversification of LNG supply sources, caused by a growing number of liquefaction plants around the world, an increase in short-term trade and a general trend toward a worldwide liberalization of gas markets, reception terminals now need to deal with a greater variety of incoming LNG qualities. Also, with the need to reduce capital and operating costs, existing storage capacity must be used to its fullest extent and the capacity of new storage tanks must be utilized to the maximum.

Consequently, mixing different grades of LNG in receiving tanks cannot be avoided. However, mixing multiple grades of LNG in a single tank immediately calls for proactive measures in view of the possible creation of stratification and the associated risk of rollover.

Background

In a paper named "Advanced LNG Storage Tank Management" [1], the various operational consequences, the LNG behavior prediction software tool named "LNG MASTER®" and the required operational actions have been described in greater detail. This paper is available on the Whessoe S.A. website at : www.Whessoe.fr in the "Liquid Natural Gas" section.

Whessoe, being the supplier of an advanced and single source designed and custom built LNG storage tank gauging instrumentation system, inclusive of the availability of the stratification predictive tool LNG MASTER®, has received multiple requests from site owners to develop an on-line, real time, roll-over predictive alarm.

Such requests may be logical, since a Level, Temperature and Density (LTD) measurement gauge is included in this system. However, to find the solution was something else.

In order to develop such a predictive alarm tool, one has to fully understand the phenomenon of rollover and by which process (site and structural conditions) it is influenced. Next, the predictive alarm tool should not be introduced on a site for operators to rely on, if its reliability has not been validated.

Roll-over itself is triggered by the evolution of two layers of different grades of LNG stored in the same tank. Both layers will take in heat from the tank's walls, however only the top layer can evacuate such heat through free surface boil-off. The bottom layer is isolated from the free surface, thus not being able to boil-off and therefore accumulates this heat. The initially different grades will eventually reach a state of density equalization, where the lower layer will be able to rapidly mix with the upper layer, thus allowing all of the accumulated heat in the lower layer to be released via free surface evaporation. This phenomenon is named roll-over and often results in excess boil-off generation, which can lead to overpressures in the tank's gas phase. Because of this phenomenon, gas companies have requested an on-line alarm system, since LTD gauges are indeed the only instruments to monitor the temperature and density throughout the LNG column in a single tank.

However, the stratification development of the tank contents, eventually leading to a roll-over, is influenced by many factors, far beyond the measurement capabilities of an LTD gauge. Factors that impact the creation and evolution of an LNG stratification are amongst others :

- the tank's physical details such as the height, diameter and insulation of the walls and the roof;
- the type of filling devices;
- the pump capacity and actual send-out rate;
- the compressor and flair capacity and its real-time load (actual boil-off);
- the capacity and availability of safety devices (are some out for maintenance or not);
- the chemical composition of LNGs;
- the total number of tanks connected to the same vapor space; and
- the total volume of vapor space, inclusive all piping etc.

Having access to LNG MASTER, a software developed by Gaz de France to assist plant managers in optimizing their LNG storages and having the ability, amongst other functionalities, to calculate the evolution of stratifications and predict roll-over, Whessoe, in collaboration with Gaz de France, developed a new software package named LNG EXPERT, to perform exactly what was requested : to generate an on-line, real-time roll-over predictive alarm.

LNG EXPERT system set-up

Since LNG EXPERT requires much more site and process related data than just the LTD's temperature and density evolution data, it needs to have direct access to the single source where all data should be available : the terminal's Distributed Control System (DCS) platform. This is the sole location where all actual data is available and also where instructions are issued to, for example, take one or more safety relief valves and/or compressors out of service (see figure 1).

Figure 1 : LNG EXPERT linked to DCS system and LNG tank instrumentation

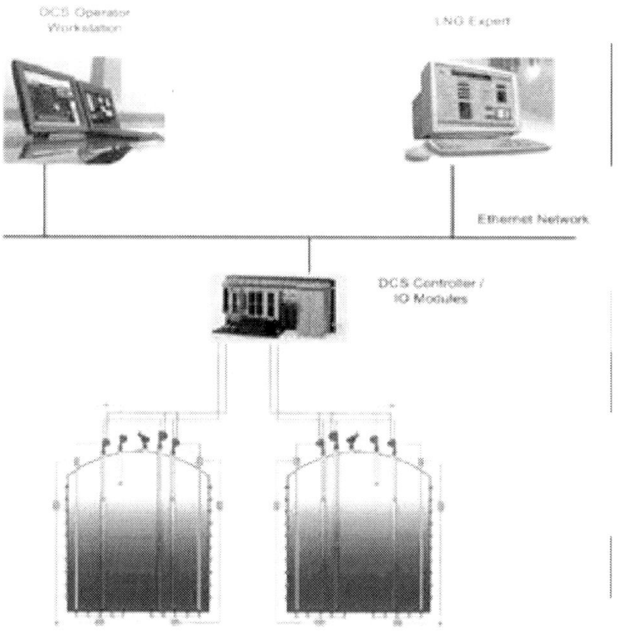

In case the site is equipped with non-Whessoe gauging instrumentation, LNG EXPERT will not run on Whessoe's instrument control platform (LNG MANAGER), but on a separate PC, linked to the DCS through an OPC link, thus minimizing the data communication back and forth. If the site is equipped with Whessoe's gauging instrumentation, then LNG EXPERT will run on LNG MANAGER. In either case, LNG EXPERT is also independent of which DCS vendor is selected.

Description of LNG EXPERT

LNG EXPERT is tailor-made for the specific site where it is intended for. The software consists basically of two major parts : the Interface Module (IM) and the Calculation Module (CM). Its data base is pre-programmed by Whessoe and stored within the CM. On a regular interval basis (typically every 4 hours), the IM acquires all variable process data and communicates such with the CM, which first verifies if site specific details have altered with respect to the data base. Based on the newly collected and verified data, the CM will then run a stratification evolution calculation, based on newly calculated chemical compositions of the in-tank LNG layers and the outcome is reported back to the DCS via the IM and OPC link.

The outcome may be any of the three following conditions :

1. No roll-over is expected within the next (maximum*) 30 days. Status : **normal**.
2. A roll-over is expected within the next 30 days. Status : **alert**, plus indication of expected Boil-Off Gas (BOG), plus roll-over onset time and date.
3. A Roll-over is expected within the next 2 days. Status : **alarm**, plus indication of expected BOG plus roll-over onset time and date.

* : the number of days over which LNG EXPERT calculates the future evolution of the stratification is user configurable from 1 to 30 days

Figure 2 : normal operation - no roll-over expected within 30 days.

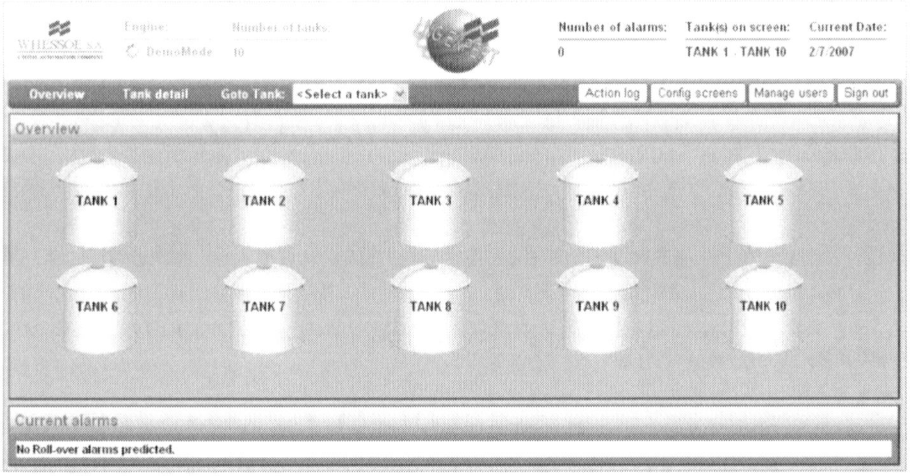

For the initial first condition described above, the operator has figure 2 on his screen, giving him an immediate overview of all tanks, connected to and supervised by LNG EXPERT and indicating that all is normal.

Figure 3 : alert generation : roll-over expected in 20 days

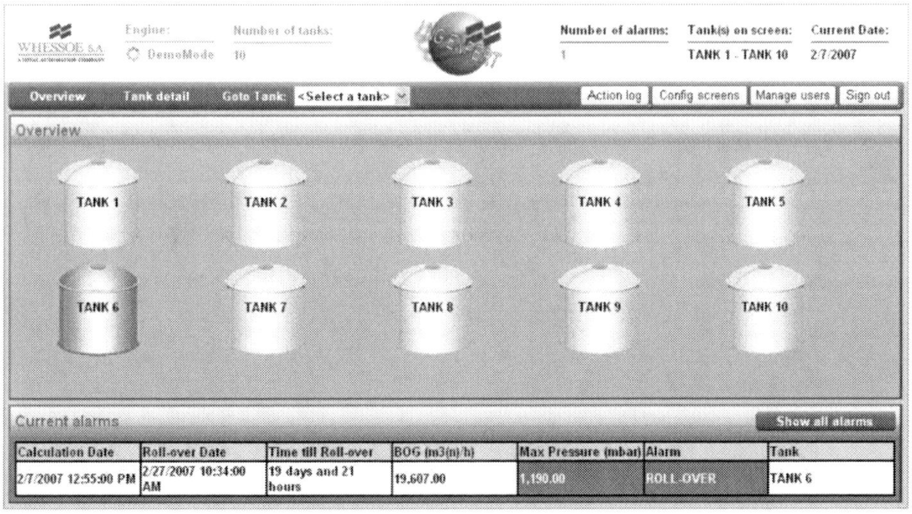

For conditions 2 and 3 described above, if a rollover is detected to occur within a configurable number of days, an alert or alarm, respectively, will be generated. Since an LNG EXPERT simulation is automatically carried out every 4 hours, LNG EXPERT can detect a rollover up to a maximum of 30 days in advance. However, it will not trigger the alarm status for a rollover if such does not occur within a pre-set period of say 2, 3 or 4 days from the date of the predictive calculation simulation. From the "alert" point onwards, the operator should have sufficient time to take corrective actions to avoid rollover. From the "alarm" point onwards, such can become difficult, if not critical.

As soon as LNG EXPERT has detected a potential roll-over possibility in any of its connected tanks, it will identify in which tank, at what date and time and what the expected BOG level and tank pressure would be (see figure 3 above). These values may be such that they fall well within the capacity of the plant's BOG compressors and that no special actions by the operators are required.

What actions the operator should take to remedy a rollover situation, whereby the expected BOG and pressure levels are indeed over and above the plant's BOG treatment capacity, is beyond the realm of LNG EXPERT, which is purely to be seen as a safety tool. Corrective actions, if any, remain the sole responsibility of the operators. In these cases, possible solutions to such a rollover situation can best be found by utilizing LNG MASTER.

LNG MASTER enables the user to run various simulations of prescribed operational scenarios and site conditions, thus being the appropriate tool to determine what would be the optimum solution to avoid or to resolve a roll-over situation, whereby excessive BOG levels may be expected.

Once an alarm has been generated by LNG EXPERT on a specific tank, the operator has the possibility of further zooming-in on the specific tank and analyzing the expected roll-over condition in greater detail (see figure 4 below).

Figure 4 : details of rollover characteristics for a specific tank

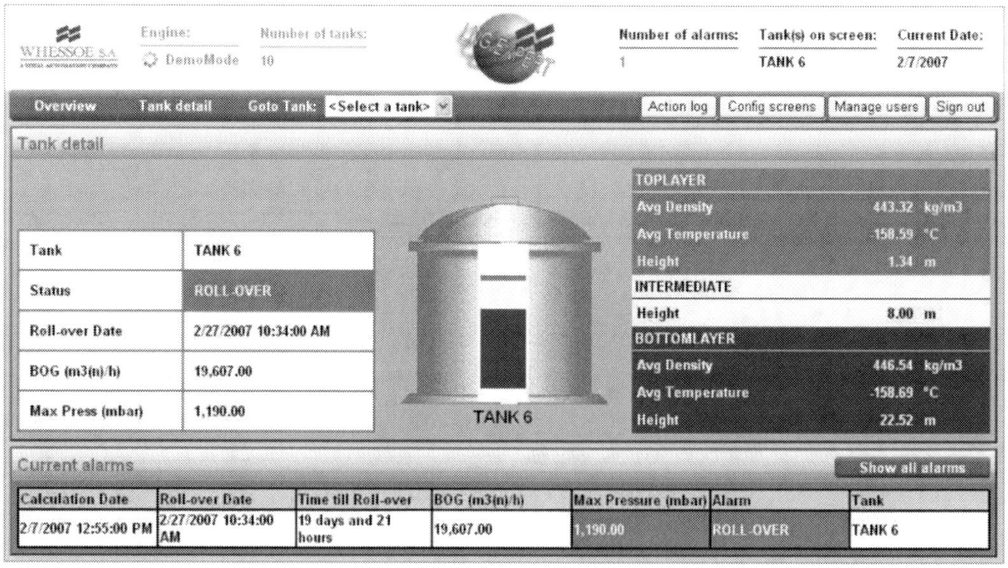

Scientific validation of the LNG EXPERT calculation module (CM)

The LNG EXPERT CM, developed by Gaz de France, is at the heart of the entire LNG EXPERT platform, as it carries out the essential role of predicting, for a given tank configuration, whether or not rollover will occur and if so, giving the occurrence date and time, as well as the intensity (BOG and pressure peaks) of the rollover event.

Therefore, a sound validation basis for the CM is an important element in ensuring that LNG EXPERT will indeed give an accurate evaluation of the rollover potential of LNG storage tanks.

Figures 5 below show the evolution of an LNG stratification created in the Nantes (France) 500 m^3 LNG storage tank during a Gaz de France experimental campaign conducted between 1987 and 1989 on its cryogenic testing station [2]. This evolution can be decomposed into 4 distinct phases, each governed by its own phenomenology :

- Phase 1 : thermal evolution of the lower layer;
- Phase 2 : interpenetration of the layers;
- Phase 3 : hydrodynamic overturning of layers (rollover)
- Phase 4 : return to thermodynamic equilibrium

Figure 5 : evolution of an LNG stratification in a 500 m^3 storage tank

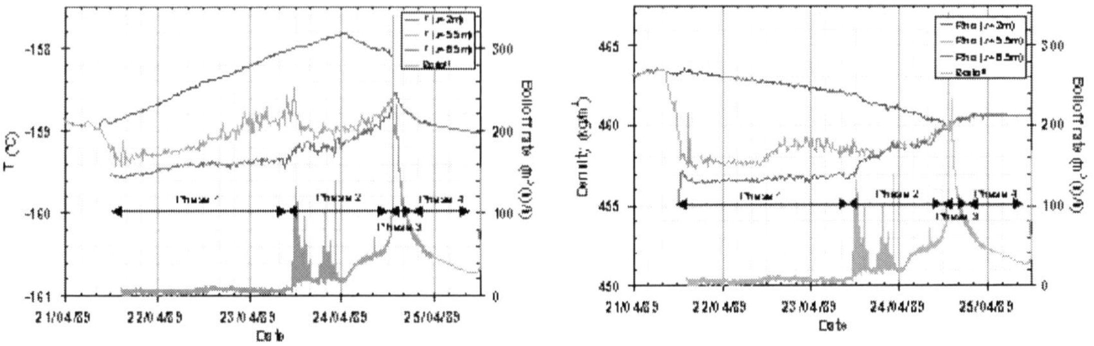

During the first phase, the stratification's layers can be considered as insulated from each other both heat and mass wise, while only the lower layer heats up progressively. The upper layer remains essentially in a thermal pseudo steady state, as the heat ingress into the upper layer can be released through free surface boiloff. This results in a progressive decrease of the density difference between the two layers over time.

During the second phase, interlayer penetration takes place between the two layers, reducing even further the layer's density difference. During the third phase, density equalization occurs, which results in a rapid hydrodynamic mixing of the two layers, producing the rollover event. The rollover itself is characterized by a sudden liberation of overheat accumulated in the lower layer which can now be evacuated at the free surface through evaporation.

During the fourth phase, which takes place after the rollover event, the homogeneous LNG resulting from the mixing of the two layers of the stratification progressively loses its superheat and returns to a thermodynamic equilibrium state. The

BOG rate decreases and the temperature and density of the LNG reach new equilibrium values.

The above figures also show the difference in nominal BOG rates between the stratified tank and the homogeneous tank after rollover and return to equilibrium have occurred. In the trial shown, the stratification reduced the nominal BOG rate by a factor of 5. This and other examples show the advantages in using stratifications whenever possible to decrease electrical power consumption for BOG gas compression during LNG aging, and thus terminal operating costs. However, LNG stratifications need to be managed in total safety and software such as LNG EXPERT and LNG MASTER can help operators achieve this goal.

The CM was validated with the help of the following experimental and field data :

- rollover trials conducted by Gaz de France in the 500 m^3 tank of the Nantes LNG testing station;
- the La Spezia rollover incident which occurred in 1971 [3];
- a stratification evolution recently recorded on a methane terminal.

In what follows, several examples of inter-comparisons of LNG EXPERT CM results with validation data are shown, as well as the dispersion of results obtained for time to rollover and peak BOG rate for the Nantes rollover trials.

The following figure shows the inter-comparison of the measured boil-off rate [3] and the calculation performed by the LNG EXPERT CM for the La Spezia rollover incident. As figure 6 shows, the CM predicts rollover occurrence about 4 hours after the actual rollover, which is quite good, given the uncertainty linked to the cargo and heel initial conditions (no traveling LTD gauge was installed in the tank at the time of the incident and therefore no recordings of the stratification's initial conditions could be made).

Figure 6 : La Spezia incident prediction by the LNG EXPERT CM

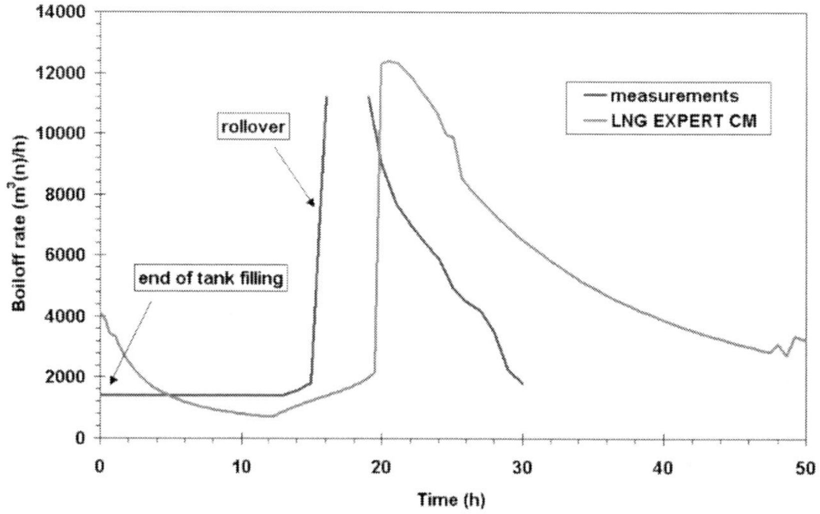

Figure 7 below shows the inter-comparison of the measured boil-off rate and the calculation performed by the LNG EXPERT CM for one of the Nantes rollover trials. As figure 7 shows, the CM predicts rollover occurrence about 8 hours before the actual rollover, which again is good, given that the time to rollover for this stratification was 71 hours.

Figure 7 : Nantes rollover trial prediction by the LNG EXPERT CM

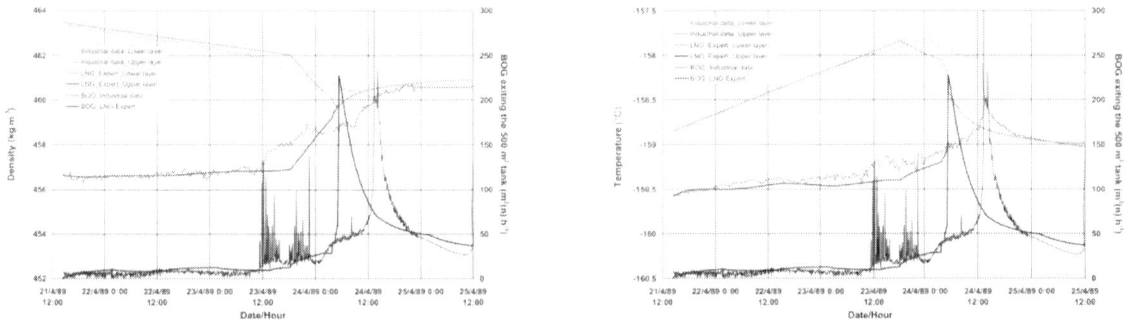

Conclusion

It is easily understood from the above that simply monitoring the temperature and density evolution as measured by an LTD gauge is far from truly "safe", if the focus is to be on roll-over prediction. An LTD gauge by itself is simply not capable of doing so. It would be like the parking radar installed in your car : it will indicate to you that you are getting closer to a wall. However, without any details on driving speed, road and brake conditions, it will be unable to predict as to when you will actually hit the wall, let alone with what impact.

Therefore, it is essential for the operator to have at his disposal a tool such as LNG EXPERT, capable of assisting him in making the right decisions at the right time, in order to manage his LNG storages in a safe and optimal way.

In view of this, Whessoe, in collaboration with Gaz de France, has developed a software named LNG EXPERT, whose stratification evolution and rollover potential calculation module was successfully validated against known incidents and well documented LNG field data in the possession of Gaz de France.

References

[1] "Advanced LNG storage tank management", Dominik Uznanski and Pieter Versluijs, AIChE Spring Meeting, 3rd Natural Gas Utilization Topical Conference, Session LNG VI – Storage & Facilities, New Orleans, Louisiana, March 30 – April 3, 2003

[2] "Rollover trials in a 500 m^3 LNG storage tank", Olivier Marcel, internal Gaz de France report (N 1153), 1993

[3] "LNG Stratification and Rollover", J. Sarsten, 1971

Biography

361 avenue du Président Wilson, BP 33, 93211 Saint Denis la Plaine Cedex, France
Phone : (33) 1 49 22 52 91, Fax : (33) 1 49 22 50 40
E-mail : dominik.uznanski@gazdefrance.com

Mr. Dominik Uznanski joined Gaz de France's Research and Development Division in 1993 as a research engineer at the Cryogenic Studies Section. Since then, he has been involved in numerous studies on the behavior of LNG in storage tanks (including research on stratification behavior, weathering and tank filling operations), on LNG vapor heavy gas dispersion and on water curtain mitigation methods and has gained extensive experience in these fields. He is also involved in the development of Gaz de France's LNG tank management software named LNG MASTER and its delivery to end users, including Gaz de France's two LNG receiving terminals and external clients. He has managed several projects dealing with LNG Technologies and Operations of LNG storage tanks. In this respect, he is involved in particular in studies on LNG behavior and in developing a methodology for the diagnosis of the aging of LNG installations.

Mr. Uznanski holds two Bachelor of Science degrees in chemical engineering (specialization in process control and optimization and numerical methods) and pure mathematics from the University of Maryland, U.S.A. (1986) and a Master of Science degree in fluid dynamics and heat transfer from the Ecole Centrale Nantes engineering school (1991).

Rentech's Fischer Tropsch Technology Development.

Harold Wright, Belma Demirel, Sergio Mohedas, Rentech Inc.

Abstract:

The pace of FT development is changing rapidly as crude prices remain high and everyone form government to coal companies look for opportunities to convert low valued resources into valuable fuels and chemicals. Rentech has developed an aggressive plan to both develop and license its proprietary brand of Fischer-Tropsch conversion technology to work with any reasonable feedstock – natural gas, coal, petroleum coke and even bio-feedstocks that can't compete on an economic scale today.

This paper will provide an overview of Rentech's technology, its current projects and its license partners that are together creating a domestic industry primed for moving from coal to bio-technology.

Introduction

Rentech has developed technology that converts synthesis gas from coal or other carbonaceous to clean economical fuel from secure sources. Persistent high crude prices and unstable oil supply have created the need for these projects. The United States has the largest proven coal reserves in the world with over 500 billion tons of proven reserves amounting to over 300 years of production. Coal is also available at stable and relatively low cost. The United States and Canada also have ever growing supply of petroleum coke from refining operations. Biomass can be another source of feedstock for these plants. The end result is that Rentech's Fischer Tropsch from solids is economically feasible.

A coal to liquids plan comprises a gasification section. Gasification can be used to convert coal, petroleum coke, or other carbon bearing feedstock is converted into syngas which is a mixture of CO and Hydrogen. The Fischer Tropsch section of the plant converts the syngas to hydrocarbon liquid. This is Rentech's proprietary technology. The FT product is then upgraded into ultra clean synthetic fuels.

Rentech's Industry Leadership

Rentech is leading Fischer Tropsch technology provider. The company was created over 25 years ago. The company has 22 US Patents. The company has operated six different pilot plants. A fully integrated solids-based 10 BPD Fischer Tropsch demonstration plant is under construction in Commerce City, CO and will be completed later this year.

Figure 1: Artists Rendition of the Commerce City Rentech PDU

This PDU in Commerce City will be the first fully integrated Coal to Liquids Production Unit in the United States

The first commercial Solids Based Fischer Tropsch plant in the United States will be at the Rentech Energy Midwest facility in East Dubuque, IL. This plant is currently a natural gas to ammonia plant that will be converted to coal as a feedstock. The syngas will be used to make ammonia and Fischer Tropsch diesel fuel.

This current plant has substantial existing operations and infrastructure. The 680,000 TPY fertilizer plant ready for immediate conversion. The plant is on the Mississippi River and is the northernmost ammonia facility on the Mississippi River.

Figure 2: The Rentech Energy Midwest facility in East Dubuque, IL. (REMC)

The current REMC facility uses natural gas and air to produce syngas that goes to Ammonia Synthesis. Figure 3 shows the block flow diagram of the current facility.

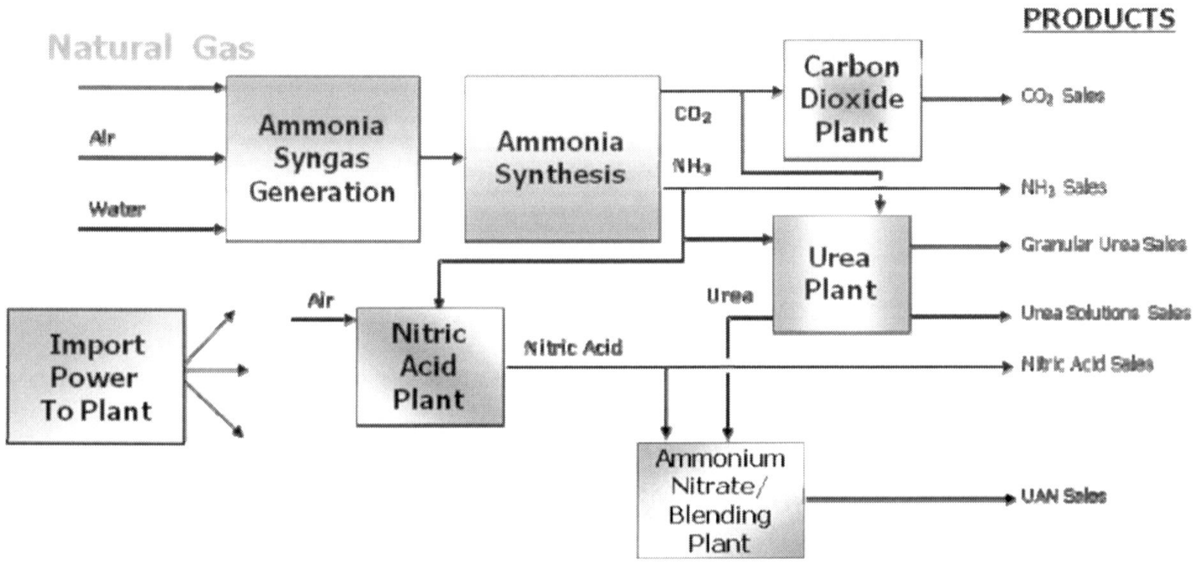

Figure 3: Current REMC Block Flow Diagram

After conversion the plant will be configured as shown in Figure 4.

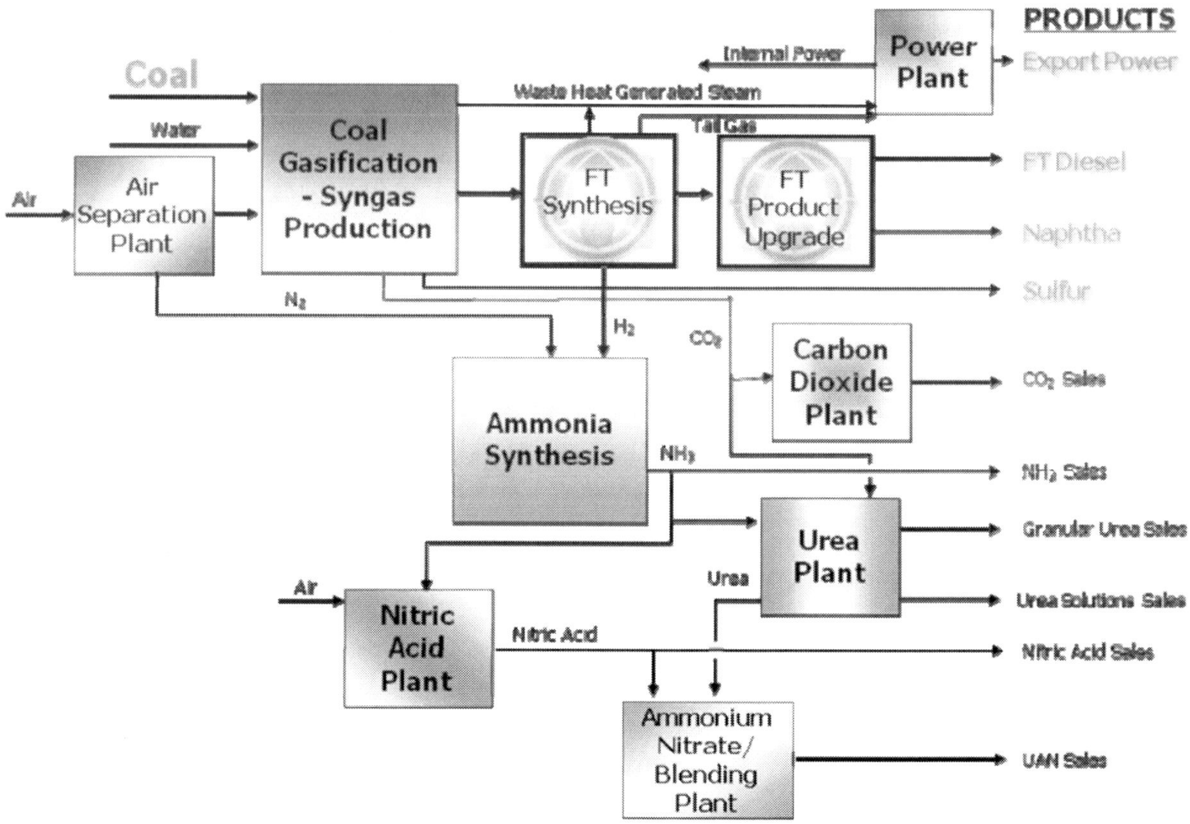

Figure 4: The Block Flow Diagram of the East Dubuque REMC facility after conversion

The project will be completed and started up in 2010.

Conclusion

With the completion of the REMC conversion, Rentech will operate the first commercial Fischer Tropsch plant in the United States. We see this plant as the first of many plant that will us coal, petroleum coke, and biomass to high quality diesel fuel.

RENTECH PROCESS – SCALE-UP AND COMMERCIALIZATION OF A COAL-TO-LIQUIDS PROCESS

Belma Demirel
12441 East 37th Avenue, Denver, CO

Rentech opened a new era by taking a major step on the way to the production of fuels and chemicals from coal, natural gas, petroleum coke and even bio-feedstocks. There has been a tremendous effort to scale up and commercialize Rentech's Fischer-Tropsch process. A Process Development Unit of 10 bpd is being built in Colorado. The nation's first commercial Coal-to-Liquids plant is being designed for Rentech Energy Midwest facility in East Dubuque. This paper will review the ongoing work at Rentech's Technology Development site and plans for the future.

INTRODUCTION

Rentech, Inc., a Colorado corporation founded in 1981, has developed a patented and proprietary Fischer-Tropsch (FT) technology that has demonstrated high productivity over a wide range of hydrogen to carbon monoxide ratios. By concentrating on the catalyst, the slurry reactor and the efficient integration of the FT module with multiple gasification technologies, Rentech's technology has been recognized as one of the leading and most cost-efficient FT technologies currently being offered in the market today. The technology has the flexibility work with various feedstocks such as natural gas, industrial off-gas, heavy crude oil, refinery byproducts, coal, biomass and petroleum coke into valuable liquid hydrocarbon products. The Technology produces a wide range of products, including clean-burning diesel fuel, naphthas, and specialty products such as waxes, petrochemical feedstocks, fuel cell feedstocks and synthetic lubricant base stock.

HISTORY OF RENTECH'S TECHNOLOGY

Rentech's technology is based upon the FT conversion process that was originally developed in Germany during the 1920s to create synthetic transportation fuels. Based on the work of Drs. Chuck Benham and Mark Bohn, Rentech's founders, the Company developed its own Fischer-Tropsch technology in the early 1980s, using a proprietary and patented iron-based catalyst in a slurry reactor.

The ability of the Rentech FT Technology to convert synthesis gas, a mixture of hydrogen and carbon monoxide into valuable liquid hydrocarbons was first established in the Company's initial pilot plant in 1982. This pilot facility was a small, skid-mounted, integrated system including a single tube steam reformer, a Benfield carbon dioxide removal system, and a slurry F-T reactor. This pilot plant was operated up until 1985, with the primary goal of providing a test facility for catalyst development. A second higher-capacity pilot plant was operated during 1989 to acquire sufficient products for characterization and further engine testing. Additional confirmation of several significant aspects of the Rentech FT Technology was obtained from tests conducted between 1991 and 1998 in a third pilot plant. The Company continues to use this third pilot plant at its Denver Technology Development facility in Denver, Colorado, to further advance the development of the technology and to develop F-T data in response to inquiries from licensees and prospective licensees.

Use of the Rentech FT Technology in a commercial-scale FT plant was successfully demonstrated in 1993. This plant, the Synhytech plant located at Pueblo, Colorado, built under a license from Rentech to Fuel Resources Development Co., part of Public Service Company of Colorado (PSCo), had a designed capacity of 235 barrels of liquid hydrocarbons per day. Rentech had responsibility for the design, fabrication, installation and operation of the F-T reactors, and provided the F-T catalyst. The slurry reactors at L/D of 9.2 were, at that time, the largest F-T slurry reactors ever constructed; and to date, still the largest ever operated outside of South Africa. Despite the severe inadequacies of the landfill gas in terms of both quantity and quality, the Rentech FT Technology, including the F-T reactors and catalyst, performed as expected. In mid-1992, due to the lack of adequate feedstock from the landfill, PSCo closed the plant.

By the terms of a negotiated settlement between PSCo and Rentech, ownership and control of the Synhytech plant, was then transferred to Rentech. In order to further evaluate performance of the Rentech FT Technology at a commercial-scale, the Company decided to operate the plant for a short period of time. Extensive modifications were made to improve the safety and reliability of several mechanical systems of the plant that did not involve the Rentech FT Technology. The plant was decoupled from the landfill gas source, and added a temporary pipeline supply of natural gas. In July and August 1993, Rentech operated the plant continuously for three weeks. The results confirmed that the Rentech FT Technology operated successfully and confirmed several factors that are keys to the use of the technology.

In October 1998, Rentech granted an exclusive license to Texaco Energy Systems, Inc. (Texaco), a division of Texaco, Inc., to use the Rentech FT Technology in plants where solid and liquid hydrocarbons are used as feedstock. Texaco also has the right to grant sublicenses for this application. The Company retained rights to grant licensees to others for natural gas feedstocks, which includes industrial off-gases. In addition, Texaco was granted a non-exclusive license, but not the right to sublicense Rentech's FT Technology in plants which use natural gas as feedstock.

In April 1999, Rentech entered into a Technical Services Agreement (TSA) with Texaco to carry out the additional development activities to confirm the performance of the Company's proprietary and patented technology and catalyst at the low hydrogen-to-carbon monoxide ratio synthesis gas produced by Texaco's gasifiers, using either coal or petroleum coke as a feedstock and to integrate Rentech's FT Technology with Texaco's gasifiers. The work carried out by the Company under the TSA has further confirmed the efficacy of the Company's Technology for a wide range of synthesis gas hydrogen to carbon monoxide ratios and the demonstration at semi-commercial scale of the Company's patented primary wax/catalyst separation technology.

In August 1999, a team led by Texaco, of which Rentech was a part, was selected by the U.S. Department of Energy (DOE) to develop the necessary data and designs for the "Early Entrance Coproduction Plant". For this project, Texaco plans to combine its gasification technology with the Rentech FT Technology to enable the production of both high quality transportation fuels and electricity from coal and petroleum coke at a coproduction plant.

The work continued for several years. The team members used Texaco's gasification technology, Rentech's FT Technology, General Electric's power generation design, Praxair's oxygen plant design and Kellogg, Brown and Root's engineering capabilities. After a feasibility study and successful completion of an integrated design, the team developed an engineering design package for a commercial scale fossil fuel plant to use the combined technologies.

Rentech completed its tasks under the first phase of the DOE contract. The work to date consisted primarily of supporting a preliminary engineering design for the FT portion of the plant that would use the Rentech FT Technology. In support of this effort in November 2000, Rentech participated in Texaco's funded program to demonstrate use of the Rentech FT Technology. The program was carried out at the U.S. Department of Energy's slurry reactor facility in La Porte, Texas.

Review of the test data confirmed the expected performance of Rentech's patented and proprietary Fischer-Tropsch catalyst. Key reactor operating parameters were also demonstrated. The preliminary data indicated that the technology for separating waxes and catalyst met expectations. Representative products were collected for use in the next phase of work on the Early Entrance Coproduction Plant project.

In the past, the R&D effort at Rentech has focused on catalyst development. This effort led to a catalyst which has the high activity and selectivity required for economic gas to liquids and coal to liquids systems using slurry bubble column Fischer-Tropsch reactors.

RENTECH'S TODAY AND FUTURE

Major capabilities of Rentech's R&D facility currently include a Bubble Column Slurry Reactor (BCR) system, a wax/catalyst separations test facility, an autoclave reactor system, a catalyst preparation laboratory, a number of process simulation software.

The BCR system includes a 1.6" id x 26' tall reactor with all supporting equipment. The reactor is designed for operation at temperatures up to 300°C and pressures up to 500 psig. It is externally heated by 14 zones of electrical heat that allows the reactor to operate isothermally and simulate a full-scale reactor. An internal cooling tube is used to transfer the exothermic heat of reaction to a cooling oil stream. The unit has the capability for extended runs by the incorporation of Rentech's patented wax/catalyst separation technology. The system nominally produces 3.5 kg/day total collected hydrocarbons.

A 2-liter Autoclave Engineers autoclave system allows engineers to screen catalyst performance before committing to testing on the BCR.

The catalyst preparation laboratory allows the preparation of small quantities of advanced F-T catalyst suitable for autoclave or BCR testing.

Rentech engineers use proprietary process simulation software to examine new process flow sheets, determine yields, product distributions, and utility requirements from the process.

While the tools available to the Rentech R&D program have been adequate in the past, the new era of commercialization requires additional equipment and personnel to meet the challenge of building on past successes.

Technology Development Center founded recently started to expand the capabilities to adequately pursue technology improvements. A large number of equipments such as spray dryers, CSTRs, fixed-bed reactors, etc. were already purchased and are being installed. Product Development Unit (PDU) of 10 bpd unit is under construction and will be in operation soon. Rentech's staff including new hires in the areas of catalyst development, process development, process design and modeling will enable Rentech to facilitate Rentech's future commercial plants.

GTL and CTL Commercialization: Status and Impact on Global and Regional Product Markets

Iraj Isaac Rahmim
E-MetaVenture, Inc.
Houston, TX 77277-1522
iir@e-metaventure.com

Abstract:

Interest in commercial gas-to-liquids (GTL) and coal-to-liquids (CTL) technologies, limited for over 70 years to countries with political rather than economic drivers, has heightened in the past decade with now-monthly announcements of technical advances and commercial agreements, studies and demonstrations and front-end engineering and design (FEED) awards, and even a groundbreaking or two. Companies with commercial GTL/CTL histories, such as Sasol and Shell, as well as others including BP, Syntroleum, ExxonMobil, Rentech, and ChevronTexaco, have made advances in hardware design, catalysts, and operating conditions in an attempt to improve GTL/CTL economics. Certain governments, such as those of Qatar, the State of California, Peoples Republic of China, and the State of Montana, have taken leading positions in developing and implementing GTL/CTL as a component of long-term strategies.

Key GTL/CTL products fall into the categories of fuels (primarily diesel but also LPG), specialty streams (lube basestocks and waxes), and petrochemicals (naphtha for steam cracking). This presentation will examine these product markets, particularly those of the automotive diesel, high-grade lubricants, waxes, and naphtha and the potential impact of GTL/CTL developments on the global and regional balance for these. Issues discussed will include projected volumes and quality, regionality, impact of regulations, and blending and transportation.

Fischer-Tropsch Fuel for Use by the Us Military as Battlefield Use Fuel of the Future

Delanie Lamprecht
Sasol Technology Research and Development
Sasolburg, South Africa
delanie.lamprecht@sasol.com

Abstract:

The United States Department of Defence (DoD) currently uses a single kerosene-type fuel for all its gas turbine and diesel engine applications. Technological changes introduced to the battle space make the single battlefield fuel philosophy of the DoD less prone to the use of crude-oil derived kerosene-type fuel. The DoD initiated a three phase Joint Battlefield Use Fuel of the Future (JBUFF) program to evaluate, demonstrate, certify and implement turbine fuels produced from alternative energy resources such as Fischer-Tropsch (FT) fuels for use in all its gas turbine and diesel engine applications. Sasol Synfuels International (Pty) Ltd, amongst others, was invited to participate in the program with the objective to supply the DoD with a FT BUFF that conforms to fuel volatility and low temperature fluidity requirements. These requirements included a freezing point of -47 °C, viscosity of 8 cSt at -20 °C and a flashpoint of 38 °C for a BUFF conforming to "Jet Propellant # 8" (JP-8) specifications and a flashpoint of 60 °C for a BUFF conforming to JP-5 specifications for the handling of fuel on aircraft carriers. Technical, economic and strategic challenges related to the introduction of FT fuels into the military fleet include the interchangeability of FT fuels with crude-oil derived kerosene-type fuels; elastomer compatibility of fuel systems already conditioned in crude-oil derived kerosene-type fuels with subsequent exposure to FT fuels containing no aromatics; demand and supply of FT fuels at a price comparable to crude-oil derived kerosene-type fuels; and the modification of existing fuel specifications to allow for general approval of FT kerosene-type fuel. Although the DoD is more interested in Coal-to-Liquid (CTL) technology, fractionator feed from the Gas-to-Liquid (GTL) Products Work-Up Demonstration Unit in Sasolburg, South Africa, was used to evaluate the possibility of producing a BUFF fraction from the Sasol Slurry Phase Distillate™ (SPD™) low temperature FT (LTFT) and Chevron ISOCRACKING® technology. The fractionator feed used to develop a BUFF from the Sasol SPD™ LTFT process contains only normal and iso-paraffins with virtually no aromatics or naphthenes. Concluded from the study was that the production of a synthetic BUFF is feasible using the Sasol SPD™ technology alongside the current Chevron ISOCRACKING® technology. Also concluded from the study was that the Sasol SPD™ LTFT BUFF product yield and the FBP of the product will be restricted by the freezing point requirement due to the low iso to normal paraffin ratio of ~1:1 of the Sasol SPD™ LTFT BUFF and not by the maximum viscosity requirement like the "Single Battlefield Fuel" of the Syntroleum Corporation (S-5) which has an iso to normal paraffin ration of ~14:1. One would therefore need to optimize the hydrocracking process conditions to increase the product yield. Sasol recently demonstrated that fully synthetic high temperature FT (HTFT) aviation turbine fuel, produced at Sasol Synfuels in Secunda, South Africa, is fit-for-purpose for commercial use. The fully synthetic HTFT aviation turbine fuel comprises of blends of hydrocracked tar liquid, light distillate refinery streams and iso-paraffinic kerosene. It differs from Sasol SPD™ LTFT kerosene in that it contains naphthenes and aromatics resulting in a comparable elastomer response with crude-oil derived aviation turbine fuel with similar aromatic content. The fully synthetic HTFT aviation turbine fuel addresses the technical challenges related to the introduction of FT fuels into the military fleet. It further conforms to all JP-5 specifications with properties and characteristics that are typical of conventional, crude-oil derived aviation turbine fuel. Modification of existing fuel specifications is therefore not required for a HTFT BUFF. A comparison between the bulk composition of LTFT and HTFT fuel for use by the US Military as a BUFF will be presented.

Synthetic Fuels: past, Present and Future

Terry Mazanec
Velocys, Inc.
Plain City, OH 43064
mazanec@velocys.com

Lee Tonkovich
Velocys, Inc.
Plain City, OH 43064
tonkovich@velocys.com

Kai Jarosch
Velocys, Inc.
Plain City, OH 43064
jarosch@velocys.com

Jeff McDaniel
Velocys, Inc.
Plain City, OH 43064
mcdaniel@velocys.com

Abstract:

Sustained high oil prices and the growing concerns about 'peak oil' have renewed interest in fuels made from alternative feedstock materials, including natural gas, coal and biomass. These processes go by the acronyms GTL, CTL and BTL for gas-to-liquids, coal-to-liquids and biomass-to-liquids respectively. Although these catchall acronyms cover all manner of processes for generating a range of liquid products, including methanol, ethanol and biodiesel, the highest hopes are pinned on Fischer-Tropsch based systems that produce synthetic transportation fuels. The history of the Fischer-Tropsch process reaches back to the 1920s when Franz Fischer and Hans Tropsch, working at the Kaiser Wilhelm Institute first converted synthesis gas, a mixture of hydrogen and carbon monoxide, into liquid hydrocarbons. Today, multiple world-scale plants are being built to convert low cost, stranded natural gas reserves into salable petroleum products. The future of Fischer-Tropsch processes looks bright with the proliferation of world-scale GTL and CTL plants located near energy deposits and smaller scale facilities to convert offshore associated gas and locally sources biomass to market ready synthetic fuels. Synthetic fuels processes consist of three basic steps: 1) gasification of feedstock material, 2) Fischer-Tropsch synthesis, and 3) hydrocracking. The gasification technology employed is dependent on the feedstock material used. However, the most economically efficient path to synthesis gas is via steam methane reforming. The Fischer-Tropsch process is a catalyzed chemical reaction where synthesis gas is converted into longer chain hydrocarbons, known as waxes. The final step, hydrocracking, is the conversion of longer chain hydrocarbons into salable liquid products. Hydrocracking has historically been performed at refineries, but synthetic fuel processes require this to be done at the production site as the longer chain products from the Fischer-Tropsch process can be a transportation challenge. Microchannel process technology can improve each of these steps by improving efficiency, reducing capital cost and shrinking facility footprints, which is especially important for offshore installations. Although the emphasis today is on world-scale gas-to-liquid

Fischer-Tropsch Synthesis: Impact of Sulfur on Iron Catalysts

Burtron H. Davis, Center for Applied Energy Research, University of Kentucky, 2540 Research Park Dr., Lexington, KY, 40511

The generation of syngas for the Fischer-Tropsch synthesis invariably produces small levels of potential catalyst poisons. One of these is sulfur, primarily as H_2S, COS and CS_2 but also small amounts of organic sulfur compounds, such as mercaptans and thiophenes. In commercial operations these sulfur compounds are reduced to low levels in order to prolong catalyst life. The removal of these sulfur compounds adds an expense to the generation of the synthesis gas and the benefit of life become an optimization based upon these two factors. In spite of the importance of sulfur removal in the FTS plant, these is limited data on the effects of sulfur on catalyst operation.

An early report by the German developer of FTS, I. G. Farben, indicated that small loadings of sulfur on a Group VIII metal catalyst could be beneficial [1]. It is not possible to learn from the patent the impact of sulfur on the activity but the data show that more liquid hydrocarbons are produced and that these products are lower molecular weight than is produced when sulfur is not present.

A more detailed report of the beneficial effect of low loadings of sulfur on the activity/selectivity of iron Fischer-Tropsch was published in 1946. Layng [2] indicated that the disadvantage of producing CO_2 by the WGS reaction could completely or largely be overcome by maintenance of the catalyst at a sulfur content sufficient to inhibit the formation of CO_2 but in an insufficient quantity to prevent the desired hydrocarbon synthesis reaction. It was claimed that when operating with the optimum sulfur level on the catalyst the yield of liquid hydrocarbons was improved and the products assumed a more unsaturated character together with a somewhat greater molecular weight. In one version of the patented operation, sulfur is added to a clean syngas while in another approach a high sulfur containing syngas is utilized until the optimum effect is obtained and then the feed is switched to a clean syngas. Alternatively, the high-sulfur feed gas can be utilized continuously provided a means is available for regenerating a portion of the catalyst in a recycle operation.

In one example, Layng indicated that when utilizing a fluid bed operation, the initial syngas feed could contain 0.5 molal-% of sulfur. It was indicated that after the initial run-in period, the production of the desired hydrocarbons will proceed, while a decreasing proportion of CO_2 in the effluent gas with a corresponding increase in the molecular weight and degree of unsaturation will be observed. The change in selectivity will continue with time on stream until the optimum sulfur level is attained after which the sulfur added after this point will serve as a catalyst poison. In the preferred mode of operation, the addition of sulfur will be terminated just before the optimum amount has been added and thereafter a rigidly purified syngas will be utilized. This mode of operation implies that the sulfur is adsorbed on the catalyst irreversibly, or at least nearly irreversibly.

In another mode of operation, a high sulfur concentration in the feed gas is tolerated by withdrawing continuously a portion of the catalyst from the reactor and subjected to a regeneration that will decrease the sulfur level of the withdrawn catalyst. The example used to illustrate this operation utilized the circulation fluid bed reactor.

An example run utilized a catalyst containing 2 to 3% potassium oxide and alumina (presumably 2 to 3% of each). The catalyst size was such that essentially all of it passed through a 200 mesh screen and 85% passed through a 325 mesh screen. The catalyst mass was maintained at 600°F (315°C) and 200 psig. The feed was composed of syngas with a H_2/CO ratio of about 2.0, about 14% diluent gases (CH_4, N_2 and CO_2) and 0.4 molal-% sulfur, determined as H_2S. At the beginning of the run the effluent from the reactor contained 30 to 40% CO_2. After a period of about eight hours the proportion of carbon dioxide contained in the effluent stream will have been reduced to a minimum, say below 5%, and further sulfur addition to the catalyst mass would be detrimental.

Sufficient details are not given in the patent to learn whether the period of sulfur addition began with the introduction of syngas to an unactivated catalyst or to an activated catalyst. Should the addition be to an unactivated catalyst then the production of CO_2 and iron carbide is expected to occur as the catalyst is activated and then CO_2 production from activation will decrease to a low level, eventually being zero. The feed gas contained hydrogen sufficient to convert a significant amount of CO_2. This was a high temperature synthesis condition where the reverse WGS reaction does occur. Thus, it appears that the low CO_2 make is due to a combination of the reverse WGS of CO_2 and its subsequent conversion of the CO to hydrocarbons, and not due to a special effect of the added sulfur. This view is supported by the fact that subsequent patents by Layng relate to the low production of CO_2, and even the negative production of CO_2, under high conversions of CO. At the temperatures where Sasol operates, the CO and CO_2 are in equilibrium concentrations due to WGS and at high carbon conversion they will have the low CO_2 production as observed by Layng. Thus, it is concluded that it is a combination of operating conditions and method of operation that led to the conclusion of a beneficial effect of sulfur, and not to a beneficial effect of sulfur.

Cooper and Montano [3] conducted studies at low temperature (100°C) with iron supported on single crystal copper and concluded that "up to 0.2 ML [ML = monolayer]of sulfur does not produce any significant changes in CO dissociation and carbonization of fcc iron" for FTS. However, the FT reaction was conducted at 250°C, $H_2/CO = 1$, and 10^{-3} torr for only 5 minutes; it is likely that the catalyst was in the early stages of the induction period.

Reports by SRI International workers indicated significant beneficial effects for the addition of sulfur in amounts that were less than a monolayer for an iron catalyst [4]. These authors conducted many surface science studies related to various sulfur coverage. While much of the work by this group was conducted in a microreactor, studies were also conducted in larger, higher pressure reactors. In many instances sulfur was added to a reduced, or reduced with CO monolayer catalyst, and then subsequently treated in oxidizing and/or reducing conditions at high temperatures. These workers usually showed a lower conversion for the sulfur treated catalyst and a significantly different product selectivity. In general, the selectivity changes included much lower methane selectivity, a shift to lighter hydrocarbons and higher olefin composition. These selectivity changes could be qualitatively explained on the basis of lower conversions of the sulfur treated catalyst.

An absolute scale of the influence of sulfur on iron Fischer-Tropsch catalysts is not available. First, the impact of sulfur depends upon the reactor type that is used. Second, there is probably an effect of the reaction temperature and the chemical and physical properties of the catalyst.

Early work at the U.S. Bureau of Mines and at Sasol indicate that sulfur effects on a catalyst in a fixed bed reactor as a function of reaction time results in uneven deposition of sulfur within the catalyst bed. Thus, the sulfur entering the reactor is deposited preferentially in the top of the catalyst bed and progresses down the catalyst bed with increasing reaction time. Recent results for operation of a pilot plant that clearly show this trend have been published [5]. Thus, operations in a fixed bed, in general, do not provide a uniform distribution of sulfur and the results will be difficult to relate to effects due to sulfur loading.

Bromfield and Coville [6] have added sulfur to the catalyst prior to loading into the reactor. The sulfided iron catalyst was oxidized and reduced prior to use. It is possible that the sulfide added to the catalyst was converted to the sulfate during calcination. It has been established that sulfate will stabilize the surface area during calcination for many metal oxides, including iron oxide, and this effect appears to be operating in this study. The authors found that the catalysts with low sulfide concentrations were up to four times as active as one that did not contain sulfur. At least some of this increased activity could be due to the reported increase in surface area. More importantly, the sulfur containing catalyst produced more oxygenates at a higher conversion than the catalyst without sulfur and this is a surprising result. In general, the higher conversion would lead to lower, not higher, oxygenate yields. This implies a role for the sulfur species in the synthesis reaction.

In general, there are sufficient indications that low levels of sulfur may have beneficial effects on the iron FTS catalyst. However, it is difficult to maintain a catalyst with the needed low levels of sulfur. Furthermore, the iron catalyst is utilized in different reactors and this will have an impact. For example, the high temperature fluid bed iron catalyst will develop significant amounts of carbon and the sulfur will have the opportunity to distribute between the carbon and the iron portions of the catalyst particle. This distribution will be difficult to define. For the low temperature operation, the distribution of added sulfur between the catalyst and the reactor components will be difficult to control and to obtain accurate loadings of sulfur on the catalyst. Depending upon the conversion level, the catalyst may be exposed to reducing or oxidizing conditions and these conditions should have an impact on the state of sulfur in the catalyst.

In preparation for the second DOE run at La Porte, Texas, four catalysts were evaluated in order to decide upon the final catalyst to be utilized. All four candidate catalysts were inactive when activated in syngas; however, the same catalysts were active when activated in CO. Since the CO activation did not reduce the sulfate present in the catalysts, H_2S was not formed whereas it was formed during the syngas operation. Subsequent exposure of the CO activated catalyst to syngas did not produce H_2S rapidly enough for it to be a severe catalyst poison. In summary, it will require very well designed and controlled experimental conditions to resolve the question of whether sulfur can be a promoter for FTS catalysts.

References

1. British Patent 322,284, Dec. 5, 1929.
2. E. T. Layng, U.S. Patent 2,446,426, August 3, 1948 (applied March 1, 1946).
3. B. R. Cooper and P. A. Montano, Sulfur effects in multimetallic Fischer-Tropsch Catalysts, Final Report, DE-FG22-84, PC 70800, 1987.
4. R. B. Wilson, Jr., T. T. Tong, Y. W. Chan, H. W. Huang and J. G. McCarty, Improved Fischer-Tropsch Catalysts for Indirect Coal Liquefaction; DOE/PC/80016-T10, February 1989.
5. D. J. Duvenhage and N. J. Coville, Appl. Catal. A: General, 298 (2006) 211-216.
6. T. D. Bromfield and N. J. Coville, Appl. Catal. A: General, 186 (1999) 297-307.

Plant Operating Strategy in Dealing with Gas Supply Shortages to the Badak LNG Plant

Rahmat Safruddin
Process Engineering Section
PT Badak NGL
Bontang-Indonesia

AIChE Spring Meeting, April 2007
7th Topical Conference on Natural Gas Utilization
Houston, Texas, April 22-27, 2007

Abstract

The shortage in feed gas supply causes PT Badak to look into its running plant strategy in achieving the optimum operating mode. With the current gas supply, the company is running the plant at a border line between seven and eight Trains capacity. Running with eight Trains will cause higher energy consumption because the refrigeration compressors are operating with recycle valves open to avoid surging due to low load. On the other hand, running with seven Trains will result in approaching the maximum production of each Train, and the Plant has very little cushion to absorb more gas if required.

Individually, an LNG Train has optimum energy consumption at approximately 88% of its maximum production. Beyond this optimum point, the energy consumption is higher per cubic meter of LNG produced. From this point of view, it is best to operate the Train at its optimum energy consumption. However, from plant-wide point of view, putting one Train on idle and pushing the LNG production a little higher on the remaining running Trains will give the best compromise in term of plant energy consumption, because there is a fixed energy requirement to run the LNG Train, regardless of its production level. Therefore, there is a certain point where the whole plant production level will reach the best energy efficiency level, even though, individually, each Train runs beyond its optimum level.

This paper describes the gas supply chain in the East Kalimantan region in relation to the function of the Badak LNG Plant as the swing plant to absorb gas supply fluctuation. The paper also discusses several parameters used to represent its efficiency. In addition, it discusses the potential risk that the gas producers may face due to higher back pressure to the upstream operations, if one of the LNG Train trips. Shutting down old wells to avoid gas venting due to excessive gas supply will put the wells at risk.

Introduction

PT Badak NGL Bontang is presently operating with eight LNG Trains on the east coast of Kalimantan (Borneo) Island. The company is capable of producing LNG at 142,000 m^3/d or 22.5 million tons per annum (mtpa). However, as some of the east Kalimantan Producers fields have entered their mature phase, the plant LNG production is now limited to 115,000-120,000 m^3/d. At this level of production, only seven trains are required to be operated. However, to maintain a high plant availability and minimize the back pressure on the upstream mature gas fields, PT Badak, in agreement with the East Kalimantan gas producers, is operating eight LNG trains. Running with seven LNG trains at this production level can save energy, therefore, a plant running strategy needs to be developed in order to optimize the energy consumption but at the same time avoid operational problems at the plant and for upstream gas fields.

Gas Supply

The gas processed by the plant is produced from several fields located in the delta of Mahakam River, East Kalimantan. There are three main gas producers under a Production Sharing Contract with Indonesia: Total E&P Indonesie, VICO, and Chevron. Total E&P Indonesie is the major gas producer with a production of around eighty (80) percent of the total gas produced. VICO is the second largest gas producer followed by Chevron. The last two companies are facing declining production as their fields have entered a mature phase.

The gas production from all fields are sent to a gathering center located at Muara Badak, 57-km south of Bontang (Figure 1). From Muara Badak, the gas is delivered using two 36-inch and two 42-inch pipelines to Bontang. The main portion of the gas, i.e. 2950-3100 MMscfd, is delivered to PT Badak NGL to be liquefied into LNG for export. The other portion, 400-450 MMscfd, is delivered to Domestic Plants. Domestic Plants is a term to represent companies consisting of Kaltim Fertilizer Plant (KFP), Kaltim Methanol Industri (KMI), Kaltim Pasific Amoniak (KPA), and Kaltim Parna Industri (KPI). The last two plants are producing ammonia. The Fertilizer Plant itself comprises four fertilizer Trains. Each unit of Domestic Plants requires feed gas in the range of 45-80 MMscfd.

Badak LNG Plant Characteristics

Badak LNG Plant consists of eight LNG Trains named alphabetically from A to H. Each Train consists of five Plants or processing units: CO_2 Removal Unit (Plant 1), Moisture and Mercury Removal Unit (Plant 2), Fractionation Unit (Plant 3), Refrigeration Unit (Plant 4), and Liquefaction Unit (Plant 5). All eight trains were designed using propane pre-cooled MR refrigeration systems licensed by APCI (Figure 2).

The Badak LNG Plant is a steam/cooling water LNG Plant compared to other LNG Plants using gas turbine and air cooling. Steam is utilized to drive the refrigerant compressors (propane & MR compressors), as a heating media in the CO2 Acid Gas

Removal & Fractionation Units and to generate the required electricity. Sea water is the cooling media of the Plant.

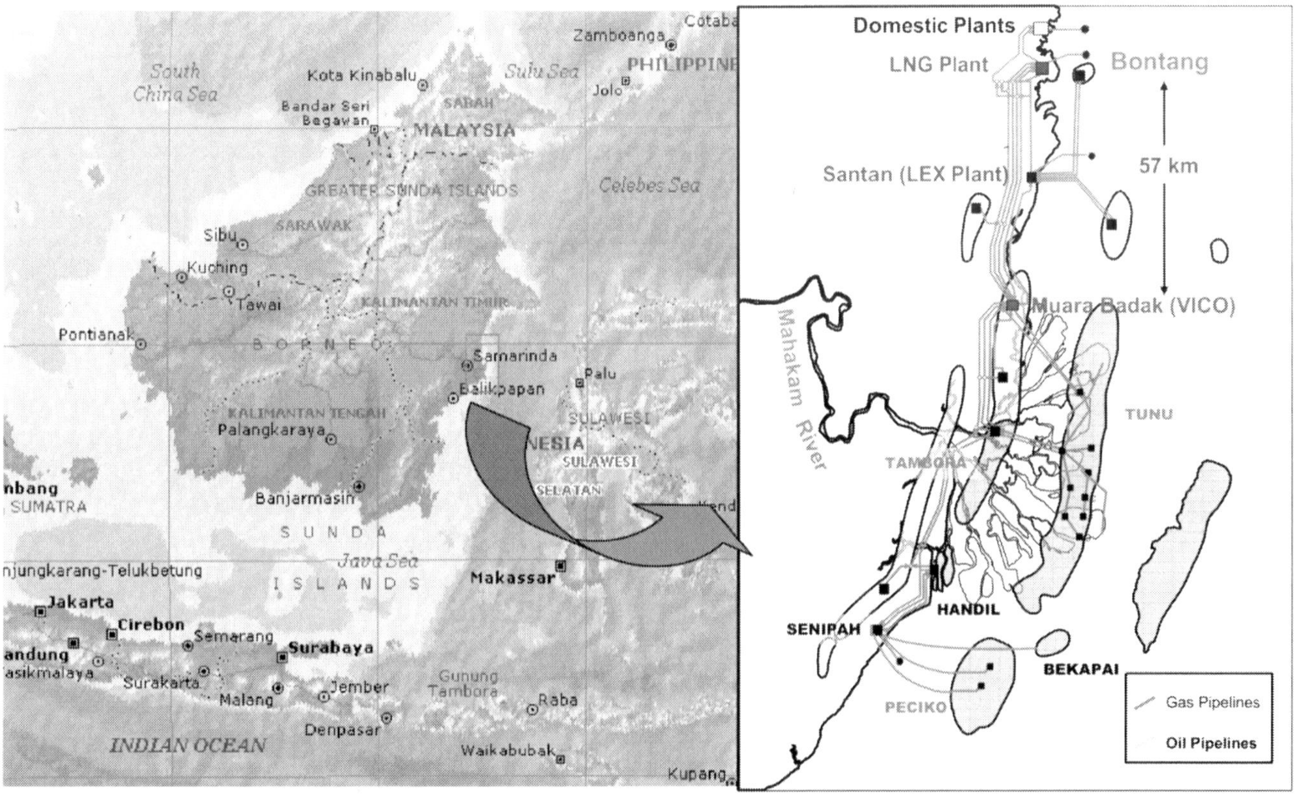

Figure 1. East Kalimantan Gas Production and supply chain

Badak LNG Plant Characteristics

Badak LNG Plant consists of eight LNG Trains named alphabetically from A to H. Each Train consists of five Plants or processing units: CO_2 Removal Unit (Plant 1), Moisture and Mercury Removal Unit (Plant 2), Fractionation Unit (Plant 3), Refrigeration Unit (Plant 4), and Liquefaction Unit (Plant 5). All eight trains were designed using propane pre-cooled MR refrigeration systems licensed by APCI (Figure 2).

The Badak LNG Plant is a steam/cooling water LNG Plant compared to other LNG Plants using gas turbine and air cooling. Steam is utilized to drive the refrigerant compressors (propane & MR compressors), as a heating media in the CO2 Acid Gas Removal & Fractionation Units and to generate the required electricity. Sea water is the cooling media of the Plant.

The production capacity of each Train is listed in the following table. About 10% of the total gas delivered to the Plant is consumed for energy to run the plant. An LNG train can be operated in two modes: with LPG production or without LPG production (all the LPG's are re-injected into the LNG).

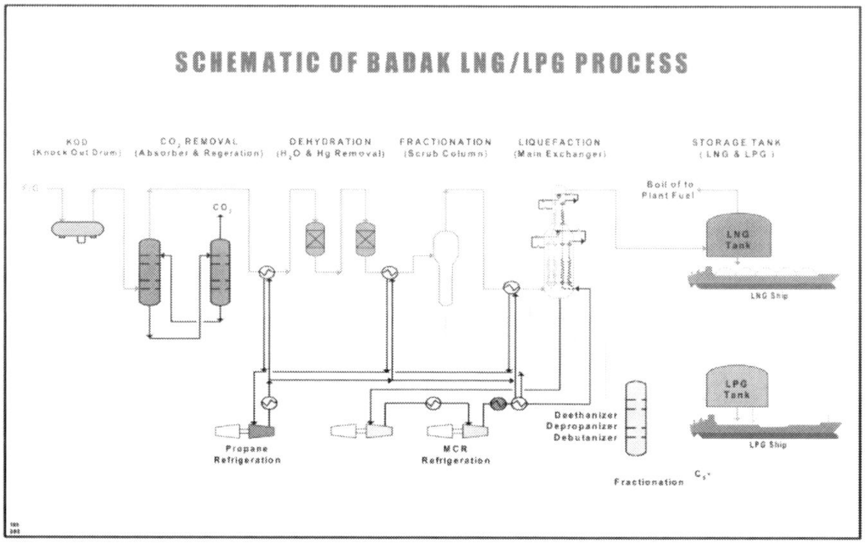

Figure 2. Process Scheme of the Badak LNG Plant

Table 1. LNG Production Capacity of Each Train with no LPG Production

Train	LNG Production		Feed Gas Equivalent (MMscfd)
	M3/h	Mtpa	
A	720	2.71	412
B	710	2.68	406
C	720	2.68	412
D	720	2.68	412
E	770	2.90	441
F	806	3.02	461
G	790	2.94	452
H	890	3.36	509

At its peak production, the Badak LNG Plant is capable of producing at 22.5 mtpa LNG, 1 mtpa of LPG Propane and Butane, and 1 million barrels of Condensate per year. At this production level, the feed gas supply to the Badak LNG Plant is approximately 3,700 MMscfd.

Currently the Badak LNG Plant receives only 2950-3,100 MMscfd as some of the upstream fields have entered their mature phase. At this level of feed gas received, PT Badak could operate seven (7) Trains. However, taking all the parameters into consideration i.e. gas deliverability from the upstream mature gas fields, the LNG Plant efficiency and the continuous high reliability of the LNG supply to its customers, PT Badak and the Gas Producers have developed an Operating Plan to decide whether to operate seven or eight LNG Trains.

In addition, at the beginning of 2006, in order to maximize the LNG production, it was decided to stop LPG production and all LPG products are recombined with the gas stream to Liquefaction unit to be liquefied as LNG product.

LNG Train Energy Consumption

Description

To operate an LNG Train, steam is utilized as the heating media and to run steam driven compressors and pumps. Steam is utilized at three levels of pressure and temperature: high pressure (HP) steam at 60 kg/cm^{2}g (850 psig) and 420°C, medium pressure (MP) steam at 17.5 kg/cm^{2}g (250 psig), and low pressure (LP) steam at 3.5 kg/cm^{2}g (50 psig) and 150°C.

LP steam is used as the heating media for the amine solvent regeneration in the CO_2 Removal Unit and in reboilers of the columns in the Fractionation Unit. About 36% of the total energy consumption by the Train is for amine regeneration. Due to the level of temperature required, High pressure (HP) Steam is also used as the heating medium for the gas drier reactivation heater.

HP steam is used to drive the refrigeration compressor turbines, the largest energy consumers in the train. There are three refrigeration compressors in a train: Propane compressor, 1st stage MR compressor, and 2nd stage MR compressor. The total design power of all refrigeration compressors are 160,000 hp (119.3 MW) or 47.25% of the total power consumed by the Train.

Another compressor driven by HP steam is the LNG Flash Gas Compressor which recovers LNG flash gas and compresses it to the fuel gas system. This compressor was designed at 3 MW power.

There are some other turbines driven by MP steam: a drier reactivation gas compressor and lube/seal oil pumps, but their energy consumption is much smaller compared to the refrigeration compressors.

In addition, electrical power is required to run motor driven pumps and aerial cooler fans. Most of the pumps in the Train are driven by electrical motors but some are driven by steam turbines. In Trains F, G, and H, amine booster pumps (266 kW) and amine circulating pumps (0.9 MW each) are driven by electrical motors while in Trains A-E they are driven by steam turbines. The electricity is produced by steam-driven turbine power generators located in the Utility area.

The most important motor driven pumps are cooling water pumps, stationed in the cooling water intake facility outside of the Train. There are 22 cooling water pumps with a power consumption of 3.5 MW each. Two cooling water pumps are required to supply a sufficient amount of cooling water for Trains A-F. For Trains G and H, three pumps are required at LNG production of higher than 720 m^3/h.

To simplify the analysis, energy consumption to run one LNG Train is divided into fixed and variable energies.

Fixed Energy

The fixed energy is the amount of energy, relatively constant, required to run a Train regardless of its production level. This energy consists of LP steam as the heating media for column reboilers, HP steam to run the LNG flash gas compressor as well as the heating medium for drier gas reactivation, MP steam to run the drier reactivation gas compressor and lube/seal oil pumps, and electricity to run all motor driven pumps.

The following equipment utilize fixed energy:

- Pumps (Plant 1~5)
- Compressors (LNG Flash Gas and Drier Reactivation Gas, in Plant 2)
- Finfan Cooler motors (Plant 1 and 2)
- Reboilers and heater (Plant 1, 2, and 3)
- Cooling water pumps

The total fixed energy per Train is 136.7 MW (3 cooling water pumps) or 133.2 MW (2 cooling water pumps)

Even though the power consumption of the abovementioned pieces of equipment varies slightly with the production rate, this power consumption is considered fixed.

Variable Energy

The variable energy, on the other hand, is the energy whose magnitude greatly depends on the Train production level. This is the energy required to run all three refrigeration compressors. Even though this energy varies with the LNG production level, there is a minimum energy to run the compressors before LNG can be produced during the startup process. The following are the designed power of the refrigeration compressors:

- Propane compressor	60,000 hp	=	44.74 MW
- 1st stage MCR compressor	50,000 hp	=	37.29 MW
- 2nd stage MCR compressor	50,000 hp	=	37.29 MW
Total:	160,000 hp	=	119.3 MW

If one LNG Train is idled because the gas supply is just enough to keep seven LNG Trains running, the plant can save the energy by running in a more efficient mode. However, to maintain the same plant production capacity, the production of the idled Train shall be shared by the other running Trains proportionally, depending on the production capacity of the Train. Therefore, the running Trains will increase their energy consumption mainly in the refrigeration compressors to increase the production. The real saving from the idled Train is only from the fixed energy.

Statistically, due to the compressor performance, the increase in steam consumption in the refrigeration compressors is not linearly proportional with the increase of the LNG production rate. The steam consumption is higher per cubic meter of LNG produced at a production rate of higher than 88% (average) of the maximum sustainable LNG capacity of the Train.

Operational Parameters

In order to identify the best operating points and decide on whether to operate seven or eight Trains, PT Badak and the Gas Producers are utilizing the following operational parameters:

- The Refrigerant Compressors Steam-To-LNG Ratio (Train Efficiency indicator)
- The Feed Gas-To-LNG Ratio (Overall LNG Plant Efficiency indicator)
- Overall Steam-To-LNG Ratio (Overall LNG Plant Efficiency indicator)

Refrigeration Compressors Performance and Steam-to-LNG Ratio

Although the power consumed by the compressor should be calculated from the process parameters, the power consumption in this evaluation is represented by the total steam consumed by all refrigeration steam turbines in tons divided by LNG production rate in cubic meters. Since the power of all refrigeration compressors are adjusted to the level of LNG production, the steam-to-LNG ratio concept is used as it can give a direct indication of the Train performance in terms of energy consumption.

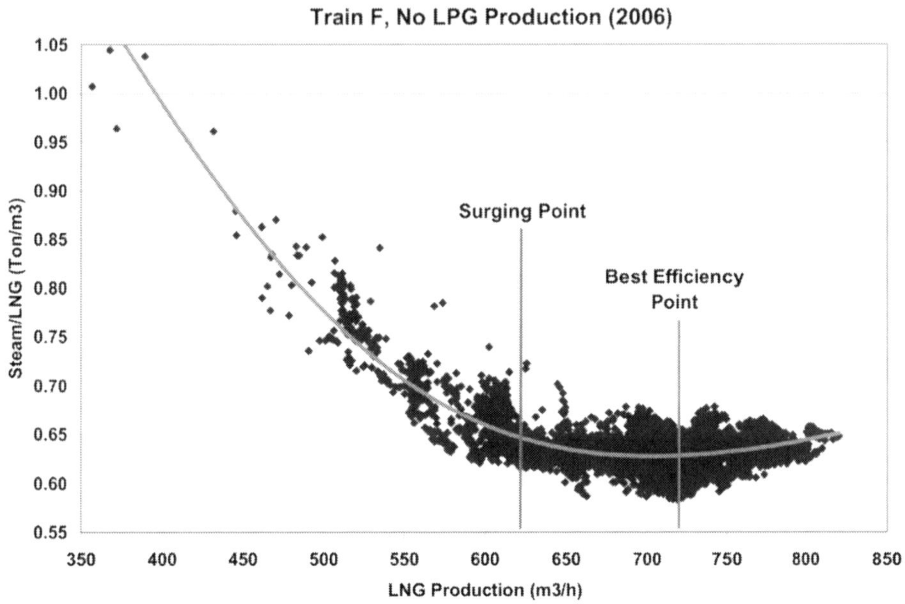

Figure 3. Steam/LNG Ratio over a range of LNG production in Train F during the no LPG production mode in 2006.

A typical curve of steam/LNG can be found in Figure 3 taken from Train F and Figure 4 for Train H. These figures are derived from the actual average hourly operating data in 2006, during which LPGs were not produced as a separate product, but re-injected into the LNG. Hence, the figures give a better approach to modeling the Train's material balance.

Figure 3 shows that the best energy efficiency level to run Train F is at 720 m^3/h LNG production. Train F is capable of producing LNG at maximum of 810 m3/h. This means that the best efficiency point of Train F is at 89% of the maximum production. Even though the Train is capable of running at higher levels of production, but the steam consumed per cubic meter of LNG produced is higher than at this best efficiency point. On the other hand, if the Train is operated at lower LNG production levels than 720 m^3/h, the refrigeration compressors will run toward their surge points. Hence, the recycle valves will open automatically to prevent compressor surging or are opened manually by the operators to avoid unstable operating conditions before the surge control opens the recycle valves automatically. From statistical data, the surging point of Train F is at the 620 m^3/h LNG production level.

Figure 3 also shows that the steam/LNG at the best efficiency point is at 0.62 tons steam/m^3 LNG. When the production is pushed to the maximum capacity, the ratio increases to 0.65 tons steam/m^3 LNG. This means that an additional 0.02 tons of steam is required per cubic meter of LNG produced at maximum production compared to the best efficiency point.

The steam/LNG curves for other Trains are typical of train F with some slight differences. Figure 4, for example, shows Steam/LNG for Train H with the same data resolution. Train H is the last Train constructed, and has a higher design LNG capacity than the other Trains. Its propane compressor was designed for a higher load while the 1st and 2nd MR compressors were designed similar with Trains E, F and G. Therefore, the best efficiency and surging points of this Train are higher than those for Train F.

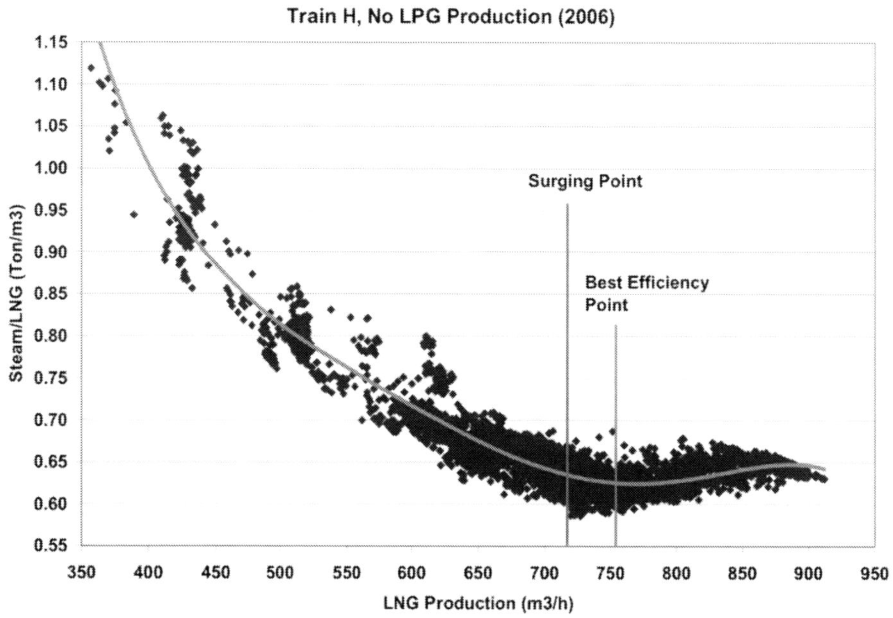

Figure 4. Steam/LNG Ratio at a range of LNG production in Train H during no LPG production mode in 2006

Data from Trains A, B, C and D cannot be acquired due to lack of flow indications in DCS control panel, however, due to typical compressor design with the ones installed in Trains E, F, and G the performance is assumed to be similar.

Feed Gas-to-LNG Ratio

The ratio of feed gas to LNG (FG/LNG) is another parameter used by the East Kalimantan Gas Producers to monitor the plant performance in term of feed gas conversion into LNG. Data is plotted daily. The feed gas is in thousand Standard Cubic Feed (Mscf) and the LNG production is in cubic meter.

The advantage of this parameter is that it can give an overall gas conversion into LNG no matter what happens to the plant such as a train trip, gas losses due to equipment leaks or purposely vented for other reasons. The Gas Producers prefer to use this parameter to establish their field production forecast based on the number of LNG cargoes to be delivered to the customers.

However, this parameter can be confusing in analyzing the plant performance if not used carefully, because some variables need to be taken into account in analyzing the plot results. For example, the fluctuation in feed gas composition such as CO_2 content or heavier hydrocarbon components can give different conversion results from feed gas into LNG. The FG/LNG ratio is plotted based on number of running Trains (Figures 5).

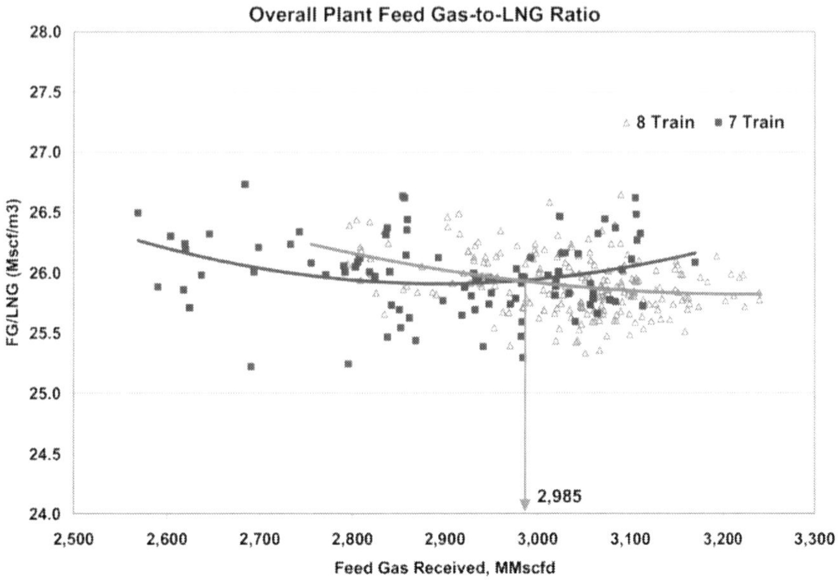

Figure 5. Feed Gas-to-LNG Ratio of the Overall Plant at no LPG production mode in 2006

Two trend lines of eight Trains and seven Trains running are drawn over the data plots. The intersection point is the FG/LNG ratio of 25.935 Mscf/m3 corresponding to 2,985 MMscfd of feed gas equivalent to an LNG production of 115,095 m^3/d. Therefore, if the feed gas received is lower than 2,985 MMscfd, the seven train mode of operation is more efficient than the eight train mode, and vise versa.

In the fourth quarter of 2006, the gas received by the Badak LNG plant was in the range of 2950-3100 MMscfd (equivalent to an LNG production of 114,000-119,000 m^3/d). This production level is on the border line between the production capacity of a seven and an eight Train operation. The gas producers prefer the Badak LNG Plant to run with eight trains. As a result, the plant is running with a high production flexibility. If one train trips, the remaining running trains will still be capable of absorbing the production loss from the Train that tripped.

Running with only seven Trains is theoretically sufficient to absorb all of the current feed gas delivered to the Plant, including a 50-150 MMscfd additional gas capacity should one of the Domestic plants trip or shutdown because the Badak LNG Plant can absorb gas up to 3,150 MMscfd when running with seven Trains.. However, there is a risk of venting gas and shutting down wells if one of the LNG Train trips.

Operating with eight trains, when seven trains would be theoretically sufficient, has a beneficial impact on the gas deliverability from the upstream fields. The current feed gas pipeline pressure of 42-43 kg/cm^2g at the Bontang receiving point is significantly low compared to the design operating pressure of 47 kg/cm^2g. As a result, the backpressure at the gas producers fields is lower allowing some of the low pressure wells to produce gas which would not be the case otherwise. Therefore, a good compromise is required to have an optimum operating scenario in terms of energy efficiency and gas production.

Steam-to-LNG Ratio

Another parameter developed to eliminate some of the inaccuracy derived from the FG/LNG plot is the steam to LNG ratio, i.e. the total steam production from all the boilers in tons divided by the total LNG production in cubic meters measured in the LNG storage per day. The total steam production represents the fuel consumption of the plant and the fact that the curve of steam production versus fuel consumption from all 12 boilers at the Badak LNG Plant is linear over a wide range of boiler capacity, the steam-to-LNG ratio parameter is directly related to the plant efficiency (running trains).

Figure 6 shows the plot of total Steam/LNG ratio versus gas received by the Badak LNG Plant. In this plot, data is distributed more consistently compared to the Feed Gas/LNG plot. The plot is made for two operating modes: seven trains and eight trains. The trend lines of both data series give a crossover point at 3,104 MMscfd of feed gas supply. This is higher than the crossover point given by the Feed Gas/LNG ratio, i.e. 2,985 MMscfd. Therefore, from this curve, at 2,985 MMscfd gas supply, operating with a seven train mode is more efficient.

For a feed gas flow rate of 2,985 MMscfd, running with seven Trains requires 0.025 Tons steam consumption less per cubic meter of LNG produced than running with eight Trains. This amount corresponds with a saving of 235,800 Nm^3/d fuel gas or equivalent with 360 m^3/d LNG or 1.05 standard LNG cargoes per year (1 standard cargo = 125,000 m^3 LNG), therefore, it would be more beneficial to run with 7 Trains at the current gas supply level to the Plant.

Discussion: Running Seven or Eight Trains?

Another consideration before making a decision in putting one Train on idle is the amount of flared gas during shutdown. Statistically, the amount of flared gas recorded during a Train shutdown ranges from 338,000 Nm^3 to 1,300,000 Nm^3 depending upon the activities carried out during the idle time.

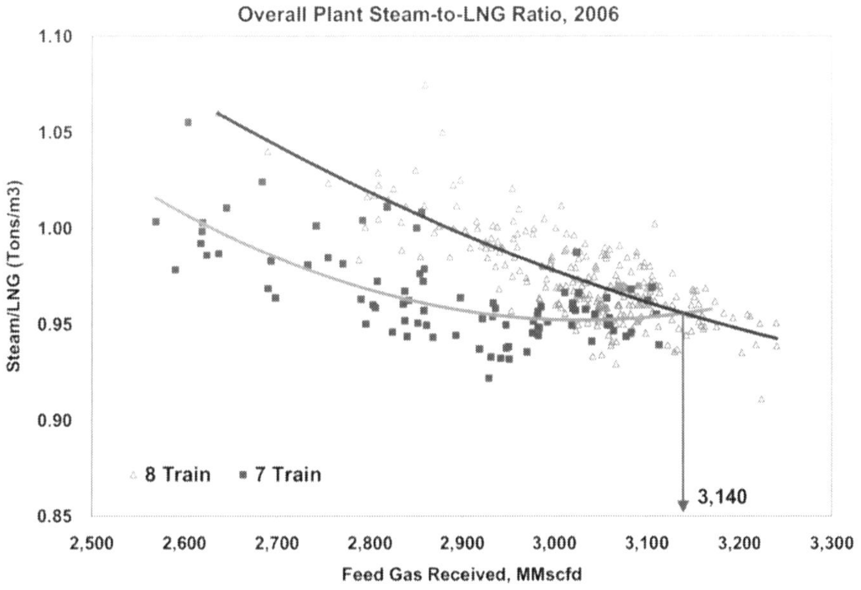

Figure 6. Steam-to-LNG Ratio of the Overall Plant at no LPG production mode in 2006

Therefore, this flared gas must be taken into account in calculating the pay back time of putting one LNG Train on idle. From Figure 6, taking the range of gas received from 2,985 (crossover point from FG/LNG plot) to 3,140 MMscfd (crossover point from Steam/LNG plot), the number of idle days is calculated and plotted for the two amounts of flared gas (Figure 7). If ten days is taken as a reasonable pay back time for putting one Train on idle, running with seven Trains is recommended at a feed gas rate of 3,070 MMscfd or lower with the flared gas of 1,300,000 Nm^3 and at feed gas rate of 3,120 MMscfd or lower with the flared gas of 338,000 Nm^3. At 3,000 MMscfd gas received, and considering the highest amount of flared gas during a shutdown, the pay back time for putting one Train on idle is six days.

If energy consumption is the main concern in operating the plant, running with 7 Trains is the recommended mode of operation at the current feed gas supply of 2950-3,100 MMscfd at is it could lead to the production of an additional 1.05 standard LNG cargoes per year. In addition, it also gives a chance to maintain a higher plant reliability because any repair or maintenance activity can be carried out during the idle time shutdown. This saving can only be achieved when the flared gas during the Train shutdown is carefully controlled and the idle time is longer than six days.

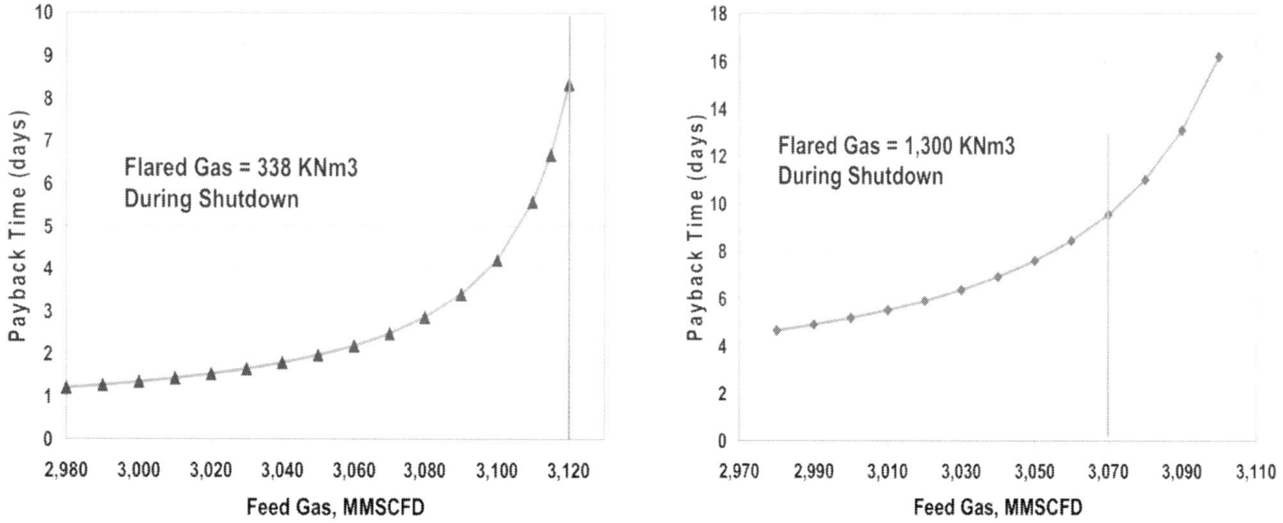

Figure 7. Payback Time of Putting One Train on Idle versus the Amount of Flared Gas

On the other hand, if the gas deliverability (Producers' concern) or LNG production reliability are the main concern, then running with 8 Trains at the current feed gas rate is recommended.

Conclusions

- Several parameters are used to describe the Badak LNG Plant efficiency versus feed gas supply, namely Feed Gas-to-LNG ratio, individual Train Steam-to-LNG ratio, and overall plant Steam-to-LNG ratio.
- The Feed Gas-to-LNG ratio parameter is preferred by the Gas Producers, as it can give a direct conversion of feed gas into LNG. However, the plot does not give a clear indication between running 7 and 8 Trains.
- The Steam-to LNG ratio is a better parameter to determine the cutover point between a 7 and 8 Train operation.
- At a feed gas rate of 2950-3100 MMscfd, the seven train operating mode is recommended for better overall efficiency. However, from the gas delivery point of view, the eight train operating mode is recommended.
- In order to have the best compromise between the gas deliverability (especially from the mature fields) and the efficiency, the Badak LNG plant is still operating eight LNG trains at the current gas supply of 2950-3,100 MMscfd. However, a further evaluation of the situation is currently underway between Badak plant personnel and the Gas Producers.

Biography

Rahmat Safruddin is the head of the Process Engineering Section in the Technical Department, PT Badak NGL, Bontang, Indonesia. He has been working for the company for over 10 years, mainly in the Process Train Group. He holds a BS degree from North Sumatera University in Chemical Engineering and a MS degree from Ohio State University in Materials Science and Engineering.

Address: Process Engineering Section, PT Badak NGL, Bontang 75324, Kalimantan Timur, Indonesia.
Phone: 62-548-553804, Fax: 62-548-21605, Cell: 62-812-552-8381.
E-mail: rahmat@badaklng.co.id

Transient Characteristics of Two- Phase LNG Expanders

Christopher D. Finley
Test Engineering Manager
Ebara International Corporation
Sparks, NV 89434, USA

Chen-Hwa Chiu
Senior Technology Advisor
Chevron Energy Technology Company
Houston, TX 77002, USA

Katarzyna Cholast
Andrzej Kociemba
Process Advisors
KRIO Polish Oil and Gas
Odolanow, Poland

Andrew Kimmel
Research Advisor
University of Nevada Reno
Department of Physics
Reno, Nevada

AIChE Spring National Meeting 2007
7th Topical Conference on Natural Gas Utilization
Houston, Texas, April 22-27, 2007

Abstract

For optimized LNG production, particularly to reduce liquefaction costs or to extend the lifetime of depleting gas fields, it was previously suggested to use two-phase LNG expanders operating in series with a single-phase liquid LNG expander. While the performance of liquid expanders depends only on mass flow and differential head, the performance of two-phase expanders depend on additional properties like LNG composition, temperature, and outlet pressure. The effect of non-linear oscillations and of transient operation on the performance of two-phase LNG expanders operating in series with single-phase LNG expanders is discussed and presented.

Introduction

A system is in a transient state when a process variable has been changed and the system has not yet reached steady state. Two-phase LNG expanders have several process variables and a variety of transient states occur as a result of process variable changes. The present paper discusses the transient behavior of oscillating liquid hammer phenomena and pressure distribution effect for expanders operating in series with de-energized electric generators.

Oscillating Liquid Hammer Phenomena

The dynamic behavior of liquid expanders is governed by the conservation laws of mass, energy and momentum [Chiu [1]].

$$H(t) = \alpha Q(t)^2 + \beta N(t)^2$$

$$H(t) = H_0 + \sigma \frac{dQ(t)}{dt}$$
$$Q(t) = Q_0 + q(t)$$
$$N(t) = N_0 + n(t)$$

The differential head H represents the static energy across the expander and the sum of the squares of the volumetric flow Q and the rotational speed N represents the kinetic energy. H, Q and N are functions of the time t, but at any moment the sum of the total energy is constant. Changes in the volumetric flow cause changes in the differential head, which is described as liquid hammer [Smith [2]]. In LNG, with σ as the speed of sound, the change of differential head is calculated by

$$h(t) = \sigma \frac{dQ(t)}{dt}$$

For the steady state the total energy is described as

$$H_0 = \alpha Q_0^2 + \beta N_0^2$$

By using this relation the transient non-steady part of the energy equation reduces to a non-linear differential equation.

$$\sigma \frac{dq(t)}{dt} = \alpha q(t)^2 + \beta n(t)^2$$

During the transient state of an expander with a de-energized electric generator, small disturbances of the rotational speed occur, given by

$$n(t) = n_0 \sin(ft + \varphi)$$

The transient energy equation transforms into

$$\sigma \frac{dq(t)}{dt} = 2\alpha Q_0 q(t) + \alpha q(t)^2 + 2\beta N_0 n_0 \sin(ft + \varphi) + \beta n_0^2 \sin^2(ft + \varphi)$$

This non-linear differential equation has no exact solution and an approximate solution is calculated using the referenced method with Fourier coefficients for the error [Kimmel [3] [4]].

The approximate solution is given as

$$q(t) = q_0 \sin(ft)$$

when substituted in the differential equation a periodic error E(t) is generated:

$$E(t) = -\sigma f q_0 \cos(ft) + 2\alpha Q_0 q_0 \sin(ft) + \alpha q_0^2 \sin^2(ft) + 2\beta N_0 n_0 \sin(ft + \varphi) + \beta n_0^2 \sin^2(ft + \varphi)$$

Calculating accurately only the main frequency by neglecting all higher harmonic frequencies, the Fourier coefficient a_1 and b_1 for the error for the main frequency are set to be zero [Kimmel [3]].

$$a_1 = \int_0^{2\pi} E(t)\cos(ft)d(ft) = 0 \qquad b_1 = \int_0^{2\pi} E(t)\sin(ft)d(ft) = 0$$

By proceeding using the referenced algorithm the amplitude q_0 for the periodic changes of the flow can be calculated, together with the phase difference φ between the fluctuations in rotational speed and flow.

$$q_0 = \frac{2\beta N_0 n_0}{\sqrt{4\alpha^2 Q_0^2 + f^2\sigma^2}} \qquad \tan\varphi = -\frac{f\sigma}{2\alpha Q_0}$$

The phase difference between the fluctuations in speed and flow depends mainly on the ratio between the frequency of the fluctuations and the steady state flow rate. For small frequencies and large flows, the phase difference is approximately 180°, therefore

$$Q = Q_0 + q_0 \sin(ft) \qquad N = N_0 - n_0 \sin(ft)$$

and the amplitude of the flow fluctuations approaches the limit

$$q_0 = \frac{\beta N_0 n_0}{\alpha Q_0}$$

A periodic increase in rotational speed causes a periodic decrease in flow. Finley[5] reported test results for the continuously transient operation of LNG expanders. Figure 1 shows the transient changes of flow and speed for low frequency changes. The changes of flow and rotational speed are delayed with a phase of approximately 180°, but for faster changes close to the maximum speed the delay in speed is reduced to less than 180°.

For high frequencies and small flows the phase angle approaches 90° with

$$N = N_0 + n_0 \cos(ft)$$

and the amplitude of the flow fluctuations approaches zero.
The flow fluctuations cause pressure waves called liquid hammer according to

$$h(t) = \sigma \frac{dq(t)}{dt} = \sigma f q_0 \cos(ft) = \frac{2\beta N_0 n_0}{\sqrt{1 + \left(\frac{2\alpha Q_0}{f\sigma}\right)^2}}$$

The amplitude of the pressure fluctuation increases with the frequency and approaches asymptotically the maximum value h_{max}.

$$h_{\max} = 2\beta N_0 n_0$$

This maximum value depends only on the product of the steady state rotational speed N_0 with the change of speed n_0 .The pressure change Δh for a liquid hammer during transient operation of liquid expanders is always smaller than h_{max}

$$\Delta h \leq 2\beta N_0 n_0$$

The maximum possible change of speed is $n_{0max}= N_0$ from full rotational speed to zero speed in case of a locked rotor due to bearing failure. Then the maximum possible pressure change for the liquid hammer is therefore

$$\Delta h_{\max} = 2\beta N_0^2$$

In case of liquid expanders the expression βN_0^2 is equal to the Euler-Head H_E for zero flow.

$$\Delta h_{\max} = 2H_E$$

In case of a locked rotor the maximum possible pressure change for a liquid hammer in liquid expander operation is equal to twice the Euler-Head. The Euler-Head is equal to the square of the tip speed of the rotational runner divided by the gravity. For multiple stages the total Euler-Head is the sum of the Euler-Heads of each single stage.

In case of a change of the rotational speed to a higher speed, expressed by using a factor ε,

$$n_0 = \varepsilon N_0$$

the maximum possible pressure change of the liquid hammer is equal to twice the Euler-Head times the factor ε

$$\Delta h_{max} = 2\varepsilon H_E$$

Transient Characteristics of LNG Expanders Operating in Series

Figure 2 shows an assembly of two LNG expanders operating in series and mounted together in a tandem configuration. The flow across both expanders during normal operation is equal, but the rotational speeds and the resulting pressure differences on each expander may be different or independently selected to achieve the desired total pressure difference.

Liquid expanders are occasionally subject to transient operation if a power failure occurs at the plant. If two expanders are operating in series then a power failure at one expander immediately effects the operation of the other expander.

Figure 3 shows the schematic of two expanders operating in series. During normal operation the LNG flow across both expanders X1 and X2 with the attached electrical generators G1 and G2. The inlet pressure P1 for X1 and the outlet pressure P4 for X2 measure the total pressure difference for both expanders and P2 measures the pressure between X1 and X2. Each expander has a bypass with a pressure relief valve PV1 and PV2, and a third bypass redirects the LNG flow across a Joule-Thomson Valve in case of any mechanical or electrical failure involving X1 and X2.

Figure 4 shows the performance characteristics of two identical expanders operating in series. The horizontal axis represents the flow Q and the vertical axis represents the differential pressure or differential head H. If both expanders are operating at the rated points R1 and R2, then the identical rated flow for both is Q_R and the rated heads are H_{R1} and H_{R2}. The rated points are located at the constant speed curves S1 and S2. If power failure occurs only on expander X1, then expander X1 is operating along the no-load characteristic N1 and expander X2, still producing power, is operating along the constant speed curve S2. Since the flow QA for both expanders has to be equal due to flow conditions in series, the operating point for expander X1 moves to A1 and for expander X2 to A2.

The differential head H_{A1} of expander X1 and the differential head H_{A2} of expander X2 add up to the total differential head H_{R1} + H_{R2} for both expanders. To maintain this condition in the event of a power loss, the expander X1 must operate at a speed above the permitted maximum speed. This is because the rotational speed is determined by the differential head at the no-load characteristic, The rotational speed is proportional to the square root of the differential head.

To avoid this effect of higher speed in the case of power failure at one expander, the other expander must be simultaneously disconnected from the power net. If both expanders are disconnected at the same time then the operating points of expander X1 and X2 move to the points B1 and B2 at the no-load characteristics N1 and N2 with the identical flow Q_B, and both expanders are operating at the permitted maximum rotational speed.

Figure 5 shows the performance characteristics of two expanders, which are not identical but operating in series. Expander X2 has a no-load curve N2 that is different from the no-load curve N1 of expander X1. The rated points R1 and R2 with the rated heads H_{R1} and H_{R2} and the rated flow H_R are the same like for the configuration shown in Figure 4. If power failure occurs on one expander and the other expander is simultaneously disconnected from the power net, then the operating point for expander X1 moves to C1 at the no-load characteristic N1 and for expander X2 moves to C2 at the no-load characteristic N2 with the identical flow Q_C. Because N1 is different from N2 in this case, and the sum of the differential heads H_{C1} + H_{C2} has to be equal to H_{R1} + H_{R2}, the expander X1 is operating at a higher speed and above the permitted maximum speed, despite that both expanders have been disconnected from the power net. The reason for this over-speed performance is the difference in the no-load characteristics.

Figure 6 shows the means to avoid over-speeding as above described. The by-pass for expander X1 with the pressure relief valve PV1 as illustrated in the schematic of Figure 3 modifies the no-load characteristic N1 to NB as soon as the by-pass valve PV1 opens. The operating point of expander X1 moves to D1 at the modified no-load characteristic NB and the operating point of expander X2 moves to the no-load characteristic N2 with the identical flow Q_D. The sum of the differential heads H_{D1} + H_{D2} is equal to H_{R1} + H_{R2}, but in this case, compared to Figure 5, the differential head H_{D1} is significantly smaller than H_{C1}, and since the differential head determines the rotational speed at the no-load characteristic, the speed will not exceed the permitted maximum speed. Analog conditions apply for the expander X2 with the by-pass and pressure relief valve PV2, if the no-load characteristics N1 and N2 are not identical.

Summary

Transient characteristics of two-phase LNG expanders operating in series are subject of the no-load characteristics of each individual expander. Due to the condition that the same mass flow rate passes through both expanders, there are certain transient conditions that require by-pass installations to avoid over-speeding of the expanders.

Any change of the rotational speed effects the mass flow and the differential head. A change in the mass flow causes a temporary pressure wave, known as liquid hammer. The maximum possible pressure difference of a liquid hammer depends on the product of the rotational speed times the change of the rotational speed.

References

1. Chiu, Chen-Hwa et al.: "Turbo-expander technology development for LNG plants" LNG-13 Conference, Seoul, Korea, May 2001.
2. Smith, D.R. et al.: "Liquid hammer for the de-energized radial inflow turbine generator" Methods and Applications of Analysis Vol.5, 1998, ISSN 1073-2772.
3. Kimmel, A. "Pressure induced non-linear oscillations in two-phase LNG pipe-flow" AIChE Spring National Meeting 2006, Orlando, FL, April 23-27, 2006.
4. Kimmel, A. "Effect of quadratic fluid damping in two-phase liquefied natural gas" AIChE Spring National Meeting 2007, Houston, TX, April 22-27, 2007.
5. Finley, C.D. "Continuously transient operation of two-phase LNG expanders" AIChE Spring National Meeting 2004, New Orleans, LA, USA, April 25-29, 2004.
6. Cholast K., Kociemba A., et al. "Improvements in nitrogen rejection unit performance with changing gas compositions" Gas Processors Association Annual Conference, Warsaw, Poland, 21-23 September 2005.

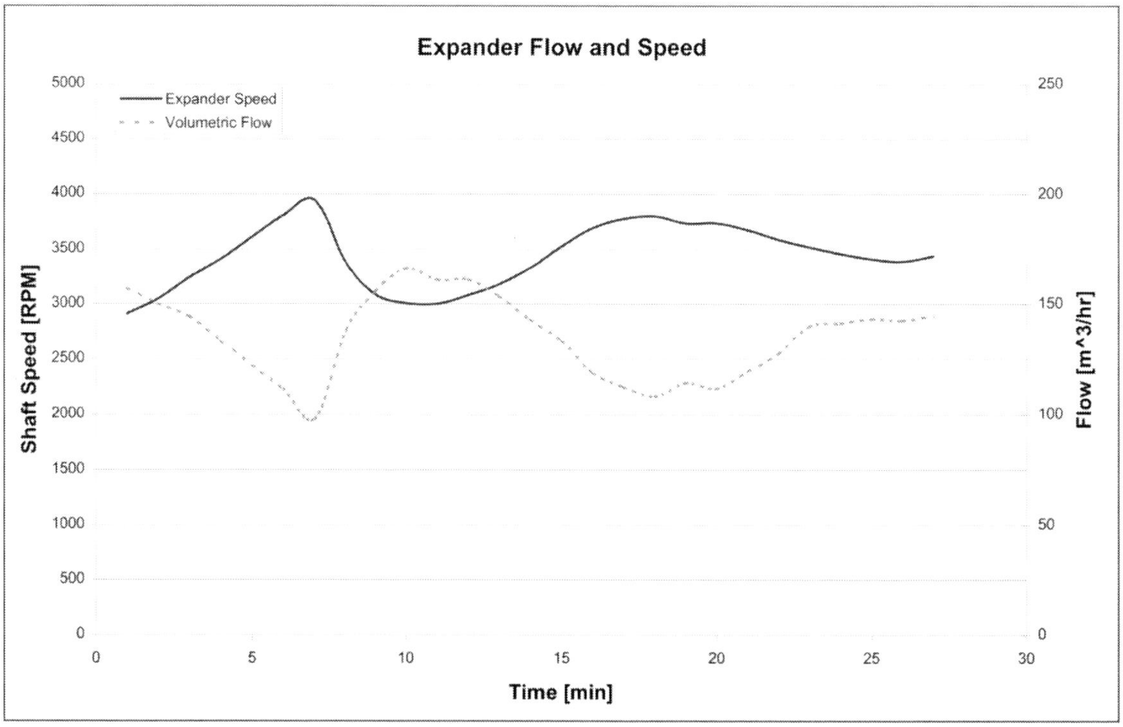

Figure 1: Transient Changes of Flow and Speed for Low Frequencies

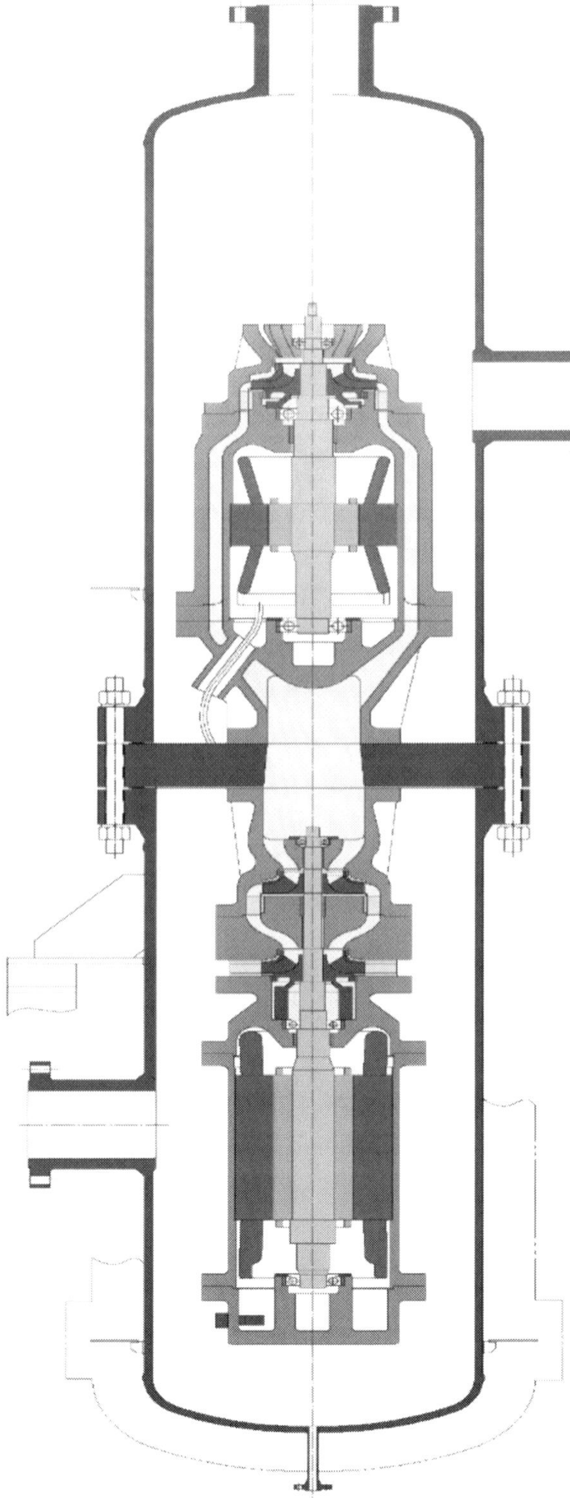

Figure 2: Assembly of Two LNG Expanders Operating in Series

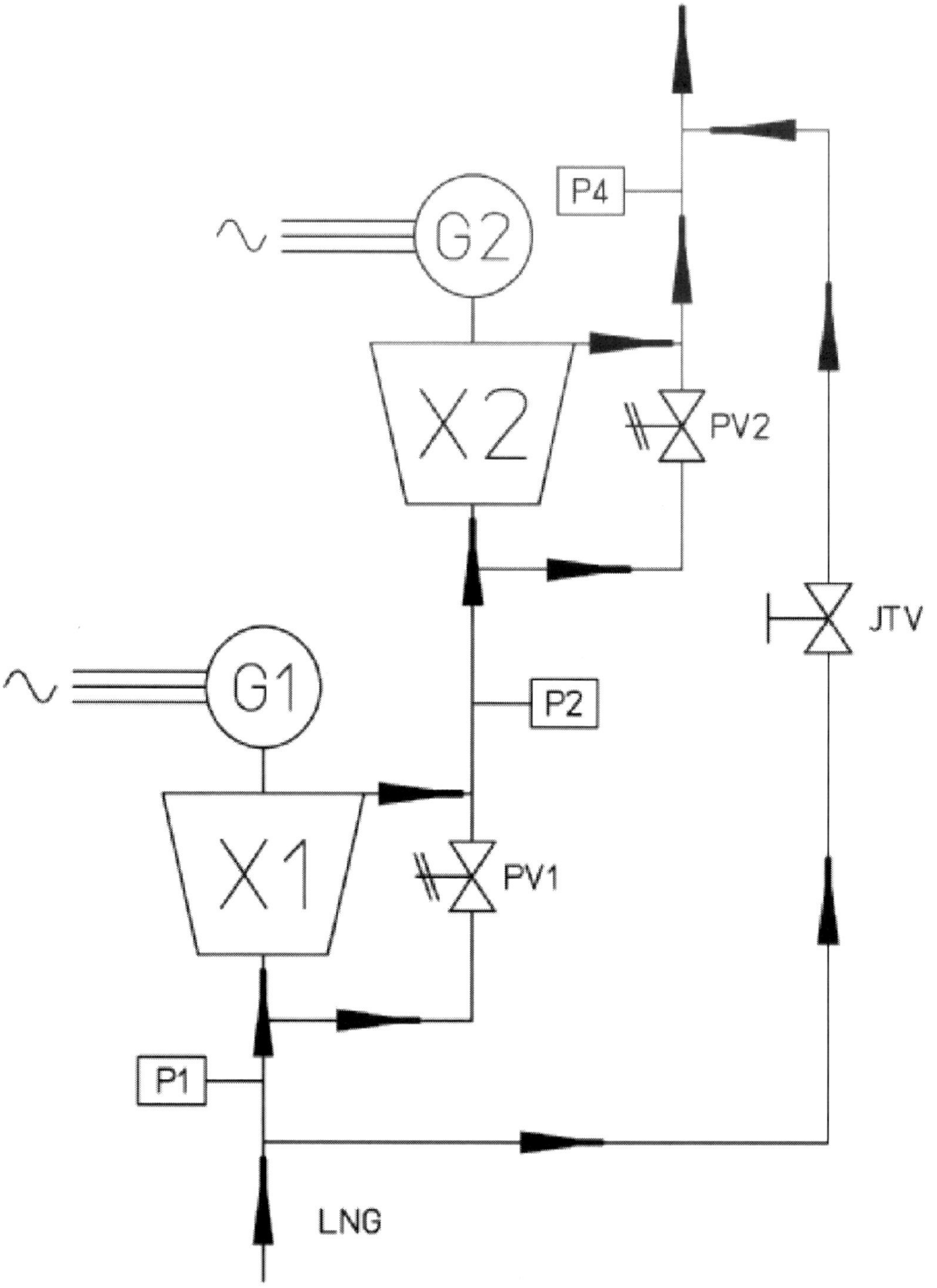

Figure 3: Schematic of Two LNG Expanders Operating in Series

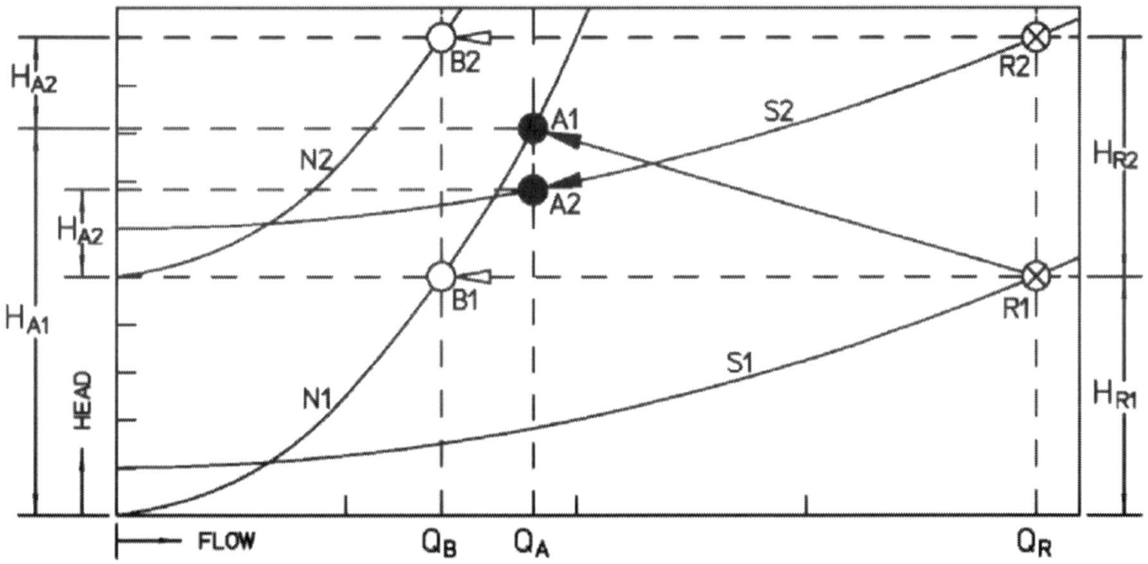

Figure 4: Performance of Two Identical LNG Expanders Operating in Series

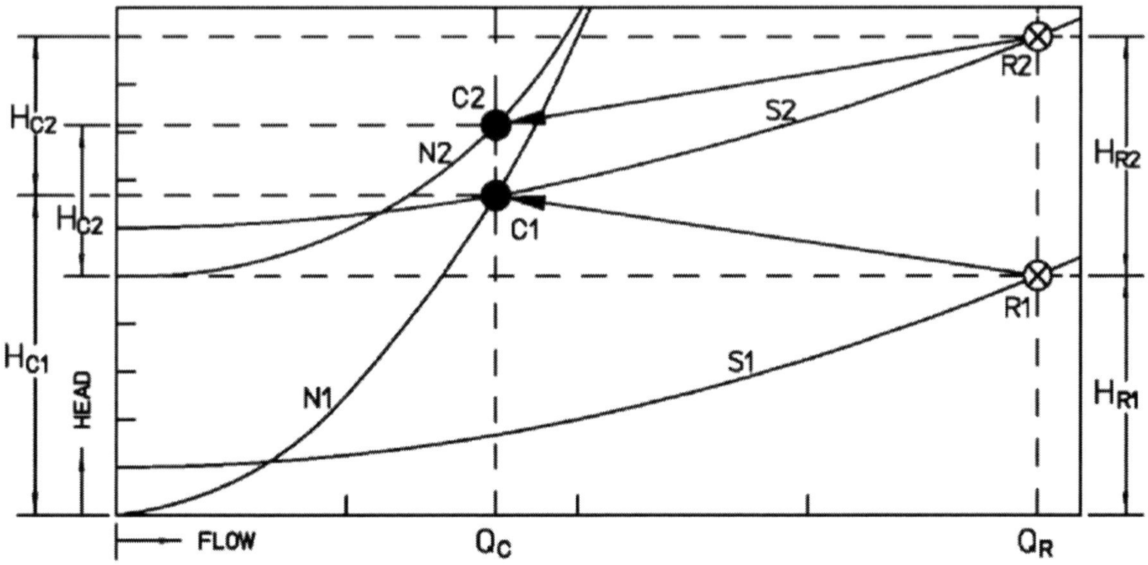

Figure 5: Performance of Two Different LNG Expanders Operating in Series

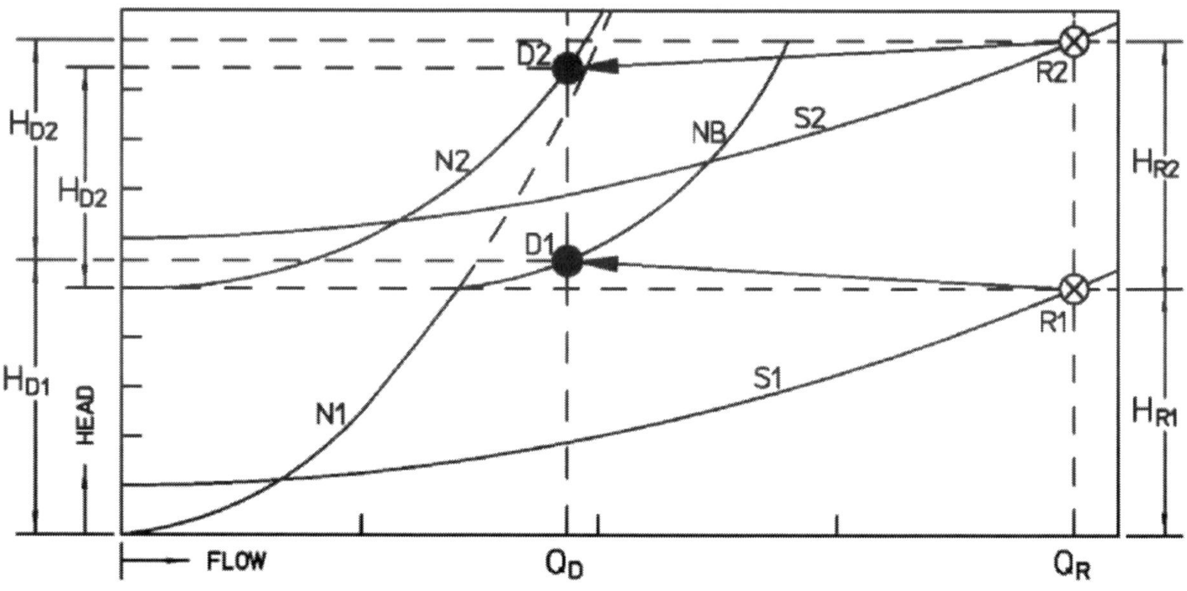

Figure 6: Two LNG Expanders Operating in Series with Bypass Installation

Biography

Christopher D. Finley is Test Engineering Manager of Ebara International Cryodynamics division. Mr. Finley earned a B.S. degree in mechanical engineering from the University of Nevada, Reno and is currently enrolled in graduate courses also at the University of Nevada, Reno.

During his 11 year tenure at Ebara, Chris has been involved in numerous research and development projects including the development of the first two-phase LNG expander. As Test Engineering Manager, Chris has lead several projects involved with the advancement of testing technology of cryogenic pumps and expanders, including novel heat exchanger designs and controls, variable speed induction motor control and development of data acquisition software.

Dr. Chen-Hwa Chiu is a Senior Technology Advisor of the Chevron Energy Technology Company. Dr. Chiu has served as a company representative to the Joint Industry Projects of Azure Floating LNG Chain, LNG FPSO Safety, and Offshore Cryogenic Loading for LNG transfer. Dr. Chiu is a Fellow of American Institute of Chemical Engineers and was a Chair of AIChE Fuels and Petrochemicals Division. He is the organizer and LNG Session chair of the AIChE Topical Conference on Natural Gas Utilization. Dr. Chiu graduated from National Taiwan University with a B.S. and earned his Master of Engineering and Ph.D. from the University of Oklahoma, all in Chemical Engineering. He is a Registered Professional Engineer in Texas and Pennsylvania.
Dr. Chiu can be contacted at 1500 Louisiana Street, Houston, TX 77002, USA. Phone: 832-854-5305, Cell: 713-397-4097, and Chen-Hwa.Chiu@Chevron.com.

Andrew Kimmel is graduating at the University of Nevada, Reno, College of Science, in Physics and Mathematics. His research contributions include analyses of non-linear oscillations, and steady state solutions for non-linear vibrations with applications in cryogenic fluid dynamics. He is a member of the German Physical Society and participates in the German Research Group for Non-Linear Phenomena in Physics and Engineering. He is also a member of the Golden Key International Honour Society.

Improving Profitability of LNG Plants using Advanced Process Control

Jaleel Valappil, Sanjay Wale and Vibhor Mehrotra
Bechtel Oil, Gas and Chemicals
Houston, Texas

Ram Ramani and Satish Gandhi
Conoco Phillips
Houston, Texas

AIChE Spring Meeting, April 2007
7th Topical Conference on Natural Gas Utilization
Houston, Texas, 22-26, 2007

1. INTRODUCTION

There are a number of LNG plants that are built and operating based on the ConocoPhillips *Optimized Cascade*SM* LNG Process. This provides many opportunities in applying advanced technologies to improve the operation of these plants. There are many benefits to be derived from application of advanced process control (APC) to an LNG process. LNG plants are characterized by operational challenges including varying ambient conditions, heat integration and product quality constraints. For LNG plant owners, the main benefit from APC is product maximization, which increases operating profit. Other benefits like increased process efficiency, increased natural gas liquids (NGL) production and reduced operator intervention are also valuable.

Use of APC for optimization has established itself as a widely used technology in process and other industries. Currently, there are hundreds of industrial units that benefit from this technology. The main reason for its popularity is the ability to run plants at optimal conditions while handling various operational constraints and disturbances. Model predictive control (MPC) is the most popular advanced control technique. MPC consists of a model capable of predicting the behavior of key variables in the plant. Adjustments are made to the corresponding manipulated variables to keep the control variables tracking close to their desired target values. These target values are obtained from an optimizer that aims to increase the profit function.

*Optimized Cascade services are provided by ConocoPhillips Company, Phillips LNG Technology Services Company and Bechtel Corporation via a collaborative relationship with ConocoPhillips Company. Optimized Cascade, the Optimized Cascade logo, ConocoPhillips and its logo are trademarks of ConocoPhillips Company. Bechtel and its logos are trademarks of Bechtel Group Inc.

There was not much interest in applying advanced control to LNG units in the early days of APC implementation. This has changed in recent years, and different applications of APC to LNG processes are being reported. Recently, Shell has reported applications to the Shell MR process based LNG plants. Application of APC to a LNG train in Qatar was reported by Aspen Technology.

More interest in the application of APC to LNG plants is expected in the future. This is due to changing market conditions, whereby operating plants will be able to sell excess LNG production in the spot market. This provides significant economic incentives in maximizing the existing plant potential. This also raises the issue of tailoring the LNG quality to the appropriate market and changing the product mix based on economics. Advanced Control is a proven tool to achieve these objectives.

2. ADVANCED CONTROL OF LNG PROCESSES

Process Description

The studies presented in this paper are based on the Optimized Cascade process. The overall plant configuration of this process has been driven from a gas plant perspective and has proven extremely reliable. The feed gas is first processed in the feed-treatment section of the plant. Treated gas from the amine system is fed to the molecular sieve dehydrators and processed through activated carbon beds to remove any mercury. The treated gas is fed to the liquefaction unit where it is cooled in stages and condensed prior to entering the LNG tanks. The liquefaction process includes three refrigeration circuits consisting of predominantly pure component refrigerants, propane, ethylene and methane. Although not required, each refrigeration circuit typically utilizes parallel compressors (up to two or three per refrigerant service) combined with common process equipment. The feed gas passes through multiple stages of chilling in the propane, ethylene and open loop methane circuits. Each successive stage is at a lower temperature and pressure. The resulting LNG product is pumped to storage where it is stored at near atmospheric pressure and –161°C.

Advanced Controller Design

The objective of the LNG APC controller is to maximize gas feed rate subject to the various process operating constraints. Other objectives such as minimizing the power usage, maximizing NGL production, or maximizing liquefaction thermal efficiency can be explicitly incorporated into this APC controller. To achieve these objectives, the APC controller is designed with seven manipulated variables (MVs), three feed forward or disturbance variables (FFs), and eighteen controlled variables (CVs).

The main manipulated variables are the plant feed rate and the various refrigeration system suction pressures. These pressures are used to balance the loads between the systems to increase the plant production. The controller will typically push the feed up against the high limits on the gas turbine exhaust temperatures. Other manipulated variables like LNG condensing pressure are also included in the controller. The main disturbance variable is the site ambient temperature, which provides the indication to balance the loads and to reduce or increase the production rate. Other process variables, such as, feed gas pressure that affects the overall plant operation, are included as disturbance variables.

There are a total of eighteen controlled variables for the LNG process. The main constraints in the process are the gas turbine exhaust temperatures and compressor speeds, which ensure that the six compressors and associated gas turbines are efficiently operated. An intermediate heat exchanger outlet temperature is controlled to ensure that there is enough sub-cooling for LNG produced. The valve positions are controlled within limits to ensure that the regulatory controllers in the Hysys simulation hold the setpoints, which is a standard practice in advanced control projects.

3. SIMULATION SETUP FOR ADVANCED CONTROL TESTING

The benefits from applying advanced process control to the LNG plant are first analyzed using dynamic simulation. There are several advantages to this including the fact that the behavior of the plant with the controller can be analyzed before the plant startup. Also, this can reduce the time required for commissioning in the field.

The plant-wide dynamic simulation model for a recently constructed LNG plant is the main simulation tool used for this study. This dynamic model includes the entire liquefaction and LNG storage areas of the plant. This model was also used as the base plant model for the Operator Training Simulator delivered to the plant site. The plant-wide dynamic simulation model is fundamental, and it is based on rigorous thermodynamics. For discussion here, a model with level of details to perform engineering studies is assumed. Unlike steady state simulation, dynamic simulation is characterized by the need for accurate pressure profile, equipment volumes and other relevant information. The dynamic model is built based on actual engineering data during the EPC phase of the project and has the actual equipment and process information.

A schematic of the system setup for the APC testing is shown in Fig 1. Here, the offline plant data (ambient temperature) is fed to the MATLAB software. This data is written every 10 seconds from MATLAB to the Hysys plant model to drive the plant behavior. The Hysys model is linked to the DMC Online using the DMC Link. The control calculations are executed in the online software every minute. The DMC Online relays this data to the Graphical User Interface.

Figure 1: System architecture for LNG plant advanced control testing.

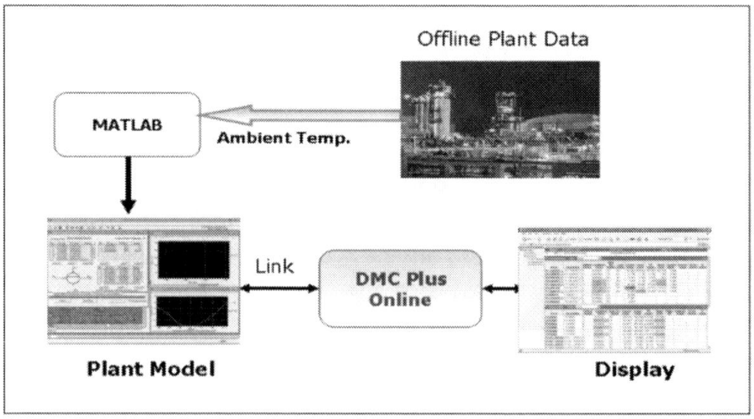

4. SIMULATION RESULTS

The simulations were performed initially to tune the advanced controller and to test its performance. It is well known that a change in ambient temperatures is the main driving factor for optimization in LNG plants. Hence, the APC controller was subjected to this disturbance to study the benefits of optimization. A typical ambient curve profile obtained from the climate data at plant site is used. The ambient temperature increases rapidly during the day. The main effect of this is the imbalance in the loading of the various refrigeration systems.

The benefits from advanced control implementation depend on the comparative performance of an operator in the LNG plant. To mimic a real operator is a very subjective task. Instead, we have tried to study three different operators with different skill levels and analyzed the performance of APC vis-à-vis these operators.

The first simulation that was performed was the closed loop operation of the LNG plant with advanced process control. The controller changes the MVs to maximize the LNG production. The ambient temperature data and the plant feed rate for this case are shown in Fig 2. As can be seen, the increase in ambient temperature causes the plant feed to decrease. The three turbine exhaust temperatures for the 24 hour period are given in Fig 3. Here, the temperatures are always at or close to the constraint. The APC controller is constantly manipulating its handles to balance the refrigeration systems and to operate the plant at its temperature constraints. This is mainly accomplished by manipulating the suction pressure setpoints of the refrigeration systems and the other handles.

The increase in LNG production achieved by using advanced process control depends on the operator actions in the plant. For a highly skilled operator, who makes frequent plant adjustment, a production increase of around 1% is achievable. For an average operator who makes less frequent changes in a day, a production increase of 2-3% is attained from advanced control. This is sufficient economic incentive to justify the implementation of APC in the LNG plant. The intangible benefits like smoother and consistent plant operation also provides additional benefits from APC.

5. CONCLUSIONS

The Advanced control scheme was found to be very effective in improving the operation of the LNG plant. The main objective of the controller is to maximize the LNG production. Other objectives like efficiency and NGL recovery could be considered. To estimate the benefits from APC, three different operator skill levels were considered. For an operator with average skills, a 2 to 3% percent LNG production increase is expected with advanced process control. It needs to be noted here that the benefits from reducing the variability in the constraints is not estimated here. This is possible, since the advanced control improves the dynamic control of variables like exhaust temperatures. This allows these constraints to be moved closer to their true limits, providing additional benefits from advanced control.

Figure 2: Plant Production and Ambient Temperature

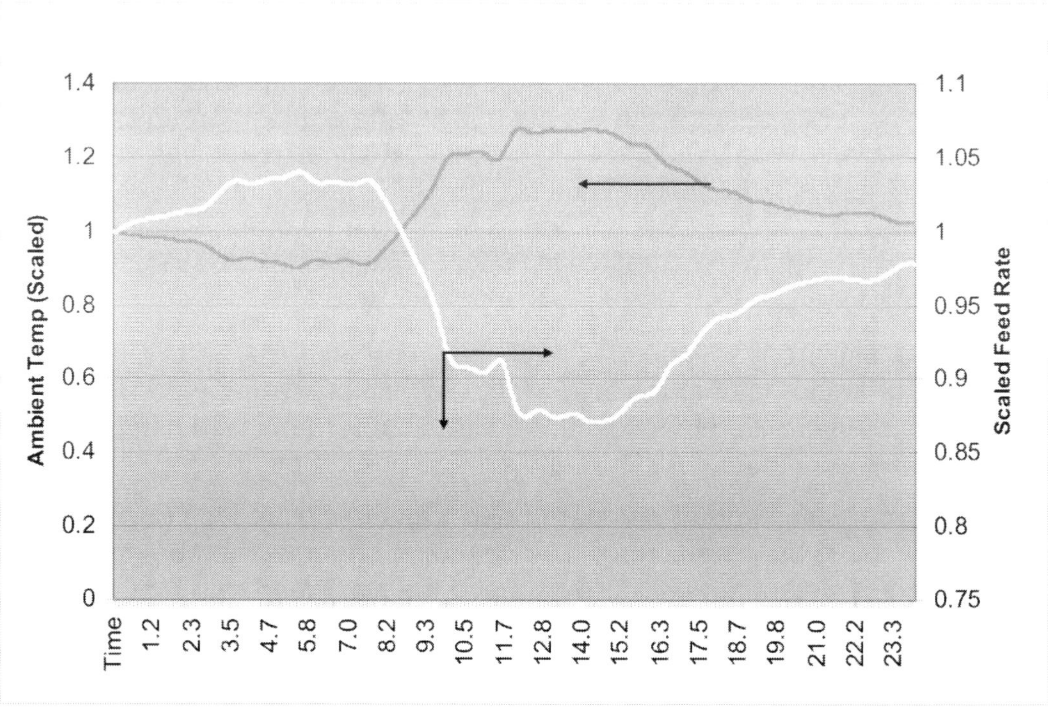

Figure 3: Turbine exhaust temperatures with APC

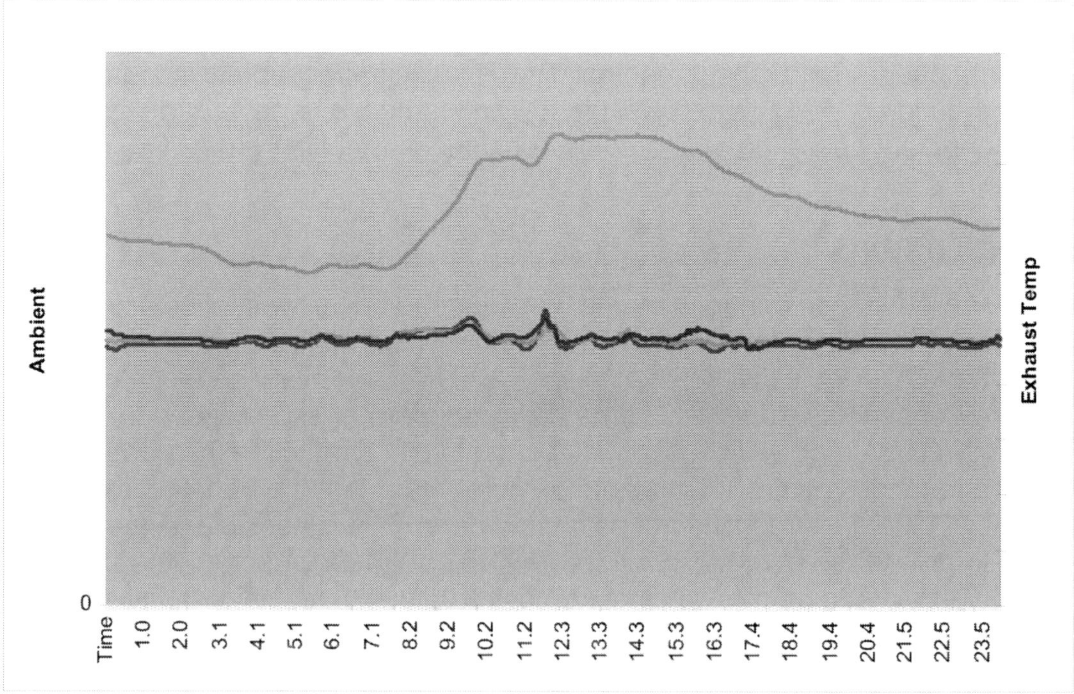

6. REFERENCES

- "LNG Lifecycle Simulation", Jaleel Valappil, David Messersmith and Vibhor Mehrotra, Hydrocarbon Engineering, October 2005.
- "Dynamic simulation and optimization of LNG plants", Jaleel Valappil, Paul Bruner and Vibhor Mehrotra, Aspenworld Conference, 2004, Orlando, FL.
- "A Step change in LNG operations through advanced process control", K. den Bakker, 23rd World Gas Conference, Amsterdam, 2006.
- "Step-test free APC implementation using dynamic simulation", Nicholas Alsop and Jose Maria Ferrer, AIChE Spring National Meeting, 2006, Orlando, FL.

Biography

Jaleel Valappil is a Senior Engineering Specialist in Bechtel's Advanced Simulation and Analysis organization. He is actively involved in building and validating dynamic plant models and utilizing simulation for process and control design improvements, operator training and optimization/advanced control of LNG plants and other units. Jaleel holds a Bachelor in Science from Indian Institute of Technology and a Doctor of Philosophy in Chemical Engineering from Lehigh University.

A Novel Method for Testing the NPSHA Level of a Submerged Centrifugal LNG Pump

Christopher D. Finley
Test Engineering Manager
Adam B. Clinton
Test Engineer, Mechanical
Ebara International Corporation
Sparks, Nevada

AIChE Spring Meeting, April 2007
7th Topical Conference on Natural Gas Utilization
Houston, Texas, 22-26, 2007

ABSTRACT

Net Positive Suction Head Required (NPSHR) is defined as the Net Positive Suction Head (NPSH) at which a pump's total discharge head (first stage head in multi stage pumps) has decreased by three percent (3%) due to cavitation within the pump. Cavitation begins as small harmless bubbles before any indication of loss of head or capacity occurs. This is called the point of incipient cavitation. Testing has shown that between two and twenty times the NPSHR is required to fully suppress incipient cavitation, depending on the impeller shape (specific speed number) and operating conditions. To keep a pump from cavitating the Net Positive Suction Head Available (NPSHA) must be equal to or greater than the NPSHR.

Current methods used for testing the NPSHR of submerged centrifugal LNG pumps cannot be performed in real time while the pump is running. This is due to the fact that current methods assume ideal methane as a working fluid and assume that the test will be performed with the working fluid in a constant, thoroughly saturated state.

This paper presents a novel method of testing NPSHR levels in real time using pure methane or hydrocarbon mixtures. This is accomplished by using custom data acquisition software that calculates the instantaneous fluid properties of the test liquid, including the vapor pressure, based on the chemical composition delivered by the LNG supplier. With an accurate vapor pressure calculation based on the known working fluid composition, more accurate NPSHR levels can be defined.

INTRODUCTION

Cavitation

Cavitation occurs when the local pressure of a liquid is dropped below its vapor pressure due to a flow condition (White). For example, as water moves past the propeller fin shown in Figure 1, the local pressure at the tip of the blade drops low enough to cause the water in that area to boil. The same phenomenon can occur inside a centrifugal pump at the leading edge of the impeller or inducer.

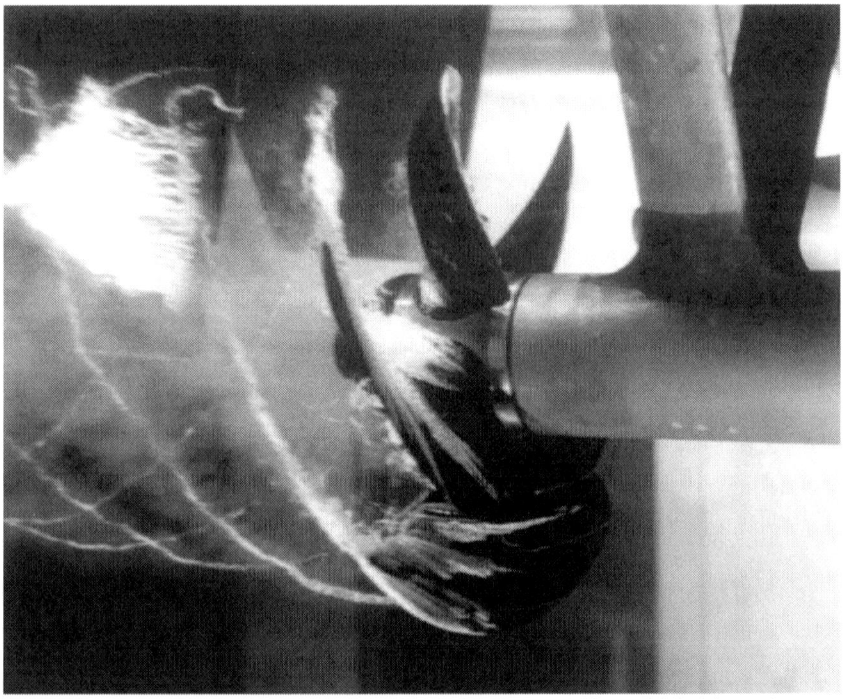

Figure 1: Cavitation of Water Propeller (US Navy)

When cavitation occurs inside a pump, there are some drawbacks that need to be considered. The first is physical damage to the pump caused by the collapse of bubbles formed by a low pressure condition of the operating fluid. Long-term exposure to the repeated collapse of cavitation pockets can erode the material of the impellers and the inducer. Another problem is the loss of discharge pressure-head that accompanies the formation of vapor within the pump. If left to cavitate long enough, the pump will begin to expel vapor along with the working fluid, decreasing the effectiveness of the pump. One problem specifically related to centrifugal pumps that use the working fluid as the cooling medium for the induction motor as well as lubrication for the main bearings, is caused by decreasing the feed fraction of liquid within the pump. Decreasing the volume of liquid flowing through the pump also decreases rate at witch the pump motor can be cooled and the bearings can be lubricated. Running the pump without sufficient cooling or bearing lubrication can lead to long term damage of the motor and bearings.

Net Positive Suction Head – NPSH

When using a centrifugal pump, which can operate upwards of 3600 RPM, to move LNG, which is sometimes pumped near its saturated liquid conditions, cavitation can be a likely occurrence. In order to help prevent cavitation the Net Positive Suction Head (NPSH) must be greater than what is required by the manufacturer's specifications.

$$NPSH = \frac{P_i}{\rho g} + \frac{V_i^2}{2g} - \frac{P_v}{\rho g} \qquad \text{Eq. 1}$$

Equation 1, from White, gives the NPSH value where P_i is the pressure at the pump inlet, V_i is the velocity at the pump inlet, P_v is the vapor pressure of the fluid being pumped, ρ is the density of the liquid being pumped and g is the acceleration of gravity. Reduction of this equation to a simpler form that can be easily measured and calculated using typical data acquisition systems is done by applying the following equations and substituting them into equation 1:

$$P_i = P_T + \rho g h_{LL} \qquad \text{Eq. 2}$$

$$V_i = 0 \qquad \text{Eq. 3}$$

Where P_T is the pressure in the pump run tank and h_{LL} is the liquid level in the pump run tank, above the pump inlet. Equation 1 is then simplified as seen in the following equation:

$$NPSH = h_{LL} + \frac{P_T - P_v}{\rho g} \qquad \text{Eq. 4}$$

NPSH Testing

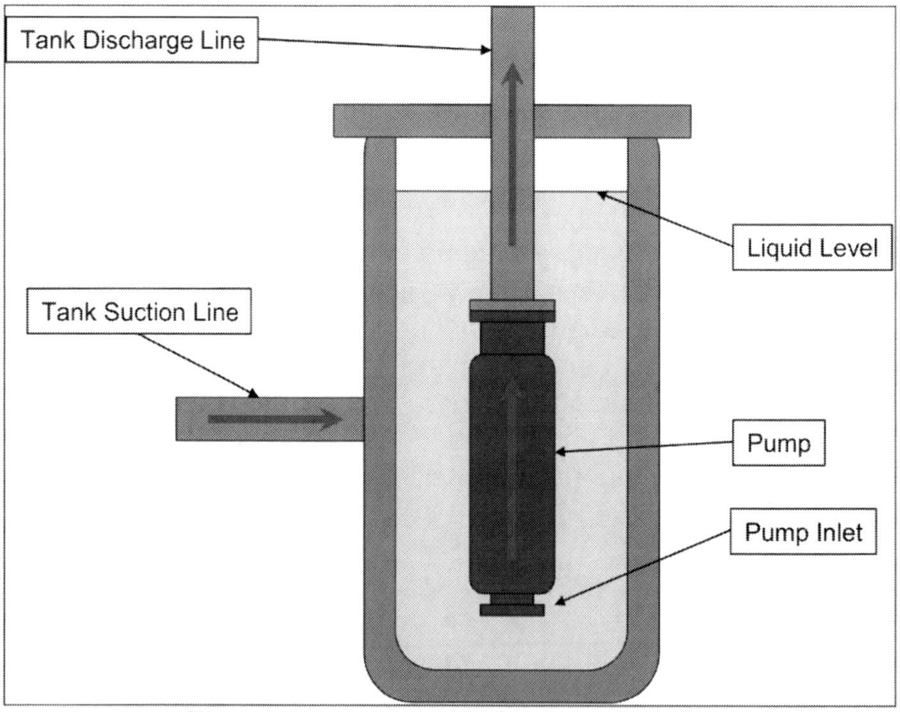

Figure 2: Typical NPSH Test Set-Up

The test method in question involves installing a centrifugal pump into a static run-tank, as depicted in the diagram of Figure 2. Throughout this test, several parameters are monitored and recorded via a data acquisition system. These parameters should include at a minimum, run tank liquid level, run tank pressure, pump suction and discharge pressure, pump suction and discharge temperature, pump flow and motor electrical. Once installed, the pump is started and brought to the necessary flow condition as required by predetermined test parameters. Once the pump flow has stabilized, the suction valve to the pump run tank is slowly closed, reducing the inlet flow to the pump vessel while keeping the pump discharge flow constant. This restriction in inlet flow causes the pump vessel liquid level to reduce. As the liquid level decreases, the pump suction pressure decreases. This decrease in suction pressure leads to the formation of gas pockets in the pumping medium that leads to a decrease in pump head. A 3% drop in head is interpreted as indicating pump performance impairment, or the NPSH available. After this point has been reached, the pump continues to operate with restricted inlet flow until a 40% drop in pump head is observed. At this time the pump is shut down to avoid damage to the motor and hydraulic components of the pump.

METHODS

Current Methods

The current method for measuring and calculating the NPSH level of a pump tested per the preceding section are typically performed using a custom data acquisition system. This method also takes into account several assumptions. The first assumption is that V_i, or the tank liquid level velocity, is negligible and is set to 0. This is due to the relatively large size of the tank as compared to the rate at which the level is reduced. The second assumption is that the run tank pressure is equal to the vapor pressure of the operating fluid. Applying these assumptions to equations 1 and 4 result in the following equation:

$$NPSH = h_{LL} \qquad \text{Eq. 5}$$

The NPSHR (Net Positive Suction Head Required) value is taken as the liquid level in the tank when a 3% drop in head is observed. Although this simplified method of measuring the NPSH level for submerged centrifugal pumps is commonly used by pump manufacturers, several problems exist. The first problem is the assumption that the run tank pressure is equal to the vapor pressure of the operating fluid. During most NPSH tests described previously, this assumption is incorrect. Due to the limited cooling and venting capacity of the run tank together with the heat input to the operating fluid from the electric motor, the fluid in the run tank cannot reach equilibrium. In most cases, this causes the run tank pressure to be higher than the vapor pressure of the operating fluid.

A further problem with applying the simplified NPSH equation is that the run tank liquid level measurement may not be accurate depending on the condition of the liquid in the pump run tank. The most common method of measuring liquid level in a tank used today is to measure the differential pressure between the top and bottom of the tank. This pressure is then converted to liquid level mathematically utilizing the estimated density of the liquid in the tank. During a typical NPSH test, the density of the operating liquid is not constant due to temperature and pressure fluctuations. These changes in density are not typically taken into account during the liquid level calculation, so the liquid level measurement during the NPSH test can be considered inaccurate.

New Method

The following section describes a novel method for testing the NPSH level of a submerged centrifugal pump that addresses the problems with the current method of testing. To apply this new method, the exact composition of the operating fluid must be determined. Generally, this composition is confirmed by the test liquid supplier but may it be determined experimentally. Once the composition of the fluid is confirmed, a saturation pressure curve plotted as a function of temperature of the mixture must be created. This can be performed using several thermodynamic software packages. For the example presented here, the National Institute of Standards and Technology's "Supertrap" software was used. Figure 3 shows this data for a typical LNG mixture.

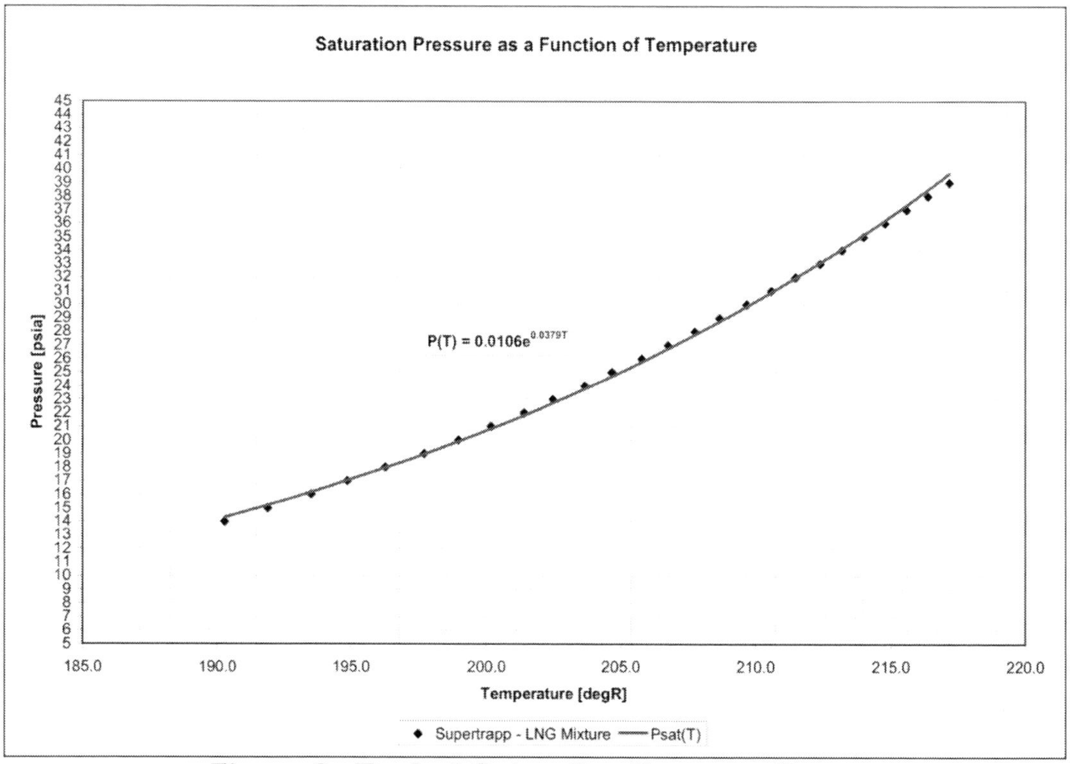

Figure 3: Typical Saturation Pressure Curve

Once the saturation curve is generated, an exponential trend line is fit to the data. For this example, this exponential function is:

$$P_{sat}(T) = 0.0106e^{0.0379T} \quad \text{Eq. 6}$$

Using equation 6, the vapor pressure of the liquid mixture can now be calculated throughout the NPSH test.

The next step in applying this new method is to calculate the density of the operating fluid mixture. Because the density of the fluid is a function of both pressure and temperature, a table of densities must be generated. For this example, the "Supertrap" software was also used for this purpose. Figure 4 shows a graphical representation of the density of typical LNG as a function of pressure and temperature.

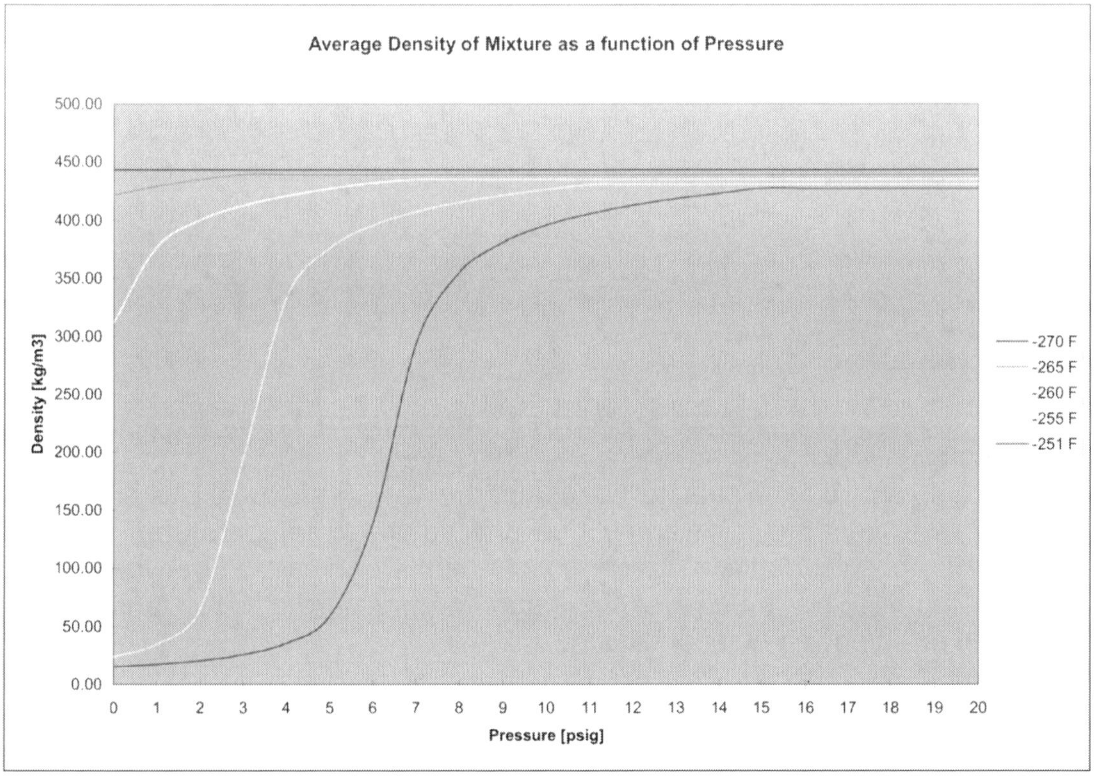

Figure 4: Liquid-Vapor Mixture Density

Once this table is generated, the density of the fluid mixture can be interpolated using the known pressure and temperature of the fluid throughout the NPSH test. Now that the vapor pressure and density are know for every data point throughout the NPSH test, equation 4 can be used to determine a more accurate NPSH level.

TEST RESULTS

With the equation for the saturation pressure as a function of temperature known for a range of typical operating temperatures, the calculations necessary to perform the new NPSH measurement were applied to old test data. The NPSH values generated with the new method were then compared with the old method to determine the amount of improvement achieved.

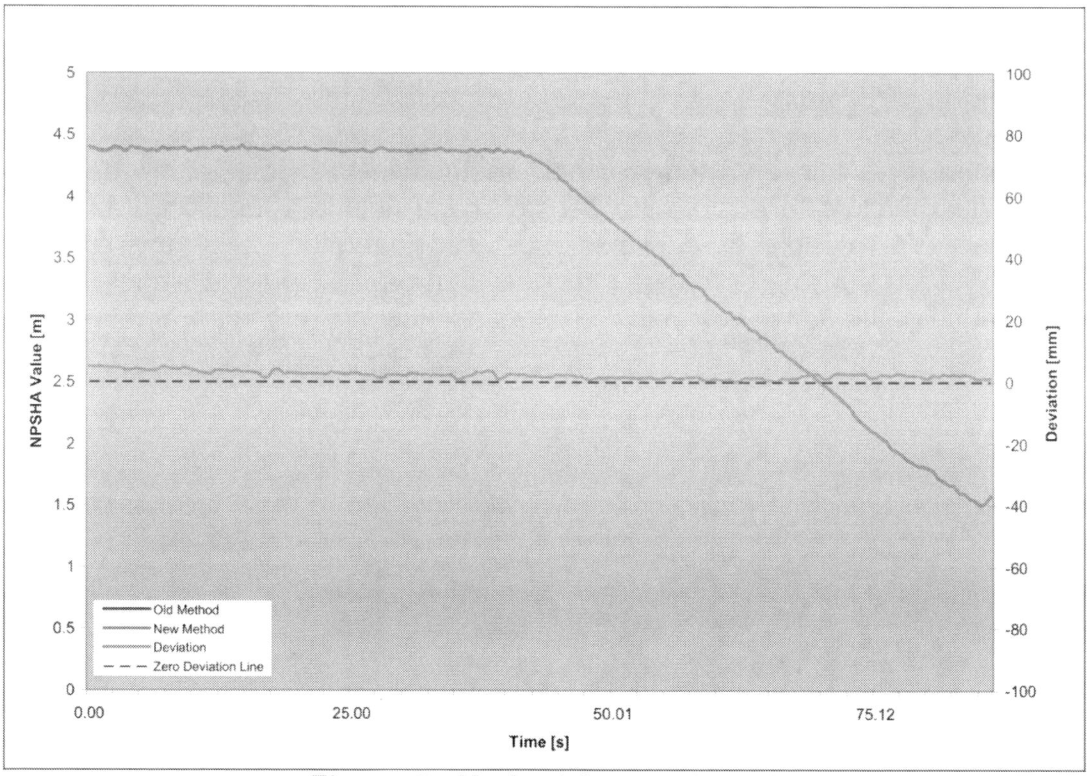

Figure 5: Method Comparison 1

The first set of test data to be compared was from a test where the suction temperature was kept slightly warmer than the saturated temperature. As evident in Figure 5, these conditions show a very small amount of deviation between the old and new test methods. This is because the saturation pressure and the tank pressure are almost equal for the given test conditions.

The second set of data that was compared was from a test where the suction temperature was approximately 5°C colder than the measured saturation temperature. For this case, the difference between both methods was much greater. As shown in Figure 6, the deviation between the two methods was, at its greatest, 48.8 mm.

The third set of data used to compare the different methods was from a test where the tank pressure and temperature fluctuated during the test. Figure 7 shows that as these fluctuations were nevertheless accounted for and corrected to their actual value with the improved test method.

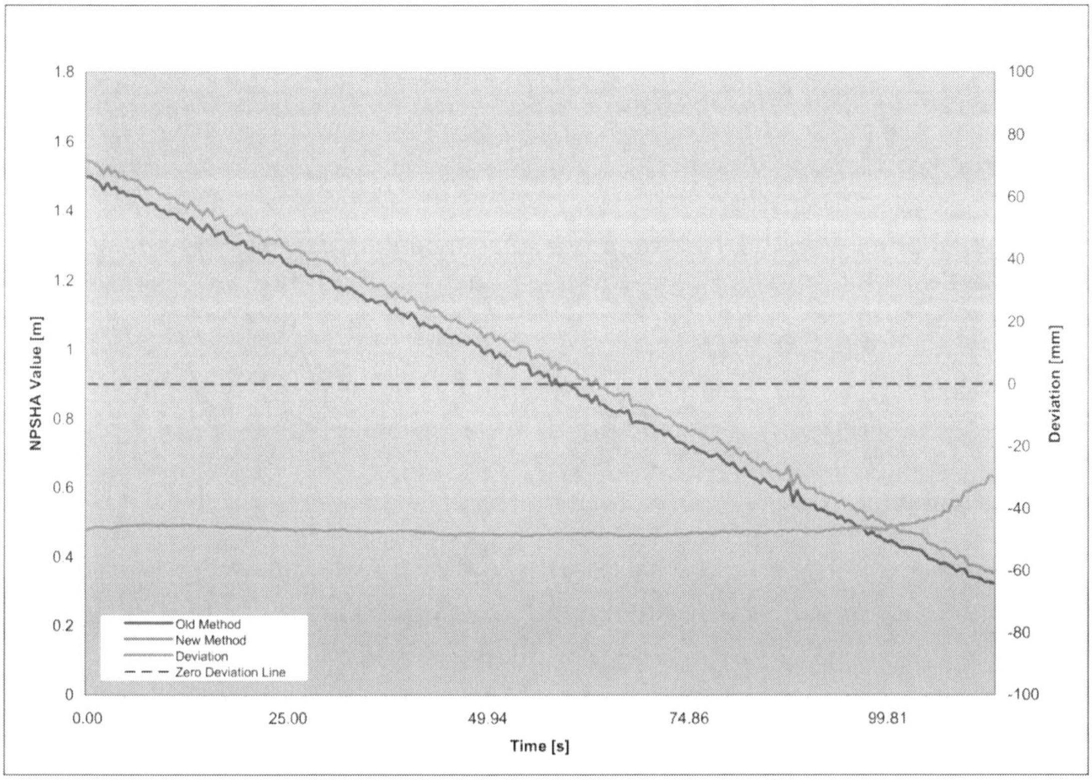

Figure 6: Method Comparison 2

Figure 7: Method Comparison 3

The third set of data used to compare the different methods was from a test where the tank pressure and temperature fluctuated during the test. Figure 7 shows that as these fluctuations were nevertheless accounted for and corrected to their actual value with the improved test method.

CONCLUSIONS

When comparing the two methods of calculation for NPSH levels, a relatively small difference was noted. For the sets of data that had the new calculation method applied, the largest observed difference between the two methods of calculation was 48.8 millimeters. Since the NPSHR values are typically given in meters, improvements in their measurement by millimeters are arguably negligible. Nonetheless, the new method of calculation is an improvement of the accuracy of the test.

FUTURE WORK

The next step to be taken will be to apply the equations to the in-house test measurement software. By rewriting the existing code to include the new NPSH calculations, more accurate NPSH levels can be measured.

REFERENCES

- API Standard 610, 8th Edition. August 1995. “Centrifugal Pumps for Petroleum, Heavy Duty Chemical, and Gas Industry Services.”
- US Navy Image. David Taylor Model Basin. Controllable-Pitch Propeller in Large Cavitation Channel. http://www.dt.navy.mil/div/about/galleries/gallery4/095.html.
- White, Frank M. Fluid Mechanics, 5th Edition. 2003. McGraw-Hill Companies Inc. New York, NY.

BIOGRAPHIES

Christopher Finley

Christopher D. Finley is Test Engineering Manager of Ebara Intl. Cryodynamics division. Mr. Finley earned a B.S. degree in mechanical engineering from the University of

Nevada, Reno and is currently enrolled in graduate courses also at the University of Nevada, Reno.

During his 11 year tenure at Ebara, Chris has been involved in numerous research and development projects including the development of the first two-phase LNG expander. As Test Engineering Manager, Chris has lead several projects involved with the advancement of testing technology of cryogenic pumps and expanders, including novel heat exchanger designs and controls, variable speed induction motor control and development of data acquisition software.

Adam Clinton

Adam Clinton is a Test Engineer for Ebara International Corporation. His responsibilities include conducting performance tests on Centrifugal Liquefied Gas Pumps as well as the research and development of improved test methods.

A recent graduate of the University of Nevada, Reno, Adam has extensive experience in the field of research and development. He worked for two and a half years as an assistant development technician at the Nevada Terawatt Facility (NTF), a research lab for the Physics Department at the University of Nevada. There, he designed and fabricated several pieces of experimental equipment to be used in conjunction with the Zebra Pulsed-Power Generator as well as the Leopard and Tomcat Laser systems. His senior year as a student, he was awarded a grant from the University of Nevada Office of Undergraduate Research and the National Science Foundation to perform his own research at the Composite and Intelligent Materials Laboratory (CIML). There, under the guidance of Dr. Faramarz Gordaninejad, he designed and conducted experiments to characterize the electrical and material properties of new "Smart Materials" known as Magneto-Rheological Elastomers (MRE's). Adam completed his education in May of 2006, graduating with a Bachelor's Degree in Mechanical Engineering.

Effect of Quadratic Fluid Damping in Two-Phase Liquefied Natural Gas

Andrew Kimmel
Research Advisor
University of Nevada Reno
Department of Physics
Reno, Nevada

AIChE Spring National Meeting 2007
7th Topical Conference on Natural Gas Utilization
Houston, Texas, April 22-27, 2007

Abstract

Non-linear oscillations occur in two-phase LNG fluid flows. Due to its compressibility the vapor bubbles act as a spring with an asymmetric non-linear characteristic. Compressing pressure fluctuations in two-phase liquid-vapor LNG cause less displacement than expanding pressure fluctuations.

The displacement of the vapor bubbles and the surrounding liquid are subject to quadratic fluid damping due to the viscosity of the LNG. The effect of the quadratic damping depends on the ratio of LNG liquid to vapor, the ratio of pressure fluctuations over average pressure and on the exciting frequency of the pressure fluctuations.

The paper presents and discusses approximate steady state solutions for pressure induced two-phase fluid oscillations with quadratic damping. Resonance criteria and damping effects of LNG two-phase flows on plant safety and operation are discussed.

Introduction

The following introduction is an extract from Kimmel[(4)]: "Pressure induced non-linear oscillations in two-phase LNG pipe-flow". Two-phase LNG is a mixture of liquefied and gaseous natural gas. It forms in LNG liquefaction plants during the condensation of the saturated natural gas or during the expansion of the already condensed pressurized natural gas. The volumetric ratio r between gaseous and liquid natural gas is defined as

$$r = \frac{V_g}{V_g + V_l}$$

with $r = 0$ for pure liquid and $r = 1.0$ for pure gas. V_g is the volume of gas and V_l is the volume of liquid. With very good accuracy, the liquefied natural gas can be assumed to be incompressible when compared to the compressible gaseous natural gas. Gaseous natural gas follows approximately the ideal-gas equation of state with the gas volume V_g inversely proportional to the gas pressure p at

$$\frac{p+\Delta p}{p}=\frac{V_g}{V_g+\Delta V_g}$$

and can also be written in the form:

$$\Delta p=-p\frac{\Delta V_g}{V_g+\Delta V_g}$$

$$V_g=r$$

and for very large Δp the gas condenses. Therefore the maximum volumetric change is $\Delta V_g = -r$ for increasing pressures (compression) and $\Delta V_g = +\infty$ for decreasing pressures (expansion) with $p+\Delta p \to 0$ approaching vacuum.

$$\Delta p= -p$$

The density of the gaseous natural gas is negligibly small compared to the density ρ of the liquefied natural gas. Therefore the oscillating mass m within the controlled volume $V_g+V_l=1$ depends only on the volume of the liquid V_l.

$$m=\rho\ V_l=\rho(1-r)$$

For pipe flows with constant cross sectional area the axial displacement z of the fluid is linearly dependent to the volumetric change

$$z\approx\Delta V_g$$

Figure 1 demonstrates the asymmetric oscillations in a two-phase pipe flow. For a given absolute pressure difference the absolute displacement $\Delta z(t)$ is smaller for compression than for expansion. Due to the fluid viscosity the movement of gaseous and liquid fluid are subject to friction forces resulting in a damping effect proportional to the square of the displacement velocity.

Steady state damped oscillations in two-phase flow

Pressure changes Δp act as a force in the axial direction of the pipe. The compressible gaseous volume and the mass of the liquid volume form an oscillating system. Friction forces generated by fluids are opposed to the velocity vector and proportional to the square of the displacement velocity multiplied by the damping factor D. The exciting force Gf^2 depends on the exciting amplitude G and the square of the exciting frequency f.

In the case for a forced oscillation with harmonic excitation and quadratic fluid damping the following non-linear differential equation describes the pressure-induced oscillation:

$$(1-r)\rho\frac{d^2 z(t)}{dt^2}+D\frac{dz(t)}{dt}\left|\frac{dz(t)}{dt}\right|+p\frac{z(t)}{r+z(t)}=Gf^2\cos(ft)$$

This non-linear differential equation cannot be solved by analytical methods. Numerical methods generate only non-algebraic solutions and a proposed approximation method described by Kimmel[4] is used to calculate algebraic approximate solutions for the steady state non-linear damped oscillation.

In case of cylindrical pipe flows r has the dimension of a length and a dimensionless displacement $y(t)$ is introduced:

$$y(t)=\frac{z(t)}{r}$$

$$r(1-r)\rho\frac{d^2 y(t)}{dt^2}+r^2 D\frac{dy(t)}{dt}\left|\frac{dy(t)}{dt}\right|+p\frac{y(t)}{1+y(t)}=Gf^2\cos(ft)$$

The following parameters are introduced to simplify the equation: the natural frequency f_N, the damping parameter C and the exciting amplitude parameter F

$$f_N=\sqrt{\frac{p}{r(1-r)\rho}}$$

$$C=\sqrt{\frac{rD}{(1-r)\rho}}$$

$$F=\frac{G}{r(1-r)\rho}$$

are substituted into the differential equation:

$$\frac{d^2 y(t)}{dt^2}+C^2\frac{dy(t)}{dt}\left|\frac{dy(t)}{dt}\right|+f_N^2\frac{y(t)}{1+y(t)}=Ff^2\cos(ft)$$

The proposed method is using Fourier coefficients for the periodic error to generate an approximate analytical solution for the steady state case of pressure induced oscillations in two-phase pipe flow, Kimmel [4].

If $y(t)$ is the exact solution then let $x(t)$ be an approximate solution for the above-described differential equation.

$$x(t) \approx y(t)$$

Then $E(t)$ is the function of the error for this approximate solution $x(t)$.

$$\frac{d^2x(t)}{dt^2} + C^2 \frac{dx(t)}{dt}\left|\frac{dx(t)}{dt}\right| + f_N^2 \frac{x(t)}{1+x(t)} - Ff^2 \cos(ft) = E(t)$$

The error $E(t)$ can be expanded in a Fourier series for periodic functions of the error.

$$E(t) = \frac{1}{2}a_0 + \sum_{k=1}^{\infty}[a_k \cos(kft) + b_k \sin(kft)]$$

The Fourier coefficients a_k and b_k are determined by the following integration:

$$a_k = \int_0^{2\pi} E(t)\cos(kft)d(ft)$$

$$b_k = \int_0^{2\pi} E(t)\sin(kft)d(ft)$$

By setting certain a_k and certain b_k equal to zero, the error for the k-th term of the Fourier series $E(t)$ becomes zero. The approximate solution $x(t)$ is then correct for the k-th term of the function $E(t)$.

Approximate solution for the steady state damped oscillating system

For the given differential equation with forced oscillation through harmonic excitation

$$Ff^2 \cos(ft + \varphi)$$

The approximate solution $x(t)$ corresponds to the same periodic function but with different amplitude A and with a phase difference φ between excitation and approximate solution.

$$x(t) = S + A\cos(ft)$$

By substituting $x(t)$ into the differential equation the Fourier coefficients a_0 , a_1 and b_1 of the error $E(t)$ can be determined by integration:

$$E(t) = -Ff^2\cos(ft+\varphi) - Af^2\cos(ft) - C^2f^2A\sin(ft)\left|-A\sin(ft)\right| + f_N^2\frac{S+A\cos(ft)}{1+S+A\cos(ft)}$$

$$a_0 = \int_0^{2\pi} E(t)d(ft)$$

$$a_1 = \int_0^{2\pi} E(t)\cos(ft)d(ft)$$

$$b_1 = \int_0^{2\pi} E(t)\sin(ft)d(ft)$$

The integration of these equations leads to the following functions:

$$a_0 = a_0(A,S) \qquad a_1 = a_1(A,S,\varphi) \qquad b_1 = b_1(A,\varphi)$$

Setting all three functions equal to zero produces approximate solutions for A, S and φ as functions of the frequency f of the exciting force. The solutions are in implicit form with S, f and φ as functions of the amplitude A.

$$S = -1 + \sqrt{1+A^2}$$

$$\frac{f}{f_N} = \sqrt{\frac{6\pi(-1+\sqrt{1+A^2})(3\pi A \pm \sqrt{9F^2\pi^2 - 64A^4C^4})}{A(64A^4C^4 + 9\pi^2(A^2 - F^2))}}$$

$$\sin\varphi = \frac{8C^2A|A|}{3F\pi}$$

The oscillating system with quadratic damping has only a steady state solution if the exciting force F meets the following condition

$$F \geq F_{\min} = \frac{8A^2C^2}{3\pi}$$

For $F=F_{min}$ the formulas for the frequency and the phase become

$$\frac{f}{f_N} = \frac{\sqrt{2}}{A}\sqrt{-1+\sqrt{1+A^2}} \qquad \sin\varphi = \frac{|A|}{A}$$

The approximate actual displacement $X(t)$ is

$$X(t) = rx(t) = r[S + A\sin(ft)]$$

The energy losses L_E for one period due to quadratic damping are calculated by

$$L_E = D\int_0^X \left|\left(\frac{dX(t)}{dt}\right)^2 dX(t)\right|$$

$$L_E = r^3 D\int_0^{2\pi} \left|f^2(-A\sin(ft))^3 d(ft)\right| = \frac{8}{3}r^3 DA^3 f^2$$

The largest resonance amplitude A_{max} occurs for $F=F_{min}$ for the critical frequency f_C:

$$A_{\max} = \sqrt{\frac{3\pi F_{\min}}{8C^2}}$$

$$f_C = f_N \frac{4C}{\sqrt{3\pi F_{\min}}}\sqrt{-1+\sqrt{1+\frac{3\pi F_{\min}}{8C^2}}}$$

Substituting the above defined expressions for C and f_N, with

$$F_{\min} = \frac{G}{r(1-r)\rho}$$

into the energy loss equation L_E generates the expression for the maximum energy loss L_{EC} at the critical frequency f_C as a function of the volumetric vapor content r

$$L_{EC} = \left(\frac{4p}{\rho}\sqrt{\frac{2\pi}{3}}GD\right)\left(-1+\sqrt{1+\frac{3\pi G}{8D^2 r^2}}\right)\left(\frac{r}{1-r}\right)$$

The maximum energy loss L_{EC} has a minimum value for the volumetric vapor content r_{EC}:

$$r_{EC} = \frac{1}{2}\left(1-\frac{3\pi G}{8D}\right)$$

The smallest damping effect or energy loss L_{ECmin} occur for the volumetric vapor content r_{EC} at the resonant amplitude for the critical frequency f_{Cmin} .

$$L_{EC\min} = \frac{8p\sqrt{6\pi^3 G^3 D}}{\rho(8D + 3\pi G)}$$

The critical frequency is a function of the vapor content r and has its minimum value f_{Cmin} at $r = r_{EC}$:

$$f_{C\min} = 4\sqrt{\frac{2p}{\rho}}\sqrt{\frac{D}{8D + 3\pi G}}$$

$$A_{C\min} = \frac{8\sqrt{\frac{3\pi G}{2D}}}{8D - 3\pi G}$$

Review of the approximate solution

Figure 2 shows the graph of the absolute amplitude A as a function of the exciting frequency f/f_N for different values of the damping parameter C and with an exciting amplitude parameter F = 0.1.

Figure 3 shows a typical example of a steady state asymmetric oscillation with quadratic damping in a two-phase pipe flow

Figure 4 shows a typical spectrum of the steady state oscillations for different values of the exciting amplitude parameter F.

Conclusions

- Oscillations in two-phase fluids with quadratic fluid damping demonstrate a complicated pattern
- The ratio between exciting frequency and natural frequency is an important parameter. The natural frequency has a minimum value if the volumes for gas and liquid are equal
- The resonant amplitude occur at a critical frequency below the natural frequency
- The damping effect and energy dissipation depends on the ratio between the volumes for gas and liquid.
- For a certain ratio between the gas and liquid volumes the damping effect or energy dissipation has a minimum value. For a small damping effect the oscillations are more critical and have to be closely monitored to maintain plant safety and reliable plant operation.

- Exciting frequencies above the natural frequency are less critical than exciting frequencies below the natural frequency.
- Due to the damping effect and the energy dissipation the two-phase LNG is heating up and increasing the amount of vapor and the oscillation patterns are changing with the ratio between gas or vapor and liquid

References

1. Henry, R.E. "Pressure Wave propagation in two-phase mixtures" Heat transfer Minneapolis, Chemical Engineering Progress Symposium Series No.102, 1970, Volume 66, American Institute of Chemical Engineers, New York.
2. Tse, Francis S. et al. "Mechanical Vibrations", Allyn and Bacon Inc., Boston, MA 02210, 1978, ISBN 0-205-05940-6
3. Rao, S.S. "Mechanical Vibrations" Pearson Prentice Hall, New Jersey 07458, 2004, ISBN 0-13-048987-5
4. Kimmel, A. "Pressure induced non-linear oscillations in two-phase LNG pipe flow", AIChE Spring National Meeting 2006, Orlando, FL, April 23-27, 2006.

Biography

Andrew Kimmel is graduating at the University of Nevada, Reno, College of Science, in Physics and Mathematics. His research contributions include analyses of non-linear oscillations, and steady state solutions for non-linear vibrations with applications in cryogenic fluid dynamics. He is a member of the German Physical Society and participates in the German Research Group for Non-Linear Phenomena in Physics and Engineering. He is also a member of the Golden Key International Honour Society.

High Expansion (Hi-Ex) Foam Application for Controlling LNG Pool Fire – Experiment Results and Analysis

Jaffee Suardin[a], Yanjun Wang[a], Mike Wilson[b], Sam Mannan[a]
[a]Artie McFerrin

[a]Department of Chemical Engineering
Mary Kay O'Connor Process Safety Center
The Texas A&M University System
College Station, TX 77843-3122

[b]ANGUS Fire
Thame Park Road
Oxfordshire, United Kingdom

AIChE Spring Meeting, April 2007
7[th]Topical Conference on Natural Gas Utilization
Houston, Texas, 22-26, 2007

Introduction

BP Sponsored LNG Research

A new LNG emergency response training facilities has been constructed as the BP Global LNG sponsorship to Texas A&M University (TAMU) Emergency Services Training Institute (ESTI). The facilities are located at Texas Engineering Extension Services (TEEX) Brayton Fire School where emergency responses training and researches have been conducted. In line with that effort, BP Global LNG has also joined forces with the Mary Kay O'Connor Process Safety Center (MKOPSC) of the Texas A&M University to perform research on LNG vapor dispersion and pool fire with the application of water curtain and high expansion foam as control and recovery measures. One of the main focuses is to study Hi-Ex foam application and other alternative method to suppress LNG vapor and pool fire. ANGUS Fire facilitated the Hi-Ex foam generator, personnel, and instrumentation needed.

The objectives of the experiment are:

- To measure the effectiveness of high expansion (Hi-Ex, 500:1) foam in controlling LNG pool fire in two (2) different types of pits.
- To validate the effectiveness of the recommended Hi-Ex foam application rate (10 L/min/m^2) and to compare it with lower foam application rate (3.5 L/min/m^2 and 7 L/min/m^2). This is to find the feasibility of having more economic Hi-Ex foam application rate.

This paper focuses on the effectiveness of the Hi-Ex foam controls LNG pool fire by measuring the heat radiation flux reduction produced by applying the Hi-Ex foam on LNG pool fire with different application rate on different types of LNG pit.

Angus fire provided its latest Turbex fixed Hi-Ex foam generator and instrumentation to measure the Hi-Ex foam application rate released by the foam generator and the heat radiation flux.

Hi-Ex Foam Application on LNG

Since LNG is cryogenic, it readily receives heat from its surroundings. When LNG is spilled into the concrete pit which is usually at ambient temperature, concrete provides heat in addition to the heat from the surroundings. This supports the LNG to evaporate. When the Hi-Ex foam is applied and enveloping all LNG pool surface, LNG vapor penetrates upward through the foam until released into the atmosphere. Cryogenic temperature of LNG provides high temperature difference with the Hi-Ex foam. The heat transfer between the LNG vapor and the water in the Hi-Ex foam creates ice tubes along the LNG vapor path. LNG vapor is warmed during the heat exchange along the frozen flow path, the vapor density is reduced and LNG vapor becomes more buoyant, thus disperses more upward instead of downwind. This is how the Hi-Ex foam application reduces the hazard distance of LNG vapor dispersion, as presented in Figure 1.

When LNG vapor finds an ignition source, LNG pool fire is started. Heat radiation from the fire maintains and increases the evaporation thus increases the fire size further. Hi-Ex foam acts as insulation on the LNG pool surface and prevents the oxygen and the heat radiation from reaching the LNG pool. This phenomenon is shown in Figure 1.

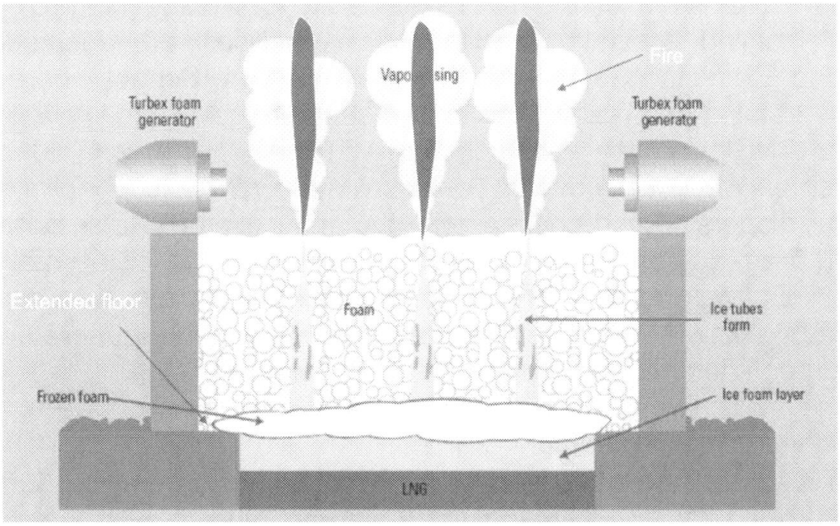

Figure 1. Hi-Ex foam on LNG liquid and pool fire (adapted from [1])

In summary, there are three method of Hi-Ex foam to be effective in suppressing LNG vapor dispersion and fire [2]:

- It reduces the radiation and convective heat input from surroundings to the LNG pool surface.
- It reduces the LNG vapor density by increasing its temperature while penetrating through Hi-Ex foam.
- It reduces the fire growth by lowering the oxygen concentration or flow to the burning LNG pool.

Experimental Setup

Five experiments have been conducted on October 2005 and April 2006. The summary of the experiment parameters is provided in Table 1.

Table 1. Foam application experiment on LNG pool fire on 2005 and 2006.

Pit Area	Foam Application Rate	Equivalent Pool Diameter	Year	Radiometer (x Pool Diameter)
m^2	$L/min/m^2$	M		
45	7	7.57	2006	4.0
65	3.5	9.10	2005	3.3
65	7	9.10	2006	3.3
65	10	9.10	2005	3.3
65	10	9.10	2006	3.0

Two pits (dikes) were applied to simulate industrial LNG spills. The bigger pit is called the "65 m^2" pit, as shown in Figure 2, while the smaller one is the "45 m^2 pit", as shown in Figure 3. The 65 m^2 pit is four feet deep, and the 45 m^2 pit is eight feet deep with a four feet wall above the ground. Side views of these two pits are presented in Figure 4.

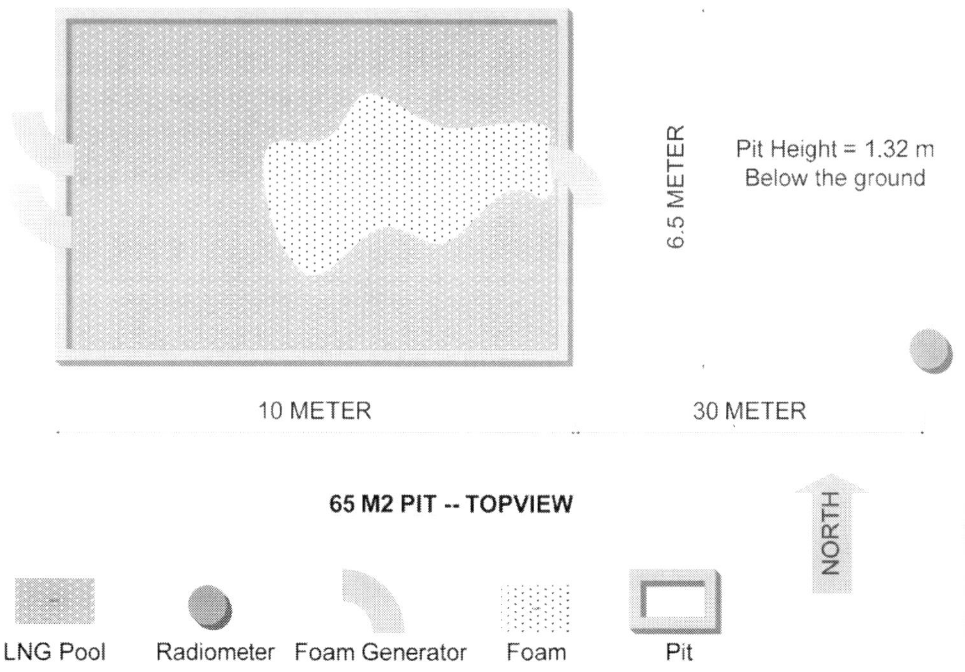

Figure 2. Experiment layout for the 65 m^2 pit

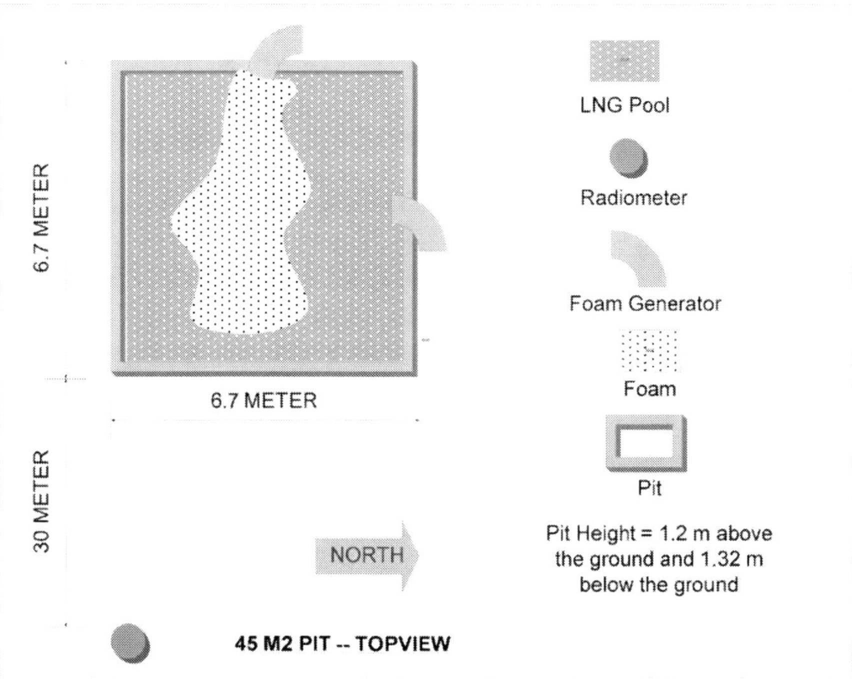

Figure 3. Experiment layout for the 45 m^2 pit

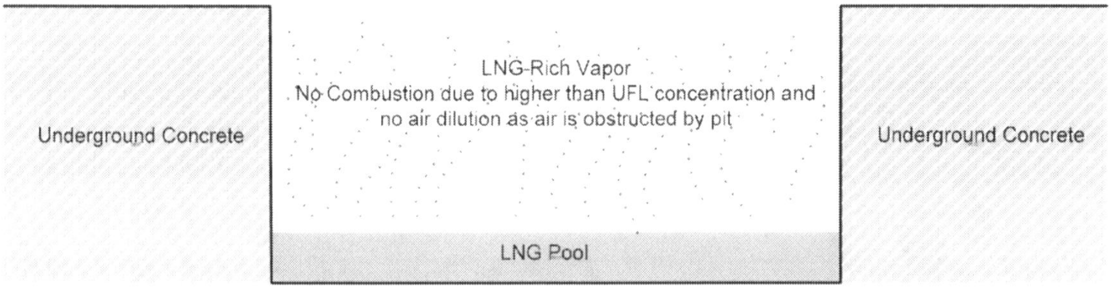

65 m^2 Pit

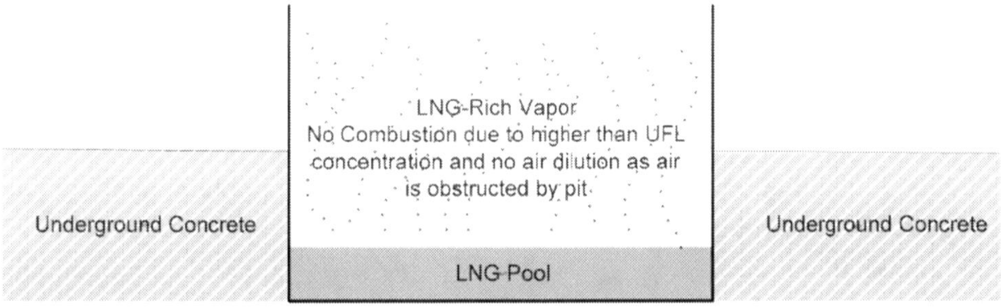

45 m^2 Pit

Figure 4. A side view of the pits

There are four scenarios performed on the 65 m^2 pit in 2005 and 2006 with application rate of 3.5 L/min/m^2, 7 L/min/m^2, and 10 L/min/m^2. The 10 L/min/m^2 is the recommended foam application rate, thus in both years experiment with 10 L/min/m^2 were tested. The experimental layout is shown in Figure 2.

There was only one experiment conducted for 45 m^2 pit which was using foam application rate of 10 L/min/m^2. It was performed on April 20th 2006. The experiment layout is presented in Figure 3.

References

1. LNG hot spot, Frontiers, August 2005, p. 15
2. Takeno, K., Ichinose, T., Tokuda, K., Ohba, R., Yoshida, K., and Ogura, K. (1996). Effects of high expansion foam dispersed onto leaked LNG on atmospheric diffusion of vaporized gas. Journal of Loss Prevention in Process Industries, 9(2), p. 125-133

Biography

Jaffee Suardin is a PhD candidate at the Mary Kay O'Connor Process Safety Center (MKOPSC) of the Artie McFerrin Department of Chemical Engineering at Texas A&M University (TAMU). His research is directed by Professor Sam Mannan, the director of MKOPSC, and he is expected to graduate on 2008.He can be reached at:

Mary Kay O'Connor Process Safety Center
The Texas A&M University System,
3122 TAMU, College Station, TX 77843-3122.
Phone: (979) 862-1310 Fax: (979) 458-1493 Cell: (979) 739-2251
Email: jaffee.suardin@chemail.tamu.edu

He started his chemical engineering background at Institut Teknologi Bandung (Bandung Institute of Technology, ITB) in Indonesia. In 2005, he obtained a master degree in chemical engineering at the MKKOPSC. His current research focuses on high expansion foam application and other alternatives methods to suppress LNG vapor dispersion and pool fire.

He has been working on the computer program development on the process safety methodologies. He developed Layer of Protection Analysis (LOPA) expert system. Currently, he is in the final steps of the development of the Dow Fire and Explosion Index and the Dow Chemical Exposure Index computer program.

In addition to research, he has gained industrial safety experiences from having summer internship at Shell International Exploration and Production (SIEP, Shell Oil Company). He developed computer program that is able to perform the fire and explosion assessment for the oil and gas FPSO facilities. He is also performed several consulting projects in the area of process design and safety engineering with the MKOPSC.

As part of his PhD curriculum, he was the teaching assistant for classes such as process safety engineering, process design, gas processing and process simulation with AspenPlus.

Bulk Temperature Profile of LNG Spill on Unconfined Water – Experiment Results and Analysis

Benjamin Cormier, Yanjun Wang, and M. Sam Mannan*
Mary Kay O'Connor Process Safety Center
Artie McFerrin Department of Chemical Engineering Department
Texas A&M University System
College Station, Texas 77843-3122, USA
*(979) 862-3985; mannan@tamu.edu

AIChE Spring Meeting, April 2007
7th Topical Conference on Natural Gas Utilization
Houston, Texas, 22-26, 2007

Abstract

An LNG experiment has been conducted at Texas A&M University's Brayton Fire Training Field in April 2006. LNG was spilled onto unconfined water and the water temperature profile was recorded at twelve different points with temperature sensors in the water.

In the 70s and 80s, many tests were performed in order to estimate the boil-off rate of LNG on water. Most of these tests were performed in boiling cell and with small quantities in a laboratory. In a confined space, the formation of ice underneath the LNG layer happens rapidly. Therefore it changes the boil-off rate and the behavior of LNG on water. In an unconfined space, the formation of ice is reduced because of the movement of the water underneath the LNG layer.

In this paper, experimental data are analyzed and presented. Bulk temperature and the estimated evaporation rate are discussed and compared with currently available models.

Introduction

Energy demands have caused a sharp increase in the construction of LNG terminals. LNG is mostly transported over seas by ship. Each ship carrier is capable of transporting more than 140 m^3 of LNG. In the case of an incident, LNG could be released onto water and pose fire or explosion hazard. Since 1970, many entities have performed experiments to analyze the behavior of LNG spills on water but still there are some uncertainties.

Once LNG is spilled onto water, it starts boil off rapidly due to its cryogenic nature. The analysis of LNG boiling off is one key component in the vapor dispersion modeling. In this paper, the experiment and analysis of LNG boil-off rate over an unconfined space of water is presented. The boiling-off rate was estimated from the temperature profile of the water underneath LNG layer. The results were compared to theoretical estimation and experimental values from smaller spills. The results were also compared to the values used in the computational modeling as well.

Background

Once LNG is spilled onto water, because of temperature difference between the LNG layer and water, the LNG layer goes through several stages of boiling. There are three types of boiling related to the temperature difference. First, film boiling occurs with large temperature difference. Then meta-boiling starts when the temperature difference decreases. Finally it goes to nucleate boiling with small temperature difference.

The estimation of LNG evaporation rate represents a critical input to the vapor dispersion modeling. Since 1970, many tests were performed. Bureau of mines [1, 2], AGA [3], and Gaz de France [4] tested the spread of LNG onto water both on confined and unconfined space. During these tests, the vapor cloud and the size of the liquid pool were measured. Temperature profile of the water was recorded only in case of confined space.

At the same time, laboratory tests were performed, and water temperature was measured in small confined space. The evaporation rate was estimated from the change in the mass. Most of the papers published on this subject are based on experiments conducted in confined space that promotes the formation of ice underneath. The formation of ice changes the heat transferred and therefore the evaporation rate.

The data gathered in the tests mentioned above have been analyzed and resulted with different evaporation rates. Many vapor dispersion software use different boil-off rates to simulate LNG spills on water. Table 1 presents the LNG evaporation rate estimation on water from our literature review.

Table 1 Evaporation rate estimation onto water from literature review for unconfined spill

	Evaporation rate	Estimation/Conditions
Bureau of Mines[1]	0.181 kg $m^{-2}.s^{-1}$	Pond scale
Bureau of Mines[2]	0.155 kg $m^{-2}.s^{-1}$	Pond scale
Shell Ltd[5]	0.024 to 0.195 kg $m^{-2}.s^{-1}$	Laboratory scale
Esso[5]	0.195 kg $m^{-2}.s^{-1}$	Matagorda bay
Burro[6]	0.085 kg $m^{-2}.s^{-1}$	Visual estimation

Ospchoor [7]	0.04 and 0.17 kg $m^{-2}.s^{-1}$	Theoretical values
DNV(PHAST)[8]	0.182 kg $m^{-2}.s^{-1}$	Software
ABS[8]	0.072 kg $m^{-2}.s^{-1}$	Software
Quest[8]	0.2 kg $m^{-2}.s^{-1}$	Software

Experimental setup

In April 2006, in a project sponsored by BP, the Mary Kay O'Connor Process Safety Center (MKOPSC) performed LNG spill tests at the Texas A&M Brayton Fire School Field. The analysis of the vapor dispersion behavior from previous tests at the fire school was published in 2006[9]. During the tests in 2006, LNG was spilled onto water. The size of the pit is twenty two feet by twenty two feet with a depth of eight feet. This pit was filled with water up to the top and LNG was spilled continuously at a constant rate.

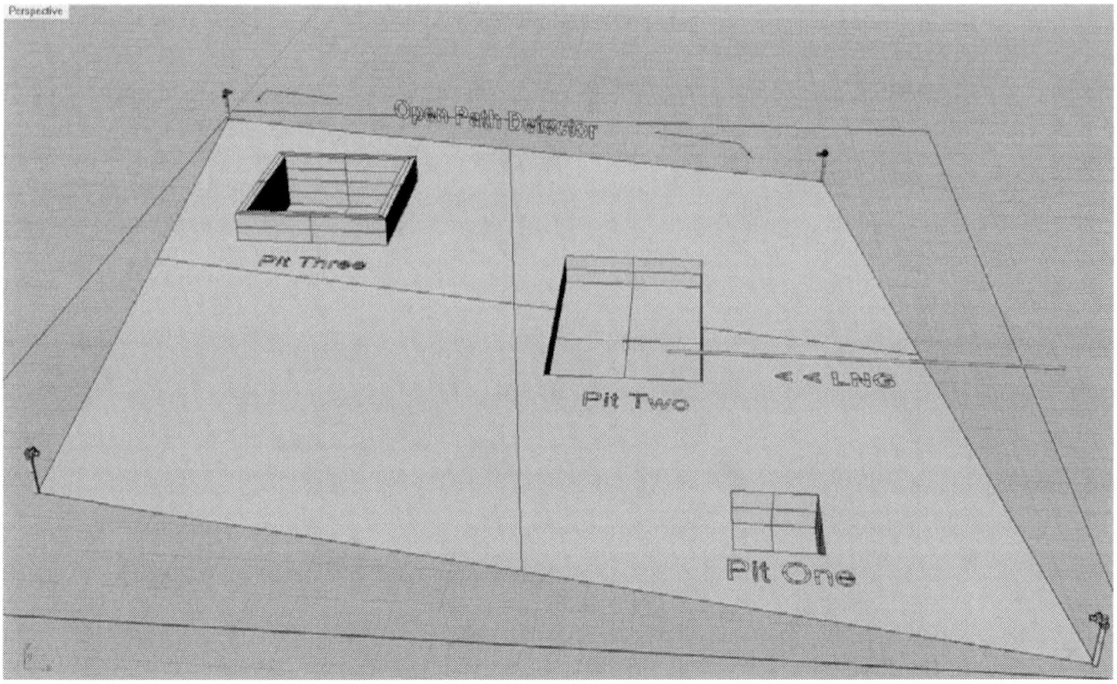

Figure 1 LNG pits at Brayton Fire School, Texas A&M

Temperature sensors were distributed in the water in order to record the temperature of the bulk. The main purpose of these sensors was to capture the heat transfer within the water while LNG was spilled.

Because of fast evaporation and vapor cloud formation, the temperature sensors within the water are able to capture the heat transferred to the LNG layer to estimate the boil-off rate. Twelve temperature sensors were used and placed in the water underneath the LNG layer. The placement of the temperature sensors is presented in

Figure 2. The LNG release point is with a 45 ° angle of 2-inch aluminum pipe. As shown in Figure 2, four sets of three sensors were placed around the release point to obtain a three dimensional representation of the water temperature profile.

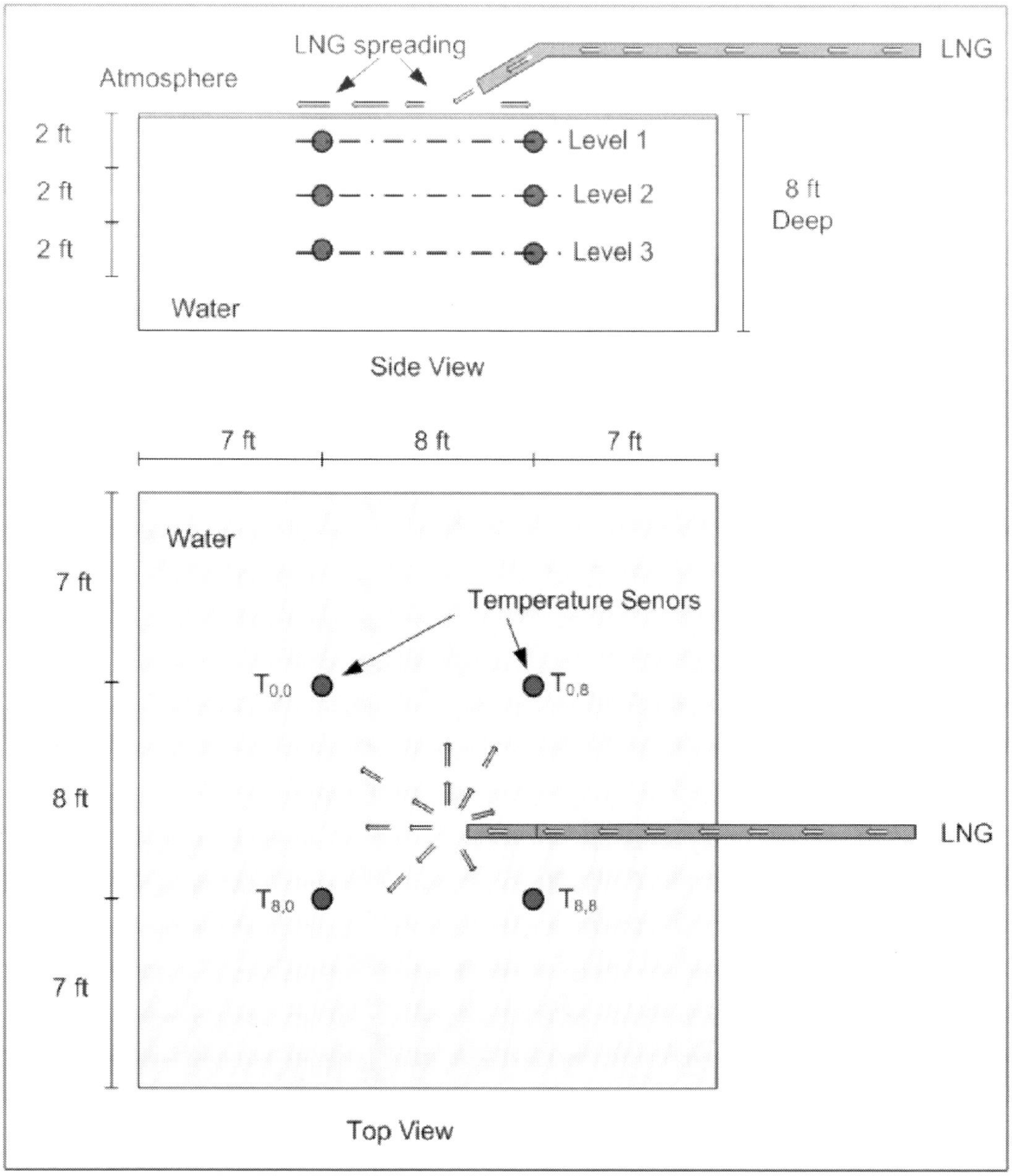

Figure 2 Placement of the temperature sensors in the water, top and side view

During the day, two runs of approximately ten and twenty minutes were performed. The LNG flow rate was estimated as 56 GPM from the release point. Some of the LNG flashes before reaching the surface of the water. LNG composition was about 99% methane.

Data analysis

Approach

The temperature data was collected every second and analyzed. The temperature profile can be used to estimate the heat transfer from and to the LNG layer. An algorithm was developed to analyze the temperature profile of the water.

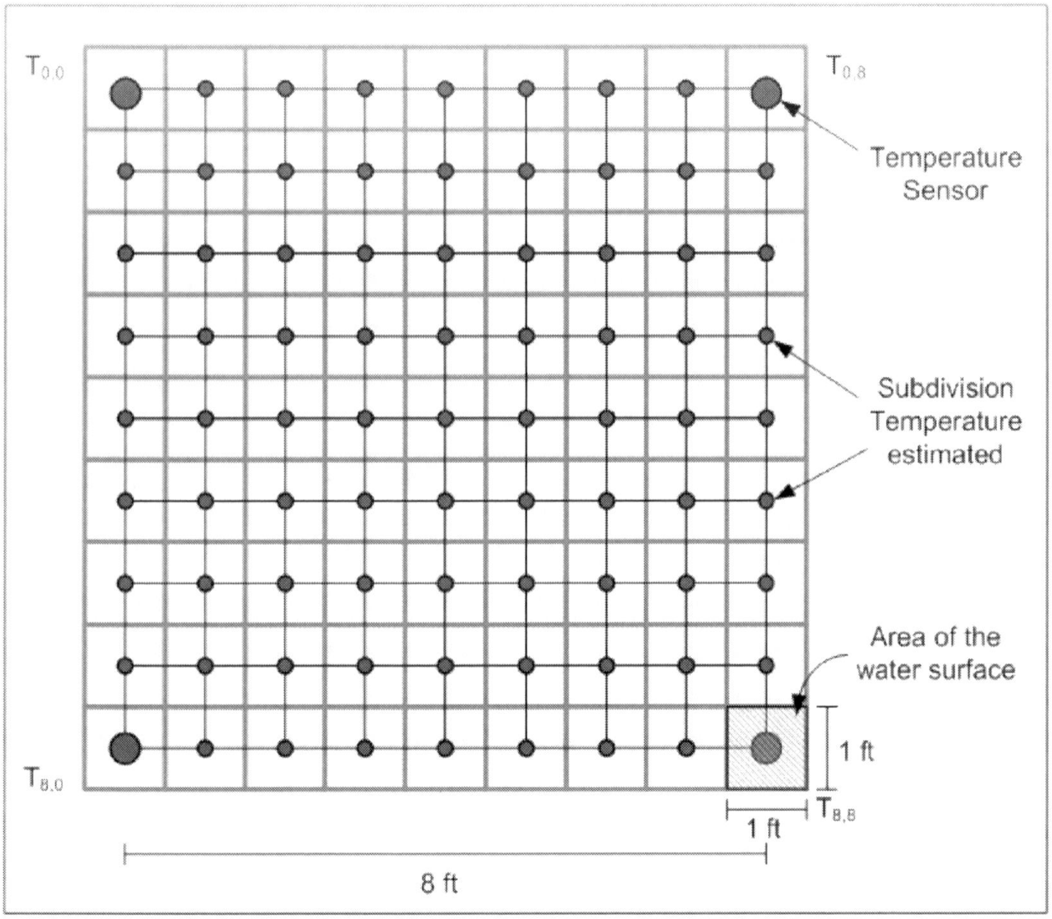

Figure 3 Configuration of the grid at level one

Figure 3 shows the configuration of the grids for level one of temperature sensors. This same grid was developed for the two other levels. The temperature profile can be determined as presented in

Figure 3. Each point can be represented algebraically by using the following notation:

$T_{j,k}(t)$ with j=0…8 and k=0…8 at instance t.

j and k represent the location on the grid. $T_{0,0}$, $T_{0,8}$, $T_{8,0}$ and $T_{8,8}$ represents the location where the thermocouples are located. The distance between each sensor was eight feet. This distance was divided into eight subdivisions. Each surface created by the subdivisions has an area of one square foot. The points from each subdivision are estimated by using a linear interpolation with respect to the thermocouples. The heat transferred to the water from the LNG layer can be formulated by the following:

$$Q_{j,k}(t) = m * C_P * \Delta T_{j,k}(t) \quad (1)$$

$$\Delta T_{j,k}(t) = T_{j,k}(t+1) - T_{j,k}(t) \quad (2)$$

m represents the mass of water present around the temperature sensors. This volume is constant over time and it is 0.056 m^3 (or two cubic feet). The heat capacity, C_p is also constant, 4.185 $kJ.kg^{-1}.K^{-1}$ for water. The only variable is the temperature change. Similarly, the heat transferred to the water at each level can be estimated. The total heat transferred to the layer is calculated with following equations:

$$Q(t) = \sum_{j=0}^{8}\sum_{k=0}^{8} Q_{j,k}(t) = m * C_P * \sum_{j=0}^{8}\sum_{k=0}^{8} \Delta T_{j,k}(t) = m * C_P * \sum_{j=0}^{8}\sum_{k=0}^{8} \left(T_{j,k}(t+1) - T_{j,k}(t)\right) \quad (3)$$

The calculation of the heat transferred was performed using an algorithm and implemented in MATLAB®. The results are presented hereafter.

Results

Heat transfer phenomenon within the water when LNG is spilled on the top is studied using the collected data. The temperature profile shows the effects of the cryogenic liquid spill onto the water. A layer of LNG was formed and some ice formation was observed on the edge. From the data analysis, LNG boil-off rate was estimated at four different locations. These results are presented in this section.

Heat Transfer Distribution

The water is constrained into a pit, the wall of the pit are made of special concrete developed by the fire school. During this analysis, the heat input was considered to come only from the LNG layer because the wall of the pit does not provide any heat input to the water. Therefore the total heat input from the LNG layer should be equal to the total heat transferred into the water. By using equation 4, the lost of heat to the surrounding was calculated over the period of the test.

$$Q_{in} - Q_{out} = Q_{surrounding} \quad (4)$$

Where Q_{in} is the negative heat input due to the LNG layer and Q_{out} is the positive heat input resulting from mixing to reach equilibrium temperature. The positive heat input is the response for the temperature change due to negative heat input. $Q_{surrounding}$ is either energy lost to the surrounding or data lost during the recording. The data shows $Q_{surrounding}$ (i.e. energy lost) is less than 4%.

Figure 4 shows the total heat transferred at level 1 using equation 4. The negative and positive peaks represent a cooling and heating cycle of the layer respectively. In other words, the negative and positive peaks show the decrease and increase in the temperature at the level 1 from the data respectively.

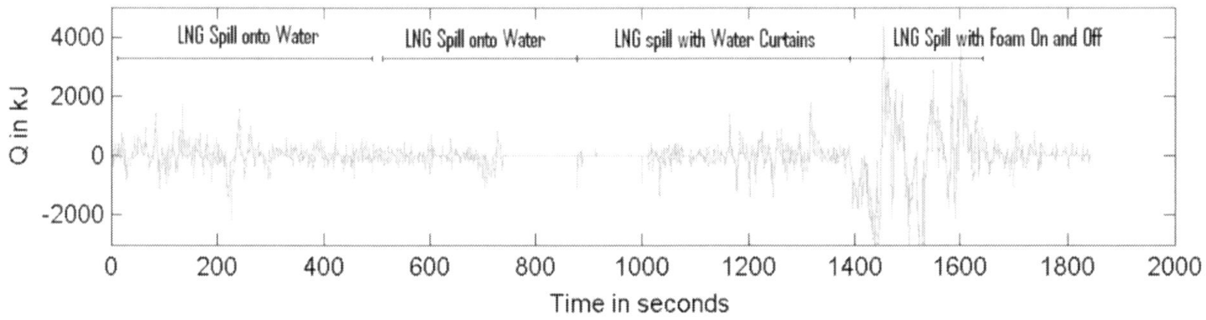

Figure 4 Heat Transfer for level 1 (L1)

Figure 4 has four segments, which is pointed out in the top of the figure. For the first and second segments, LNG is released onto water with no mitigation measures. For the third segment, LNG is released onto water with water curtain application which did not affect the evaporation rate since it is independent from LNG boiling phenomenon. For the fourth segment, LNG is released onto water with foam application. Between 720 and 1000 seconds, the data was lost due to circuit short. The negative and positive peaks are studied further in the following sections.

Heat transfer due to Convective Current

The positive heat transfer peaks were extracted and presented in Figure 5. This figure shows the distribution of this positive heat transfer during the test.

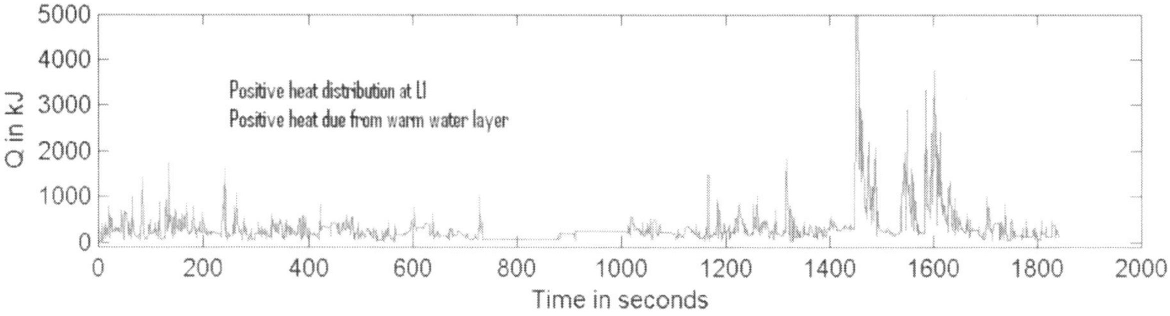

Figure 5 Positive Heat Transferred due to Convective Current

The behavior of the positive heat transferred seems to be corresponding to the negative energy input. While LNG layer is applied onto water, heat is provided from water surface close to level 1. Temperature gradient changes the water density and creates a negative buoyancy effect. In other words, the colder water sinks to the bottom of the pit. While the chilled water sinks, it pushes the warm water underneath upward. The warm water is pushed up along the level 3, on to level 2 and then level 1. Figure 6 illustrates this effect.

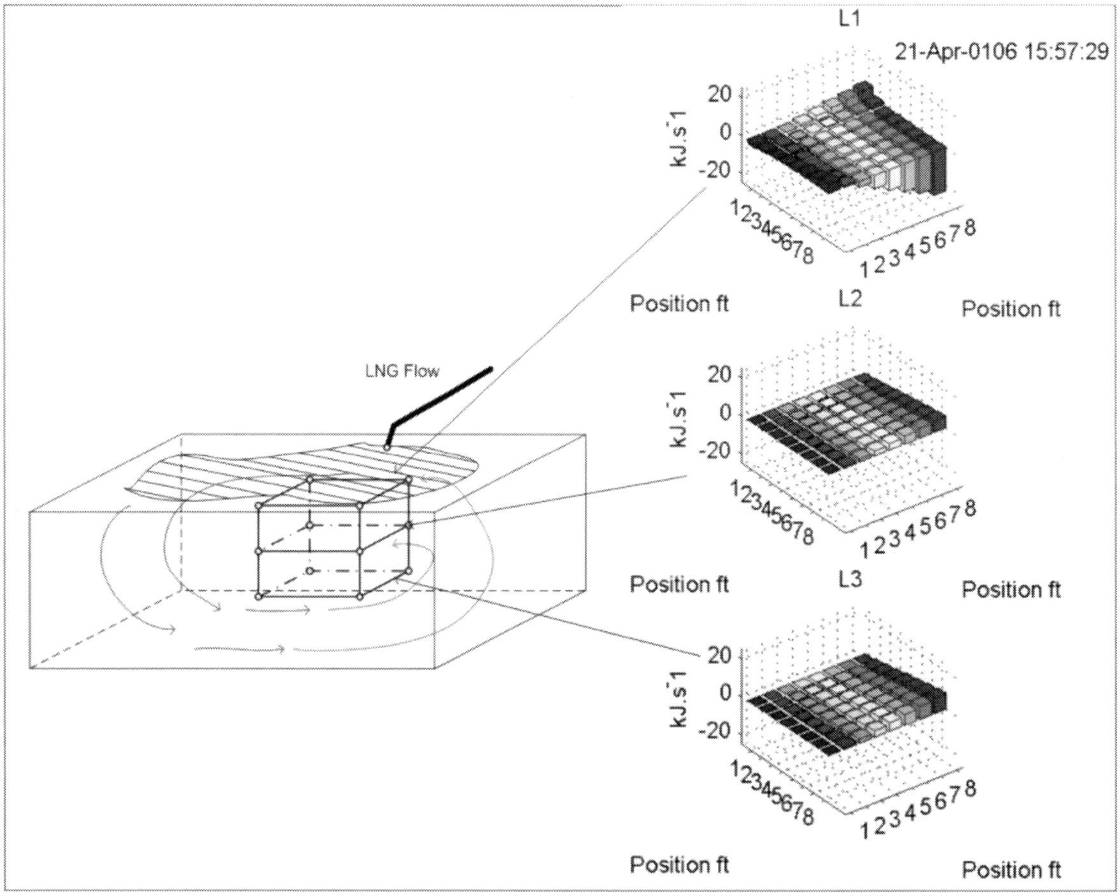

Figure 6 Heat Transferred for Each Layer and Current Effects

Figure 6 shows the convective current effect and heat transferred to level 1, level 2, and level 3 for an instance. Blue and red arrows represent cold and warm water currents in the pit, respectively. Because of the orientation of the LNG layer, not only vertical convective current is created but also a horizontal current as well. On the left of the figure, the bar-graphs present the heat input at each level and illustrate the effect of the convective current due to the cold LNG layer. While negative heat transfer occurs at level 1, a positive heat transfer is introduced at level 2 and level 3.

Boil-off Rate

From the heat transfer analysis, LNG boiling-off rates were estimated. Each vertical array of three temperature sensors was placed at different position relative to the incoming LNG hose. The pipe opening was situated about one and half above the water surface. The end of the pipe was open and the LNG was shot directly onto the water surface. $T_{8,8}$ was the closest to the LNG source and $T_{8,0}$ was the next closest. $T_{0,8}$ and $T_{0,0}$ were further down the LNG layer as shown in

Figure 2. Because of this configuration, several observations were made from the data analysis.

- LNG was spread with an initial momentum due to the height between the water surface and end of the pipe. This initial momentum creates a lot of turbulence near the landing area, located next to $T_{8,8}$.
- Because of the uneven distribution of the poles compared to the incoming LNG, the estimated boil-off rates at the spots (0,0), (0,8), (8,0) and (8,8) are not exactly equal to each other.

By using equation 3 and the latent heat of methane, the boil-off rate was estimated and presented in Table 2.

Table 2 Summary of Estimated LNG Boil-Off Rate

	$T_{8,8}$	$T_{0,8}$	$T_{8,0}$	$T_{0,0}$
Layer State	Unsteady	Steady	Meta steady	Steady
Boiling Rate Calculated, $kg.m^{-2}.s^{-1}$	0.50 to 0.80	0.25 to 0.30	0.40 to 0.50	0.25 to 0.30

Table 2 presents the boil-off rates at different location in the pit. $T_{0,8}$ and $T_{0,0}$ represent the steadiest results, while $T_{8,8}$ shows the most instability due to the high turbulence created by the initial momentum of the spill. This behavior is also illustrated in Figure 7.

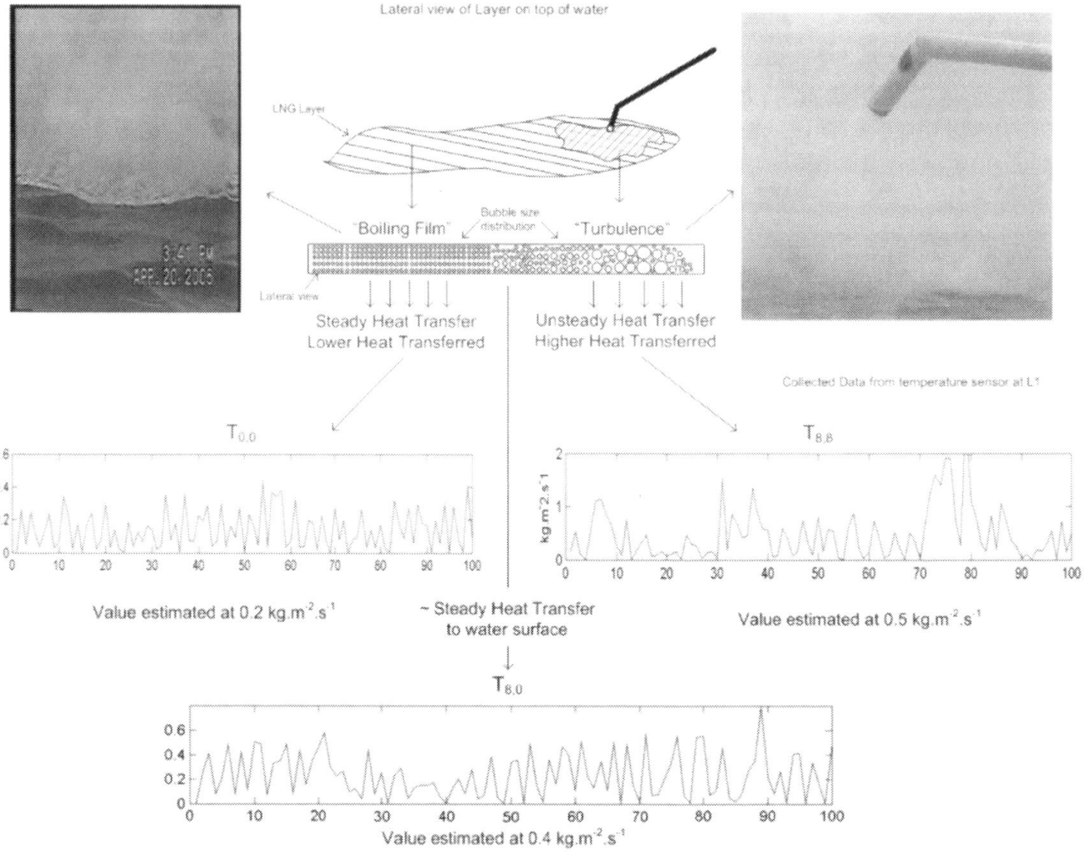

Figure 7 Lateral View of LNG layer and Behavior Description of Boil-off Rate

Figure 7 shows the lateral view of LNG Layer and the change in the boil-off rate with respect to the states of the layers. The interface between the LNG layer and the water surface is characterized by the bubble formed (bubble being methane gas vaporized at the contact with water). There are three main regions noticed:

- The unsteady region (i.e. landing area near the LNG source), the momentum created a stronger mixing effect between the water and the LNG, thus leading to the formation of larger bubbles and a high fluctuation in the estimated boil-off rate, as shown in the graph $T_{8,8}$.
- The meta-stable region, the bubble size distribution is spread out better and the LNG layer starts to stabilize. Therefore the boil-off is more steady as shown in graph $T_{8,0}$.
- The steady region, the bubble size is small and uniform, therefore the boil-off rate is more stable as shown in the graph $T_{0,0}$.

Conclusion

In April 2006, LNG was spilled onto water and the temperature profile of the water was recorded. This paper presents the analysis of the temperature profile to study and estimate the LNG boil-off on water. In this test, two runs were performed. The following observations and conclusions were made from the data analysis.

The cooled water underneath the LNG layer sinks and pushes up the warmer water to the surface of the pool. This effect creates a current inside the pit and prevents the formation of ice underneath the LNG layer. In the case of confined spill, ice formation slows down the boiling rate because the heat transfer is reduced from water to LNG layer.

In previous unconfined spill tests, the boiling rate was estimated through visual control of the pool diameter. In LNG spill, the condensate vapor makes this measurement really difficult and inaccurate. During April 2006 test, the temperature profile helps to estimate the boiling rate instead of a visual measurement of the pool diameter. The evaporation rate was estimated by calculating heat transferred from the LNG layer to the water. The evaporation rate was estimated to be between 0.2 $kg.m^{-2}.s^{-1}$ and 0.8 $kg.m^{-2}.s^{-1}$ depending on the region of turbulence. 0.8 $kg.m^{-2}.s^{-1}$ was estimated almost underneath of the pipe where the LNG was shoot in. 0.2 $kg.m^{-2}.s^{-1}$ was estimated at the tail of the LNG pool where the LNG is clamed onto the water surface. The previous reported values from the laboratory tests and theoretical calculations agree only when the LNG layer is spreading quietly onto water and the film boiling is dominant. The momentum of the incoming LNG increases the interaction between gas and water. Therefore the evaporation rate is greater than the values presented in Table 1, which is from the open literature. Once the LNG is spreading quietly, there is a good agreement with the values estimated during previous tests within the uncertainties of these experiments. This only represents the tail of the LNG layer. This paper shows the uneven evaporation rate along the LNG layer due to the initial momentum. In a case of accident scenario, this should be included in the simulation. Most of the software use an evaporation rate of 0.2 $kg.m^{-2}.s^{-1}$ that is the evaporation rate once the LNG is calmed onto the surface of the water.

In case of a large spill, the variation in the boil-off rate could have a significant impact on subsequent vapor dispersion modeling. Today, most of the simulations assume a constant evaporation rate. The use of uniform values does not represent the real behavior of LNG is spilled onto water. A more realistic source term should be integrated into vapor dispersion modeling.

Acknowledgements

LNG spill experiments and this project are sponsored by BP. During the test, data was collected with the participation of ABB, Honeywell and Detection & Measurement Systems, Inc. The authors would also like to thank the input provided by the late Dr. Harry West and other researchers at the Mary Kay O'Connor Process Safety Center.

Reference:

1. Burgess, D., J. Murphy, and M.G. Zabetakis, *Hazards Associated with the spillage od Liquified Natural Gas on Water*. 1970, Bureau of Mines.
2. Burgess, D., J. Biordi, and J. Murphy, *Hazards of spillage of LNG into water*. 1972, Bureau of Mines: Pittsburg, Pennsylvania. p. 93.
3. *LNG safety program: Phase II*. 1974, American Gas Association.
4. Humbert-Basset, R. and A. Montet. *Dispersion dan l'atmosphere d'un nuage gazeux forme par epandage de G.N.L. sur le sol*. in *Third International Conference & Exhibition on Liquified Natural Gas*. 1972. Washingtion D.C.
5. Hightower, M., et al., *Guidance on Risk Analysos and Safety Implications of a Large Liquified Natural Gas (LNG) Spill over Water*. 2004, Sandia.
6. Hanna, S.R., D.G. Strimaitis, and J.C. Chang, *Hazard Response Modeling Uncertainty; A Quantitative Method. Volume II: Evaluation of Commonly-Usede Hazardous Gas Dispersion Models*. 1991, Sigma Research Corporation: Westford, MA.
7. Opschoor, G., *Investigations into the spreading and evaporation of LNG spilled on water*. Cryogenics, 1977. **17**(11): p. 629-33.
8. Baik, J., V. Raghunathan, and H. Witlox, *Consequence Modeling of LNG Marine Incidents*. American Society of Safety Engineers, 2006.
9. Cormier, B., Y. Wang, and M.S. Mannan, *LNG Mitigation Experiments - Preliminary Results*, in *Mary Kay O'Connor Process Safety Center 2006 International Symposium*. 2006: College Station, TX.

Biography of speaker

Benjamin Cormier is a PhD candidate at the Artie McFerrin Department of Chemical Engineering, Texas A&M University. He attended Virginia Tech where he obtained his Bachelor of Science degree in Chemical Engineering in May 2004. He came to Texas A&M University for a Master of Science degree in Chemical Engineering and completed his Masters in December 2005. Since summer 2005, he has been involved with Mary Kay O'Connor Process Safety Center under the direction of Dr. Sam Mannan. Benjamin Cormier is also responsible for teaching Aspen Plus to the senior class at Texas A&M University.

Benjamin Cormier has been involved with LNG since the beginning of January 2006 and he is sponsored by BP technology. His research involves LNG tests which are usually carried out at the Brayton fire school at Texas A&M University. He has been focusing on the computational fluid dynamics on source modeling for LNG evaporation.

Email: Benjamin.Cormier@chemail.tamu.edu

Safety Assessment and Design Code Development for Membrane Type LNG Storage Tank

Seung-Rim Lee, Boo-Kil Kwon, Seung-Hyun Lee
Korea Gas Safety Corporation (KGS)
Siheung, Kyounggi, Korea

Young-Soon Lee
Seoul National Univ. of Technology (SNUT)
Seoul, Korea

AIChE Spring Meeting, April 2007
7th Topical Conference on Natural Gas Utilization
Houston, Texas, April 22-27, 2007

ABSTRACT

Liquefied Natural Gas (LNG) is becoming one of the important parts of the world's energy sources. Since the imported LNG is only the source of natural gas in Korea, huge storage volumes are required to effectively support the seasonal fluctuations. To carry this huge inventory of LNG, ensuring safe design of the storage tanks is as important as achieving cost effective construction of the storage tanks.

The paper discusses the safety assessment of a membrane tank developed by Korea Gas Corporation (Kogas). The following areas are specifically evaluated prior to approval of the tank as a 'full containment tank without spill containment':

- Failure frequency of an external and internal containment system;
- Comparative risk assessment of dike function of the outer concrete tank of membrane LNG storage tank.

This paper also discusses the highlights of Korean safety verification codes for above ground membrane storage tank.

INTRODUCTION

In recent years, the majority of new LNG storage tanks have been designed on the basis of Full Containment Tank concept, which provides an inner vessel to contain the LNG and an outer, gas tight, concrete vessel which can also self-contain the LNG in the event of an emergency. A significant variation on this concept is the membrane type tank developed by KOGAS where the inner vessel is a thin stainless steel membrane which is supported by structural insulation and the concrete outer vessel.

The objective of the study is to carry out a comparative Quantitative Risk Assessment (QRA) of two KOGAS tank designs using a fault tree methodology, a standard "Full Containment" tank and a "Membrane" tank. For the membrane tank, both the initial Kogas design and 4 Alternative Kogas designs have been investigated and assessed, presenting six separate cases.

The modified membrane tanks are as follows:

1) Option 1: An absorber structure to prevent dropped pumps penetrating the membrane.
2) Option 2: In addition to the absorber structure, a secondary barrier consisting of a 0.5mm aluminum sheet sandwiched within the insulation is included
3) Option 3: A pump catcher is added to the pump lifting device.
4) Option 4: In addition to the absorber structure and pump catcher a secondary barrier is included.

Expected event frequency has been assessed for the three types of release: 1) external LNG liquid leak to the environment, 2) internal liquid leak (retained within the concrete vessel); and 3) vapor leak to the atmosphere. The external LNG leak is the critical safety concern. The internal and vapor leaks are largely operational issues.

In this paper, the frequencies of expected risk are quantified using a Fault Tree Approach (FTA). For clarity in the analysis, and to ensure consistency, all cases have been quantified using the same fault tree. Logic within the fault tree is used to select each of the cases. Full quantification of risks is often difficult, owing to a lack of relevant failure data, but the aim of this study is to provide a fully transparent failure information as quantitative as possible. According to the result of FTA, Korean domestic standard for above ground LNG storage tank was developed and published. Thermal protection system and absorber structure were made mandatory to reinforce weak points of membrane tank. These items were reflected in its standard.

STUDY BASIS

Fault trees are a means of quantifying the frequency (or probability) of a Top Event based on the combinations of failures which are needed for the Top Event to occur. The Base Events, for which failure rates (or probabilities of failure) are provided, are combined together using Logiv Gates that reflects the combinations of failures which are needed for the Top Event to be realized. Using Boolean algebra, the combination of events can be reduced to produce combinations of events, known as Cutsets, which can result in the Top Events. By studying the Cutsets and the failure results within the tree, the main contributors to the failure of a system can be determined.

In general, events which are common to all the tank types are quantified with a single number, while events which are quantified differently for each tank require a switch in the fault tree. The fault tree for external leakage consists of three main branches:

1) Failures starting with a failure of the outer concrete vessel walls or base,
2) Failures starting with the concrete roof, and
3) Failures starting with a failure of the inner vessel wall (whether nickel steel or membrane).

In each case, the causes of the initial failure are further divided as given below. For example, failures of the outer concrete wall or base can be caused by:

1) External fire
2) Aircraft impact
3) High winds/hurricane damage
4) Flood
5) Structural failures from
   i) Ground movement, due to ice heave of piling failures.
   ii) Latent defects in the concrete construction or failures of pre-stressing.

Safeguards (e.g. the hydrotest of the concrete vessel for the membrane tank) are included where appropriate. Many events are the same for both the full containment and membrane tanks. Where there is a difference, different events or branches of the tree are used for each case.

Initial failures of the roof are divided between:

1) Extraneous causes (snow loading and lightning strike).
2) Structural failures due to a defect in roof construction.
3) Operational failures due to overpressure, overfilling and vacuum.
4) Impact from a dropped pump during removal.
5) An explosion in the external plant.

Initial failures of the inner vessel are divided between as follows:

1) Operational failures from dropped pumps and internal missiles split between:
   i) Events that directly fail the outer vessel as well.
   ii) Events that fail the outer vessel indirectly
2) Operational failures, as above, but with an indirect cause of failure of the outer vessel.
3) Seismic failures (which are not fully quantified)
4) Structural defects (split between failures on initial filling and failures in normal operation).
5) Failures due to detachment of the membrane due to a differential pressure across the membrane.

RESULTS OF FAULT ANALYSIS

Top Event Frequencies

The top event frequencies are summarized in the Table 1. Because the seismic events were assessed nominally only, the frequencies are presented with and without seismic events included.

The vapor leak frequencies are the same with and without seismic events.

The external (LNG) leak frequency is the main measure used to compare the safety of the tank designs. The internal and vapor leaks are considered more as operational issues. An internal leak has no external safety implications, but would require repairs to the tank. A vapor leak will release flammable vapor external to the tank, but the pressure and flow rate are expected to be low and have only very limited safety implications.

Table 1 : Top Event Frequencies

		FT	MT	MT: Opt 1	MT: Opt 2	MT: Opt 3	MT: Opt 4
Including seismic events	External Leak	1.63×10^{-6}	4.47×10^{-6}	1.73×10^{-6}	1.73×10^{-6}	1.46×10^{-6}	1.45×10^{-6}
	Internal Leak	1.08×10^{-4}	1.33×10^{-4}	1.05×10^{-4}	1.05×10^{-4}	1.03×10^{-4}	1.03×10^{-4}
	Vapor Leak	1.80×10^{-4}	8.02×10^{-5}	8.02×10^{-5}	8.02×10^{-5}	8.02×10^{-5}	8.02×10^{-5}
Excluding seismic events	External Leak	6.32×10^{-7}	3.47×10^{-6}	7.29×10^{-7}	7.29×10^{-7}	4.55×10^{-7}	4.55×10^{-7}
	Internal Leak	8.05×10^{-6}	3.27×10^{-5}	5.33×10^{-6}	5.33×10^{-6}	2.64×10^{-6}	2.64×10^{-6}
	Vapor Leak	1.80×10^{-4}	8.02×10^{-5}	8.02×10^{-5}	8.02×10^{-5}	8.02×10^{-5}	8.02×10^{-5}

The external LNG leak frequency should not be taken as directly equivalent to measures of individual and societal risk. These measures include additional effects due to the vaporization of the liquid, distance the vapor travels, how much the vapor cloud spreads, the probability of the cloud being ignited, the location of people, etc. The estimate of individual risk of fatality (often quoted in risk criteria) will always be lower than the estimate of external leak frequency made here and will depend on distance and wind direction from the tank. An assessment of individual (or societal) risk levels for comparison with criteria would require these additional factors to be taken into account by including consequence analysis for the identified leaks.

However, we would expect all the factors in the consequence analysis to be the same for each tank. This would only change if the modes of failure of the different tanks were very different, and would result in releases of different sizes, or if differences in design (e.g. including a dike for one tank) would alter the consequences between the tanks. This is not the case here, with the dominant failure modes being similar and expected to result in similar sizes of releases.

The external (LNG) leak frequency therefore provides a good comparison between the safety performance of the full containment and membrane tank, without introducing the additional complexities of the consequence of a release.

The spread of the results between the full containment tank and the membrane tank with Options 1~4 is only 16%. This is small compared to the uncertainties in the analysis. These differences are considered minor and, if individual risk contours were plotted for these different release frequencies (based on additional consequence analysis), they would be expected to be almost indistinguishable. The results for the modified membrane tank span the results for the full containment tank.

For the membrane tank, without any modification, the external LNG leak frequency is predicted to be significantly higher than for the full containment tank (a factor of 2.7 higher or

170% increase). This difference is due to the different predictions for the failure probabilities of the membrane with only thicker plates for protection and the nickel steel tank base in the event of a dropped pump. Modifications 1 and 3, for the membrane tank, address this issue to provide an overall external LNG leak frequency that is comparable to the full containment tank.

Comparison between Designs

The relative magnitudes of the types of the causes of failure are shown in the following Figs.1~3.

Fig. 1 shows the difference in vapor release frequencies between the tanks. The vapor release frequency is predicted to be higher for the full containment tank, and is dominated by concrete defect failures. These are reduced for the membrane tank due to the hydrotest of the concrete wall. The internal release frequencies shown in Fig. 2 are predicted to be almost the same for all the tanks except for the unmodified membrane tank, where the failures are dominated by dropped pump.

Fig. 3 shows the predicted external release frequency for all the tank designs. The unmodified membrane tank shows a greater external leak frequency than the full containment tank. This is because, even with the 20mm reinforced plates above and below the membrane, the dropped pump could still penetrate. The addition of the absorber structure (Option: MOD in below Figs.) reduces the dropped pump frequency to a similar level to the full containment tank. The addition of a pump catcher (MOD3) reduces the dropped pump frequency further bringing the overall leak frequency of a membrane tank to a marginally lower level than predicted for the full containment tank.

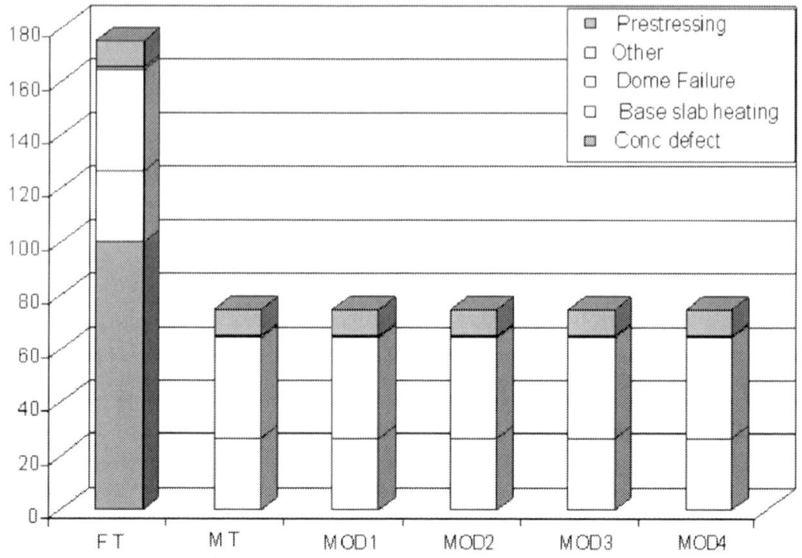

Fig. 1 External Vapor Leaks

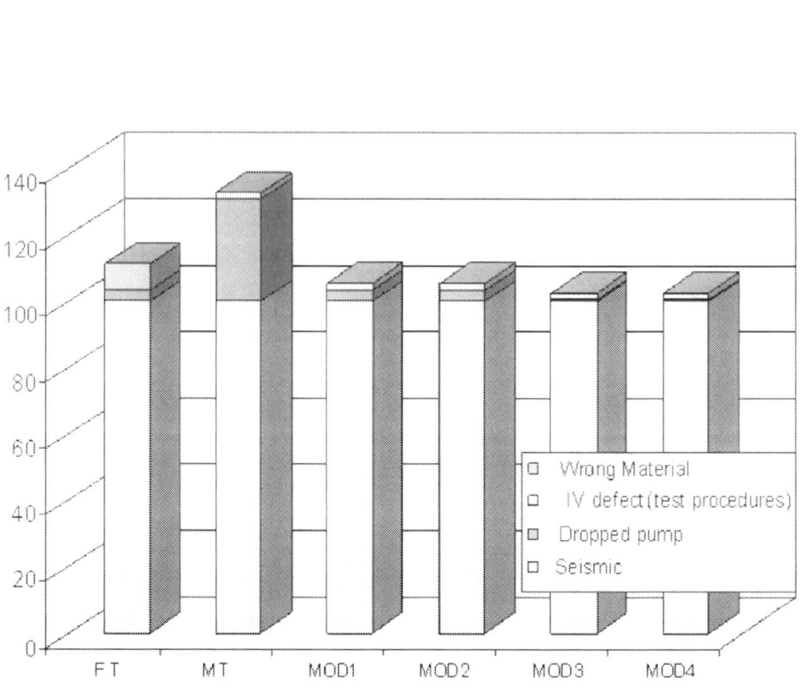

Fig. 2 Internal Leaks

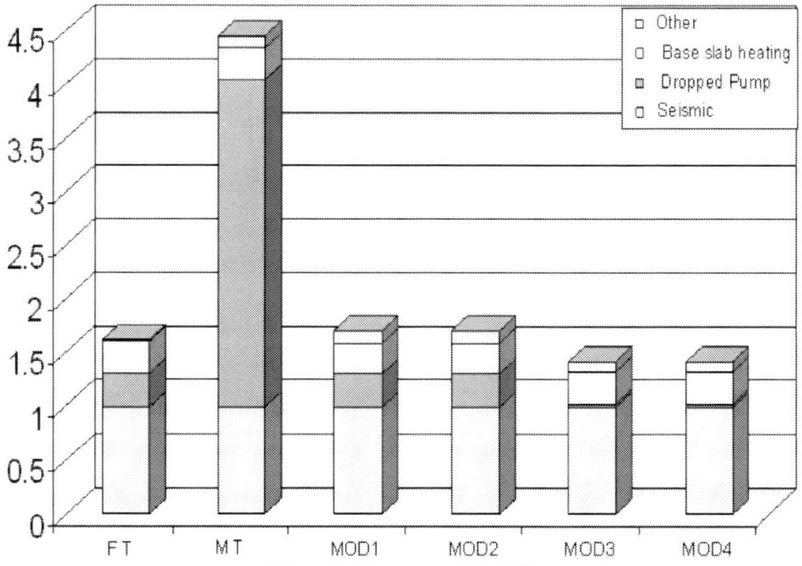

Fig. 3 External LNG Leaks

It is possible that the absorber structure may perform better in avoiding penetration of the tank base than the thicker nickel steel plate, but the uncertainties involved are larger and a more conservative approach is appropriate for the membrane absorber design. However, modification 1 is predicted to remove the additional susceptibility of the unmodified membrane tank to this type of incident. Careful attention to lifting equipment and lifting operations remains important to avoid damage to the tank (which would require repair of the pump well at least) even when the potential for an external LNG release has been reduced. Whilst a "pump catcher" has been included in the membrane tank analysis, it is noted that this could also be included for the full containment tank and that there is no particular reason for including it in only one tank once the absorber structure has been included. The "pump catcher" would have the potential advantage (for both tanks) of avoiding significant damage in the event of a lifting equipment failure. However, the additional equipment could potentially cause operational issues if any of the pivots seize whilst in service. This could cause difficulties in removing a pump from the tank, which have not been assessed here.

Seismic failures dominate the external LNG leak frequency for the full containment tank and the modified membrane tanks. However, seismic failures have only been nominally quantified and rely on the tanks all being qualified to the same seismic standard. There is no reason why the membrane and full containment tanks cannot be designed to resist the same seismic loads, but this may include greater strength in the outer tank wall for the membrane tank due to the greater static loading. It should be remembered that the magnitude of the earthquake required to cause failure is also expected to be sufficiently large to cause widespread destruction of buildings and loss of life, independent of the failure of an LNG tank. However, it remains clear that the seismic design of the LNG tank is critical in ensuring that risk levels are kept as low as reasonably practicable.

Failures of the base slab heating system are predicted to be the next most significant cause of external LNG leaks, due to ice heave. Ice heave will take a long time to occur after failures of the base slab heating system and time should be available to correct the failure or empty the tank. A base event has been included to quantify the probability of the tank continuing to be operated long term without an adequate base slab heating system. This allows for the uncertainty in the level of the potential problem and the difficulty in correcting failures within the base slab. It should be noted that ice heave will not be critical if a tank is not operated for a long time (typical timescale of months) without effective base heating. If this can be guaranteed then an external LNG leak due to a base slab heating system failure cannot occur. This is a matter of good operational practice and effective regulation.

The second barrier included for modifications 2 and 4 is predicted to make a negligible difference to the external LNG leak frequency. The barrier has very little structural strength and will make essentially no difference to the dominant causes of seismic failure, dropped pumps and base slab failure. The construction of the barrier also requires extensive bolting between panels and there is no test of the integrity of these joints. For the limited cases where the barrier could be effective, the probability of there being a leak at the joints is predicted to be high. This second barrier also performs no additional useful function if the foam insulation is liquid tight. Overall, the inclusion of a second barrier, of this type, does not appear to make any useful contribution to the integrity of the membrane tank against external LNG leakage.

Risk Analysis for Kogas Membrane Tank

The comparative risks of leakage from the Kogas design of membrane LNG tank and a full containment LNG tank have been analyzed using a fault tree approach. The following conclusions are apparent:

1) The predicted frequencies of an external LNG leak for the full containment and modified membrane tanks are very similar.
2) Failures due to dropped pumps are predicted to be significantly greater for the membrane tank with only a thickened bottom plate than for the full containment tank. The impact

absorbing structure (modification 1) is expected to result in a similar level of risk to the full containment tank. The pump catcher (modification 3 - which is not assumed to be present in the full containment tank) reduces this risk further.

3) The addition of a secondary barrier to the modified membrane tank (modification 2 and 4) is predicted to make very little difference to the overall LNG leak frequency.
4) The frequency of vapor leaks is greater for the full containment tank. This is because the hydrostatic test on the membrane tank is assumed to detect defects on the pre-stressed concrete wall, while the pneumatic test on the full containment tank is only a leak test for the dome vapor barrier.
5) Most significant general cause of external LNG leaks is predicted to be a seismic event, which has been quantified nominally. An improvement on this would require a more detailed assessment (e.g. finite element). However, if the membrane tank is seismically qualified to the same level as the full containment tank (as assumed in this study) and designed to be equally protected against a dropped pump (as for modifications 1-4), there is little to differentiate the tanks in terms of the risk of an external LNG leak.
6) Failure of the base slab heating system and subsequent ice heave is potentially a serious problem for both tanks. The failure of both coils within the base slab is quite likely during the tank lifetime and could present a major problem as they are not replaceable. There will be little risk if it is ensured that a tank does not operate for an extended period without adequate base-mat heating.

STANDARD DEVELOPMENT

After validation of safety for Kogas membrane tank, Korean standard for above ground membrane LNG storage tank which is meeting domestic circumstances was developed and applied for tank construction in 2006. Technical committee composed of experts from design, construction, and operation companies, inspection authority, etc. was formed for developing and reviewing from an early stage draft to a final draft. Up to now several standards related to membrane LNG tank have been developed and used as follows:

1) BS EN 14620 "Design and manufacture of site built, vertical, cylindrical, flat-bottomed steel tanks for the storage of refrigerated, liquefied gases with operating temperatures between -5°C and -165°C", 2006
2) BS 7777 "Flat-bottomed, vertical, cylindrical storage tanks for low temperature service", 1993
3) BS EN 1473 "Installation and equipment for liquefied natural gas (Design of onshore installations)", 2007
4) JGA RPIS (Recommended Practice for LNG In-ground Storage), 2002

Standard for above ground membrane storage tank was developed with some directions as below:

1) Developing performance based standard suitable to domestic circumstances

2) Benchmarking existing overseas standards which is used in Korea
3) Developing standard covering all kinds of membrane developed
4) Reinforcing the safety equipment
   - Thermal protection system for outer concrete tank
   - Bottom reinforcement device against pump drop
   - Pressure relief and purge system for Inner Barrier Space (IBS)
5) Developing a real consensus based standard in the process of all related party attendances

Standard developed was composed all 4 parts and main contents are shown in Table 2.

Table 2 : Composition and main contents of standard

Sections	Contents
Part 1 General	- Design conditions - Storage concepts - Loading cases and design considerations - Pressure relief system - Commissioning and decommissioning
Part 2 Membrane inner tank	- Design conditions - Materials - Design of membrane inner tank - Fabrication and Assembly - Insulation system - Inspection
Part 3 Concrete outer tank and foundation	- Design conditions - Materials - Prestressed Concrete outer tank - Liner - Foundations
Part 4 Testing, drying, purging and cool- down	-Hydrostatic and pneumatic testing - Drying, purging and Cool-down

CONCLUSIONS

Korean domestic standard for Kogas membrane tank was developed. According to the results of risk analysis and structural analysis, membrane tank got a government approval for dike function of outer concrete tank. Some articles for safety devices which were deduced from FTA for membrane tank contained in the standard. Three safety devices are as follows:

1) Thermal protection system for outer concrete tank
2) Bottom reinforcement device against pump drop
3) Pressure relief and purge system for Inner barrier space (IBS)

Commercialized application of Kogas membrane tank has a good prospect due to lower construction cost and the same level safety comparing with traditional full containment LNG tank.

REFERENCES

1. AEAT-KGS, Comparative Risk Assessment of LNG Tank Designs, 2004.
2. N Ketchell, R Robinson and P Genoud, Quantification and comparison of the risks of LNG storage concepts – membrane and full containment, LNG12, Perth, Australia, 1998.
3. TGC-TGE, Quantitative Risk Assessment of LNG aboveground Tanks based on past operating records of LNG regasification terminals and life cycle assessment, 2003.
4. EN 14620-1-5: Design and manufacture of site built, vertical, cylindrical, flat-bottomed steel tanks for the storage of refrigerated, liquefied gases with operating temperatures between -5 and -165 deg C.
5. EN 1473: Installation and equipment for liquefied natural gas - Design of onshore installations, CEN, 2007.
6. BS7777: Flat-bottomed, vertical, cylindrical storage tanks for low temperature service; Parts 1 – 4, British Standards Institution, 1993.
7. Unified Partial Method Version 3, "A Practical Approach to Dependent Failure Assessment", AEA Technology plc.
8. Theofanous, T. G., "On the Proper Formulation of Safety Goals and Assessment of Safety Margins for Rare and High-Consequence Hazards," Reliability Engineering and System Safety, 54, 243 (1996).
9. Welker, J. R. and Schorr, H. P., "LNG Plant Experience Data Base," American Gas Association 1979 Operation Section Proceedings, A.G.A., Arlington, VA, USA (1979).
10. Han, S. H., "KAERI Integrated Reliability Analysis Code Package (KIRAP)," Korea Atomic Energy Research Institute, Korea (1999).
11. Johnson, E.M. and Welker, J. R., "Development of an Improved LNG Plant Failure Rate Data Base," GRI-80/0093, Gas Research Institute, Chicago, IL, USA (1980).

BIOGRAPHY

Seung-Rim Lee is a Senior Researcher in the Institute of Gas Safety R&D at Korea Gas Safety Corporation. He has been working for Korea Gas Safety Corporation for eleven years. He has been involved research and development projects related with domestic standardization in LNG industry. Especially, he has been participated in development of Korean Standards related with LNG storage tank and LNG terminal from 1999. He obtained a Master of Science in field of Safety Engineering in 2001 and he is under the doctor´s course in field of Gas Engineering.

Email: srlee@kgs.or.kr

Understanding and Managing Business Interruption Risk

Gøran Andreassen, Bryce Levett
DNV Energy
Houston, Texas

AIChE Spring Meeting, April 2007
7th Topical Conference on Natural Gas Utilization
Houston, Texas, 22-26, 2007

Introduction

As the LNG market continues to expand, bringing with it both technical and commercial challenges, the key to successful implementation will be understanding and management of risk. This can be risk in terms of project cost overruns, HSE, security, reputation and business interruption. Focusing on the latter, a typical LNG import terminal project is exposed to business interruption risk in the form of project delays, regulatory shutdown, infrastructure failure, equipment failure, supply delay/diversion, and environmental disruption, all which could result in significant financial losses if not properly managed.

Risks can generally be mitigated in the following ways: technically, organizationally, contractually and commercially. In order to determine which options provide the best solution, one must understand the full impact of risk in terms of overall business interruption.

This paper will examine the use of risk management as a pro-active, strategic tool to evaluate business interruption costs and provide the basis for decision making. Methodology will be discussed that includes ranking and valuation of risk to optimize resources, maximize opportunities and enhance business performance.

The LNG Industry uses several standards for HSE-related risks and works systematically to achieve acceptable HSE risk levels. Less standardization exists for managing the financial risks due to business interruption. This paper will also discuss how project risk management can be applied to maximize the odds of meeting project objectives like reduced business interruption.

Why Focus on Business Interruption Risk?

Until recently economic models used to evaluate new LNG development projects have focused on optimizing the balance between potential revenue, CAPEX and operational expenditures (OPEX). A proportional amount of effort was not spent assessing the magnitude of reduced revenue due to business interruption and cost associated with unwanted incidents in the form of property damage and tarnished company reputation. Such costs can be considered Risk Expenditures (RISKEX). The value of including

RISKEX in the early evaluations of LNG development projects is apparent as large amounts of money are invested up-front with an offset revenue stream and a long payback period. It takes a balanced and mature appraisal of the uncertainties and risks involved to accurately consider front-end (CAPEX) cost savings that may have detrimental consequences on immediate and long term revenue streams.

Managing HSE related risks are mainly driven by regulatory requirements and company HSE standards. Many HSE type risks can also cause significant business interruptions. A vapor cloud explosion at a LNG facility would not just cause potential fatalities but also result in significant property damage and worst case re-build of the facility with major financial impact in terms of business interruption and loss of reputation. Hence proper management of HSE risks will also result in reduced RISKEX.

However, most business interruption risks such as equipment failure or LNG carrier delays due to transport issues en-route do not represent significant or any HSE impact and will not be managed through the regulatory driven HSE risk management process.

Figure 1 - RISKEX over project life cycle

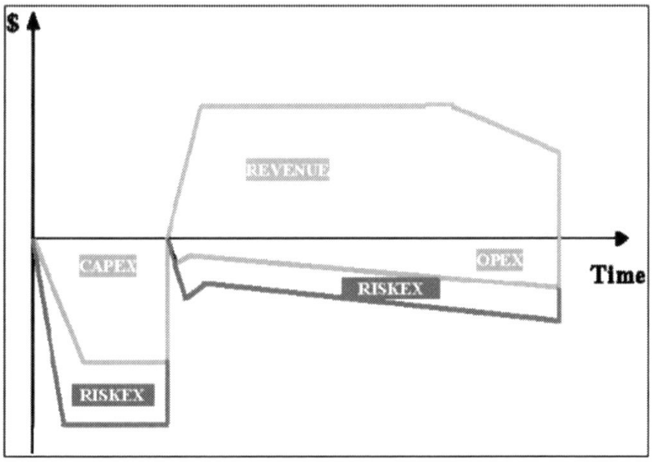

RISKEX in terms of business interruption represents a significant threat to LNG development projects and to the contractors involved. The CAPEX overrun is only one element of project risk. Business interruption further increases the lifecycle RISKEX and needs to be addressed at early project stages to reduce significant revenue losses.

Impact of Risk Management

Financial impact due to business interruptions can be calculated as the product of frequency of disruption and value of business interrupted

Figure 2 - RISKEX Calculations

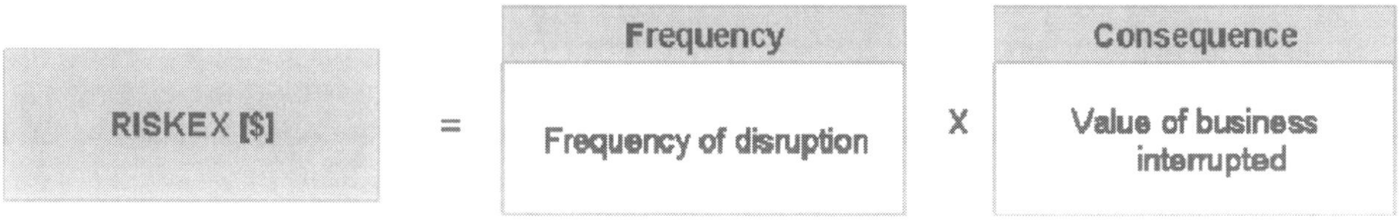

RISKEX can be reduced by reducing the frequency of occurrence, the consequences or a combination of the two as illustrated below.

Figure 3 – Impact of Risk Management

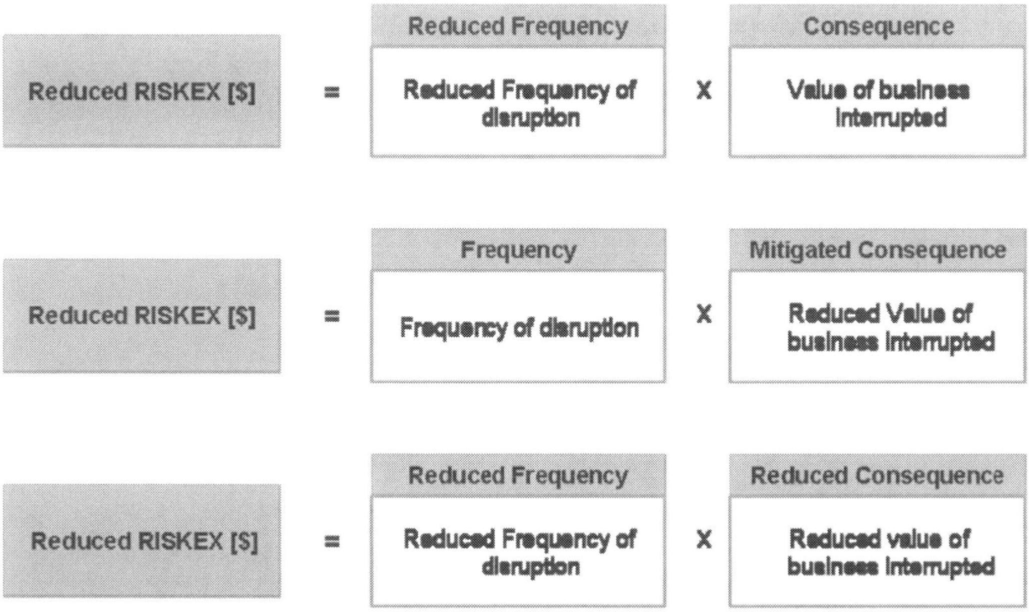

Frequency reducing measures can be:

- Technical
- Organizational
- Managerial

Consequence reducing measures can be:

- Contractual
- Technical
- Organizational
- Managerial
- Commercial Pricing

Risk Management Effect

Effective risk management will ensure that all significant business interruption risks are identified and prioritized based on criticalities (RISKEX). This will aid developers in the decision making process to ensure that the real risks are given the right attention and resources are optimized to reduce the risk to acceptable levels.

Figure 4 – Effect of Risk Management

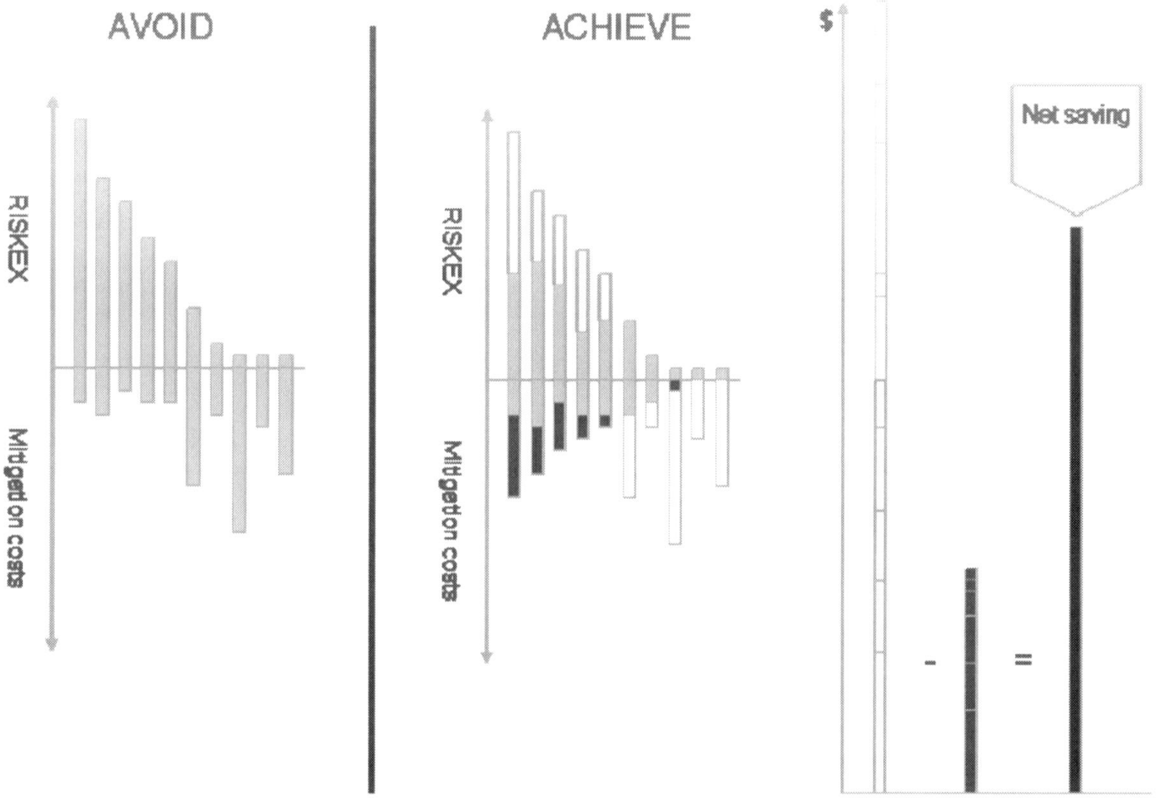

What Approach should be used to Manage Business Interruption Risk?

The LNG Industry uses several standards for HSE-related risks and works systematically to achieve acceptable HSE risk levels. Less standardization exists for managing the financial risks due to business interruption. The burden of determining the approach to take is placed largely on the risk owners. The risk owners consist mainly of the operators, the partners and the contractors. Nevertheless, lack of systems and tools to manage business interruption risk could cause significant financial losses throughout the project lifecycle.

How can then business interruption risks be given the right attention and be managed effectively? The answer to this question is project risk management. The next two sections will discuss how any project risk like business interruption risks in the LNG industry could be effectively managed as part of a structured project risk management process.

Project Risk Management

Project Risk Management (PRM) is a systematic approach for analyzing and managing threats and opportunities associated with a specific project. The use of PRM enhances the understanding of the major risk drivers and how these affect the project objectives. The objectives are typically, cost, schedule, technical performance, revenue over the project life cycle and may also include HSE. Through this understanding, decision makers can develop insight and suitable risk strategies and action plans. These actions plans would manage and mitigate potential project threats and exploit potential opportunities. This increases the likelihood of attaining project success.

Uncertainties affecting project objectives can come from a wide range of areas and disciplines:

- Technical
- Financial
- Organizational
- Contract / Procurement
- Sub-contractors
- Political / Cultural

While vastly different in nature, the uncertainties listed above all contribute to the overall risk in the planning and execution of the project, and hence control the success of the final product.

The term project risk management encompasses a wide range of activities each occurring at different intervals during a project:

1) Day-to-day identification, assessment and control of uncertainties, i.e. risk management activities related to understanding and controlling the most important risks threatening the achievement of well-defined project objectives. This type of risk management typically uses a qualitative approach where each risk is assessed separately.
2) Periodic assessment of the performance against project objectives, i.e. assessing the probability of achieving well defined project objectives such as schedule, budget technical performance but also revenue over the project lifecycle. The periodic assessments typically use a quantitative approach based on the aggregation of all (or at least the most critical) uncertainties.
3) Ranking of a set of alternative decision options / system solutions based upon their desirability measured in terms of the corresponding project objectives. This ranking uses cost-benefit analyses and is typically performed at major decision points during the conceptual design stage, where it can be of greatest benefit.

The first step in a PRM process is to identify project risks using formal brainstorming sessions and during daily work processes. Experience transfer from previous projects is also a key source for identifying potential project risks. It is important to revisit risk identification throughout the project to place less emphasis on those risks that have diminished, and to capture new risks, or those that now appear to be more critical than they did in the initial assessment.

The next step is to define probability of occurrence and consequence categories. Risk matrices can be utilized to help visualize criticalities, since these are the product of probability and consequence.

The last step, but by far the most important step, is to take actions to mitigate or prevent the various project risks. The criticality assessment enables actions to be prioritized, and maximizes the likelihood of successfully achieving the project objectives. It is this feature that distinguishes project management, which is carried out in most projects, and effective PRM, which is executed by only a few.

Success Factors for Effective Project Risk Management

PRM is often perceived as an unnecessary burden by project members. It is therefore important to create an organizational culture that supports the PRM process.

The core of the PRM process is the identification, prioritization and closure of risk reducing actions. While the completion of actions is clearly what delivers project success, it is common to expend a vast amount of resources on the steps in the process *except* for the vital action step. In such cases, PRM does nothing more than "make work", and is an unnecessary burden on the team. The greatest factor for success is therefore a culture and a system that are focused on the completion of risk-reducing actions.

Normally, a lot of good efforts are paid in identifying risks and in defining their criticalities, for example by placing them in risk matrixes. It is also normal to consider the criticality assessments as the most difficult step in the project risk management process. However, this is not the stage where the process fails.

The most difficult and the most important step in PRM is to define actions that mitigate or prevent risks effectively in a SMART manner. The SMART acronym is illustrated below:

Figure 5 - SMART

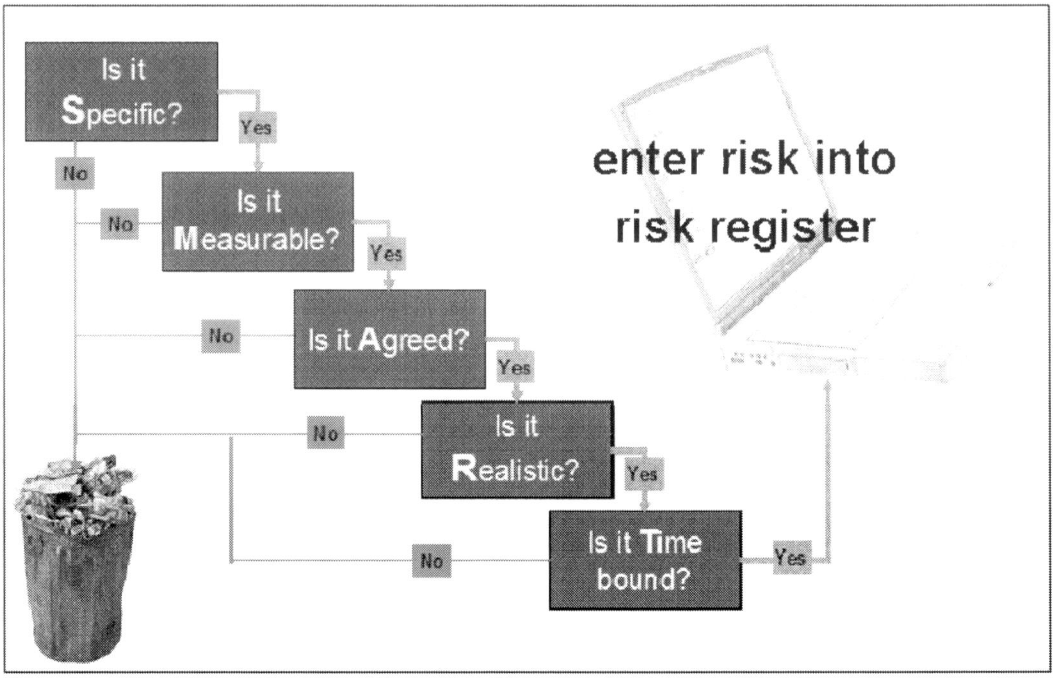

Any action that does not fulfill all these requirements is likely to have limited impact on the project risk.

A typical project risk register for a complex LNG development project would comprise hundreds of risks (including several business interruption risks) and actions not fulfilling these SMART requirements. This results in an unwieldy and often unmanageable process that serves only to undermine its own value. Quite justifiably, this frustrates project personnel.

For PRM to succeed, it is essential that the following are in place prior to project commencement:

- Management commitment to the project risk process
- Clear expectations from project organization about the role of PRM
- Organizational consensus on working routines and the PRM instrument
- Defined project targets and agreed costs for not meeting them
- Functions in the project organization that can manage the risk register and filter input.
- Action follow-up system
- Possibility to report risk trends

Having the above will enable the project team to define, prioritize and close risk reducing actions throughout the entire organization.

Conclusion

Effective risk management will ensure that significant business interruption risks are identified and prioritized based on criticalities. This will aid developers in the decision making process to ensure that the real risks are given the right attention and resources are optimized to reduce the risk to acceptable levels.

This paper has examined the use of risk management as a pro-active, strategic tool to evaluate business interruption costs to provide the basis for decision making. Methodology has been discussed that includes ranking and valuation of risk to optimize resources, maximize opportunities and enhance business performance.

This paper has also discussed how project risk management can be applied to maximize the odds of meeting project objectives like reduced business interruptions

References

- "The Need to Factor Asset Risk Into LNG Terminal Agreement Negotiations", Pedram Fanailoo and Steven Sparling, to be presented at Offshore Technology Conference (OTC), Houston, May 2007
- "The Need to Manage Risk in Floating LNG Developments", Ernst A. Meyer, Offshore West Africa Conference (OWA), March 2005

Biography

16340 Park Ten Place, Suite 100, Houston, TX 77087, USA.
Phone: 281-721-6694, Fax: 832-854-5325, Cell: 281-221-0692.
E-mail: goran.andreassen@dnv.com.

Mr. Andreassen is a Senior Consultant with 7 years experience from Quantitative and Qualitative risk assessments, HSE management, performance forecasting and technical safety from several major oil and gas engineering projects including offshore and onshore LNG receiving terminals, offshore production platforms, Mobile Offshore Drilling Units and onshore gas receiving terminals. He commenced his career with DNV in 2005, after 5 years with Aker Kvaerner Engineering and Technology in Norway. Gøran has been DNV's project manager for several LNG risk assessment studies in the US, including continuous support on Safety and Security concerns towards regulators. He holds a Master of Science Degree in Mechanical Engineering from The Norwegian University of Science and Technology.

The Use of Quantitative Risk Assessments for (QRA) LNG Terminals

Baris Arslan
Consultant
Relcon Scandpower AB
Uppsala, Sweden

Anders Wiik
Principle Consultant
Scandpower AS
Bergen, Norway

AIChE Spring Meeting, April 2007
7[th]Topical Conference on Natural Gas Utilization
Houston, Texas, 22-26, 2007

Abstract

Application of quantitative risk assessment (QRA) becomes more common during early phases of LNG projects instead of deterministic methods, such as worst-case or maximum credible event approaches. Among other reasons, it is mainly due to the fact that deterministic methods are heavily relied on "over-sized" scenarios, and may result in wasting sources on over-confident LNG-terminals in some cases. Deterministic methods are also unfavourable from the viewpoint of project owners who attempt to build LNG terminals more compactly with minimum sufficient separation distances, but still complying with applicable codes and regulations and without compromising on safety aspects. QRA is an integrated approach and has many advantages in comparison to deterministic methods. Evaluating up to hundreds of scenarios, it provides an in-depth understanding and valuable input to decision-makers. QRA also indicates the risks related to components or sections in an LNG plant dominating the overall risk picture and thus enabling that the resources for risk-reducing measures are allocated correctly. Another benefit of QRA is that, once quantifying the risks for a specific LNG plant, the change of risk level due to a future expansion project or any modification can easily be implemented and communicated. However, QRA has some limitations. An important limitation of QRA is encountered while identifying and treating the uncertainties, introduced during different stages of the risk assessment. The objective of this paper is to describe the application of QRA for LNG terminals and discuss various types of uncertainties in QRA for LNG terminals. This is achieved partly based on a case example from an ongoing Swedish LNG terminal project in the Stockholm region. It is the authors opinion that today, the QRA method, notwithstanding the problems regarding mapping out and treating uncertainties, is the most powerful tool available for safety evaluation of LNG terminals. When applied properly and the uncertainties treated well, QRA provides valuable input to decision-makers to plan cost-effective LNG terminals with very high level of safety.

Introduction

There is a tendency that LNG terminals, either proposed or under construction, are located near populated areas. This is due to the proximity to end users, access to infrastructure and utilities etc. It results in the attempts to build LNG terminals more compactly and minimize the separation distances, but still comply with the applicable codes and regulations without compromising on safety (Arslan, 2005). In this context, performance of quantitative risk assessments becomes more common during early phases of LNG-projects instead of deterministic methods, such as worst-case or maximum credible event based approaches, because deterministic methods are heavily relied on "over-sized" scenarios, and may result in wasting sources on over-confident LNG-terminals.

The purpose of this paper is to illustrate how a quantitative risk assessment (QRA) is applied for LNG terminals and discuss advantages and disadvantages compared to deterministic methods and uncertainties, arising during different steps of QRA. A case example is also given, from an ongoing Swedish LNG project, to give the reader a more clear understanding regarding both the application of QRA and the introduced uncertainties.

Deterministic approaches for safety evaluation of LNG terminals

There are various approaches, applied for safety evaluation of LNG terminals. The choice of the method is dependent on a number of factors, such as regulations, requirements from local authorities, application of standards such as EN1473, NFPA59A etc., available site-specific information prior to performance of safety evaluation and other resources (budget, competency, etc.) Based on evaluation of these factors, either a deterministic or a probabilistic approach is applied for LNG terminals. Traditionally, worst-case and maximum credible approaches are applied for LNG terminals as deterministic methods while quantitative risk assessment is applied as probabilistic method.

A deterministic risk evaluation is based on evaluation of the consequences which can occur during a "pre-defined" accident. Such accidents can be of "worst-case" or "maximum credible event" type as described below:

Worst Case Approach

The worst case event approach is based on evaluation of the most serious incident, focusing solely on the consequences. This approach enables decision-makers to evaluate several site alternatives by considering proximity to people or infrastructure. A typical conclusion from such an approach can be "In the worst-case, none of the identified hazards can lead to consequences, threatening third-parties" etc. (Pitblado, 2004)

Maximum Credible Event Approach

Maximum credible event approach is based on evaluation of consequences for a credible incident. The argumentation for the choice of such a "maximum credible" incident must be both powerful and plausible. Maximum-credible event approach takes into consideration both technical and organizational risk-reducing measures. Determination of "maximum-credible" events is a well-defined task, since the judgement is based on consideration of an incident which reasonably can occur, with regards to both active and passive safety measures. Therefore, it can be concluded that this deterministic approach has actually a probabilistic side to a certain extent. (Davidsson, 2005).

QRA-Approach for safety evaluation of LNG terminals

At a fundamental level, a risk assessment based approach, such as QRA, can be described as a structured way of identifying and evaluating the most important contributions to the overall risk that an activity exposes to society or environment. A typical QRA consists of the steps, shown in Figure 1.

As seen in Figure 1, QRA combines the likelihood of occurrence and potential consequences for all identified hazards. Thereafter, a so-called risk picture is estimated, representing all analysed risks. It can be done by various methods, such as F/N-curve (F=Frequency/N=Number of casualties), potential loss of life (PLL) value etc. Further in a QRA, a risk evaluation is done in the light of applied acceptance criteria. There are a number of factors, playing role while determining acceptance criteria, such as local- or company-specific requirements.

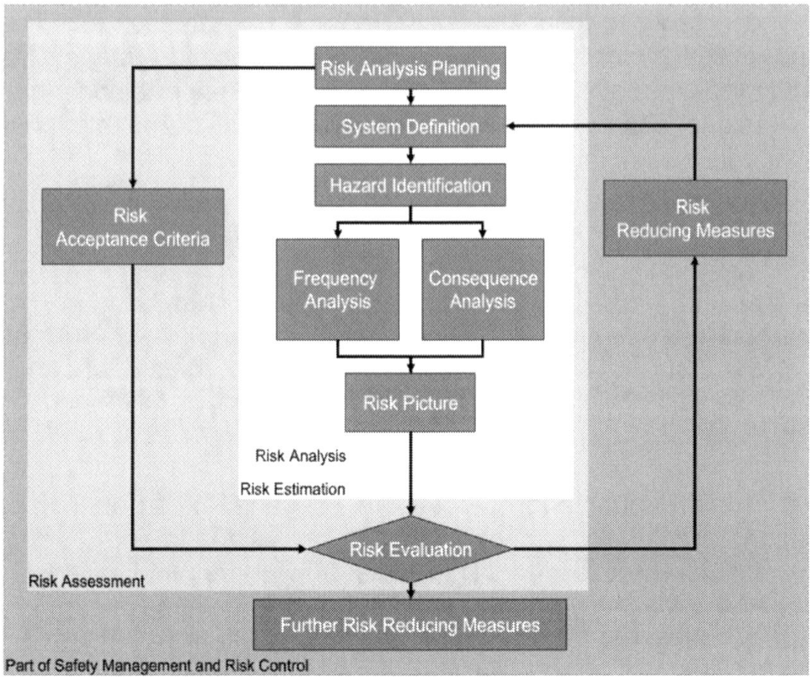

Figure 1 Risk Assessment Methodology

As an integrated approach, QRA is a very powerful tool and has many advantages in comparison to traditional deterministic methods. Evaluating up to hundreds of scenarios, it provides an in-depth understanding and provides valuable input to decision-makers. Moreover, QRA also highlights the incidents dominating the whole risk picture and hence the resources for risk-reducing measures are allocated correctly.

An important strength of the QRA is that it enables decision-makers to find cost-effective solutions. (Kirchsteiger, 1999) This statement is valid also in case of LNG terminals, because both maximum-credible approaches and worst-case approaches usually are heavily relied on "over-sized" cases, and may result in wasting sources on over-confident LNG-terminals. Another benefit of QRA is that the results can easily be communicated with the decision-makers who take necessary measures in order to reduce the risks to a tolerable level. Such measures vary remarkably from slightly changes on plant design to cancellation of the whole project. Additionally, once quantifying the risks for a specific LNG plant, the change of risk level due to a future expansion project or any modification can easily be evaluated, providing valuable input to decision-makers. A typical reason for upgrading an existing QRA is capacity enhancements where several new components, including storage tanks, unloading arms etc. are installed in an LNG terminal. In this case, decision-makers can simply study the change in overall risk picture and identify the major risk contributors.

The use of QRA may possess significant limitations when several items / steps are either not managed well or not treated at all. One of the main problems with quantitative risk assessments is that it involves a degree of uncertainty in different steps, which is not always addressed and treated properly in the risk analyses. These uncertainties can, under some circumstances, imply that the credibility and practical usability of QRA:s are questioned. The main uncertainties in QRA are introduced during identification stage, frequency analysis and consequence modelling. Dealing with uncertainties may be a challenging task in case of safety evaluation of LNG terminals because of several factors discussed later in this paper.

Application of QRA: LNG Terminal in Stockholm Region

A more refined objective of this paper is to illustrate how QRA is applied for LNG terminals and highlight uncertainties introduced during different stages of the risk assessment. A case example is also given, based on an ongoing Swedish LNG project, to clarify further the application of QRA and produced uncertainties.

The authors of this paper performed a QRA on behalf of an energy company which is planning to build and operate a terminal for import-, storage and distribution of LNG in Sweden. The LNG import terminal will be built in Brunsviks holme, an island, neighbouring an existing refinery located approx. 70 km south of Stockholm. The planned LNG terminal will have a storage capacity of 20 000 m^3 LNG and approximately 20 ship supplies a year. Approximately 2/3 of the LNG will be transported by LNG-trucks to the various end-users in the region while the rest of LNG will be regasified in the terminal and transported to the neighbouring refinery via a pipeline. (Arslan, 2006)

Figure 2 Illustration of LNG Terminal in Nynäshamn, Sweden, (Osbrink, 2006)

At the early stages of the QRA, a HAZID session (Hazard Identification) is normally held with participation of people, having extensive experience about LNG-terminals and LNG safety. A typical approach in such a HAZID is that the whole LNG-plant is divided into a number of sections and evaluated separately on a structured way. For this LNG import terminal, five sections, listed below, are used:

- Unloading of LNG-ships to storage tanks (LNG transport to terminal-usually excluded)
- Storage and daily operation of the storage facility
- Regasification and export through pipeline to boundary limit (downstream pipeline transport not included)
- Loading of LNG-trucks from storage tanks and export by road (road transport outside facility -usually not included)
- External hazards not discussed in previous sections (sabotage etc.)

Upon completing the HAZID and making a preliminary evaluation of identified hazards, a number of models were built covering all sections. The models are altogether representative for the whole LNG-terminal. Thereupon, these models were analyzed with respect to their consequences and probabilities. To further exemplify the model-based approach, one of the evaluated models is shortly described below:

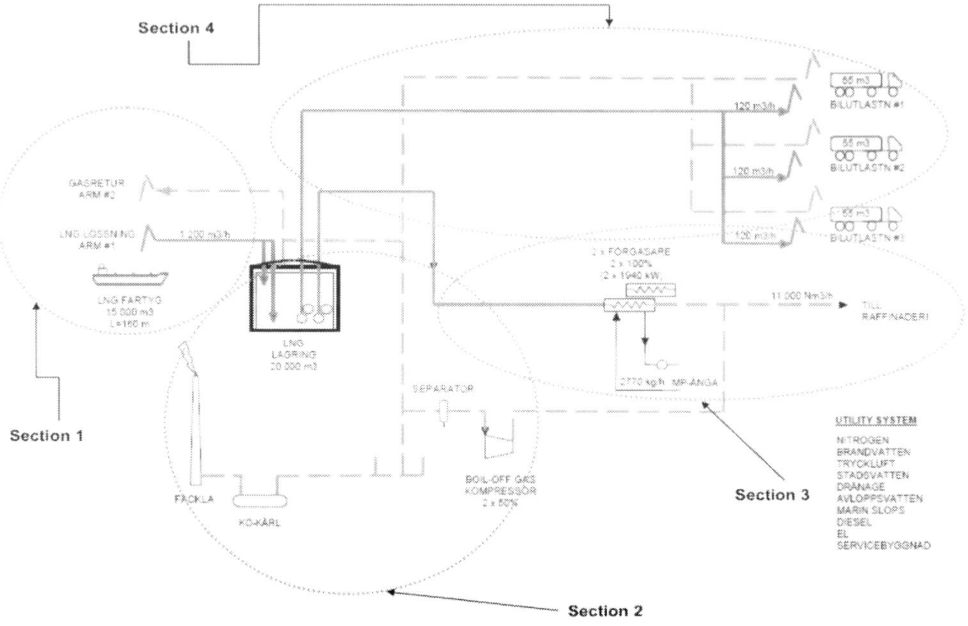

Figure 3 Illustration of different sections, used in the QRA

Case: LNG spill during loading of LNG-trucks in Section 4:
One of the evaluated models was based on the assumption that a large-scale LNG spill occurs during loading of LNG-trucks. Both loading and unloading operations at LNG plants are usually considered to be safety critical moments requiring special attention, but focus is usually put on the ship transport. Yet the loading of LNG-trucks may get less attention at times, notwithstanding the potential hazards of the operation.

The planned facility in the Stockholm region will have a filling station for LNG-trucks with a capacity to load two trucks at the same time. In the used model, a large LNG spill was assumed to take place due to a rupture of loading hose (flexible) during loading of LNG-trucks. Two alternatives are studied:

A pool fire occurs upon immediate ignition of LNG in the filling area
A flammable gas cloud is formed, resulting in a fire in another section of the facility due to delayed ignition

Having constructed the model, a series of simulations was done in order to estimate potential outcomes of above-mentioned cases. Based on the results of consequence modelling and other available information, probabilistic models were built up in order to estimate the expected number of fatalities due to activities in the section. In general probabilistic models includes a great number of parameters, containing components such as, valves, flanges, pipelines, compressors, vaporisers, pumps, tanks, supply system and fault frequencies regarding small, medium and large-scale LNG leakages, weather conditions, potential ignition sources etc. Moreover, while estimating the PLL (potential loss of life) value, information regarding people present in the facility during certain period of times, such as truck drivers, visitors, maintenance workers, security guards are included in the model.

It should be noted that the probabilistic models have taken into consideration leakage frequencies for all sections. By that way it could be ensured that none of the leakage sources were excluded from the analysis while focusing on the evaluation of special models. All of the evaluated models, including above-mentioned, constitute a whole system. The whole analysis is actually a go-through evaluation of the models and also their relationships with their underlying structure.

For the QRA of the Stockholm LNG terminal, it was very important to evaluate escalation risks in detail due to the fact that the terminal is relatively compact and located nearby a refinery where large quantities of flammable chemicals are handled. An accident, initiated in the LNG terminal, may escalate to the refinery area, leading to devastating consequences. In this case, the capability of QRA for studying underlying logical relations both inside and outside the LNG terminal enabled the risk analysts to evaluate escalation effects on a structured way. Such a detail analysis of escalation risk would by no means be possible without quantifying the risks. (Cp. deterministic approaches).

Uncertainties in QRA for LNG Terminals

At a fundamental level, there are two main types of uncertainties in QRA's, namely: knowledge-based (epistemic) and aleatory (stochastic) types. The difference between these two types is that knowledge-based uncertainty is reducible while aleatory uncertainty is not. Because the aleatory uncertainty reflects the randomness in natural it is evidently irreducible (Abrahamsson). The rest of the paper and the discussions below focus only on knowledge-based uncertainty, because this is the only uncertainty that might be reduced. The aleatory type is an inherent part of a QRA.

As shown in Figure 1, a traditional QRA consists of different steps. The uncertainties, as well, arises in each step. Particularly, the uncertainties are introduced mainly during the identification of hazards, estimation of likelihoods and consequences, the uncertainties are discussed below.

Hazard Identification

As previously mentioned, a so-called HAZID (Hazard Identification) session is usually held prior to the performance of QRA. The main purpose of this session is to identify all hazards related to the activities in the LNG-facility. The quality of findings, gathered during the HAZID, is of vital importance for the completeness of the model to be built. Participants, lacking an in-depth understanding of LNG terminals, or a HAZID-leader - not facilitating the session well-balanced, can by no means provide necessary input for further analysis.

Estimation of Consequences

Uncertainties occur also during estimation of consequences. At this stage, large knowledge-based uncertainties can arise due to both limitations of simulation tools while simulating LNG-related physical phenomena and the user's ability to simulate and finally interpret the results correctly. Furthermore, incompleteness of the consequence models due to receiving insufficient input from HAZID-session can introduce knowledge-based uncertainties. Incompleteness of consequence models can become a major issue while describing escalation risks due to fires and explosions. Mostly, consequence models focus only on the source accident without investigating potential escalation effects. However, it is a fact that most major accident can be limited if the escalation effects of the initiating event are prevented.

Estimation of Likelihoods

The uncertainty related to employment of probabilistic models is a major concern in case of LNG terminals. Due to its nature, probabilistic models, based on either historical records or event/fault tree methods, include always uncertainties with regards to accuracy and applicability of used information.

While using historical records for estimation of likelihoods, both the accuracy and applicability of information are important to keep the uncertainties as low as possible. In case of LNG terminals, accuracy is a major problem due to the fact that there is no large database, reflecting experienced failures in LNG-terminals worldwide. The uncertainties are significant, especially within the area of small- and medium-scale LNG spills (often not reported).

For LNG-terminals, applicability of historical information is also a critical issue. For instance, estimation of the likelihood for occurrence of classical "rollover" phenomena is not as straightforward as just dividing the number of rollover incidents by total tank year. For each tank and rollover incident, there are various underlying causes which also should be evaluated. To conclude, the available information is rarely directly-applicable. It can be highly time-consuming to estimate an applicable likelihood with tolerable uncertainties.

Another important source of uncertainties is application of event/fault tree method for probabilistic models. Attempts to create such models without sufficient information, may force a risk analyst to make a great number of assumptions, subsequently introducing large uncertainties in QRA. In case of LNG terminals, these assumptions might concern the parameters like; the number of operative and administrative personnel, maintenance periods, truck traffic inside the facility, fire prevention and mitigation procedures, number of visitors, security guards, type of used components including storage tanks, unloading arms, truck filling equipment etc... (knowledge-based)

Discussions

Since a large part of this paper focus on description of uncertainties, the reader may think that QRA is not a straightforward method for LNG terminals. Contrary to this belief, it is one of the main strengths of QRA that it enables risk analyst to map out the uncertainties, introduced in different levels of risk assessment. It is the only way to reduce the uncertainties to a tolerable level.

As previously stated, the main difference between aleatory and knowledge-based uncertainties is that the latter one is reducible. A closer look at the uncertainties, described in previous section, shows that majority of uncertainties during application of QRA on LNG terminals are of knowledge-based type.

The uncertainties, introduced in hazard identification stage, are exclusively knowledge-based and can be reduced easily by handpicking suitable participants and a highly-qualified facilitator. The uncertainties, related to estimation of consequences are partly introduced due to randomness of parameters, reflected in the consequence models. Another source of uncertainties, as described previously, is the limitations regarding numerical methods, model validations etc. Constant development is observed in tact with progress within the area of numerical methods and computational capacity and the uncertainty is reduced continuously.

However, the uncertainties introduced during evaluation of probabilistic models still remain as a major concern due to the lack of applicable and accurate information. When used wisely, an experience database has a great potential to serve as input to the process of probabilistic modelling. In particular, a database, focusing on experienced leakage frequencies due to various activities in LNG terminals would give a significant reduction of knowledge-based uncertainties in QRA's for LNG terminals. Obviously, it is by no means easy to prepare databases, satisfying all the needs of all risk analysts. However, a more general experience-based database, complying with the criteria's below, would enhance the usefulness of QRA for LNG terminals significantly.

- Covering most of the LNG terminals in the world
- Reliable and regularly updated.
- User-friendly and publicly accessible.

Conclusions

The experiences from the performed QRA's show that, in the light of limited LNG-specific statistical data and considering a great number of combinations to be evaluated, QRA may become a time-consuming method for LNG terminals. Adding the efforts to identify and treat the uncertainties, QRA appears to be a costly method in comparison to deterministic approaches. However, it should be noticed that the uncertainties exist regardless of which approach we do apply. Today, the QRA method, notwithstanding the problems regarding mapping out and treating uncertainties, is the most powerful tool, available for safety evaluation of LNG terminals. When applied properly and the uncertainties treated well, QRA provides valuable input to decision-makers to plan cost-effective LNG terminals with very high level of safety. Hopefully, this paper will contribute to an increase in knowledge and understanding of the appropriate choice and application of risk analysis for LNG terminals.

References

- Arslan, B., Wiik, A. (2006) Quantitative Risk Assessment of LNG Import Terminal in Nynäshamn, Scandpower
- Arslan, B. (2005) Design and Safety Aspects of LNG Import Terminals, MSc Thesis, Lund Institute of Technology, Sweden

- Osbrink, Peter (2006) Technical Description of LNG Terminal in Nynäshamn, SWECO
- SIGTTO, LNG Operations in Port Areas - essential best practices for the industry
- European Standard (1997) EN1473:1997 Installation and equipment for liquefied natural gas - design of onshore installations
- NFPA International (2001), NFPA 59A - Standard for the Production, Storage, and Handling of Liquefied Natural gas
- Pitblado, R & Baik, J. & Raghunathan, V. (2004) LNG Decision Making Approaches Compared, DNV Consulting
- Kirchsteiger, C. (1999) On the use of probabilistic and deterministic methods in risk analysis, European Comission, Systems Modelling and Assessment Unit
- Apostolakis, G. (2004) How Useful is Quantitative Risk Assessment?
- Abrahamsson, M. (2002) Uncertainty in Quantitative Risk Analysis - Characterisation and Methods of Treatment
- Davidson G & Haeffler L (2005), Värdering av Risk, Swedish Rescue Agency

Biography of Authors

Name: Arslan, Baris
Organization: Relcon Scandpower AB
Position/Title: Consultant/ MSc Mechanical Engineer
Mailing Address: Baeverns Graend 21, P.O. Box 1825, S-751 48 Uppsala, SWEDEN
Phone: +46 18188650
Fax: +46 18695416
E-mail: baa@scandpower.com

Consultant Baris Arslan holds a MSc. degree in Mechanical Engineering from Lund Institute of Technology in Sweden. Specialized in the field of design and safety of Liquefied Natural Gas (LNG) terminals, Arslan works primarily with risk analysis and technical safety in LNG industry.

Name: Wiik, Anders (co-author)
Organization: Scandpower AS
Position/Title: Principle Consultant/Manager, MSc Chemist
Mailing Address: Nygardsgt. 114, N-5008 Bergen, NORWAY
Phone: + 47 55 30 05 48
Fax: + 47 55 30 05 56
E-mail: awi@scandpower.com

Manager and Principal Consultant Anders Wiik has 10 years of experience from the oil & gas and chemical industry. In his career, he has mainly worked with gas processing, gas explosions, risk and safety analysis and emergency preparedness for gas facilities onshore and offshore. For 3 years he has been responsible for the natural gas and LNG market in Scandpower. At present he holds the position as Manager at the Scandpower Bergen office. Wiik has a Master degree in organic chemistry from the University of Bergen and supplementary education from the Norwegian School of Economics and Business Administration. Before he joined Scandpower he has worked for Borregaard Industries Ltd, Christian Michelsen Research and Gexcon.

Wind Tunnel Studies of LNG Vapor Dispersion from Impoundments

Jerry Havens, Tom Spicer and Wendy Sheppard
Chemical Hazards Research Center
Ralph E. Martin Department of Chemical Engineering
University of Arkansas
Primary contact: jhavens@uark.edu

ABSTRACT

If LNG vapor dispersion effects due to vapor holdup by impoundments or obstacles are to be determined as currently required by 49 CFR 193, FEM3A must be used, either to determine directly the downwind distance to the ½ LFL (endpoint) concentration, or to determine the rate and concentration of LNG vapor (in air) at which the vapor holdup volume (impoundment) overflows – for subsequent input to DEGADIS to determine the exclusion zone distance. The applicant may opt to use DEGADIS (without FEM3A), but must then disregard vapor holdup effects. After outlining the development of the regulatory requirements for dispersion models since 49 CFR 193 was promulgated, this paper presents data from recent wind tunnel experiments designed to determine experimentally, using carbon dioxide gas with density similar to LNG vapor, the effects upon dispersion of an impounding dike, with or without an LNG tank centered inside the dike, and compares these dispersion scenarios with that which would result in the absence of either a tank or dike. The results, while confirming an expected reduction in downwind exclusion zone requirements due to the presence of a dike and tank, indicate that the reduction in these tests is primarily due to the tank rather than the dike. However, the reduction in exclusion zone distance in this data set, due to the presence of a tank and dike, appears to be masked by the atypically high value of surface roughness used in the wind tunnel, suggesting that the relative effect of the tank would be even greater if more typical, smaller, site roughness were present. Surprisingly, the measured distance to the endpoint concentration was shown to be greater when the dike (alone) is present than when it is absent. This finding is explained as resulting from the dike's restriction of gravity spreading, which decreases the overall planar area for vertical mixing of the vapor cloud, thus increasing the downwind distance to the endpoint concentration.

BACKGROUND

Federal requirements for vapor cloud dispersion exclusion zones around LNG storage and transfer facilities direct the use of dispersion models specified in 49 CFR 193. The dispersion model required in 49 CFR 193 (promulgated in 1980) was a line-source Gaussian (passive) dispersion model ("the MTB model"; Arthur D. Little, Inc., 1974) which did not account for LNG vapor negative buoyancy (LNG vapor, because of its low temperature, can be denser than air). Neither did the MTB model provide for consideration of (earth) surface-to-cloud heat transfer, holdup in vapor detention systems (such as by impoundments, dikes, and vapor fences), wake turbulence, or directional diversion of vapor clouds by topography. It was recognized that modeling tools that would take into account all of these factors were required in order to address important issues facing the industry in determining safety exclusion zones to protect the public from vapor clouds that might be formed from LNG releases, particularly at land-based facilities where such site specific factors are most important.

The Gas Research Institute (GRI) sponsored research to develop and verify alternative dispersion modeling approaches which might be substituted for the initially prescribed passive dispersion model (Havens and Spicer, 1990; Havens et al., 1994, Spicer and Havens, 1997). GRI defined two major objectives for the development of alternative vapor dispersion methods for regulatory use (specifically in 49 CFR 193):

- The first objective was to define "short-cut" methods for determining exclusion zones for site selection screening, tentative compliance evaluations, and applications where minimization of exclusion area is not as crucial (such as for small plants located in rural areas). This objective was accomplished when 49 CFR 193 was amended in 1992 to require the use of DEGADIS (DEnse GAs DISpersion) in place of the originally prescribed model (Havens and Spicer, 1990). DEGADIS accounts for LNG vapor negative buoyancy as well as for consideration of the effect of (earth) surface-to-cloud heat transfer on atmospheric turbulent mixing. However, being a "flat-earth" model, it does not provide for vapor cloud liftoff or effects on vapor cloud dispersion of terrain or obstacles of any kind.
- The subsequent objective was to define methods for regulatory applications where topographical effects, wake turbulence, vapor holdup in detention systems, and energy addition (heat transfer and moisture condensation) to the cloud, including cloud liftoff due to positive buoyancy, considered separately or in combination, might be expected to importantly influence exclusion zone requirements. This objective was accomplished when 49 CFR 193 was amended in 2001 to allow the use of the FEM3A model as an alternative to DEGADIS <u>if the applicant desired to consider the effect upon dispersion of man-made or natural obstacles (to the flow) such as dikes, tanks, or terrain features</u> (Spicer and Havens, 1997).

Reports by Havens, Spicer and Walker (1996), submitted to the Department of Transportation (DOT) as supporting documentation for the amendment of CFR 193 to allow use of FEM3A, present the results (prior to 1996) of the extensive research program conducted to develop accurate, repeatable laboratory (wind tunnel) data sets which begin to effectively address realistic LNG vapor dispersion scenario issues and which could be used to evaluate the predictability of LNG vapor dispersion mitigation with FEM3A or other computational fluid dynamics (CFD) models. Data sets were provided to evaluate mathematical models applicable to the prediction of the following laboratory (wind tunnel) scale dense gas dispersion scenarios:

A. Isothermal dense gas (carbon dioxide) released continuously (steady state) with near-zero momentum from a floor-level area source.
B. Isothermal dense gas (carbon dioxide) released continuously (steady state) with near-zero momentum from a floor-level area source in the annular space within a dike surrounding a storage tank.

These experiments in the CHRC wind tunnel to validate the FEM3A model for prediction of the effect of the presence of tank and dike structures utilized a smooth wind tunnel floor. Further evaluation of this early work indicated that the presence of the smooth floor combined with the low wind speeds required to simulate the dense gas effects involved in LNG vapor dispersion can result in laminarization of the boundary layer near the wind tunnel floor.

Although such effects were considered in the evaluation of FEM3A to allow the data to be used for model validation at wind tunnel scale, such conditions make suspect the scaling of the wind tunnel flows to field scale (because field conditions are normally fully turbulent). There were strong indications that the experimental data from the wind tunnel would be more applicable to field conditions, and therefore more useful for model validation, if the floor were artificially roughened. The resulting research program, reported here, repeated the previous experiments with uniform roughness elements covering the wind tunnel floor, and added a third experiment:

C. Isothermal dense gas (carbon dioxide) released continuously (steady state) with near-zero momentum from a floor-level area source in the annular space within a dike, but without the storage tank.

WIND TUNNEL EXPERIMENTAL TEST PROGRAM

Experiment Design

The experiments were performed in the Chemical Hazards Research Center's Ultra-Low Speed Wind Tunnel. The wind tunnel design details, as well as a description of the methods used to measure gas concentrations and turbulence parameters, are described in Havens and Spicer, 2006.

The 150:1 scale model experimental conditions assumed Reynolds Number independence, and, consequently, are based on Froude Number scaling, as follows:

$L_m = L_f/150$ $\qquad$ $U_m = U_f/150^{1/2}$ $\qquad$ $Q_m = Q_f/150^{5/2}$

where L_m = model length scale
L_f = field length scale
U_m = model velocity scale
U_f = field velocity scale
Q_m = model volumetric gas flow rate
Q_f = field volumetric gas flow rate

It follows that the model time scale (T_m) and the field scale (T_f) are related as $T_m = T_f/150^{1/2}$.

The tank/dike configuration was a 1:150 scale model intended to represent a typical peak shaving facility with a nominal 1 BCF storage tank. The tank was located in the center of the dike, on a platform topped with a mesh screen (flush with the tunnel floor surface), that covered a gas source box mounted below the floor. A steady flow of 33.4 slpm carbon dioxide traced with 0.5 slpm (~1.5%) propane was introduced through the wind tunnel floor into the impoundment area. 150:1 scaling dictates that the equivalent field scale LNG release would be the steady evaporation, at the spill rate, of a continuous release of 0.6 m^3/s LNG, here considered to be pure methane. The experiment design is summarized:

	FIELD	TUNNEL
Release Rate	0.6 m^3 LNG/sec	33.9 l CO_2/min
Dike side length (inside)	95.4 m	0.636 m
Dike side length (outside)	169.5 m	1.13 m
Tank Wall Height	32.5 m	21.6 cm

Tank Top Height	42.5 m	28.3 cm
Tank Diameter	46.5 m	31 cm
Dike height	5.6 m	3.7 cm
Wind Speed	4.9 m/s at 10 m	40 cm/s at 6.7 cm
Atmospheric stability	neutral	neutral

The wind tunnel experiments were releases of carbon dioxide at ambient temperature (typically 23 ^{o}C) and atmospheric pressure (typically 14.2 psia) - for several reasons. First, the density of carbon dioxide (at ambient temperature and pressure) is approximately equal to the density of LNG vapor (considered here to be pure methane) at its normal boiling point. Second, the experiments were designed to be isothermal (no heat transfer effects), as it is not possible to simultaneously scale-model turbulent mixing processes and surface-to-cloud heat transfer effects in the wind tunnel. Finally, the isothermal releases, uncomplicated by heat transfer effects, allow for more direct study of the validity of the Reynolds Number independence assumption upon which direct scaling of the lab scale experiments to field conditions would be based. The wind tunnel gas release source area was 0.3341 m^2 for all three cases.

Results

Data are presented here for three experimental configurations, all with the roughened wind tunnel floor surface:

Case A. Low momentum area source CO_2 release without obstacles

Case B. Low momentum area source CO_2 release with dike and tank

Case C. Low momentum area source CO_2 release with dike only

Flow visualization tests were conducted to demonstrate the symmetry of the flow in the tunnel. Visualization of the clouds was accomplished by adding Rosco fog (theater) fluid to the carbon dioxide at the entrance to the source box under the tunnel. Fog fluid addition was at the same rate, as nearly as could be approximated, for all tests. A video camera was mounted near the center of the tunnel ceiling so as to give an oblique view of the gas flowing in the downwind direction. Camera position and angle of view were the same for all tests. Figures 1-3 show typical video frames from Cases A-C tests, respectively. We will return to these visualization test results in the Discussion section. Initially they were used primarily to demonstrate that the vapor cloud moved down the wind tunnel center line as desired.

Measurements were made of the vertical profile of mean velocity and turbulence statistics on the centerline of the wind tunnel immediately upwind of the gas source box. Figure 4 shows measurements of the upwind, approach flow, velocity profile over the roughened wind tunnel floor. These data were used to determine the mean velocity at the (model) elevation corresponding to 10 meters elevation at field scale (1000 cm/150 = 6.67 cm) as well as to determine the friction velocity, 0.035 m/s, and the surface roughness, 0.00072 m. All of the experiments reported here were for this approach flow velocity profile.

Lateral profiles of gas concentration were made at pre-assigned downwind locations. All of the lateral profile measurements were made at 0.5 cm elevation, the closest practicable measurement position above the wind tunnel floor. The lateral profiles spanned the gas cloud with a 10 cm lateral interval between measurement locations. Downwind distances at which the lateral profiles were measured varied in order to ensure accurate determination of the distances associated with the concentrations representing the upper flammable limit (UFL), the lower flammable limit (LFL), and the one-half of the lower flammable limit (LFL/2). In some cases the location of the measurement distances had to be varied slightly to avoid the concentration sensor touching a roughness element or the dike. Figures 5-7 are plots of the lateral profiles of concentration measured for Cases A, B, and C, respectively. Below are summarized the maximum carbon dioxide concentration values measured at the specified downwind locations for Cases A - C.

Case A: No tank or dike, with roughness

Downwind location (cm)	55	88	175	236	357
Maximum concentration (%)	24.2	18.9	5.1	2.8	1.4

Case B: Tank and Dike, with roughness

Downwind location (cm)	45.5	84	175	236	357
Maximum concentration (%)	11.23	4.0	2.3	1.6	0.90

Case C: Dike only, with roughness

Downwind location (cm)	45.5	84	175	236	357	410	447	535
Maximum concentration (%)	28.1	12.8	5.5	2.6	1.6	1.4	1.2	1.0

The 45.5 cm downwind measurement location is at the top downwind edge of the downwind dike.

DISCUSSION AND CONCLUSIONS

Cases A – C all show a slight shift of the plume to the right side of the tunnel. This result is highly repeatable and is believed to be the result, at least partly, of a very slight slope in the wind tunnel floor to the right hand side - it has been observed throughout the fifteen-year working history of the wind tunnel. The wind tunnel floor was leveled to the nearest 1/100 inch during construction, but experience with dense gases indicates that slopes within this magnitude of uncertainty could cause the shift observed. It is noted that slight position errors in centering the tank and elevations of the dike surfaces could also contribute to the shift observed to the right in the wind tunnel, as could possible slight pressure variations that persist due to minor differences in blockage in the area of return flow (it is impossible to place air/conditioning equipment and electrical conduits and raceways so that the blockage is completely uniform in the return space over, under, and around the tunnel cross section). Nevertheless, we do not consider this shift to be important, as it is rather slight, as long as the maximum concentration at a given downwind position is used for model evaluation, as has been done here.

Cases A - C all show the expected more rapid dilution in the near field followed by less rapid dilution in the far field.

Case B clearly demonstrates the dilution in the near field that would be expected to result from the presence of obstacles, showing the rapid dilution that occurs in the wake of the tank as a result of vertical mixing in the cloud over the length scale (height) of the tank. The tank causes bifurcation of the cloud, with near zero concentrations occurring at ground level on the centerline downwind.

Case C demonstrates that an assumption that any obstruction (such as the dike in Case C) will result in greater dilution and corresponding shortening of the exclusion zone is by no means certain – Case C indicates a greater downwind travel distance with the dike than without (Case A). Although this result was not expected, further analysis suggests that the dike restricts the gravity spreading of the cloud until the dike overflows, thus narrowing the cloud and reducing the overall top surface area of the cloud. As it is known that lateral gravity spreading can decrease downwind maximum distances traveled (due to a resulting wider cloud with greater surface area for air entrainment), we believe that Case C indicates that restriction of gravity spreading (by the dike) explains the greater travel distance observed with the dike. Comparison of the flow visualization experiments (Figure 1) and the lateral measured gas concentration profiles (Figure 3) appear to confirm this observation. We note that CFD models should be able to provide for this effect, but bulk parameter models (similar to DEGADIS) would not be expected to cope.

This work demonstrates the importance of the site roughness. The field scaled roughness value corresponding to the wind tunnel roughness would be about 11 cm, which is significantly higher than the 3 cm value typically recommended for a grassy site surrounding an LNG terminal. Comparison with earlier measurements on the CHRC tunnel smooth floor suggest that this large surface roughness alone is decreasing the downwind travel distance by roughly the same amount as the presence of the tank and dike (on the smooth floor) alone. Consequently, the large roughness effect, particularly for Case C, should not be considered typical. It is noted that the spikiness in the intermediate downwind distance lateral gas concentration profiles shown for Cases A and C is very repeatable and correlates with the position of the gas sampling probe being in or out of the wake of a roughness element. As expected, the greater dilution that occurs at these distances in Case B results in smoothing of this effect, and the spikes are not observed, even though the gas concentrations were made in the same locations as for Cases A and C.

FEM3A or DEGADIS should be used with the roughness appropriate to the site under consideration. The large roughness here is designed to result in a wind tunnel boundary layer that is scalable to field conditions. We believe that the smooth floor wind tunnel dense-gas dispersion data previously reported (Havens and Spicer, 2003 reference?), although useful for limited mathematical model validation, should not be scaled to field conditions, for the reasons stated earlier in this report. However we are confident that the rough floor wind tunnel data reported herein does not suffer that weakness. The CHRC can provide interested parties extensive background on the development of the surface roughness used in this work, as its development and testing was undertaken in a joint program by the University of Arkansas and the Environmental Protection Agency's Environmental Wind Tunnel in North Carolina for the Petroleum Energy Research Foundation (Havens et al., 2001: Briggs et al., 2001).

REFERENCES

Arthur D. Little, Inc., "Evaluation of LNG Vapor Control Methods," Report to the American Gas Association, October 1974.

Briggs, G. A., R. E. Britter, S. R. Hanna, J. A. Havens, A. G. Robins, and W. H. Synder, "Dense Gas Vertical Diffusion Over Rough Surfaces: Results of Wind-Tunnel Studies," Atmospheric Environment, 35, 2001.

Havens, J. A., and T. O. Spicer, "Development of an Atmospheric Dispersion Model for Heavier-than-Air Gas Mixtures," Final Report to U.S. Coast Guard, Contract DT-CG-23-80-C-20029, May 1985, Volumes I, II, and III.

Havens, J. A., and T. O. Spicer, "Evaluation of 3-D Hydrodynamic Computer Models for Prediction of LNG Vapor Dispersion in the Atmosphere," Series of reports (including final report) to the Gas Research Institute, GRI Contract No. 5083-252-0788, 1983-88.

Havens, J. A., and T. O. Spicer, "Modification and Extension of LNG Vapor Dispersion Simulation Methods for Regulatory Application," series of reports to Gas Research Institute, GRI Contract No. 5086-252-1287, 1988-present.

Havens, J. A., and T. O. Spicer, "LNG Vapor Dispersion Prediction with the DEGADIS Dense Gas Dispersion Model," Gas Research Institute Report No. 89-0242, 1990.

Havens, J. A. and T. O. Spicer, "Feasibility Assessment of a Conjunctive Modeling Approach for LNG Vapor Dispersion Prediction," Gas Research Institute Report, 1991.

Havens, J. A., T. O. Spicer, and H. L. Walker, "Regulatory Application of Wind Tunnel Models and Complex Mathematical Models for Simulating Atmospheric Dispersion of LNG Vapor," Gas Research Institute Report No. 92-0257, August 1994.

Havens, J. A., T.O. Spicer, and H. L. Walker, "Evaluation of Mitigation Methods for Accidental LNG Releases: Volume 1/5--Wind Tunnel Experiments and Mathematical Model Simulations to Study Dispersion of a Vapor Cloud Formed following LNG Spillage into a Diked Area Surrounding a Storage Tank," Topical Report for Gas Research Institute, November 1996.

Havens, J. A., T. O. Spicer, and H. L. Walker, "Evaluation of Mitigation Methods for Accidental LNG Releases: Volume 2/5--Wind Tunnel Experiments and Mathematical Model Simulations to Study Heat Transfer from a Flat Surface to a Cold Nitrogen Cloud in a Simulated Atmospheric Boundary Layer," Topical Report for Gas Research Institute, November 1996.

Havens, J. A., H. L. Walker, and T. O. Spicer, "Wind Tunnel Study of Air Entrainment Into Two-Dimensional Dense Gas Plumes at the Chemical Hazards Research Center," Atmospheric Environment, 35, 2001.

Havens, Jerry and Tom Spicer, "Vapor Dispersion and Thermal Hazard Modeling," Final Topical Report to Gas Technology Institute, October 2006.

Spicer, T. O. and J. A. Havens, "Evaluation of Mitigation Methods for Accidental LNG Releases: Volume 5/5--Using FEM3A for LNG Accident Consequence Analysis," Topical Report for Gas Research Institute, April 1997.

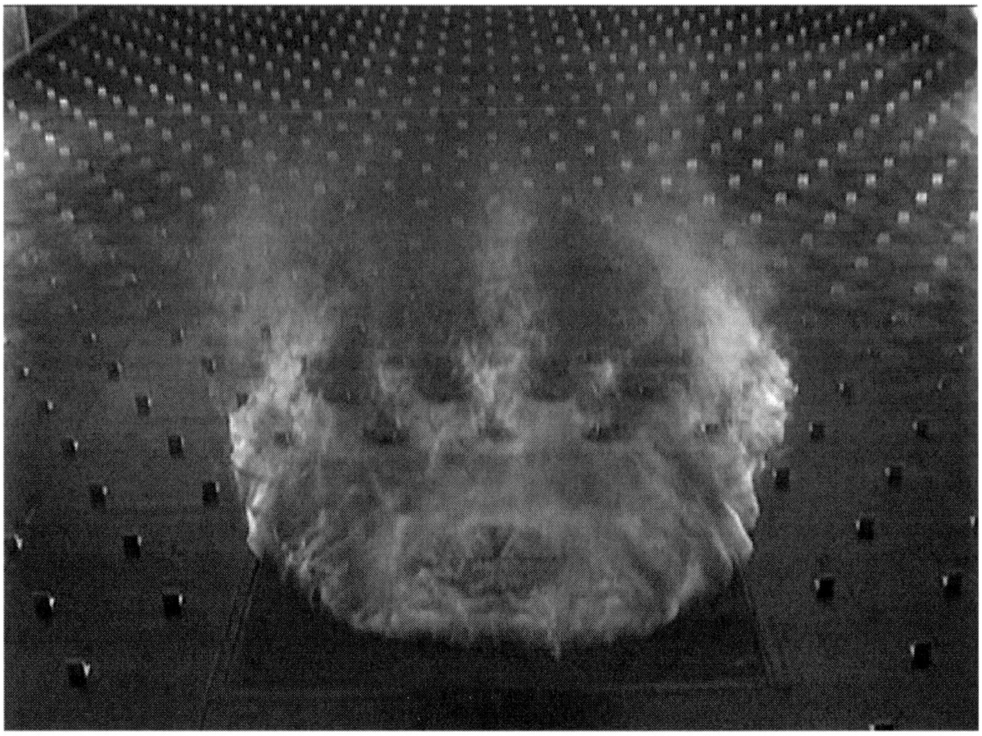

Figure 1. Case A – No Tank or Dike

Figure 2. Case B – Tank and Dike

Figure 3. Case C – Dike only

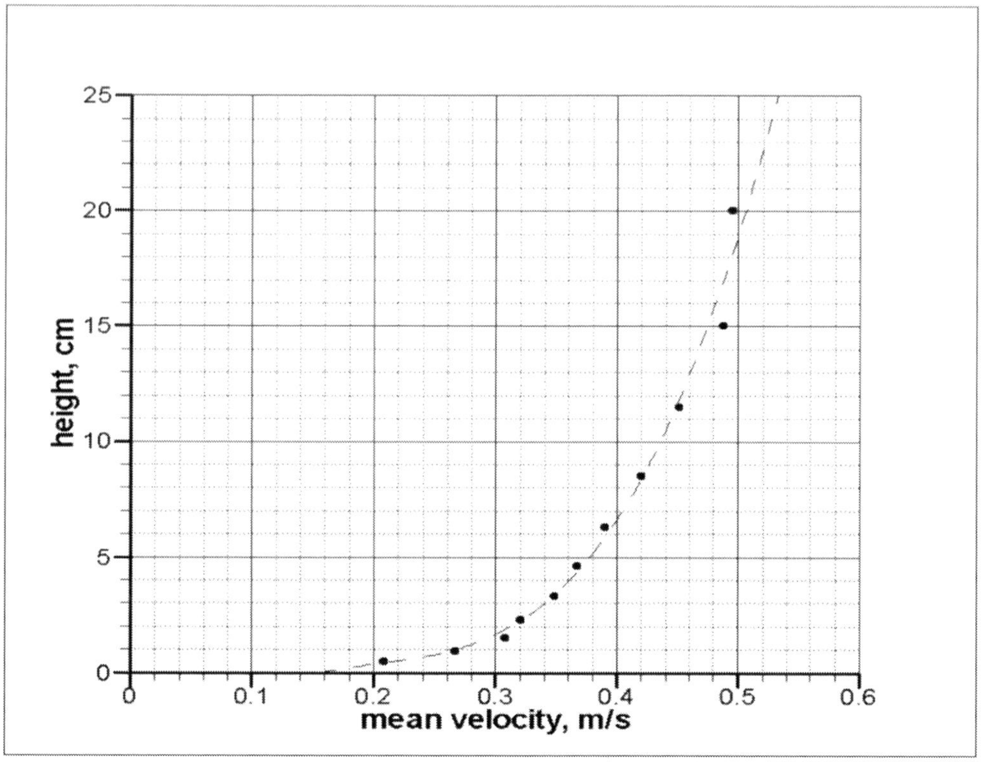

Figure 4. Mean vertical velocity profile over rough floor

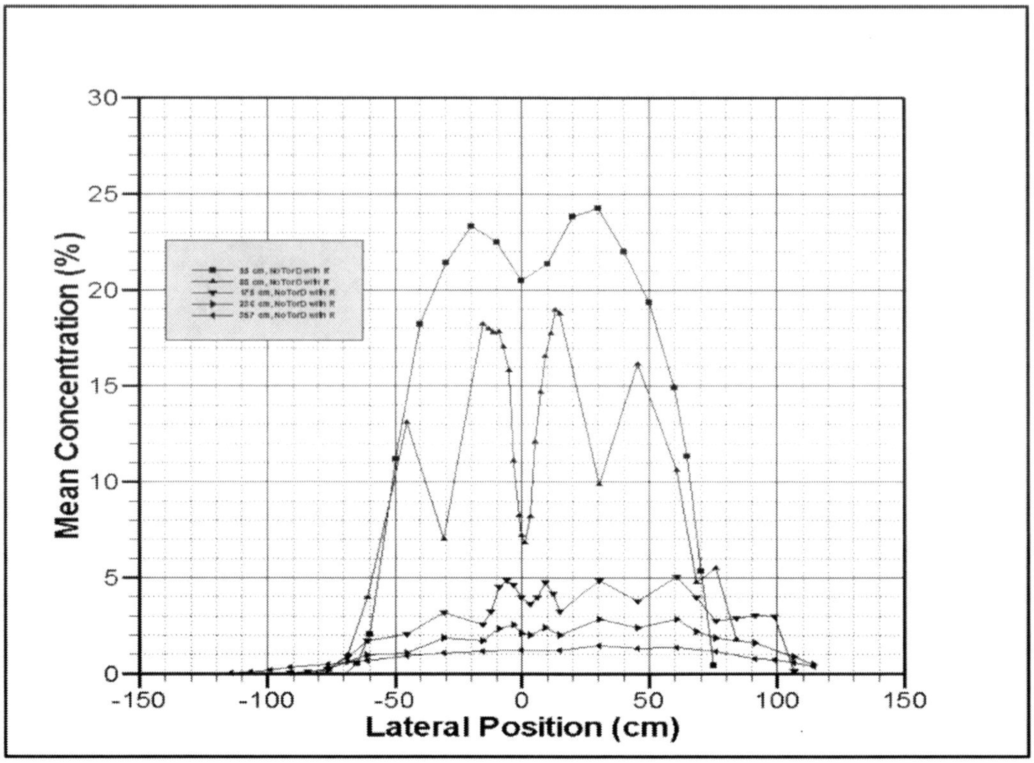

Figure 5. Case A – Lateral carbon dioxide concentration profiles

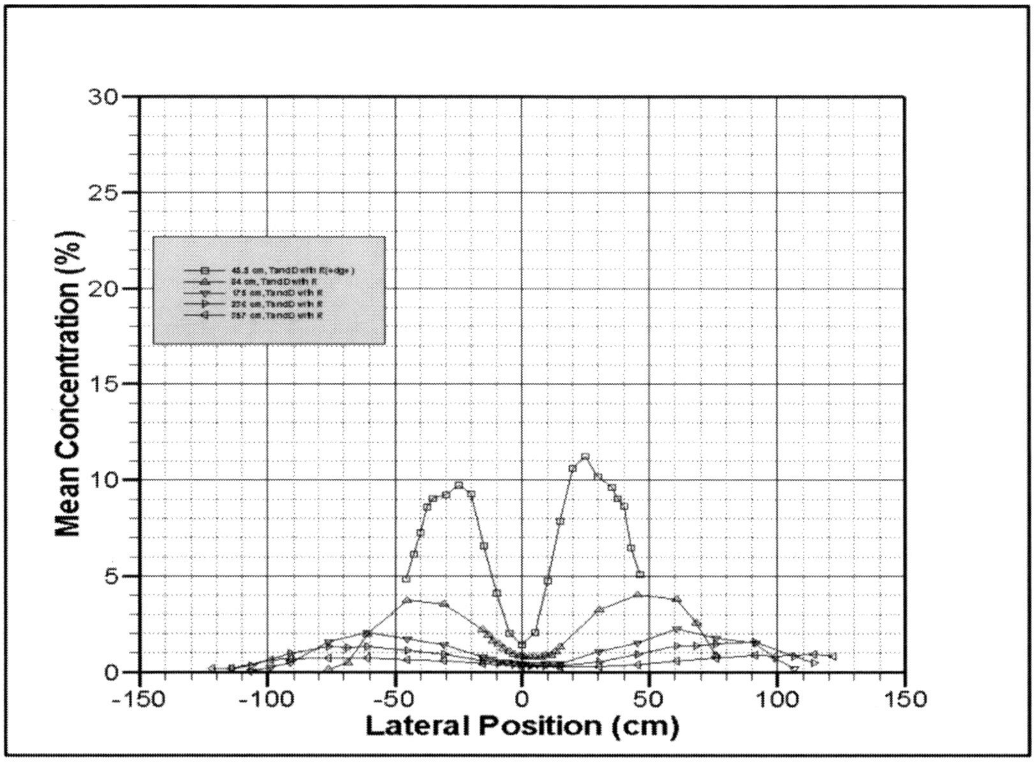

Figure 6. Case B – Lateral carbon dioxide concentration profiles

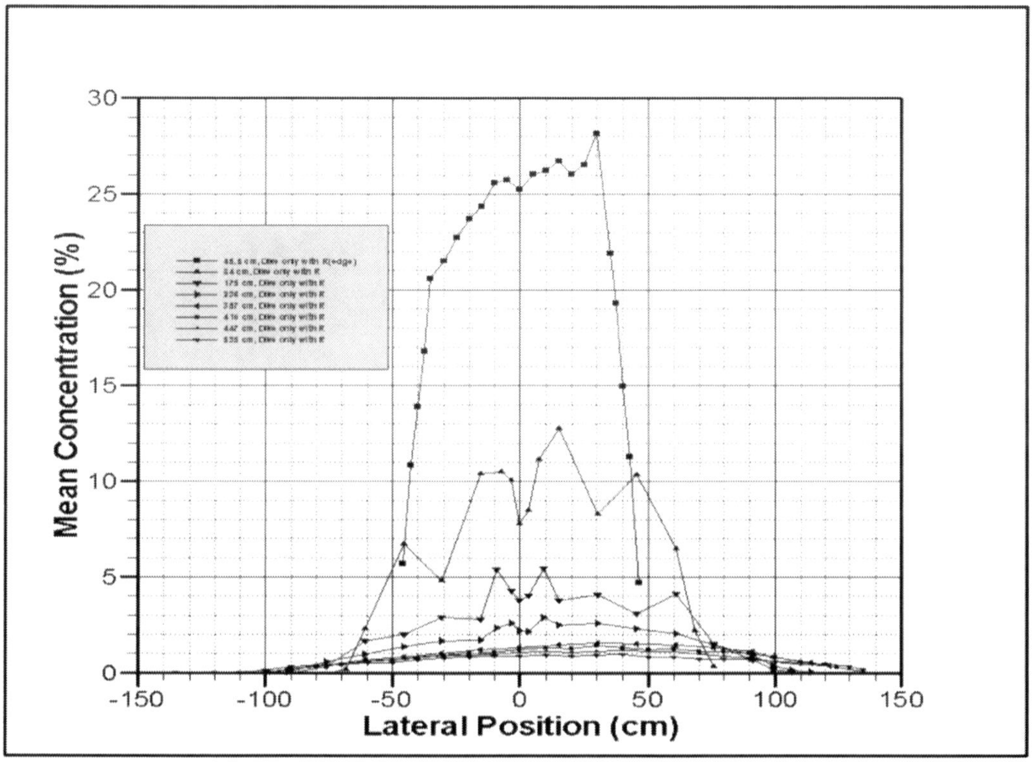

Figure 7. Case C – Lateral carbon dioxide concentration profiles

BIOGRAPHY

Dr. Jerry Havens

Maurice E. Barker Distinguished Professor
Chemical Hazards Research Center
University of Arkansas
700 Research Center Boulevard
Fayetteville, AR 72701

Phone: 479-575-2055
Fax: 479-575-8718
Email: jhavens@uark.edu

Professor Havens holds the Maurice E. Barker Chair in Chemical Process Safety and the Environmental Fate of Pollutants at the University of Arkansas. He is recognized internationally for his expertise in atmospheric dispersion of hazardous chemicals and fire/explosion phenomena. The DEGADIS and FEM3A dispersion models developed and or validated by Dr. Havens and his colleague Dr. Tom Spicer against wind tunnel experiments at the University's Chemical Hazards Research Center are specified for use in LNG plant siting applications in the United States.

Vapor Dispersion Exclusion Zones for Pressurized LNG Releases

C. Dale Zinn
Shawnee Engineers
Houston, Texas

AIChE Spring Meeting, April 2007
7th Topical Conference on Natural Gas Utilization
Houston, Texas, April 22-27, 2007

Abstract

The estimation of vapor dispersion exclusion zones as required by the DOT LNG regulations have been modeled assuming that the pressure has no effect and the released LNG drops MAGICALLY into the spill containment pit. The effect of pipe break orientation as well as pressure level has been evaluated using commercial consequence models.

CFD-Modeling of LNG Dispersion Experiments

Olav Roald Hansen, Jens A. Melheim and Idar E. Storvik
GexCon AS
Bergen, Norway

AIChE Spring Meeting, April 2007
7th Topical Conference on Natural Gas Utilization
Houston, Texas, 22-26, 2007

Introduction

The CFD-tool FLACS was developed from 1980 to model hydrocarbon gas explosions in process facilities. In order to also predict the gas build-up, very, the functionality very soon was extended to include dispersion. During three decades extensive validation against experiments has been given high priority in the FLACS development. FLACS is now considered to be a leading CFD-tool for gas explosion modeling, the modeling of hydrogen dispersion and explosion, dust explosion modeling and for dispersion modeling in offshore and onshore congested process areas.

With support from Norsk Hydro, Total, Statoil and Petroleum Safety Authority Norway (PSA), and in cooperation with DNV, a LNG-development and validation project for FLACS has been carried out recently. The first phase of the project was a validation activity in which 8 tests from Burro and 3 tests from Coyote experiments performed at China Lake (USA), and 4 tests from Maplin Sands (UK) have been simulated. Predicted downwind maximum gas concentrations at various distances have been compared to experiments showing good correlation. This validation activity will be presented in the current paper. Models have been further developed to extend the pool simulation capabilities to allow the formation and spread of a pool with local evaporation rates based on heat from ground, wind speed, local vapor pressure. These models will be described, and simulation examples will be shown. The combination of a good description of physical models and the ability to take into account buildings, terrain or process plants is essential for a proper consequence modeling of LNG-releases.

LNG-Vapor Dispersion Validation Study

The first part of this paper describes a validation study of FLACS simulating the dispersion downwind of LNG-releases. During previous FLACS development projects there has been a massive validation against explosion experiments, but also ventilation and dispersion studies both with natural gas (Phase 3B [1]), releases of buoyant or dense gas [2], and tracer gases [3]. In the study described here, releases of up to 100 kg/s or more LNG was released into a pond in Burro/Coyote tests [4] or river Maplin Sands [5]. In the modeling a simplified approach for the LNG pool has been assumed, in which natural gas is released with low momentum above ground at boiling point temperature. The ground is modeled flat, for Burro & Coyote the pool location is slightly lower than the terrain, in addition to this two buildings used for instrumentation and gas storage were also modeled. Burro & Coyote geometry model can be seen in Figure 1.

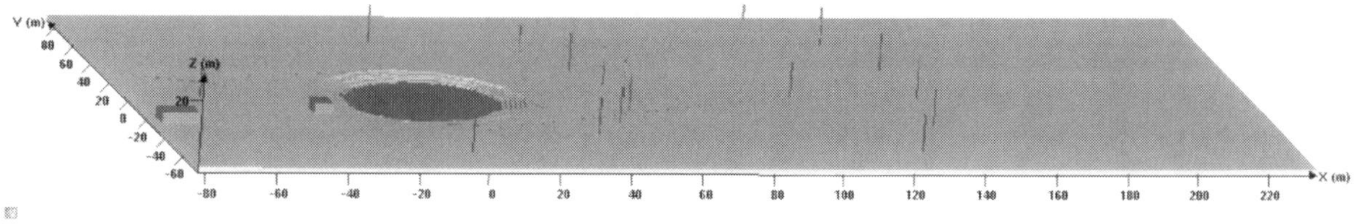

Figure 1 Burro & Coyote geometry model in FLACS [4]

In the simulations the proper wind direction was simulated, with assumption of some meandering of wind. In Figure 2 simulated gas plume (> 1%) is compared with observations in Burro test 7. In Figure 3 the observed temperature reduction due to the LNG release at 57m monitor point is compared to simulation.

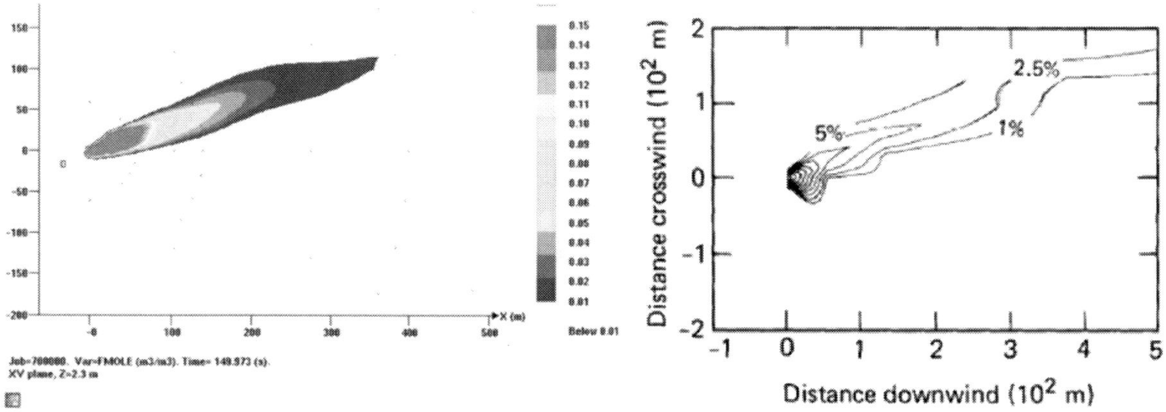

Figure 2 Example of plume comparison Burro 7 simulation (left) versus contour plots based on experimental recordings [4]

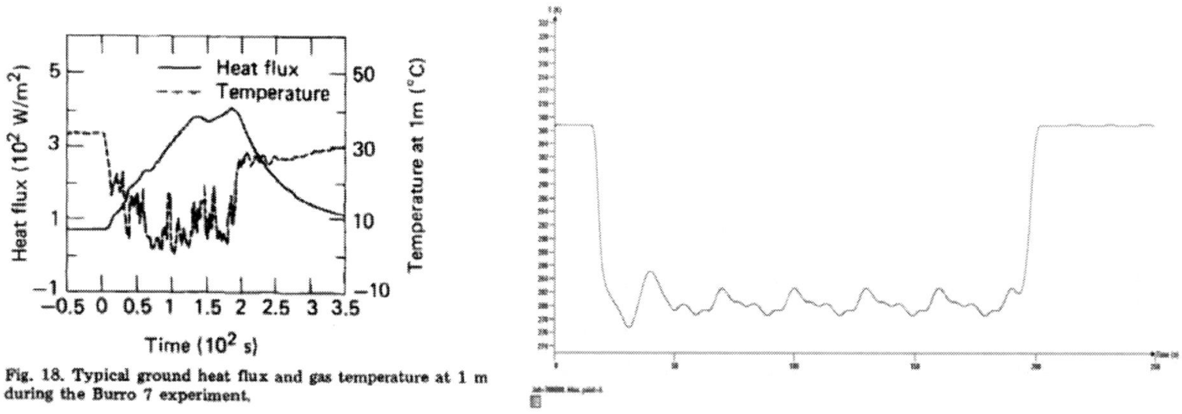

Figure 3 Example of observed temperature reduction at 57m sensor in Burro 7 experiment (left) versus simulation (right) [4]

In the Burro & Coyote validation study a very detailed curve-by-curve comparison for sensors at each monitor arch and elevation (1m, 3m and 8m) was carried out. One example of such comparison plots are shown in Figure 4, where curve comparisons for Coyote 6 experiment and simulation are compared a 1m elevation 140m, 200m and

400m from the release. Similar plots are produced for all 11 Burro and Coyote tests for all monitor point elevations. A numerical comparison of maximum reported concentrations at each sensor and arc was also carried out, with different averaging times 0s, 10s, 30s and 60s. The parabola plots in Figure 5 are using this information with 10s averaging for Burro & Coyote, each point in the plot is a geometric variance versus geometric mean observation taking into account maximum averaged observation at each of 2-3 downwind arcs (each with measurements at 5-7 different positions and 2-3 elevations). In Figure 6 a similar parabola plot is shown for simulation of Maplin Sands experiments, here each point in the parabola is the result of 2-7 downwind observations. For MS 29 observations at 7 different distances were available, and these are compared with FLACS simulations on a coarse grid (4m x 4m x 0.5m) and fine grid (2m x 2m x 0.25m) resolution.

In general good simulation results are seen. Wind conditions in experiment are not controlled, and tend always to be more chaotic than assumed in simulation model. One consequence of this is that it is difficult to compare average concentrations at downwind arcs since significant wind fluctuations are seen. When using short averaging times, and considering maximum reported concentration at each downwind arc, for two reasons some underprediction in simulations should be expected. The first reason is that the higher variability of the wind in the experiment will also make the reported concentrations oscillate more than in simulations, and with short averaging time this gives a somewhat high concentration reported from experiment. Second, with the lower wind fluctuations assumed in the simulations, we will to a greater extent fail to hit measurement points with the highest concentrations (plume falls between sensors). For these reasons, the weak underprediction of concentration level shown in Figure 5 is considered fully acceptable. For the one scenario with low wind (Burro 8, 1.8m/s) a factor of 10 underprediction of concentration are seen at measurement points. However, when comparing downwind extent of flammable gas concentration, very good results are seen. The explanation to this deviation is simply that in simulations the dense gas cloud was mainly below the 1m sensors, whereas in the experiment sporadically higher concentrations reached the sensors likely due to much more local variability in the wind.

Development of LNG pool models and humidity effects

In 2002 an LNG pool evaporation model was developed for FLACS [6]. The model took into account several parameters like heat flux from ground, solar influx, wind parameters and more. Limitations of the model were that the pool had a fixed size and shape (circular or ring), and would not include LNG added to the pool during the simulation. In the current work a further development of these models are performed.

A pool will spread until it reaches a steady state where the evaporation balances the leak or obstacles hinder the pool spread. A LNG pool is driven by gravitational forces in form of differences in the pool height and the shape of the ground and friction resist the spread. For LNG, it is assumed that the surface tension can be neglected. In several pool-spread models, a circular or semi-circular pool is considered [7]. It is, however, easy to imagine scenarios where a more general description of the pool spread will be more realistic. Woodward [8] described the pool spread by transport equations for the pool height, radial velocity and temperature on an annular grid, and it is straightforward to write the governing equations in [8] for any grid.

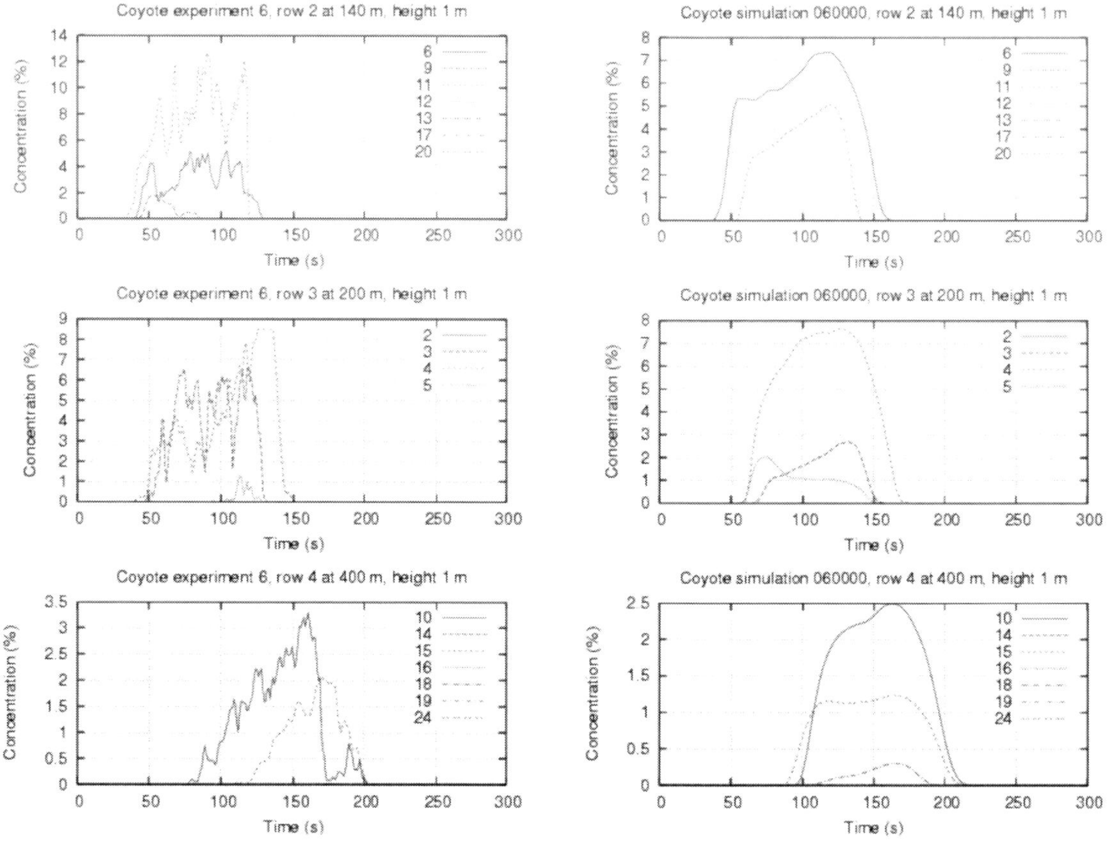

Figure 4 Example of monitor arc comparisons, observed concentrations for Coyote test 6 measurement arcs at 140m, 200m and 400m compared to simulated concentrations.

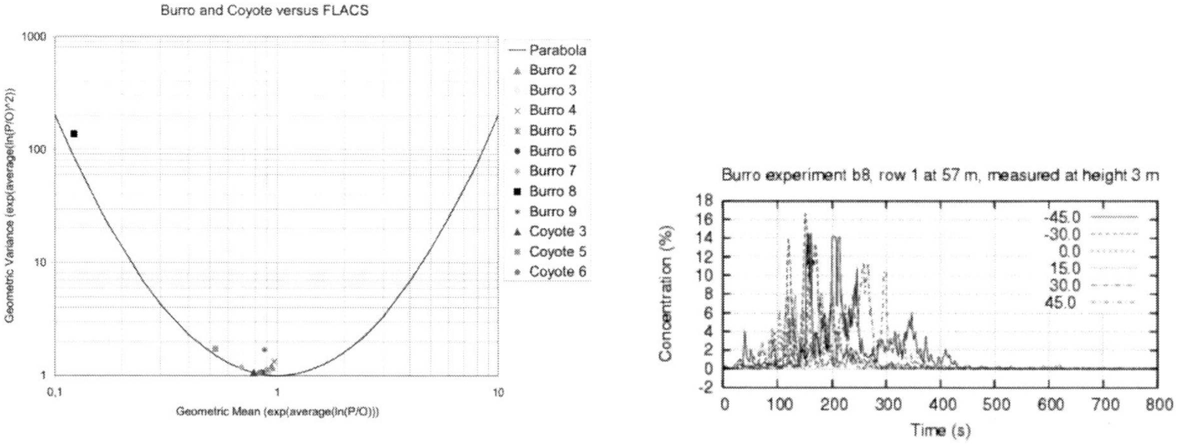

Figure 5 Geometric variance versus geometric mean for all 8 Burro and 3 Coyote experiments, each point is based on maximum observations at 6 different arc distances/elevations. In the right picture an example of sensor observation from Burro 8, with high concentrations only seen sporadically.

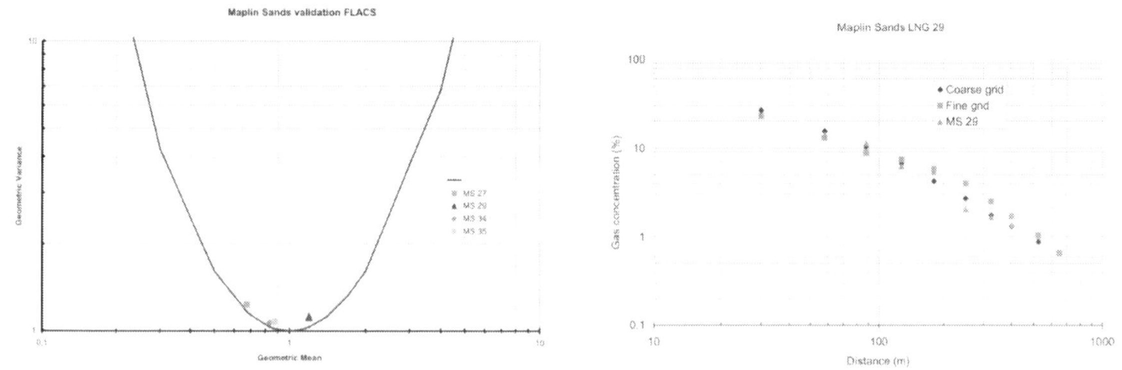

Figure 6 Geometric variance versus geometric mean for all 4 Maplin Sands experiments simulated (left plot), observations [from Hanna] at 2-7 different downwind distances are compared. For Maplin Sands test 29 (29 kg/s LNG with 7.4 m/s wind) 7 observations with increasing distance were available and compared to simulation results (right plot) with fine grid and coarse grid.

In real life, the source of the leak, for instance a storage tank or a LNG tanker, will setup a non-uniform flow field above the pool. Expressions based on boundary layer theory are therefore used to calculate the convective heat and mass transfer in grid cell separately. The equation solved for the pool height is given by:

$$\frac{\partial h}{\partial t} + \frac{\partial h u_i}{\partial x_i} = \frac{\dot{m}_L - \dot{m}_V}{\rho_l} \qquad (1)$$

and the momentum equation is written as:

$$\frac{\partial u_i}{\partial t} + u_j \frac{\partial u_i}{\partial x_j} = F_{g,i} + F_{\tau,i} \quad (2)$$

The gravity term is written as:

$$F_{g,i} = g\Delta \frac{\partial (h+z)}{\partial x_i} \Bigg/ \sqrt{\left(\frac{\partial (h+z)}{\partial x_i}\right)^2 + 1} \qquad (3)$$

where also the height of the ground above a reference level *z* has been included. Δ equals 1 for solid grounds and $\Delta = (1 - \rho_l / \rho_w)$ for spill on water. The shear stress can be given by the general formula:

$$F_{\tau,i} = \frac{1}{2} f_f u_i |\vec{u}| \quad (4)$$

where the friction factor is follows the expression for flows on fully rough surfaces. The equation for the specific enthalpy is given by:

$$\frac{\partial \theta}{\partial t} + u_i \frac{\partial \theta}{\partial x_i} = \frac{\dot{m}_L}{h}(\theta_L - \theta) + \dot{q}_c + \dot{q}_{rad} + \dot{q}_{ground} + \dot{q}_{evap} + \dot{q}_{boil} \quad (5)$$

The first term on the right hand side is due to the leak. The model for the convective heat and mass transfer is based on boundary layer theory, dividing the boundary layer into a viscous region and a log-region. The convective heat transfer is given by [9]:

$$\dot{q}_c = \frac{\rho_g C_\mu^{1/4} k^{1/2} c_{p,g} T_g \ln \frac{T_l}{T_g}}{T^+} \quad (6)$$

where $T^+ = \Pr y^+$ in the viscous region and $T^+ = E^+ \Pr + \Pr_t / \kappa \ln\left(y^+ / E^+\right)$ else.
Analogue to the heat transfer, the expression for the mass transfer is written as [10]:

$$\dot{m}_c = \frac{\rho_g C_\mu^{1/4} k^{1/2} \frac{P_0}{RT_l} \min\left\{ -\ln\left(\frac{1-x_{sat}}{1-x}\right),\ 3.1534 \right\}}{x^+}$$

where the *Stefan flow* has been taken into account. Stefan flow might be important for large mass fluxes [11]. $x^+ = \mathrm{Sc}\, y^+$ in the viscous region and $x^+ = E^+\mathrm{Sc} + \mathrm{Sc}_t / \kappa \ln\left(y^+ / E^+\right)$ outside. The modeling of the radiative heat transfer, heat from ground and heat loss due to evaporation and boiling is described in [6]. The governing equations for the pool are discretized on a non-uniform Cartesian staggered grid in two dimensions with a finite volume method. An upstream scheme is employed for the convective terms in the momentum equation, while a central difference scheme is used for the enthalpy equation. The equations are solved explicitly in time.

In Figure 7 results from a simulation using the improved pool models are shown. A 1000 kg/s LNG release from a cylindrical tank onto the ground is modeled, and the LNG-pool spreads on the ground falls into the sea. The development of methane vapor cloud volume above stoichiometry is also shown.

When modeling vapor cloud dispersion from releases of cryogenic liquids, the temperature effects when cooling the air must be taken into account. Air humidity may gradually condensate to mist when temperature is reduced, and thus influence the efficient heat capacity of the air. In Figure 8 an example is shown of the modeled heat capacity of air with humidity taking into account the condensation of water vapor. When implementing the models, we had access to a report from DNV describing how they used a similar approach when modeling with CFX [11].

Further development tasks

Despite a newly developed pool model with potentially advanced possibilities, there will always be modeling aspects that can be improved. Through previous validation work it has been demonstrated that FLACS dispersion modeling is good once the LNG evaporates. Some elements that could be improved are the understanding of heat transfer phenomena, of the arbitrary atmosphere and stability classes (in particular Pasquill A, B and C), and more details on pool behavior e.g. when mitigation is applied. For most of these phenomena it is however difficult to describe reality very detailed.

When modeling with FLACS there are however limitations using Cartesian grids. Whereas process plants and buildings are generally well described using the distributed porosity concept of FLACS, dominating sloping elements like terrain will be less good modeled. In the presented validation study, most predictions were carried out with a flat terrain assumption. Still some calculations were performed with a realistic representation of the terrain. The sloping terrain will be represented as steps on a Cartesian grid giving a too high momentum loss. With large grid cell sizes in the vertical direction, this will influence the downwind predicted hazard distance. Future development work will aim at implementing numerical schemes that will minimize the momentum loss for sloping terrain. In the meantime a fine vertical grid resolution (much finer than height of gas cloud) will be recommended when modeling dense gas dispersion in a sloping terrain.

Since FLACS is primarily an explosion simulator, the explosion effects of the dispersed clouds from LNG-pools should be well taken care of. Another hazard in connection to LNG release scenarios will be fire. FLACS has in the past not been able to model fire scenarios (with precision). There is however an ongoing development which will improve the fire modeling capabilities of FLACS in future versions of the tool.

Acknowledgments

The LNG validation and development project in which the described work has been carried out was supported by Norsk Hydro, Norwegian Petroleum Safety Authorities (PTIL), Statoil and Total, in addition to this DNV provided technical support to GexCon. The authors are grateful for access to detailed test results from Burro and Coyote experiments, which was facilitated by Steve Hanna and Joe Chang.

Conclusion

FLACS is established as a very powerful CFD tool for gas dispersion, in particular for situations where detailed description of geometry details is of importance (process facilities, urban areas and more). Numerous validation activities have been carried out in the past, the current paper describes a validation activity simulating the dispersion from LNG-release experiments Burro, Coyote and Maplin Sands. Results are generally good, FLACS predicts downwind natural gas concentrations with reasonable precision assuming the released LNG quickly evaporates from a pool with boiling point temperature. A further improvement of pool models is described in the paper which hopefully will lead to a more realistic description of accident scenarios simulated in connection to risk assessments.

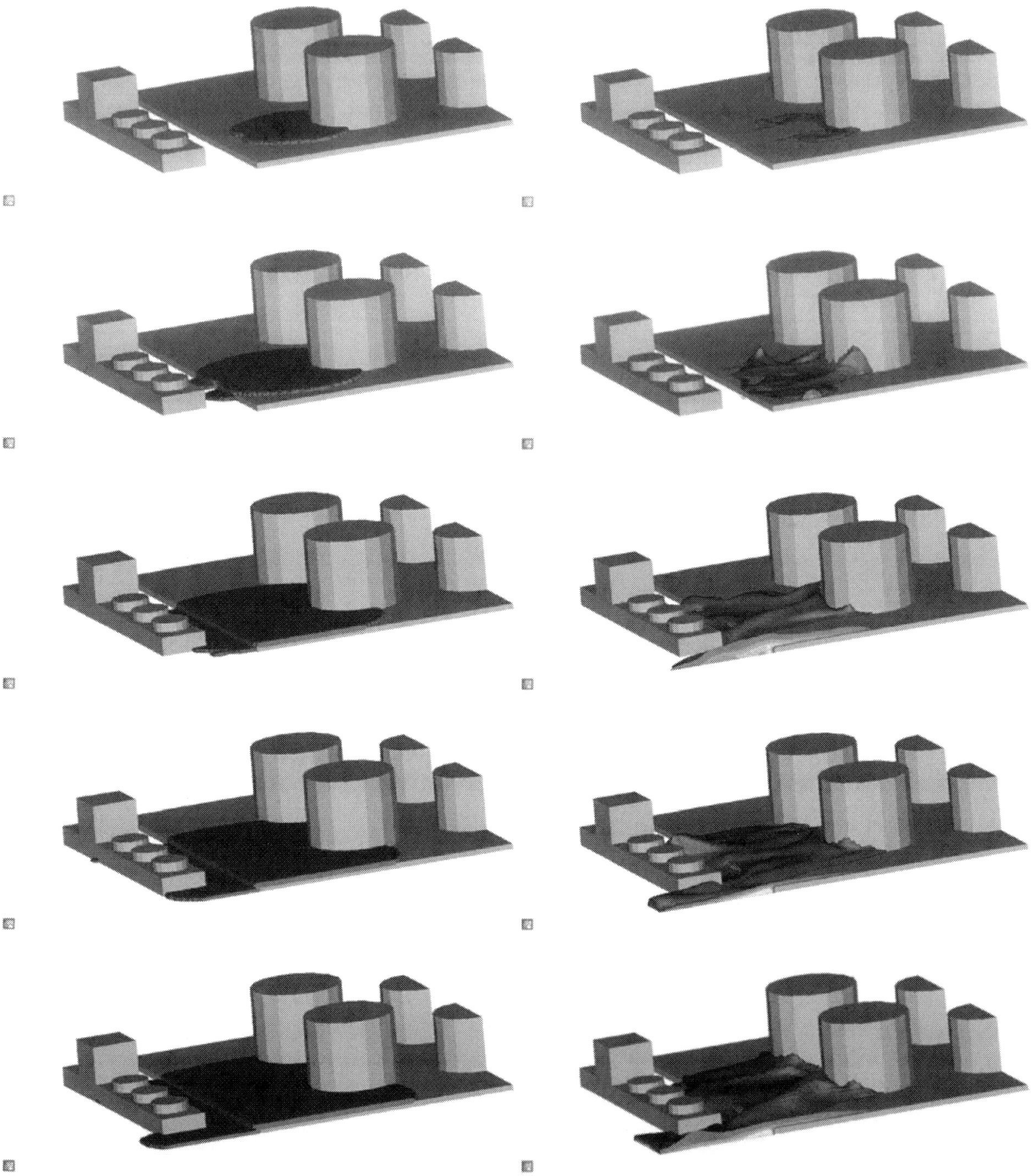

Figure 7 Simulated 1000 kg/s LNG leak from tank with liquid pool development (left) and vapor cloud above stoichiometry (right), simulated wind speed is 5 m/s, pictures show 25s, 50s, 100s, 150s and 200s after start of release. A "simplified" LNG-carrier is included in the simulation.

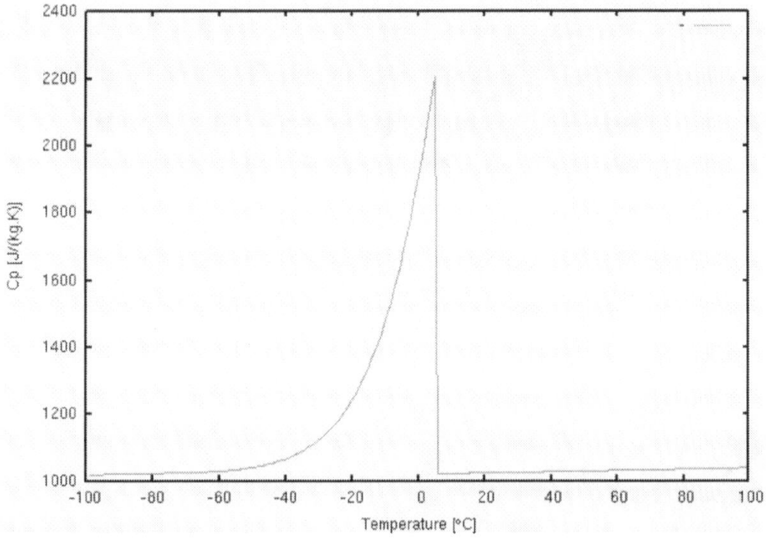

Figure 8 **Efficient heat capacity of air containing some water vapor, according to newly implemented models in FLACS. For certain release conditions, some additional humidity of air may make the evaporated cold natural gas buoyant.**

References

1 Johnson, D.M. et al. "Investigation of Gas Dispersion and Explosions in Offshore Modules", OTC-paper 14134, Houston 6-9 May 2002

2 Steven R. Hanna, Olav R. Hansen, Seshu Dharmavaram, FLACS CFD air quality model performance evaluation with Kit Fox, MUST, Prairie Grass, and EMU observations, Journal of Atmospheric Environment, Volume 38 (2004) pages 4675-4687.

3 Hanna, S.R., M.J. Brown, F.E. Camelli, S.T. Chan, W.J. Coirier, O.R. Hansen, A.H. Huber, S. Kim, and R.M. Reynolds, 2006: Detailed simulations of atmospheric flow and dispersion in downtown Manhattan. Bulletin Am. Meteorol. Soc., 87:12, 1713-1726.

4 Koopman, R.P, Cederwall, R.T., Ermak, D.L., Goldwire jr., H.C., Hogan, W.J., McClure, J.W., McRae, T.G., Morgan, D.L., Rodean, H.C. and Shinn, J.H., Analysis of Burro Series 40-m^3 LNG Spill Experiments, J. Hazardous Materials, 6 (1982), 43-83.

5 Puttock, J.S., Blackmore, D.R. and Colenbrander, G.W., Field Experiments on Dense Gas Dispersion, J. Hazardous Materials, 6 (1982), 13-41.

6 Won K. Kim and Hans-Christen Salvesen, A Study for Prevention of Unconfined Vapor Cloud Explosion from Spilled LNG Confined in Dike, CCPS Conference Proceedings, Jacksonville Florida, 2002

7 M. Hightower et al. Guidance on risk analysis and safety implications of a large Liquefied Natural Gas (LNG) spill over water. Report SAND2004-6258, Sandia National Laboratories, Livermore, 2004

8 J. L. Woodward. An integrated model for discharge rate, pool spread, and dispersion from punctured process vessels. J. Loss Prev. Process Ind., 3: 33–37, 1990.

9 Z. Han and R. D. Reitz. A temperature wall function formulation for variable-density turbulent flows with application to engine convective heat transfer modeling. Int. J. Heat Mass Transfer, 40:613–625, 1997
10 A. F. Mills. Heat and mass transfer. Irwin, 1995.
11 Huser, A. and Bjørkedal, I.E. 1999: Dispersion of heavy gas from hydrocarbon leaks, model development and survey of problem, DNV report No 99-3247, rev 01.

Biography

Postbox 6015 Postterminalen, NO-5892 BERGEN, NORWAY
Phone: +47-5557-4318, Fax: +47-5557-4331, Cell: +47-9117-1787
E-mail: olav@gexcon.com

Olav Roald Hansen is GexCon R&D Director and Manager of GexCon Software department responsible for development and sale of the FLACS CFD-software. He has an MSc in Physics from Norwegian Institute of Technology (NTH) from 1992. Hansen started working with CFD modeling and FLACS at CMR in 1993, initially with main focus on explosion modeling, but gradually also with more focus on dispersion modeling. When the explosion modeling activities of CMR was transferred to GexCon in 2001, Hansen became the GexCon R&D Director. Topics of particular focus in his work has been gas and dust explosion modeling (and experimental work), hydrogen safety, blast propagation, risk assessments, explosion mitigation, dispersion modeling (flammable and tracer gases).

Validation of a CFD Model for Vapor Dispersion from LNG Spills into an Impoundment

Filippo Gavelli[1], Melissa Chernovsky[1], Edward Bullister[2], and Harri Kytomaa[1]
Exponent, Inc.
[1]Bowie, Maryland
[2]Weston, Massachusetts

AIChE Spring Meeting, April 2007
7th Topical Conference on Natural Gas Utilization
Houston, Texas, April 22-27, 2007

Abstract

Recent discussions on the fate of LNG spills into impoundments have led to the conclusion that the commonly used combination of SOURCE5 and DEGADIS to predict the flammable vapor dispersion distances is not physically accurate, as it does not account for air entrainment into the evaporating gas and it does not allow for heating of the LNG vapor cloud within the impoundment. An alternative approach to predict the flammable vapor dispersion distance is to use a computational fluid dynamics (CFD) model. CFD codes can take into account the physical phenomena that govern the fate of LNG spills into impoundments, such as air entrainment and ground-to-vapor heat transfer. Before a CFD code can be proposed as an alternate method for the prediction of flammable vapor cloud distances, it has to be validated with proper experimental data. This paper describes the validation effort of Fluent, a widely-used commercial CFD code, for the prediction of flammable vapor cloud dispersion. Fluent was used to simulate the "Falcon" series of LNG spill tests, which is the only series of tests performed specifically to address the effects of vapor barriers and building obstructions on the vapor cloud dispersion. A comparison of CFD predictions and experimental data will be made, both graphically and by means of statistical performance measures.

Biography

Dr. Filippo Gavelli is a Managing Engineer in Exponent's Thermal Sciences practice. He received his undergraduate degree in Nuclear Engineering from the University of Bologna, Italy, and his Ph.D. in Mechanical Engineering from the University of Maryland.

Dr. Gavelli specializes in the engineering analysis and investigation of fires and explosions and applies CFD modeling tools to simulate fires and the atmospheric dispersion of dense gas releases. He has performed LNG hazard analysis studies and risk assessments for several LNG terminals, both onshore and offshore.

Prior to joining Exponent, Dr. Gavelli held the position of Assistant Professor in the Department of Mechanical Engineering at The Catholic University of America in Washington, DC, where he taught the thermal sciences and performed research with support from federal agencies including the U.S. Nuclear Regulatory Commission and the National Institute of Standards and Technology.

Spill Containment for Liquid Hydrogen Storage located at Public Fueling Stations

Dave Farese, Elizabeth Lutostansky and Peter Steiner
Air Products and Chemicals, Inc.
Allentown, Pennsylvania

AIChE Spring Meeting, April 2007
7th Topical Conference on Natural Gas Utilization
Houston, Texas, 22-26, 2007

Study for Liquid Hydrogen Diking

This study was conducted to determine the construction parameters for an area of containment to be provided to meet the requirements of Section 14.2.3 of the 2006 Edition of NFPA 52 *Vehicular Fuel Systems Code* which requires the design and installation of an impound area to contain a liquid hydrogen spill.

Background

The Vehicular Fuel Systems Code (NFPA 52, 2006 edition) was recently revised, and for the first time establishes provisions for the use of hydrogen as a vehicle fuel.[A] A newly established requirement in the code requires the retention of any spilled liquid hydrogen within the limits of the plant property [fueling station site] (14.2.3). The impound area is required to "have a minimum volumetric capacity equal to the greatest volume of LH2 … that could be discharged during a 10-minute period…" (14.2.3.3). Annex A.14.2.3.3 limits the potential volume of a spill from either a stationary tank or delivery vehicle as follows:

A.14.2.3.3 The potential spill from a fixed container equipped with the valves

---

[A] NFPA 50B *Standard for Liquefied Hydrogen Systems at Consumer Sites,* NFPA, Quincy, MA. Established for a period of over 37 years. Preparation of the Standard was initiated and materially assisted by a committee of the Compressed Gas Association, Inc., which submitted a text to the NFPA Committee on Industrial and Medical Gases in 1966. It was tentatively adopted by NFPA in 1967 and the First Edition was adopted in 1968. Subsequent editions were adopted in 1971, 1973, 1978, 1985, 1989, 1994 and 1999. It was integrated into NFPA 55 *Standard for the Storage, Use, and Handling of Compressed Gases and Cryogenic Fluids in Portable and Stationary Containers, Cylinders, and Tanks* in the 2005 Edition of that Standard.

discussed in Section 14.6 is limited to the volume of the piping system. For transfer operations where the source of a spillage is from a truck unloading, the spill volume is either the 10-minute flow from the truck, assuming design flow, or a lower volume if the truck is equipped with automatic shutoff.

The use of spill containment is not required by long standing precedent codes and standards addressing the storage or use of liquid hydrogen. Spill tests conducted by NASA indicate that it may be preferred not to impound a liquid hydrogen spill to take advantage of the dispersion mechanisms provided by spill induced turbulence that allows for the rapid warming and dispersion of the product.[B]

Determination of spill volume

This purpose of this study was to determine the spill volumes established by the code in order to provide design criteria for the construction of a containment area. In order to provide design criteria the following determinations were required:

a) Determine if pooling would occur during a liquid hydrogen spill under the requirements of Section 14.2.3.1 (see below), and

   14.2.3.1 Site preparation shall include provisions for retention of spilled LH2 within the limits of plant property and for surface water drainage

b) Determine the required volumetric capacity of the containment given the specific design parameters specified by the code under the requirements of Sections 14.2.3.5.3 and 14.2.3.5.3.1 (see below).

   14.2.3.5.3 The containment design shall include calculations and shall be installed to prevent overflow due to spill wave action or pressurized/saturated LH2.

   14.2.3.5.3.1 The containment design shall include the prevention of broken piping, and so forth, projecting LH2 or cold gas beyond the containment area.

The liquid hydrogen release was modeled using Phast, version 6.42 which is a commercially available consequence model available from DNV (Det Norske Veritas). PHAST is a comprehensive tool for modeling the consequences of accidental releases of toxic or flammable materials and the program is capable of modeling cryogenic fluids. The dispersion model in PHAST has been extensively verified and has been validated against a

---

[B] *Safety Standard for Hydrogen and Hydrogen Systems, Guidelines for Hydrogen System Design, Materials, Selection, Operations, Storage, and Transportation,* Office of Safety and Mission Assurance, National
Aeronautics and Space Administration, NSS 1740.16, February 12, 1997, pp. 9-5 to 9-6

large number of field experiments. These field experiments include continuous, elevated two phase and vapor releases, ground level liquid spills and unpressurized instantaneous releases. [C]

The following assumptions were used regarding the release for this study:

a) The release was from a 2" ID diameter pipe. The pipe size is representative of the size of the transfer hose used to off-load product from a liquid hydrogen tanker. This is the largest line on the system and represents the greatest potential to create a liquid hydrogen pool. The maximum line size on a typical piping system on a stationary tank is also 2" so this serves as a representation of a spill from the storage system, in the event a line ruptures.

b) A horizontal release at 2 feet above grade was used to simulate a release from the pump or trailer discharge connection on a liquid hydrogen tanker.

c) The quantity of liquid released was 10,000 gallons and based on the following:

    a. The typical trailer contains between 13,000 and 15,000 gallons. A quantity of 10,000 gallons represents up to two thirds of the trailer's contents being released.
    b. The design pump discharge rate is 90 gpm resulting in a 900 gallon release over 10 minutes. Therefore a 10,000 gallon release is approximately ten times the minimum volume specified in NFPA 52.
    c. A non-return valve is normally installed in the liquid fill line on the receiving vessels to prevent backflow. Therefore, there is no additional liquid to be considered other than that in the piping system that will contribute to a liquid release.
    d. In addition to the spill occurring through the pump discharge, we modeled a scenario using the tank or trailer pressure to drain the vessel through an open line. In this instance, the 10,000 gallons was released in approximately 5 minutes. It should be noted that the withdrawal lines of both the trailer and stationary tank are equipped with remotely actuated emergency shutoff valves, as referenced in A.14.2.3.3, to reduce the potential quantity of released hydrogen

d) Four different scenarios or cases were selected that were representative of various temperature and pressure conditions. The scenerios were selected based on trailer pressures that could be expected during a normal delivery. Although there is some stratification of the hydrogen, the model cannot account for varying saturation pressures for a single release. To ensure we modeled the worst case, we modeled the

[C] Model Evaluation Report on UDM Version 6.0, Ref. No. SMEDIS/00/9/E, Version 1.0, 21 January 2002 scenario where the hydrogen was saturated at the trailer delivery pressure and at zero psig saturation pressure. The four (4) different pressure and temperature conditions are described as follows:

  a. Case 1. Hydrogen at 50 psig (saturated liquid at -412°F). This case represents liquid contained in a trailer utilizing a pump to off load where the hydrogen is warmed to the saturation pressure.
  b. Case 2. Hydrogen at 50 psig at -423°F boiling point (normal boiling point of 0 psig saturated liquid). This case represents liquid contained in a trailer utilizing a pump to off load where the hydrogen is not allowed to warm to the saturation pressure.
  c. Case 3. Hydrogen at 150 psig (saturated liquid at -402°F). This case represents liquid contained in a trailer utilizing pressure to off load where the hydrogen is warmed to the saturation pressure.
  d. Case 4. Hydrogen at 150 psig at -423°F boiling point (normal boiling point of 0 psig saturated liquid). This represents liquid contained in a trailer utilizing pressure to off load where the hydrogen is not allowed to warm to the saturation pressure.

e) The analysis was conducted using three (3) different weather conditions to determine if the weather would affect the potential for pooling of liquid hydrogen. The weather conditions used were as follows:
  a. Atmospheric Stability F, Wind speed 0.6m/s (2.0 fps)
  b. Atmospheric Stability D, Wind speed 5.0m/s (16.4 fps)
  c. Atmospheric Stability D, Wind speed 9.0m/s (22.5 fps)

Analysis of the hydrogen modeling

Three (3) different weather conditions were chosen to be modeled. The atmospheric stability of F Stability with a wind speed of 0.6 m/s was chosen since the Environmental Protection Agency (EPA) requires F stability with a 1.5 m/s wind when calculating the worst case release scenario.[D] A low wind speed of (0.6 m/s) is conservative thereby providing the greatest probability of pooling due to less mixing and heat transfer. In addition to using a worst case weather represented by an F stability, two (2) additional weather conditions were modeled. These include a D stability with both a 5.0 m/s and 9.0 m/s wind speed. These wind speeds were chosen to give a representation of different conditions that may be encountered during a spill.

When liquid hydrogen is contained in a tank or vessel there is a gradient of temperatures and densities present throughout the liquid phase depending on the location in the mass of fluid where temperature and/or density is measured. The warmer, and less dense, fluid is normally found at the top of a tank and the colder, more dense, fluid is found at the bottom. When saturated liquid is assumed, the model assumes that the temperature and density of the liquid released is homogeneous and the liquid is released at the saturation pressure and temperature.

[D] 40 CFR Part 68, section 68.22 (b)

For cases 1 and 3 above, the model results in liquid temperatures of -412°F and -402°F respectively (pressures of 50 psig and 150 psig). Since these temperatures are above the normal boiling point of - 423°F for hydrogen, the released product resembles an aerosol. The liquid in the aerosol rapidly flashes before it can reach the ground. Under these conditions, it was predicted that the formation of any liquid was unlikely due to the flash loss caused by the boiling liquid. The model showed that twenty-one percent (21%) of the liquid flashed immediately for case 1 and that fifty-two percent (52%) flashed immediately for case 3.

In cases 2 and 4, assuming that all of the liquid was at its normal boiling point of -423°F results in a worst case scenario, as the colder the liquid, the greater the potential for pooling due to less liquid flashing when it is released.
From a practical standpoint any liquid within a storage tank or delivery vehicle is above its boiling point of -423°F due to the heat added during the trailer filling process and heat leak into the product during transport to the delivery site. Nevertheless, cases 2 and 4 are used to model 100% liquid hydrogen released from an open pipe in order to test the "worst case" condition.

Results (liquid pooling): The results of the modeling show that pooling of liquid hydrogen did not occur under any of the cases or conditions modeled even though gross spills in excess of the parameters established by the code were analyzed.

Graphical plots illustrating each of the four releases cases are enclosed as Attachment "A". Two plots are shown for each case to show the maximum boundary of cloud travel for concentrations at the LFL. One plot, in each case represents a side view along the cloud centerline and the other plot depicts the cloud in plan view. The time, expressed in seconds for the maximum flammable cloud, is shown in the legend and correlated with the corresponding weather condition.

The release conditions and results regarding pooling are summarized in the table below:

Hydrogen Release Conditions Modeled

Case Number	Product Pressure	Product Temperatre	% of product flashed on release	Pooling occur
Case 1	50 psig	-412°F	21%	none
Case 2	50 psig	-423°F	0%	none
Case 3	150 psig	-402°F	52%	none
Case 4	150 psig	-423°F	0%	none

Methane modeling comparison and analysis

Provisions in Chapter 12 of NFPA 52 for Liquefied Natural Gas (LNG) require the construction of spill containment or impound areas at points of transfer and storage.[E] In order to determine whether pooling occurred for LNG systems similar releases to those developed for hydrogen were also modeled using methane to represent the LNG product. The product temperatures were modified for methane to correspond to saturated liquid at the specified pressure for cases 1 and 3 and to the normal boiling point in cases 2 and 4. A 10,000 gallon inventory was also assumed for methane. No attempt was made to determine if a methane trailer could contain 10,000 gallons. Although larger than a typical LNG trailer, the release was for comparison and the quantity could be from a tank or any other storage vessel. Graphic plots illustrating each of the methane releases are enclosed as Attachment B. The plots shown for the methane release include the maximum boundary for the cloud within the flammable range as was depicted for the hydrogen release. In addition, where pooling resulted in the methane release, plots included indicating the radius and depth of the pool. The release conditions and results regarding pooling are summarized in the table below.

Methane Release Conditions Modeled

Methane Case Number	Product Pressure	Product Temperature	% of product flashed on release	Pooling occur
Case 1	50 psig	-220°F	15%	2.6 foot radius
Case 2	50 psig	-259°F	0%	14 foot radius
Case 3	150 psig	-186°F	29%	none
Case 4	150 psig	-259°F	0%	2.0 foot radius

Conclusions

The study determined that the release of liquid hydrogen will not create a pool when released under the conditions specified by the code, as developed under the cases described above. There are three (3) physical properties that contribute significantly to the ability of methane to pool while pooling for hydrogen does not occur.

[E] NFPA 52, *Vehicular Fuel Systems Code,* 2006 Edition, NFPA Quincy, MA, Section 12.2.3.

1. There is a temperature difference of approximately 150°F between the boiling points of liquid hydrogen (-423°F) and liquid methane (-259°F) which results in a more rapid boiling of the hydrogen.
2. The latent heat of vaporization is significantly higher for methane than for hydrogen, 3530 BTU/lb-mole (778 BTU/gallon) vs. 389 BTU/lb-mole (101 BTU/gallon) respectively. As a result, a significant greater amount of heat is needed to vaporize a given volume of methane compared to hydrogen.
3. Hydrogen has significantly higher flash losses (21% to 52%) than methane (15% to 29%) under the conditions that were modeled

Since it can be demonstrated that hydrogen pooling will not occur under the conditions required by the code, it is recommended that provisions for spills or the potential for condensation of liquid air at transfer sites be controlled by a spill pad or surface of noncombustible construction. There has been a long standing requirement in codes to include containment for methane storage and with the expansion of the vehicle fueling code to include hydrogen; the containment was unnecessarily included for hydrogen as well.

BIOGRAPHY

Mr. Steiner is a Lead Safety Engineer of the Process Safety Group at Air Products and Chemicals and has over 32 years experience with Air Products and Chemicals in the industrial gas industry and over 25 years have been in the process safety area.
Mr. Steiner has extensive experience in the hydrogen area and has led safety reviews on the systems used in the hydrogen fueling area. He was a member of the team that conducted the risk reviews for underground liquid hydrogen storage as well as for the ocean transport of liquid hydrogen.
He provides process safety support for the Americas and Asia
He is an instructor for the hazard review course for Air Products employees and a member of the team that provides hydrogen safety training and safety reviews to companies outside of Air Products. He obtained his BS in Mechanical Engineering from Gannon University. He is the Chairman of the Hazardous Materials Code Committee for the Compressed Gas Association, a Member of various Compress Gas Association committees including Atmospheric Gases, Bulk Distribution and Hydrogen Technology. Mr. Steiner is a Principal Member of the Industrial & Medical Gases Technical Committee of NFPA (NFPA 55, 51, 51A, 318 & 560). Serves on various task forces within NFPA.

Liquid Hydrogen Spill Study
Hydrogen Spill
Attachment A

February 26, 2007
Rev. 2

Case 1
Hydrogen (saturated liquid) @ 50 psig, -412F
Horizontal release from 2" diameter pipe @ 2' above grade
Cloud boundaries for LFL for 3 weather conditions in legend
79% liquid after release – no pool formed

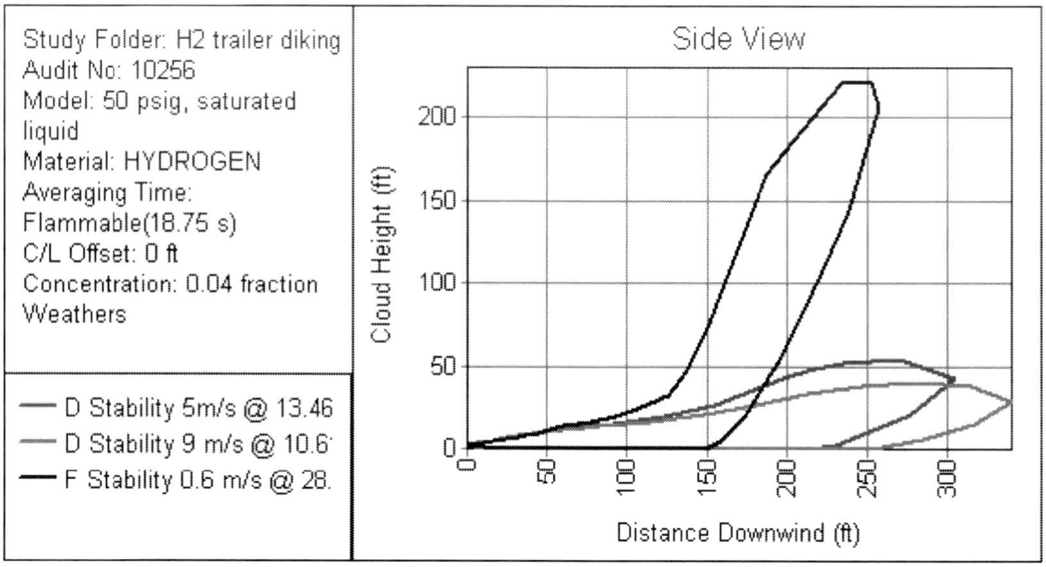

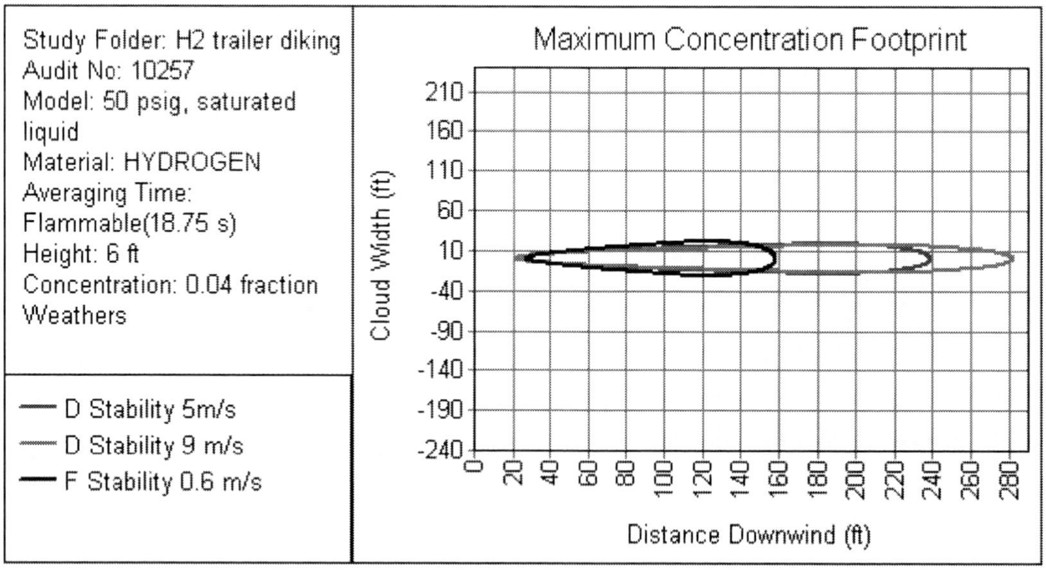

Liquid Hydrogen Spill Study
Hydrogen Spill
Attachment A

February 26, 2007
Rev. 2

Case 2
Hydrogen @ 50 psig, -423F (saturated at 0 psig)
Horizontal release from 2" diameter pipe @ 2' above grade
Cloud boundaries for LFL for 3 weather conditions in legend
100% liquid after release – no pool formed

Study Folder: H2 trailer diking
Audit No: 10257
Model: 50 psig, -423 F
Material: HYDROGEN
Averaging Time:
Flammable(18.75 s)
C/L Offset: 0 ft
Concentration: 0.04 fraction
Weathers

— D Stability 5m/s @ 16.46
— D Stability 9 m/s @ 14.2
— F Stability 0.6 m/s @ 28.

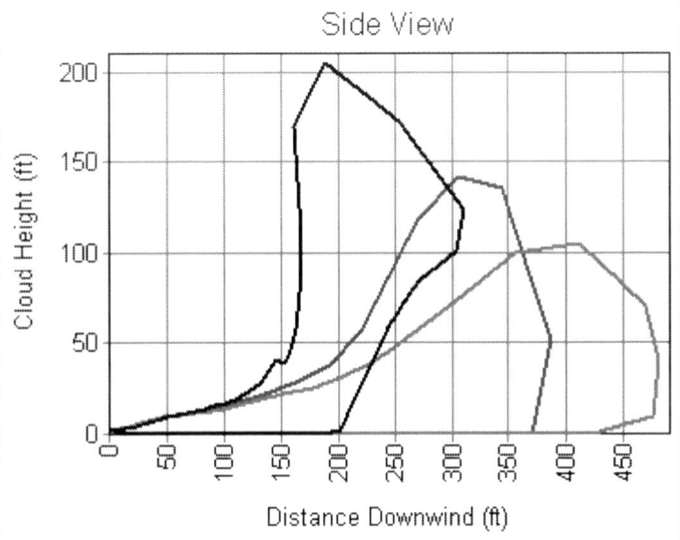

Study Folder: H2 trailer diking
Audit No: 10257
Model: 50 psig, -423 F
Material: HYDROGEN
Averaging Time:
Flammable(18.75 s)
Height: 6 ft
Concentration: 0.04 fraction
Weathers

— D Stability 5m/s
— D Stability 9 m/s
— F Stability 0.6 m/s

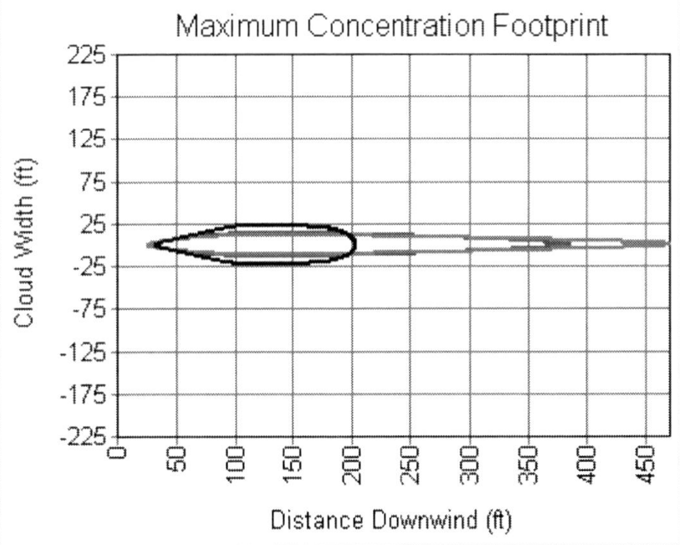

February 26, 2007
Rev. 2

Case 3
Hydrogen (saturated liquid) @ 150 psig, -402F
Horizontal release from 2" diameter pipe @ 2' above grade
Cloud boundaries for LFL for 3 weather conditions in legend
48% liquid after release – no pool formed

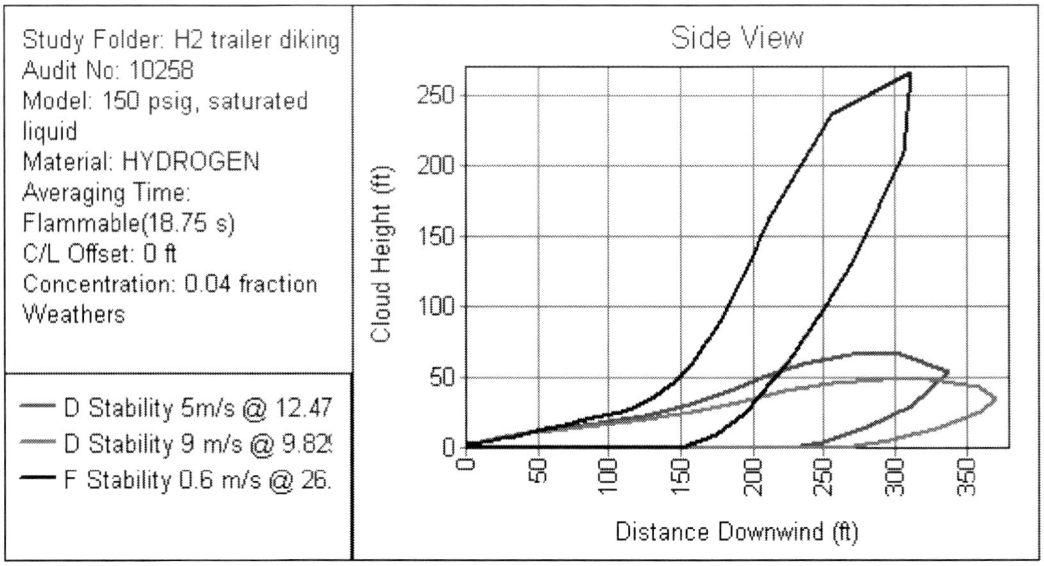

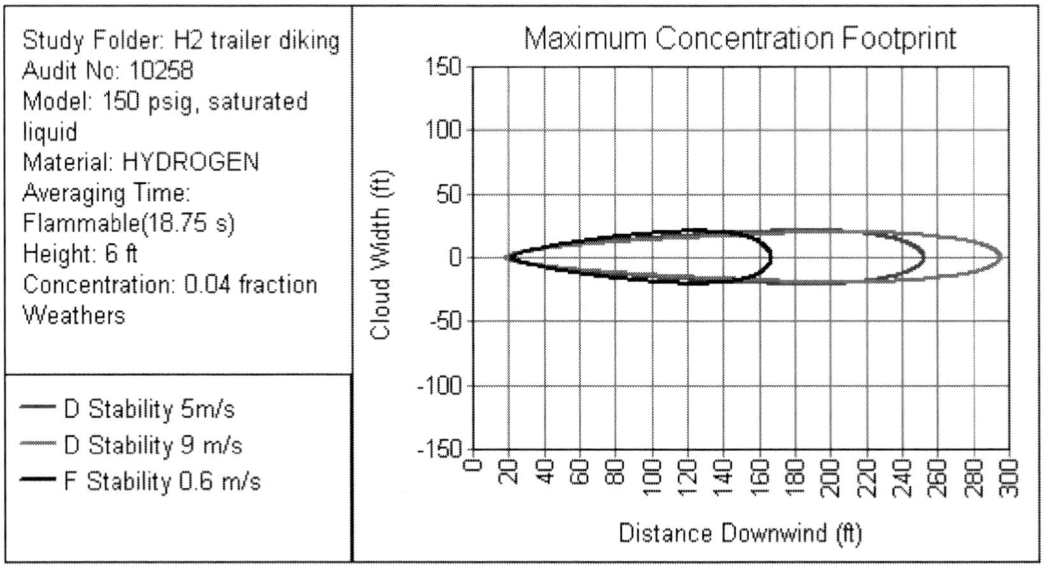

Liquid Hydrogen Spill Study
Hydrogen Spill
Attachment A

February 26, 2007
Rev. 2

Case 4
Hydrogen @ 150 psig, -423F (saturated at 0 psig)
Horizontal release from 2" diameter pipe @ 2' above grade
Cloud boundaries for LFL for 3 weather conditions in legend
100% liquid after release – no pool formed

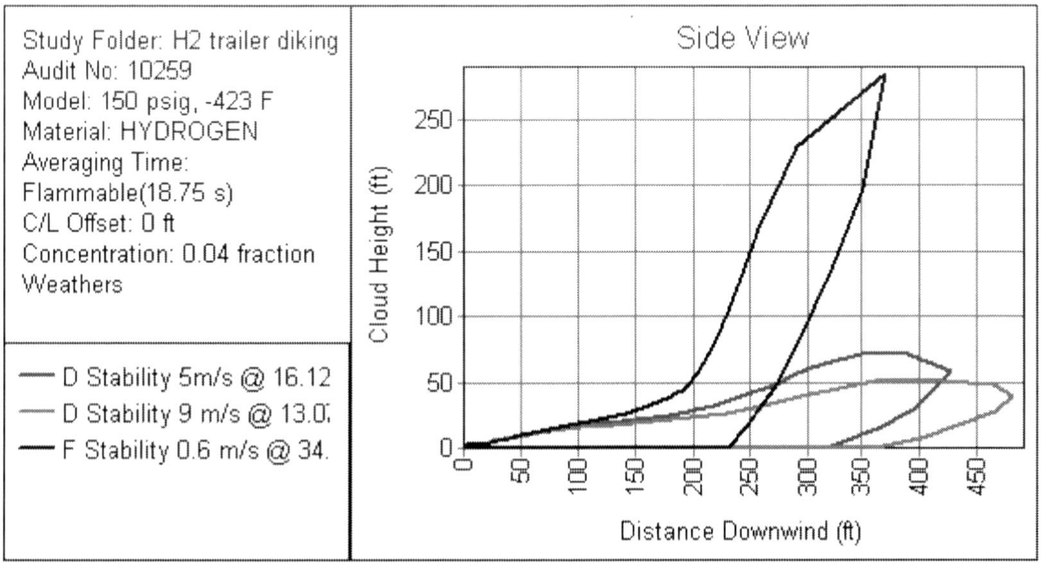

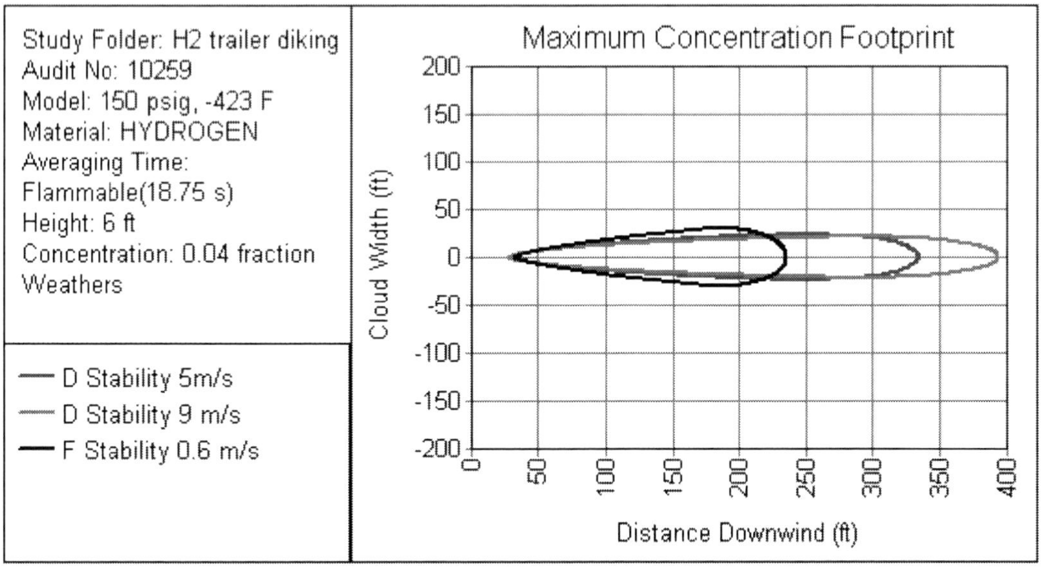

Liquid Hydrogen Spill Study
Methane Spill
Attachment B

February 26, 2007
Rev 2

Case 1
Methane (saturated liquid) @ 50 psig, -220F
Horizontal release from 2" diameter pipe @ 2' above grade
Cloud boundaries for 4.4% methane for 3 weather conditions in legend
85% liquid after release
Pool formed with F weather stability

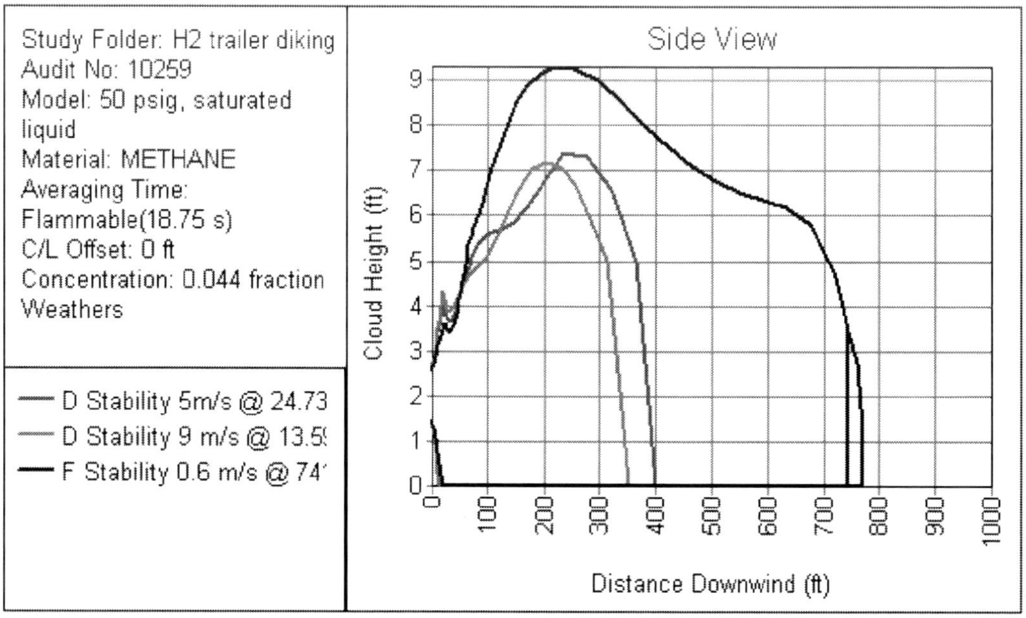

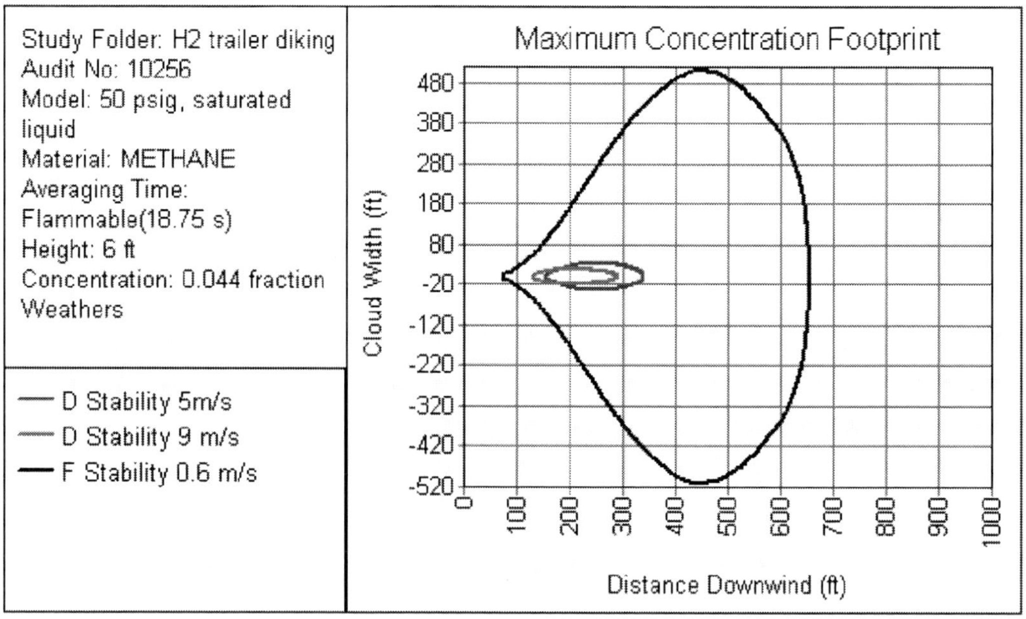

Liquid Hydrogen Spill Study
Methane Spill
Attachment B

February 26, 2007
Rev 2

Case 1
Methane (saturated liquid) @ 50 psig, -220F
Horizontal release from 2" diameter pipe @ 2' above grade
Cloud boundaries for 4.4% methane for 3 weather conditions in legend
85% liquid after release
Pool formed with F weather stability
Pool radius & depth below

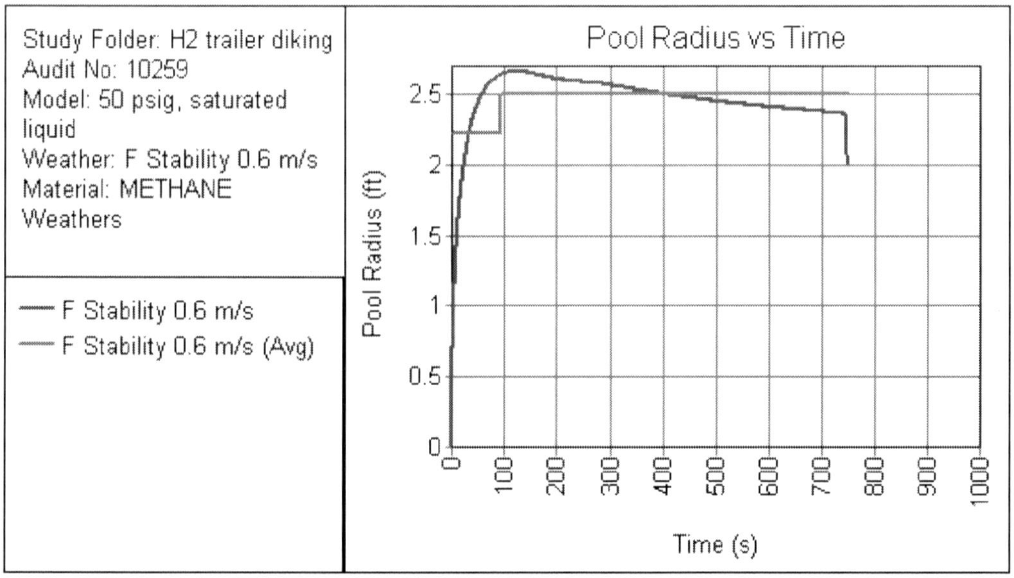

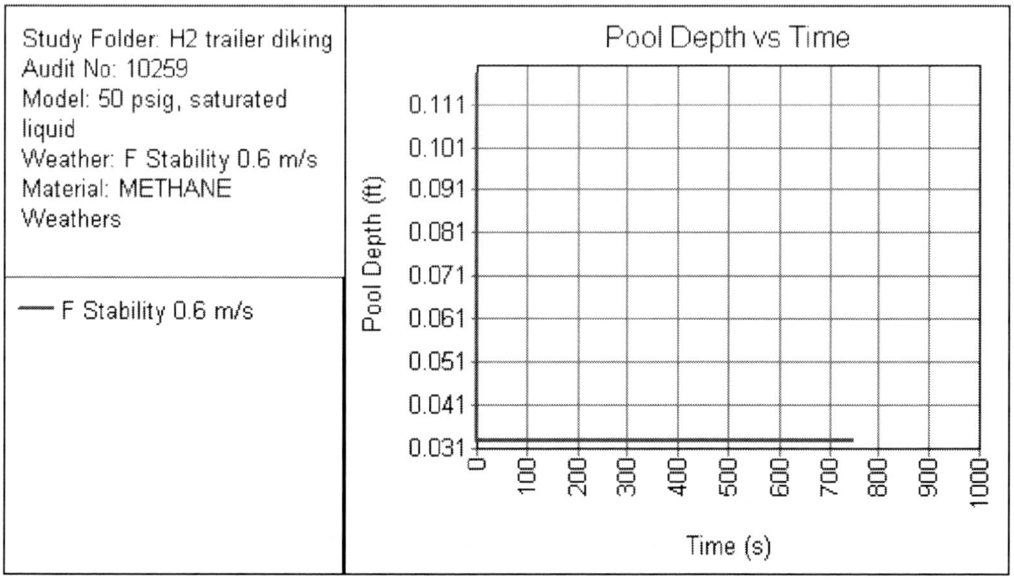

Case 2
Methane @ 50 psig, -259F (saturated at 0 psig)
Horizontal release from 2" diameter pipe @ 2' above grade
Cloud boundaries for 4.4% methane for 3 weather conditions in legend
100% liquid after release
Pool formed with F weather stability and D weather stability at 5 m/s

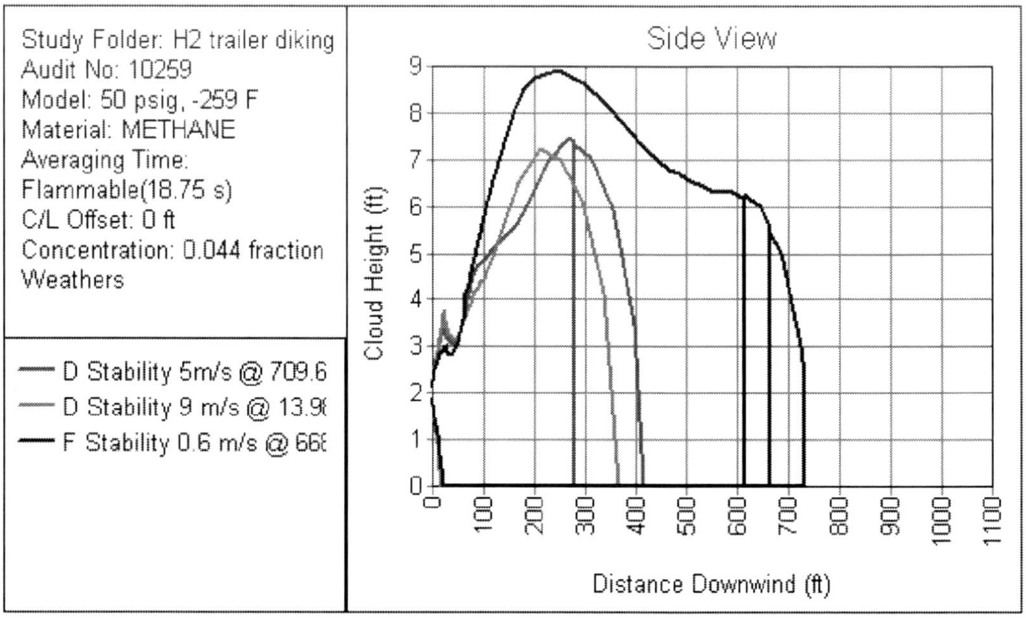

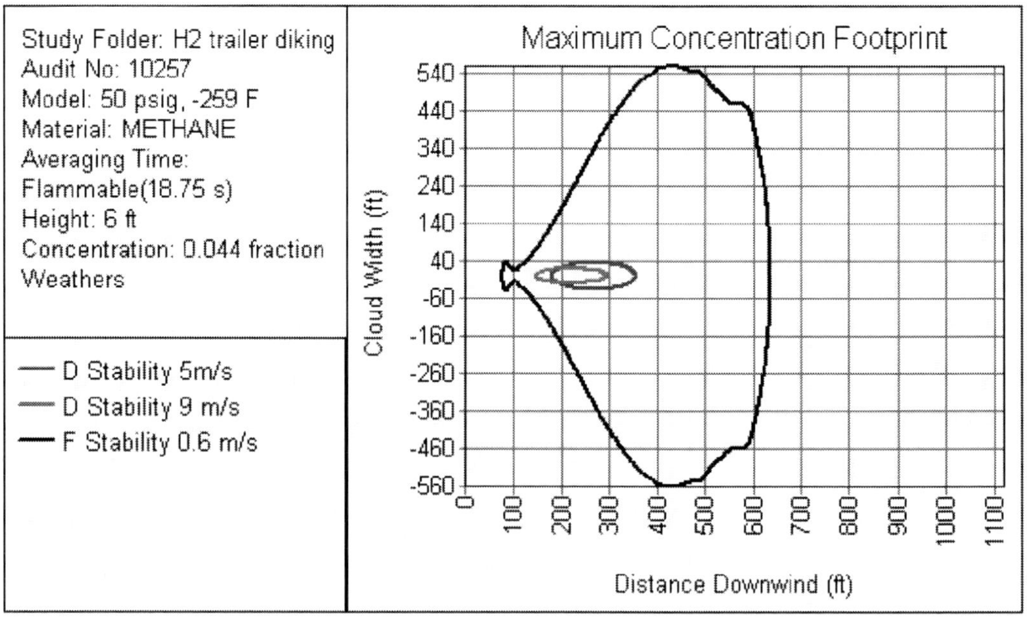

Case 2
Methane @ 50 psig, -259F (saturated at 0 psig)
Horizontal release from 2" diameter pipe @ 2' above grade
Cloud boundaries for 4.4% methane for 3 weather conditions in legend
100% liquid after release
Pool formed with F weather stability and D weather stability at 5 m/s

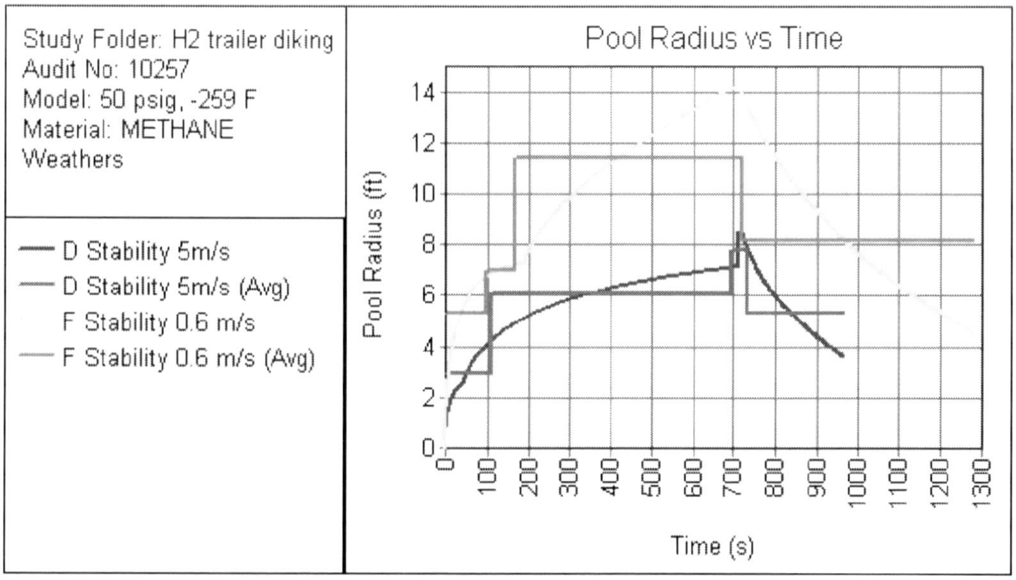

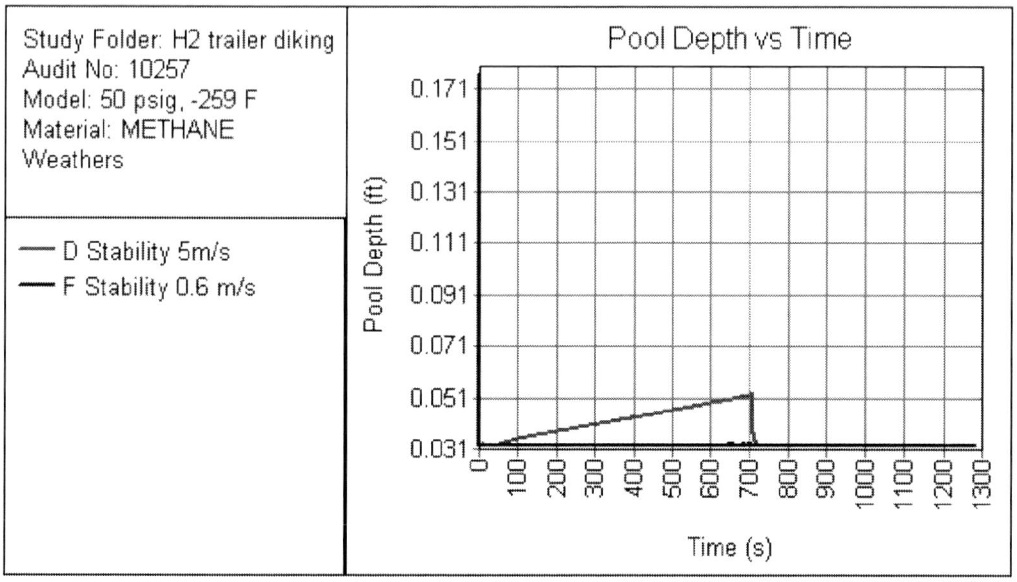

Case 2
Methane @ 50 psig, -259F (saturated at 0 psig)
Horizontal release from 2" diameter pipe @ 2' above grade
Cloud boundaries for 4.4% methane for 3 weather conditions in legend
100% liquid after release
Pool formed with F weather stability.
Radius and depth shown below

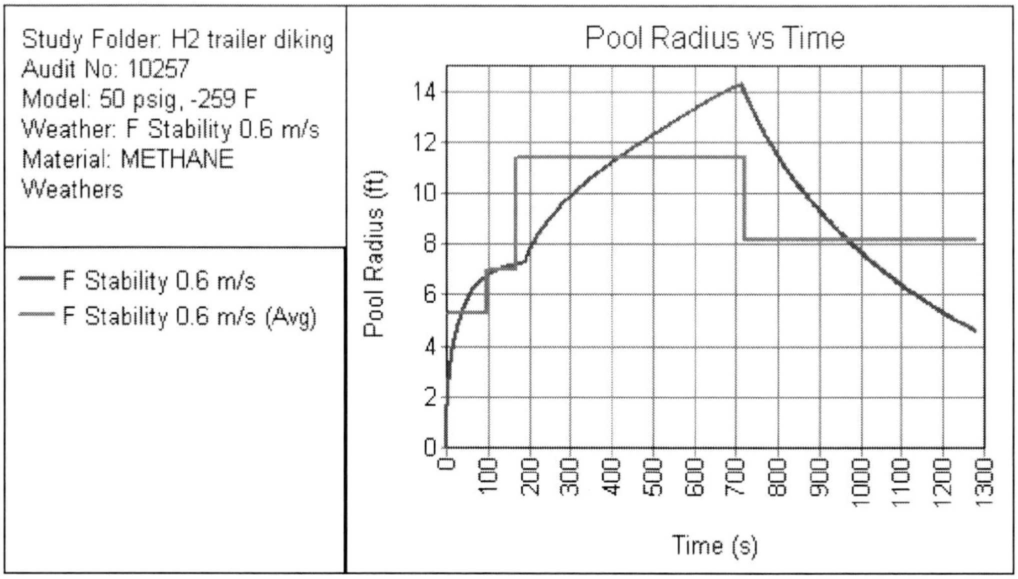

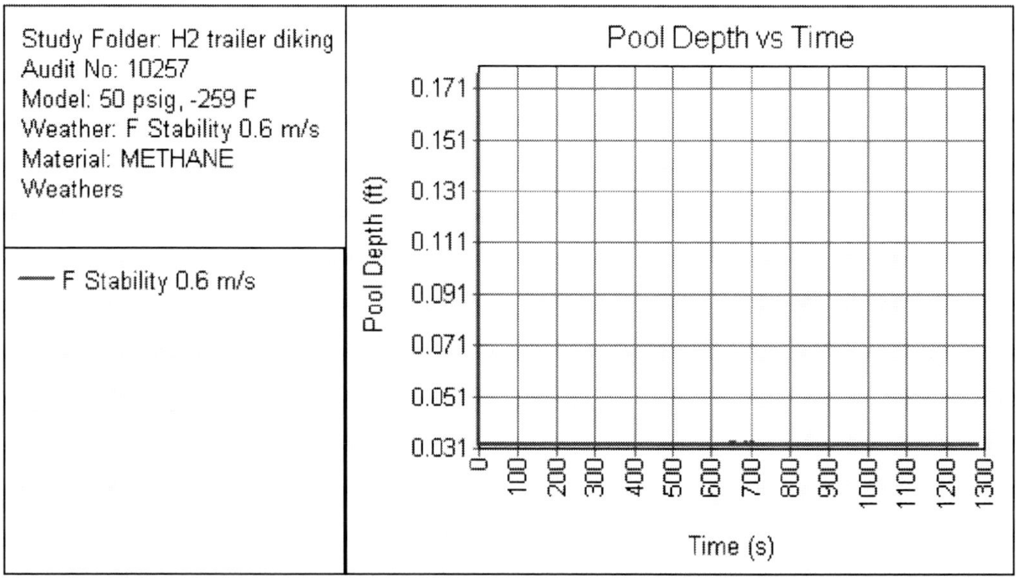

Liquid Hydrogen Spill Study
Methane Spill
Attachment B

February 26, 2007
Rev 2

Case 2
Methane @ 50 psig, -259F (saturated at 0 psig)
Horizontal release from 2" diameter pipe @ 2' above grade
Cloud boundaries for 4.4% methane for 3 weather conditions in legend
100% liquid after release
Pool formed with D weather stability at 5 m/s
Radius and depth shown below

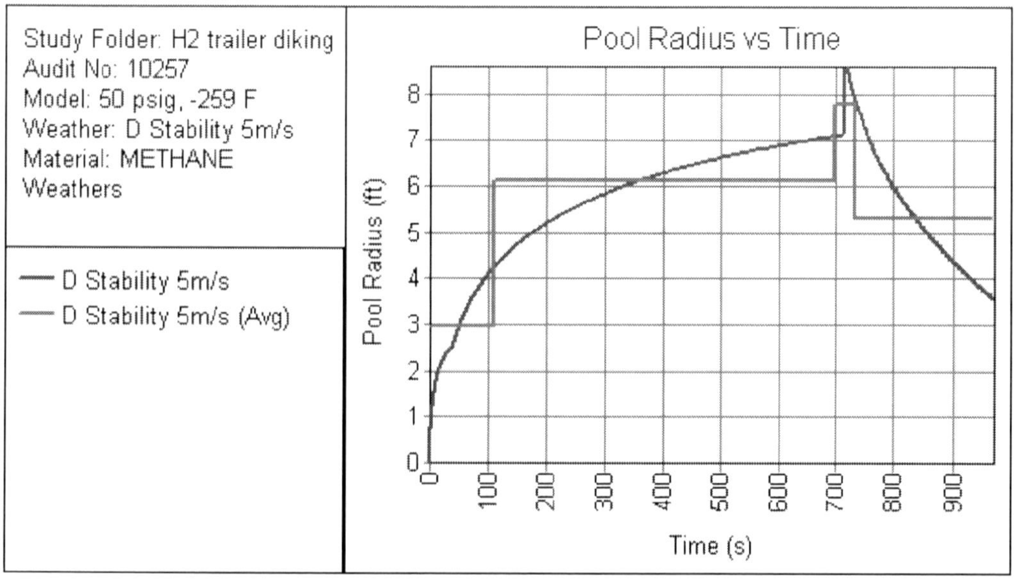

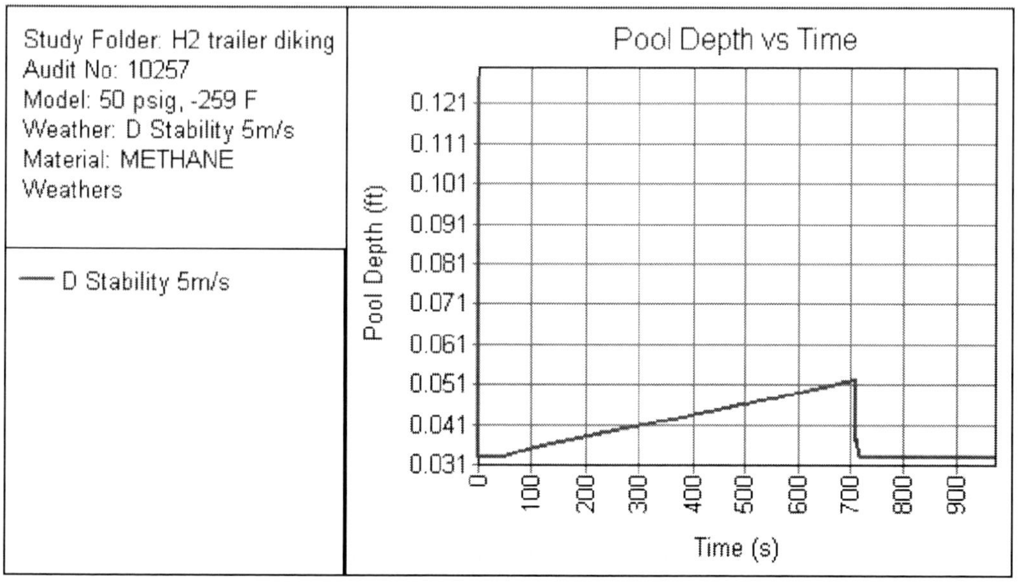

Liquid Hydrogen Spill Study
Methane Spill
Attachment B

February 26, 2007
Rev 2

Case 3
Methane (saturated liquid) @ 150 psig, -186F
Horizontal release from 2" diameter pipe @ 2' above grade
Cloud boundaries for 4.4% methane for 3 weather conditions in legend
71% liquid after release
No pool forms

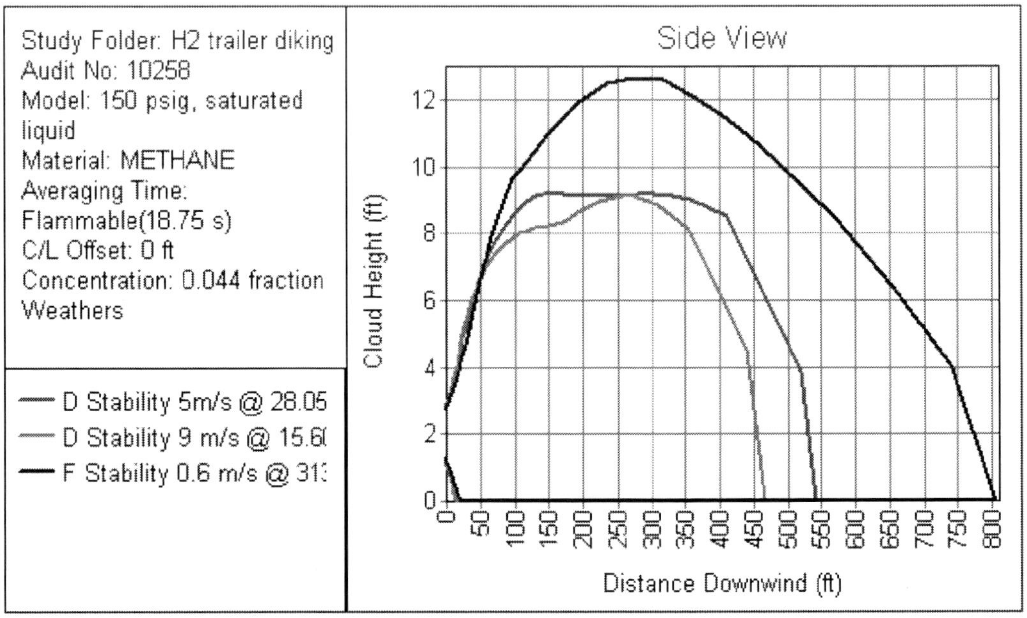

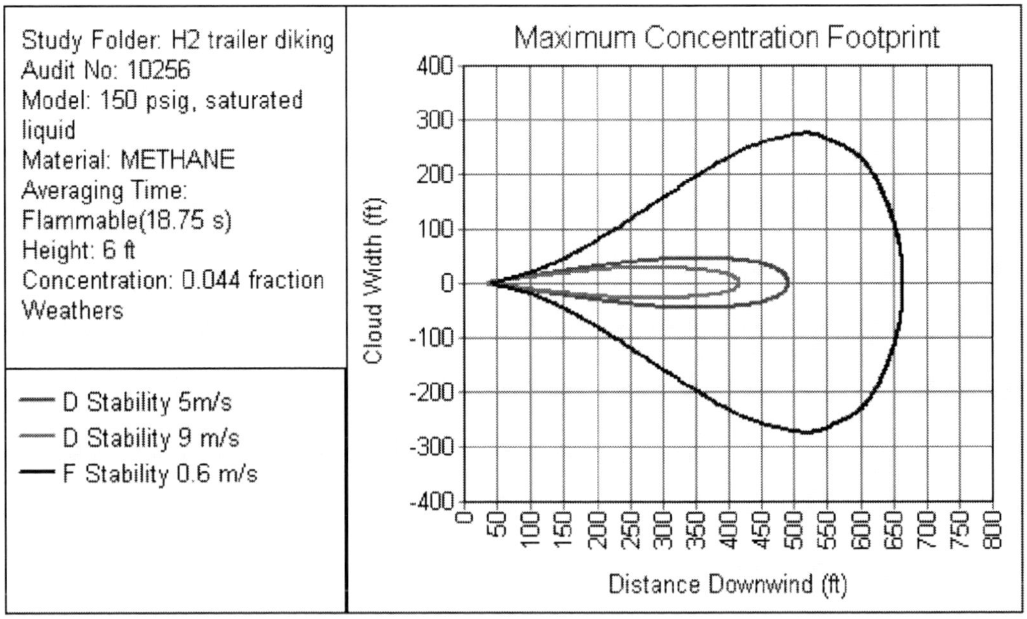

Liquid Hydrogen Spill Study
Methane Spill
Attachment B

February 26, 2007
Rev 2

Case 4
Methane @ 150 psig, -259F (saturated at 0 psig)
Horizontal release from 2" diameter pipe @ 2' above grade
Cloud boundaries for 4.4% methane for 3 weather conditions in legend
100% liquid after release
Pool formed with F weather stability

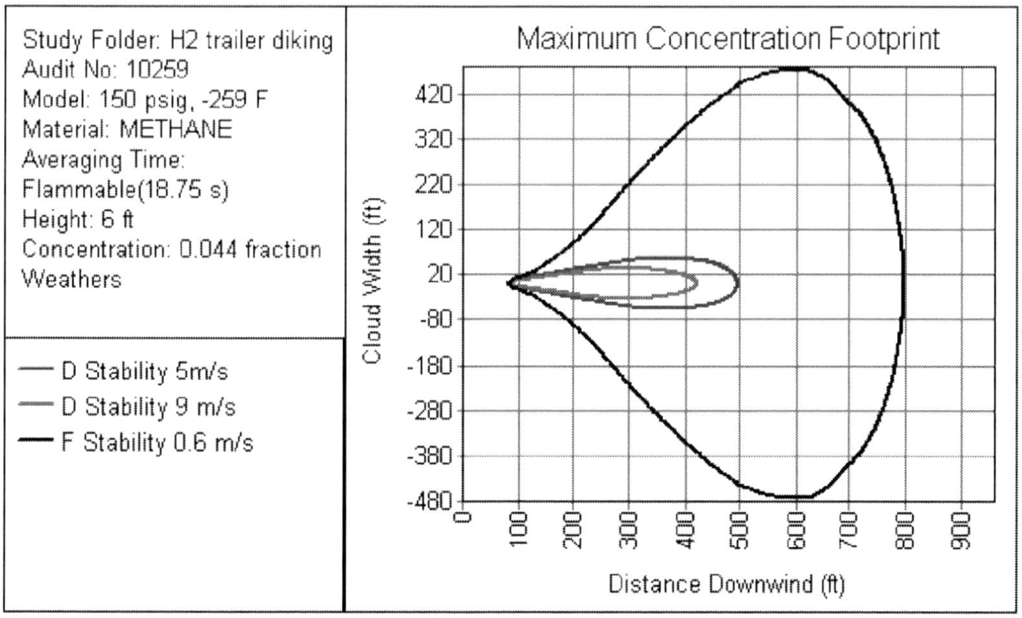

Case 4
Methane @ 150 psig, -259F (saturated at 0 psig)
Horizontal release from 2" diameter pipe @ 2' above grade
Cloud boundaries for 4.4% methane for 3 weather conditions in legend
100% liquid after release
Pool radius and depth shown below

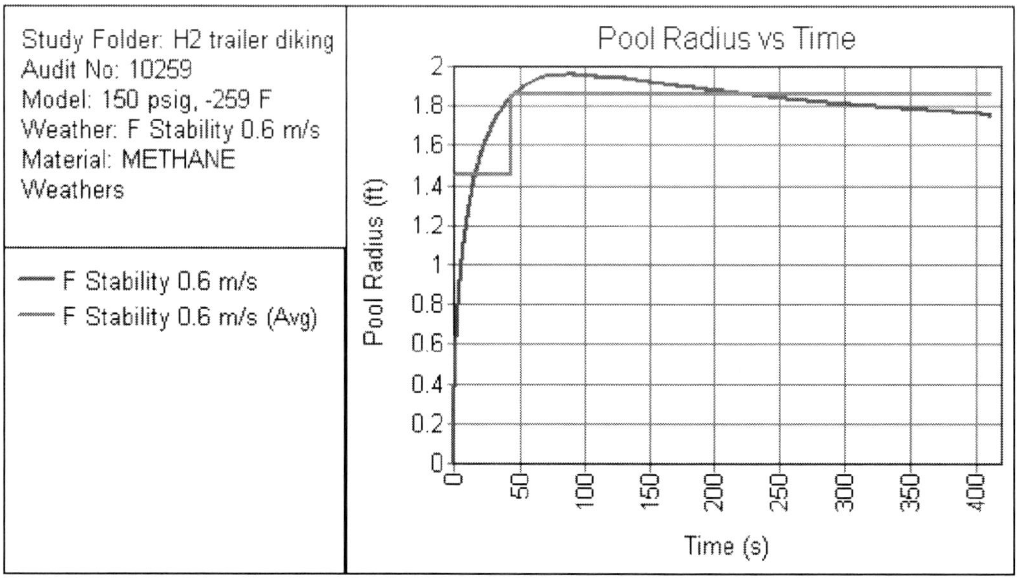

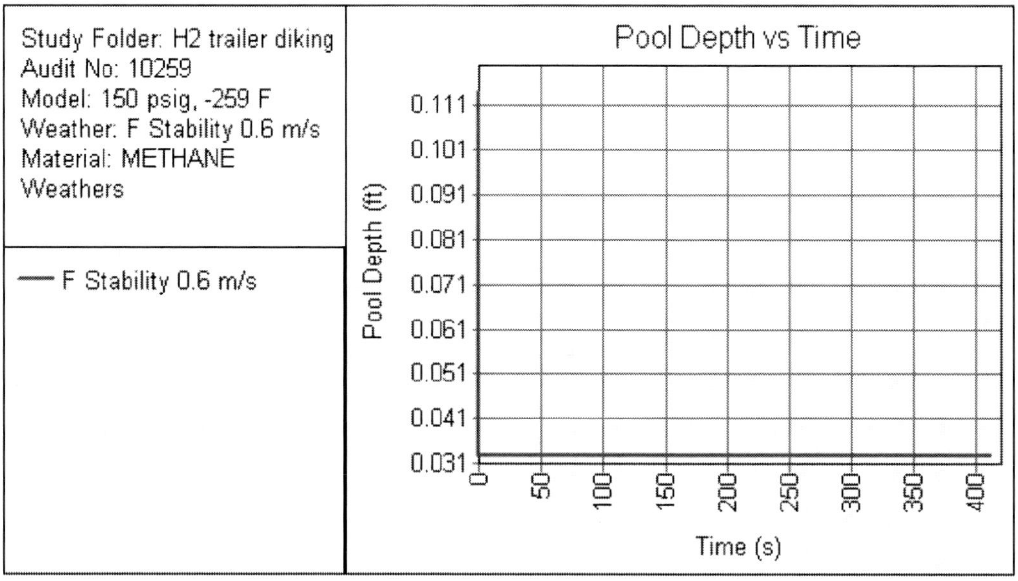

Comparison of Approaches to Managing LNG risk

*Caroline J. Deetjen, Cindy Wei, Tatiana Brown
Det Norske Veritas (USA) Inc.
Houston, Texas

AIChE Spring Meeting, April 2007
7th Topical Conference on Natural Gas Utilization
Houston, Texas, 22-26, 2007

Abstract

This paper compares the United States' consequence based approach to managing the LNG risk with other approaches taken by the rest of the world. Regardless of the country, the risk management methodologies remain largely unchanged but experiences in each individual country remarkably mirror their form of government or regulatory administration. The different risk based approaches applied to the LNG industry in different regions are outlined fully in this paper. The risk based approach is the only way to effectively manage the LNG risks that can potentially cause harm to workers, third parties and to the asset.

1. Introduction

Global demand for LNG has surged since the last decade because of the growing market in energy consumption and the environmental friendly property of natural gas. There is a need to effectively manage risks across the LNG supply chain, with the increasing activity of LNG development worldwide. In the United States, it is generally recognized by policymakers, producers, and the public that increasing LNG imports is crucial to help meet domestic energy demands and that LNG terminals can be developed in a safe and secure fashion. In the process of approving a new LNG facility, many factors including safety, environment, and economics need to be taken into account. This paper will summarize the current regulatory frameworks in the United States and other countries in terms of approaches to managing LNG risks. Areas of improvement will be isolated in order to develop a comprehensive regulatory framework or risk management program for the LNG industry.

2. LNG Safety Record

The LNG industry has a relatively good safety record comparing to other industries. LNG vessels, which have made more than 35,000 ocean voyages covering more than 60 million miles, are among the safest and most reliable cargo fleets at sea today. Since the industry's creation in 1959 there has been no record of fatalities, loss of cargo containment, or serious safety incidents on any LNG ship. The incidence of injury to LNG workers is also extremely low by industrial standards.

Worldwide, there are over 40 LNG receiving terminals, 17 LNG liquefaction export terminals, and over 240 peak-shaving and LNG storage facilities. In over half a century of operations, the world has witnessed only two deadly incidents in LNG facilities. During the inception stage of LNG industry, an LNG storage tank rupture claimed 128 lives in Cleveland, Ohio in 1944. The industry then learned a hard lesson about the proper construction material of storage tank and site selection.

Another tragic incident occurred in 2004 at Algeria's Skikda LNG production plant. The vapour cloud explosion that originated in a steam boiler killed 27 people and injured 74. Early reports suggested that it occurred when plant personnel attempted to restart the boiler, whose furnace may not have been properly purged. The explosion triggered a fire that escalated to nearby process equipment and other trains in adjacent liquefaction units. The Skikda incident highlighted the importance of safe operation and maintenance as well as enough separation spaces between equipment.

These two deadly incidents raised industry's awareness of the importance of managing public perception risks because both public perception and government regulations play an important role for LNG development.

3. Comparisons of LNG Risk Management Programs

3.1 Introduction

This section compares the major LNG regulations and standards in the North America, especially in the United States, with those in Europe. Generally speaking, consequence analysis is required by North America regulations, while risk based approaches are covered by European regulations.

3.2 49 CFR 193 and NFPA 59A

The most important LNG safety issues are to protect persons and property near an LNG facility from thermal radiation caused by an immediately ignited fire or dispersion and delayed ignition of a natural gas cloud arising from a major spill of LNG. In order to achieve the safety goals, practices need to be taken to reduce the potential for a catastrophic spill of LNG. The federal regulations and industry standards have been developed to achieve LNG process safety objectives by focusing on preventing accidental LNG containment failures and minimizing the impact of fires resulting from an accidental LNG containment release.

The major United States regulations and standards governing LNG plant siting, design, construction, inspection, testing and process safety issues are 49-CFR-193, "Liquefied Natural Gas Facilities; Federal Safety Standards" [1] and NFPA-59A, "Production, Storage and Handling of Liquefied Natural Gas (LNG)" [2]. The 49-CFR-193 standard is the U.S. federal government regulation, while NFPA-59A is an industry consensus standard issued by the National Fire Protection Association.

The approach utilized to determine adequate siting separation distances and to specify hazard control system is based upon the design spill concept. This concept hypothesizes accidental LNG releases which could occur at the facility based upon historical experience, regulatory requirements and the specific features of the facility. Potential hazards associated with the spills are then quantified using consequence analysis predictive techniques, which are widely accepted throughout the LNG industry.

3.3 Process Safety Management and RMP Rules

In the US, most other petroleum processing facilities, like a gas processing plant or a refinery that handles large quantities of flammable liquids or gases, are covered under the Occupational Safety and Health Administration (OSHA) process safety management (PSM) standard (29 CFR 1910.119) and the Environmental Protection Agency (EPA) risk management program (RMP) rule (40 CFR 68).

The PSM standard emphasizes the management of hazards through a comprehensive program that integrates technologies, procedures, and management practices. The standard has fourteen elements that are fundamental to running a safe chemical operation. The EPA Risk Management Plan builds on the OSHA PSM regulation, adding a "worst case consequence analysis" and also requires reporting of incidents.

However, both of those regulations specifically exempt LNG facilities that are regulated under the DOT regulations in 49 CFR Parts 193, being precluded by the concept of sufficient regulation by another (DOT) federal agency. It is interesting to note that the offshore petroleum production industry is also exempt due to their regulations under the Minerals Management Service (MMS).

3.4 Consequence Analysis for Exclusion Zone Determination

Unique features of NFPA 59A and 49-CFR-193 are the siting "exclusion" zones. The concept of a "design spill" was defined in order to evaluate the consequences of an accidental LNG release, with particular attention to the influence of safety protective measures. In addition to specifying the magnitude of a potential unplanned release, the design spill also specified some of the parameters for estimation of the downwind LNG vapor dispersion and the extent of the radiant heat from an LNG pool fire. Another unique feature of the USA LNG rules is the specification of distinct consequence software models. The use of alternate models which take into account the same physical factors and have been validated by experimental test data could be permitted, subject to the Administrator's approval. The federal siting requirements for onshore LNG facilities (49 CFR 193) specify the use of the DEGADIS program for the determination of dispersion distances. DEGADIS accounts for dense gas effects and was originally developed for simulation of cryogenic flammable gas dispersion, particularly for LNG. It has been validated against a wide range of laboratory and field test data. Like DNV software PHAST, it is a similarity model and does not account for complex flow patterns induced by tank and dike structures. When complex terrain and gas flow obstructions are concerns, the regulation allows dispersion distances to be calculated with the FEM3A, a CFD model.

3.5 USCG Exclusion Zones and SANDIA Report for LNG Carriers

US Coast Guard published a Navigation and Vessel Inspection Circular (NVIC) No. 05-05 entitled "Guidance on Assessing the Suitability of a Waterway for Liquefied Natural Gas (LNG) Marine Traffic" in June 2005 [3]. A NVIC provides detailed guidance about the enforcement or compliance with a certain Federal marine safety regulations and Coast Guard marine safety program. While NVICs are non-directive, meaning that they do not have the force of law, they are important tools for complying with the law.

A summary of "Zones of Concern" was provided in NVIC 05-05, and three concentric zones around the LNG carrier were based on the findings in the Sandia Report [4] for intentional breach scenarios.

Zone 1: This is the area with the most severe consequences around the LNG tanker, where an LNG spill could pose a severe public safety and property hazard and could damage or significantly disrupt critical infrastructure and key assets located within this area. Zone 1 is considered to extend about 500 m (0.3 miles) for an intentional breach of an LNG tanker.

Zone 2: This is an area with less severe consequences than Zone 1 and is considered to extend from 500 m (0.3 miles) to 1,600 m (1 mile) for an intentional breach of an LNG tanker.
Zone 3: This is an area with the least likelihood of severe consequences and is considered to extend from 1,600 m (1 mile) to a conservative maximum of 3,500 m (2.2 miles) from the LNG tanker, in the unlikely event that 3 cargo tanks were breached and a vapor cloud disperses without an initial ignition.

When performing LNG carrier safety assessments, site specific conditions such as carrier specifications and site specific meteorological data need to be used. Normally, the hole sizes (a hole area of 0.5-1m^2 for accidental breaches and a hole size of 2-12m^2 for intentional breaches) specified in the Sandia Report will be used to establish the exclusion zones.

Recent conversations with Sandia and USCG indicate that it is unrealistic to establish hazard Zone 3 based on vapor dispersion results due to 3 breached cargo tanks because of the high probability of the presence of an immediate ignition source. The intentional acts resulting in the breach of three cargo tanks would have to be inherently strong ignition sources. It is extremely unlikely that each of the three devices needed to make three holes would be so finitely timed so as to miss an opportunity to ignite a release from one of the three tanks. Also all marine traffic including LNG carriers have the potential to ignite the LNG vapor before it reaches the full LFL extent.

3.6 Canadian Regulation – Z276-01

The Canadian Standards Association (CSA) published the first edition of standard Z276, Liquefied Natural Gas (LPG) – Production, Storage, and Handling in 1972. Since then, several editions have been published with the latest seventh edition in 2001. For the seventh edition, the CSA and NFPA Technical Committees attempted to harmonize their standards while maintaining their respective countries' regulatory requirements.

3.7 Mexican Standard - NOM-013-SECRE-2004

In 2004, the Mexican Ministry of Energy published the official Mexican Standard NOM-013-SECRE-2004 entitled "Safety requirements for the design, construction, operation, and maintenance of terminals for liquid natural gas storage which include systems, equipment and facilities for the reception, transportation, vaporization and delivery of natural gas".

NOM-013 consists of four sections: the first section applies to LNG storage terminals with fixed facilities in firm ground. The second section applies to LNG storage terminals installed off-shore, from the LNG point of reception to the point of delivery of such natural gas in gaseous form to a transportation system through pipelines, and including systems for the reception, transportation, storage, and vaporization of LNG and delivery of natural gas. The third section of NOM-013 refers to the design of the underwater pipeline for the transportation of natural gas from the storage terminal out to sea to the littoral. Finally, the fourth section of NOM-013 deals with the Conformity Evaluation Procedure (PEC), which sets out the methodology to determine the licensee's level of compliance with NOM-013 in connection with the design, construction, operation and maintenance of the different kinds of terminals for liquefied natural gas storage, which include systems, equipment and facilities for the reception, transportation, vaporization and delivery of natural gas.

3.8 European LNG Safety Assessment Regulations

EN 1473, Installation and equipment for liquefied natural gas – Design of onshore installations [5]. The objective for this European Standard is to give functional guidelines for LNG installations. It recommends procedures and standards for i.e. Hazard Assessments, and requires that a hazard assessment shall be carried out.

EN 1473 opens for both probabilistic and/or deterministic hazard assessment. The probabilistic approach consists of:

- collection of failure rate data
- list of potential hazards of external and internal origin.
- determination and classification of the probability of these hazards
- determination of the consequences of each hazard
- classification of accidents in accordance with their consequences and probability criteria
- verification that no hazard comes within the "Unacceptable Risk" level
- justification of measures necessary to limit risk.

Seveso II, Council Directive 96/82/EC, is entitled "On the control of major-accident hazard involving dangerous substances" [6]. The Directive is aimed at the prevention of major accidents which involve dangerous substances, and the limitation of their consequences for man and the environment, with a view to ensuring high levels of protection throughout the Community in a consistent and effective manner. LNG facilities are covered in the Directive.

The Directive requires that the operator produce a Safety Report for the purpose of i.e.:

- demonstrating that a major-accident prevention policy and a safety management system for implementing it have been put into effect
- demonstrating that major-accident hazards have been identified
- demonstrating that internal emergency plans have been drawn up

4. Risk Based Approach vs. Consequence Based Approach

A risk-based approach is an alternative to a pure consequence-based approach. A risk-based approach, taking into account both the event likelihood and the effect of mitigation measures, is a superior decision-making basis, compared to a consequence-based approach. An example of risk-based regulation is the European Norm Standard EN 1473 that requires risk analysis in design and modification decisions. This standard provides requirements for all types of LNG-related equipment. Events are evaluated based on their frequency and consequence, and risk mitigation measures are required for events that do not meet established acceptance criteria.

In the US, due to the absence of risk based process safety management program and risk criteria, decision making has been performed on the basis of calculating expected consequences. These worst case results with low probability and very high uncertainty can inappropriately drive operational decisions and system designs, or cause costly project delays and even non-approval of projects. This consequence based approach may also inhibit opportunities to shape and implement a true risk based emergency response system by the states. Expensive and demanding means may be implemented for high consequence, but low risk scenarios while high risk events, may be ignored. The risk based approach is the only way to effectively mitigate against those incidences that can cause harm to workers, third parties and to the asset.

5. References

1 NFPA-59A, Production, Storage, and Handling of Liquefied Natural Gas (LNG) 2006 Edition, National Fire Protection Association, Quincy, Massachusetts.

2 Title 49 Code of Federal Regulations Part 193, Liquefied Natural Gas Facilities: Federal Safety Standards, U.S. Government Printing Office, Washington, D. C., October 23, 1980. (http://www.washingtonwatchdog.org/documents/cfr/title49/part193.html)

3 U.S. Coast Guard Security and Operational Updates - Navigation and Vessel Inspection Circular (NVICs) (http://www.maritimedelriv.com/Port_Security/USCG/USCG_nvic.htm).

4 Sandia National Laboratories Report, "Guidance on Risk Analysis and Safety Implications of a Large Liquefied Natural (LNG) Spill Over Water," SAND2004-6258, December 2004.

5 European Committee for Standardization CEN, EN 1473, Installation and equipment for liquefied natural gas – Design of onshore installations, 1997.

6 Council Directive 96/82/EC, "On the control of major-accident hazard involving dangerous substances", 1999.

Biography

Caroline holds a first class masters degree in Chemical Engineering from Imperial College, London. Her pervious positions for Shell International included HAE Supervisor for the LNG terminal construction at Altamira.

Caroline is currently Market Sector Leader for the Upstream, Pipeline and LNG sectors for DNV Energy Solutions in Houston. She continues in her role as Project Manager for risk management projects, focused on LNG, and works in the MSL team to develop the market.

Fluency in 6 languages means that she can communicate effectively with a wide range of clients. Her key projects this year include an LNG QRA in South Africa for a French client, another for an El Salvadorian client together with a risk assessment for a Sour Gas Development in the Middle East.

Direct Methane Conversion to Methanol

Yongchun Tang, Zaiwei Li, Weiqun Chen and Jihong Cheng

PEER Center, California Institute of Technology
738 Arrow Grand Circle, Covina, CA 91722

Finding economical ways to utilize the enormous worldwide reserves of methane (CH4) is of great interest towards increasing this clean energy resource. Low temperature direct methane conversion has shown significant promise in the past decade. Using ionic liquids could be the key to a commercial process for converting natural gas to methanol. We will present our recent findings of using ionic liquid to facilitate methane conversion in new ternary systems of inorganic Pt salts and oxides, and ionic liquids under different solution condition. One of our new ionic liquid systems shows significant improvement over the current methane to methanol process.

The direct conversion of natural gas to liquid fuel at low temperatures (<300 ^{o}C) has attracted considerable interest since the 1970s. Among many significant developments the Shilov chemistry in aqueous solution was the earliest reported method, and more recently in 1998 the Periana catalyst in oleum has been purported to be the best catalytic system so far in terms of selectivity and conversion. However, the practical application of the Periana catalyst is hindered due to its rapid deactivation by water and methanol that are produced *in situ* during the reaction. Shilov system, though is very tolerate to water, but suffers from catalyst precipitation problem and also can not be utilized for industry.

In searching for more effective catalysts for methane oxidation we found that ionic liquids could dissolve almost every solid Pt compound that we examined. Such we could fast screen many potential Pt compounds as the catalyst candidate. However, at the very harsh oxidation condition (strong acid, strong oxidation Environment, higher Temperature (150 – 250 C) and a powerful Pt Catalyst), most commonly commercial available ionic liquids will not surviving. After screening hundred commercial and synthesized ILs we finally found 3 class of new ionic liquids which are very stable in those conditions. (Fig.1 and Fig.2)

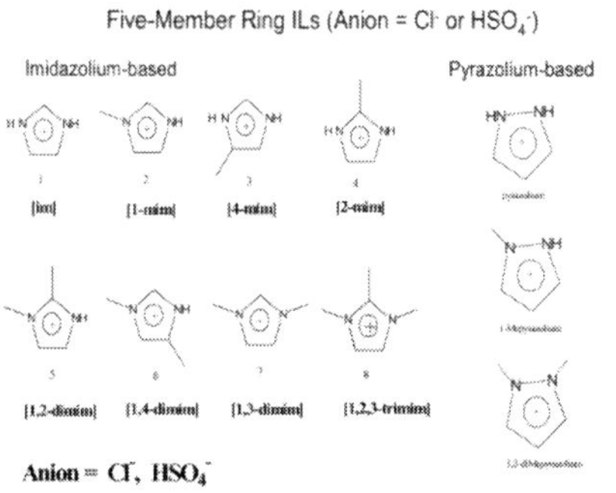

Fig.1

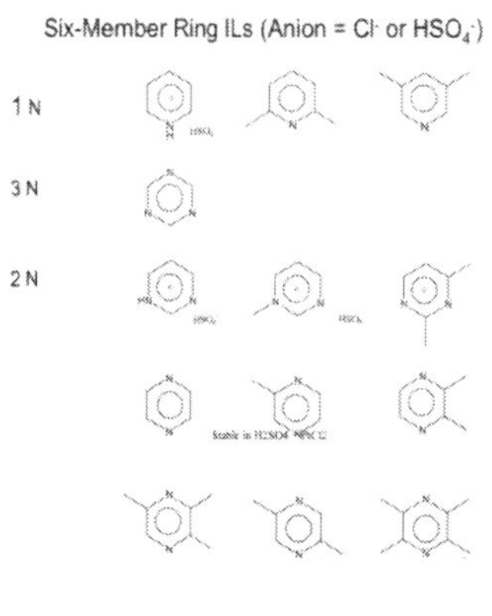

Fig.2

In the course of improving Periana's system, we found a ternary systems from inorganic Pt salts or oxide, ionic liquids, and concentrated sulfuric acid are effective in catalyzing direct, selective oxidation of methane to methanol and appear to be more tolerant with water than the Catalytica reaction (Fig.3, Fig.4).

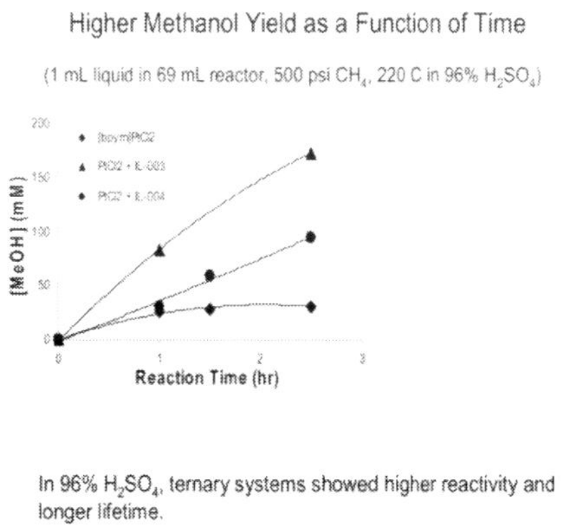

Fig. 3

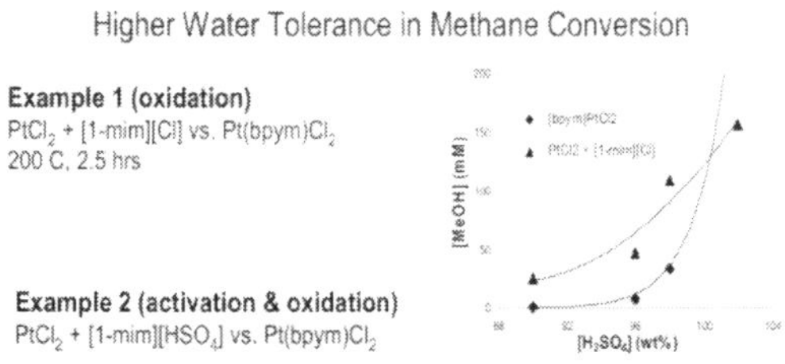

System	$[H_2O]$	Activation at 150 C		Oxidation at 200 C	
		$[CH_xD_{4-x}]$ (mmol)	TON	[MeOH] (mmol)	TON
Catalytica	2%	0.751	15.0	0.047	0.94
Catalytica	6%	0.652	13.0	0.016	0.31
$PtCl_2$ + [1-mim][HSO_4]	2%	0.843	16.9	0.065	1.31
$PtCl_2$ + [1-mim][HSO_4]	6%	0.783	15.7	0.037	0.73

Fig.3

In another direction, we also recently found ionic liquid could greatly stabilized the Shilov catalyst at aqueous solution.

Reference

1. A. E. Shilov and G. B. Shul'pin, *Chem. Rev.*, 1997, **97**, 2879
2. R. A. Periana, D. J. Taube, S. Gamble, H. Tabue, T. Satoh and H. Fujii, *Science*, 1998, **280**, 560.
3. J.Cheng, Z. Li, Y. Tang, M. Haught, *Chem. Comm.* 2006, 4617.
4. Z. Li, Y. Tang, J. Cheng, Provisional Patent Application, Caltech, 2006
5. Z. Li, Y. Tang, US Patent Application 20060069169.

Direct Methane to Methanol Conversion

Jonas Oxgaard, Yongchun Tang, Zhitao Xu, William A. Goddard, Zaiwei Li, Jihong Cheng and Weiqun Chen

The world's resources in methane are currently grossly underutilized, mainly due to a lack of efficient low temperature catalytic processes to convert the methane to transportable liquid forms. The chemistry of C-H functionalization, i.e. the activation and subsequent transformation to a C-X bond, has not yet been advanced to a commercially viable system. Nevertheless, considerable progress has been made, particularly with the introduction of the Shilov and Periana-Catalytica (PC) systems.[1-3] Shilov systems are aqueous solutions involving square-planar Pt(II) complexes $\{[PtCl_{4-n}(H_2O)_n]^{n-2}$, n=0 to 2$\}$ as the catalytic center for C-H activation and using (expensive) $PtCl_6^{2-}$ as the stoichiometric oxidant to convert methane to methanol or chloromethane. The PC system uses dichloro(bpym)Pt(II) (bpym=η^2-{2,2'-bipyrimidyl}) to convert methane to methyl bisulfate in fuming H_2SO_4, where H_2SO_4 serves also as the oxidant. The PC system displays the highest efficiency and selectivity reported to date, but the catalyst is deactivated by water or by the target product, i. e. methanol, preventing commercial exploitation .

Recently, Cheng et al. reported a ternary protocol for the effective conversion of methane to methanol, containing inorganic Pt salts, concentrated sulfuric acid and organic ionic liquids.[4] The activity of this protocol is somewhat lower than the PC system (~50% slower in 102% H_2SO_4 [2% oleum]) but also significantly more tolerant to water than the PC system (~300% faster in 96% H_2SO_4 [4% H_2O]).[4] This system, while similar in some respects to Shilov and in other respects to PC, differs mainly through the introduction of ionic liquids as additives. While the amount of ionic liquid is catalytic (6 equivalents with respect to Pt), exactly what role it plays in enabling this chemistry is unknown, prompting us to explore the mechanism for this transformation using theoretical means.

According to reported theoretical studies C-H activation by Shilov systems ($[PtCl_{4-n}(H_2O)_n]^{n-2}$, n=0 to 2) occurs either via a σ-bond metathesis {when $PtCl_2(H_2O)_2$ is the reactive species} or through oxidative addition {when $PtCl_4^{2-}$ is the dominant catalyst},. In an early report on $PtCl_2(H_2O)_2$, Siegbahn and Crabtree concluded that the rate determining step is a sigma-bond metathesis, although the authors did not rule out the possibility of a competitive oxidative addition.[5] The second coordination sphere of the solvent, mimicked by the hydrogen bonding interactions between $PtCl_2(H_2O)(CH_4)$ and a second water molecule, was found to play an important role in facilitating this chemistry. Furthermore, no studies on the generation of the prerequisite $PtCl_2(H_2O)(CH_4)$ were presented. The barrier for this ligand exchange was found later by Zhu and Ziegler to be the rate-determining step for all three species $[PtCl_{4-n}(H_2O)_n]^{n-2}$ (n=0 to 2).[6] The $\Delta H^{\ddagger}$ for the Cl-/CH_4 exchange in $PtCl_4^{2-}$, the most pertinent for this study, was determined to 27.5 kcal/mol. In all three cases, this was followed by an oxidative addition to generate a Pt^{IV} methyl hydride.[5] QM studies[7,8] on the PC system also found that the rate-determining step is CH_4 uptake, although with a higher energy of activation than $PtCl_4^{2-}$ (40.7 kcal/mol). Furthermore, the C-H activation in the PC system occurs via an electrophilic substitution, in which the reacting hydrogen is subtracted by a Cl^- or HSO_4^- anion, while oxidative addition and σ-bond metathesis were found to be higher in energy.[9]

Here, we report QM studies on the C-H activation catalyzed by $PtCl_4^{2-}$ in H_2SO_4 with imidazolium (HIM) based ionic liquid additives. We first consider a double-charged double ion pair $[PtCl_4^{2-}][HIM]_2$ (Structure **A**, Figure 1) as the possible active catalytic species. According to our computational results, the formation energy of this strong ion pair (from $PtCl_2$ + 2[HIM][Cl]) is -68.3 kcal/mol, which is sufficient for dissolving solid $PtCl_2$ in concentrated H_2SO_4 (calculated ΔE is 29.4 kcal/mol uphill for $1/6(PtCl_2)_6 \rightarrow PtCl_2$ in a continuum solvent, suggesting that forming $[PtCl_4^{2-}][HIM]_2$ from solid $PtCl_2$ and 2[HIM][Cl] is exothermic by -38 kcal/mol) and forms a homogenous solution. This result is consistent with experimental observation, which also shows that $PtCl_2$ cannot dissolve in neat concentrated H_2SO_4.

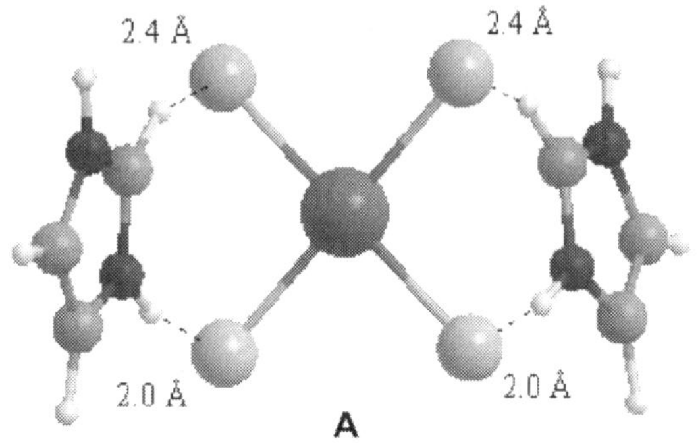

Figure 1: Structure of $[PtCl_4^{2-}][HIM]_2$ (A)

We also calculated the energies required to replace one of the Cl ligands in $[PtCl_4^{2-}][HIM]_2$, with either HSO_4^- or H_2O. The resulting complexes are 3.5 and 3.2 kcal/mol higher in energy, respectively. This is consistent with the previous theoretical results for the $PtCl_4^{2-}$ system,[6] and, together with the reported exothermic Cl-/H_2O exchange for the PC system,[4] could explain why the $PtCl_2$/IL/H_2SO_4 system is more resistant to water poisoning.

We now focus on the methane activation by $[PtCl_4^{2-}][HIM]_2$. Figure 2 shows the calculated energy profile and related structures of the methane activation by $[PtCl_4^{2-}][HIM]_2$. We see that the CH_4 uptake step, from reactant **B (A + CH_4)** to the σ-CH_4 complex **C**, has to overcome a higher barrier (**TS1**, ΔE_{solv} = 27.7 kcal/mol) than for the C-H cleavage step (**TS2**, ΔE_{solv} = 23.3 kcal/mol), an oxidative addition. This result is consistent with the previous computational results on both the PC[7,8] and Shilov systems ($PtCl_4^{2-}$),[6] and in general agreement with the qualitative potential energy profiles based on experimental observations for similar reaction processes for L_2PtCH_3X systems.[8]

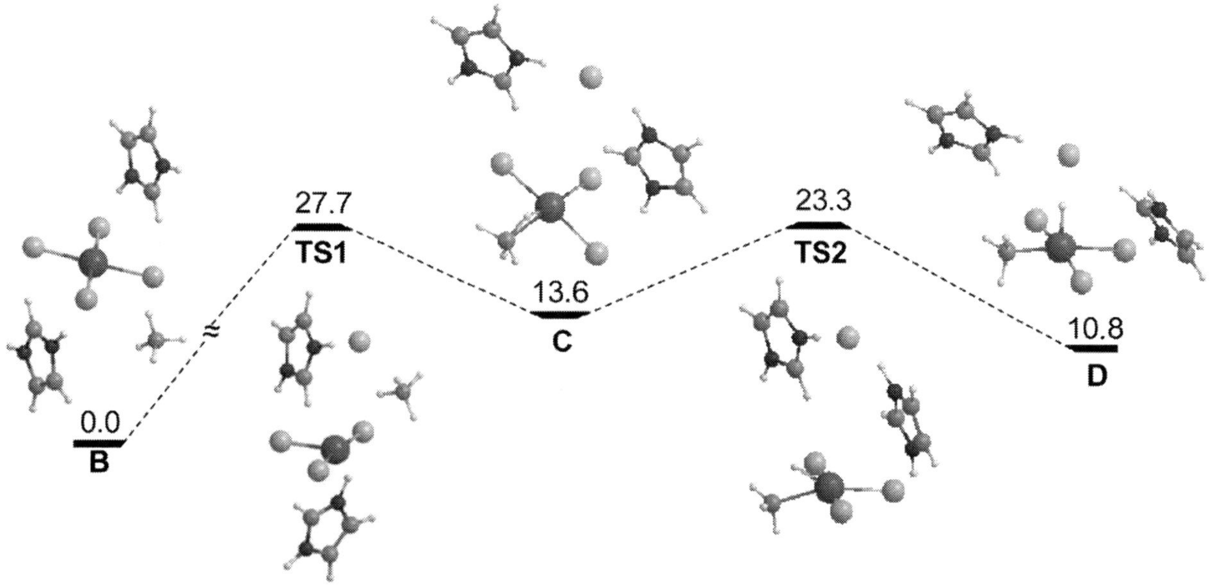

Figure 2: The energy profile and related structures for C-H activation of methane catalyzed by $PtCl_4^{2-}$ in ionic liquid solution

The overall reaction barrier for the C-H activation in the $PtCl_2$/ionic liquid by $[PtCl_4^{2-}][HIM]_2$ system is lower than that of the PC system (at least 30kcal/mol[7,8]). The C-H activation mechanism in the $PtCl_2$/ionic liquid $[PtCl_4^{2-}][HIM]_2$ system is different from that in the PC system, and can be distinguished by their different hydrogen transfer processes – oxidative addition in the current system, and electrophilic substitution in the PC system. We examined the Mulliken atomic charges on the three atoms Pt, H, and Cl in **D,** and found that both Pt and Cl are negatively charged (-0.34e and -0.43e, respectively) while the atomic charge on the H atom is positive (+0.21e). The atomic charge distribution on the Pt-H-Cl unit here is significantly different from its counterpart $\mathbf{D^{PC}}$ in the PC system (see Figure 3) where the atomic charge distribution are 0.15e, 0.27e and -0.19e on Pt, H and Cl, respectively. We see that the Pt center is indeed quite electrophilic (positive charged) in $\mathbf{D^{PC}}$ but not in **D**. We believe that the difference in charge distributions is responsible for the difference in C-H activation mechanisms catalyzed by the current ternary catalytic system and the PC system.

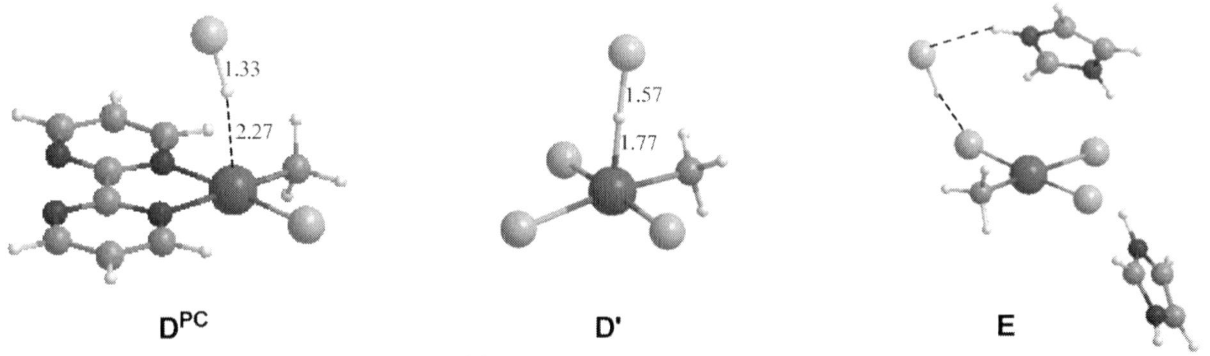

Figure 3: Structures and bond lengths of $\mathbf{D^{PC}}$, D' and E

The reaction mechanism of the catalytic system in ionic liquid solution is similar to that presented for the $PtCl_4^{2-}$ Shilov system,[6] except for the strong hydrogen bonding interactions between HIM^+ cations and the $PtCl_4^{2-}$ complex, indicated by the short Cl-H(N) distances (1.83 – 2.14 Å) and found in all reaction species. In order to identify whether ionic liquids imparts a function beyond enabling the solvation, we compared the $PtCl_4^{2-}$ systems with and without HIM^+ cations. We focused on the influence of the HIM^+ on the structures and energies of **TS2** and **D** since similar barriers (ΔH = 23.9 and 27.5 kcal/mol for dissociative and associative mechanisms, respectively)[6] have been reported for the uptake step, but the C-H cleavage step was reported in the absence of the dissociated Cl^- ligand. By removing the HIM cations from **MD2** and re-optimizing the structure (**D'**, see Figure 3) we can compare the ionic liquid system with the $PtCl_4^{2-}$ system more directly, and we find that the leaving Cl^- group is re-associated in **D'**. **D'** is 12.7 kcal/mol higher in energy than the reactant ($[PtCl_4^{2-}][CH_4]$), indicating that it is less accessible than **D** (10.8 kcal/mol) in ionic liquid solution, although this is well within the margin of error for the solvation methods used. Mulliken population analysis shows that charge distribution in the Pt-H-Cl unit in **D'** (-0.23, 0.32 and -0.51e, respectively) is similar to that of **D**. However, the Pt-H distance in **D'** is much longer (1.77 Å) but H-Cl significantly shorter (1.57 Å) than in **D** (1.57 and 2.04 Å, respectively), suggesting that **D'** might not be considered a Pt^{IV} species, but rather a molecular complex of $Pt^{II}(Cl)_3(CH_3)^{2-}$ and HCl. This could, however, also be a result of the solvation method used. Nevertheless, the mechanism of the Shilov $PtCl_4^{2-}$ system should most likely be modified to include the intermediate (**D'**) found here.

In addition to the above mechanism, we explored another possible C-H activation product (**E**) that can be considered as the product of a mechanism similar to electrophilic substitution (i.e.

a hydrogen atom transferring directly to the leaving Cl^- ligand). However, we found that **E** is 9.3 kcal/mol higher in energy than **D**. Therefore, we did not further explore the reaction barrier for the pathway.

We also tested the possibilities of the imidazolium participating in the inner coordination sphere of the catalytic center, through three different coordination modes, i.e. through **F**, **G** and **H**, as shown in Figure 4. Based on our computational results, only the C^2-carbene coordination mode (**G**) leads to a relatively stable complex (calculated reaction energy for $PtCl_4[HIM]_2 \rightarrow PtCl_3(C^2\text{-IM})[HCl][HIM]$ is -0.7 kcal/mol downhill) while the other two result in less stable complexes ($PtCl_4[HIM]_2 \rightarrow PtCl_3(\eta^2\text{-HIM}) + [HCl][HIM]$ and $PtCl_4[HIM]_2 \rightarrow PtCl_3(N^1\text{-IM}) + [HCl][HIM]$ are +16.5 and +9.5 kcal/mol uphill, respectively). Further examination on the stabilities of the two possible intermediates following the C-H activation step, structures **I** and **J** (Figure 5), shows that this reaction path can be safely ruled out due to their high instabilities (>29kcal/mol). However, it is quite likely that either F or G are involved in the breakdown of the ionic liquid over time, as a first step towards oxidation of the ring.

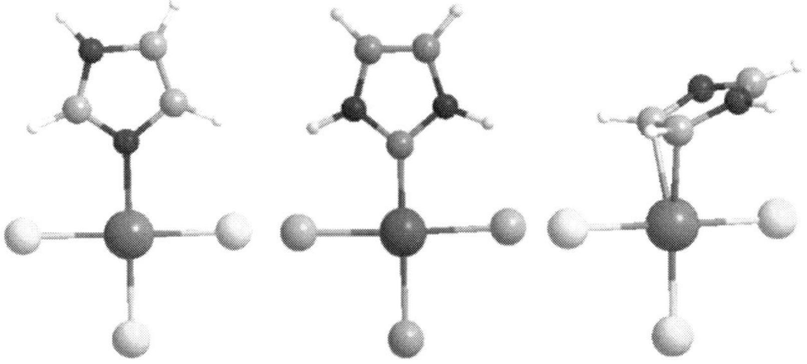

F G H

Figure 4: Possible modes of the imidazolium participating in the inner coordination sphere of the catalytic center

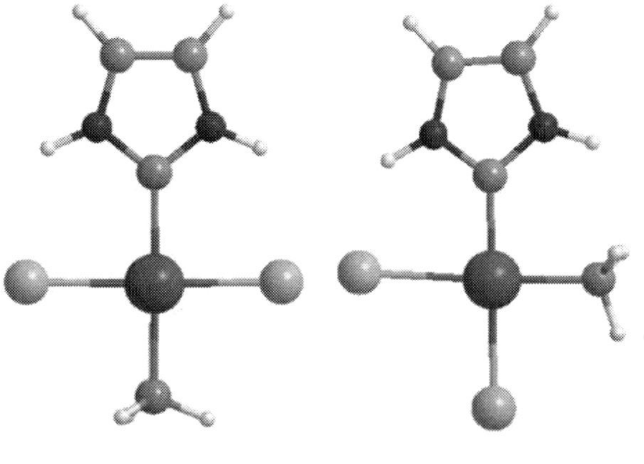

I J

Figure 5: Intermediates from reaction G + $CH_4 \rightarrow$ X + HCl

In summary, the results presented herein thus suggests that the role of the ionic liquid is to enable the solvation of the $PtCl_2$ catalyst, through the formation of a $[PtCl_4][HIM]_2$ ion pair. The C-H activation of this complex is very similar to that found for the aqueous Shilov system, with almost identical barriers. However, by coupling the more active Shilov C-H activation step with the oxidizing media of the Periana-Catalytica system, the ionic liquid system provides an important addition to the field of C-H *functionalization*. Thus, we expect that ionic liquids can be used to enable the coupling of other previously incompatible chemistries.

Computational details

The hybrid density functional theory (DFT) of the Becke three-parameter functional[12] (B3) with the correlation functional of Lee, Yang, and Parr [13] (LYP) implemented in the Jaguar 6.5 program package[14] and lacv3p**++ basis sets were used in all calculations. B3LYP is known to produce good descriptions of reaction profiles for transition metal containing compounds.[15,16] All geometries were optimized in the gas phase and transition states were verified by the proper vibrational mode of the imaginary frequency computed using the analytic Hessian from vibrational frequency calculations. Solvent effects were evaluated self-consistently using the Poisson-Boltzmann (PBF) continuum approximation as implemented in Jaguar v6.5 [17,18] (which uses the solvent accessible solute surface), through the use of single point solvation energy calculations at the B3LYP/lacv3p** level using the geometries obtained for the gas-phase optimizations.

References:

(1) Shilov, A. E.; Shul'pin, G. B. *Chemical Reviews* **1997**, *97*, 2879-2932.
(2) Lersch, M.; Tilset, M. *Chemical Reviews* **2005**, *105*, 2471-2526.
(3) Periana, R. A.; Bhalla, G.; Tenn, W. J.; Young, K. J. H.; Liu, X. Y.; Mironov, O.; Jones, C. J.; Ziatdinov, V. R. *Journal of Molecular Catalysis a-Chemical* **2004**, *220*, 7-25.
(4) Cheng, J. H.; Li, Z. W.; Haught, M.; Tang, Y. C. *Chemical Communications* **2006**, 4617-4619.
(5) Siegbahn, P. E. M.; Crabtree, R. H. *Journal of the American Chemical Society* **1996**, *118*, 4442-4450.
(6) Zhu, H. J.; Ziegler, T. *Journal of Organometallic Chemistry* **2006**, *691*, 4486-4497.
(7) Kua, J.; Xu, X.; Periana, R. A.; Goddard, W. A. *Organometallics* **2002**, *21*, 511-525.
(8) Paul, A.; Musgrave, C. B. *Organometallics* **2007**.
(9) Gilbert, T. M.; Hristov, I.; Ziegler, T. *Organometallics* **2001**, *20*, 1183-1189.
(10) Heyduk, A. F.; Driver, T. G.; Labinger, J. A.; Bercaw, J. E. *J. Am. Chem. Soc.* **2004**, *126*, 15034-15035.
(11) Romeo, R.; D'Amico, G. *Organometallics* **2006**, *25*, 3435-3446.
(12) Becke, A. D. *Journal of Chemical Physics* **1993**, *98*, 5648-5652.
(13) Lee, C. T.; Yang, W. T.; Parr, R. G. *Physical Review B* **1988**, *37*, 785-789.
(14) Jaguar, v., Schrodinger, LLC, New York, NY, **2005**.
(15) Baker, J. M., M.; Andzelm, J.; Scheiner, A. *ACS Symposium Series 629* **1996**, *Laird, B. B., Ross, R. B., Ziegler, T., Eds. American Chemical Society: Washington, DC.*
(16) Niu, S. Q.; Hall, M. B. *Chemical Reviews* **2000**, *100*, 353-405.
(17) Marten, B.; Kim, K.; Cortis, C.; Friesner, R. A.; Murphy, R. B.; Ringnalda, M. N.; Sitkoff, D.; Honig, B. *Journal of Physical Chemistry* **1996**, *100*, 11775-11788.
(18) Tannor, D. J.; Marten, B.; Murphy, R.; Friesner, R. A.; Sitkoff, D.; Nicholls, A.; Ringnalda, M.; Goddard, W. A.; Honig, B. *Journal of the American Chemical Society* **1994**, *116*, 11875-11882.

Development of Membranes for Natural Gas Dehydration

Dilip Kalthod
Air Products and Chemicals Inc.
Saint Louis, MO 63146
kalthodg@apci.com

Ray Malon
Air Products and Chemicals Inc.
Saint Louis, MO 63146
malonrf@apci.com

Terry Franks
Air Products and Chemicals Inc.
Saint Louis, MO 63146

Mimi Schaefer
Air Products and Chemicals Inc.
Saint Louis, MO 63146

Keith Murphy
Air Products and Chemicals Inc.
Saint Louis, MO 63146

Al Brooks
Air Products and Chemicals Inc.
Saint Louis, MO 63146

Wayne Bollinger
Air Products and Chemicals Inc.
Saint Louis, MO 63146
bollinwa@apci.com

Charles Page
Air Products and Chemicals Inc.
Saint Louis, MO 63146
pagect@apci.com

Abstract:

Membranes have many potential advantages over conventional technologies for the dehydration of natural gas, including lower methane and hydrocarbon loss, lower energy consumption, reduced maintenance costs due to fewer components, no spent glycol/ mole-sieve/ deliquescent wastes, amenability to unmanned operation, smaller footprints, lighter weight packages, minimal electrical power duty, and amenability to expansion due to system modularity. However most commercially available membranes for natural gas upgrading involve high capital cost and methane loss and performance degradation from operational upsets, which are barriers to their widespread adoption by the industry. Promising leads for a new polymeric membrane which can enable superior, robust, low-cost separator systems have been developed. Predicted system size and weight, capex, and opex for several membrane process schemes will be presented. These will be compared versus technical and commercial requirements for success compiled from interviews with several oil and gas companies.

AN IMPROVED AND ENERGY EFFICIENT NATURAL GAS LIQUIDS (NGL) REMOVAL PROCESS

Howard S. Meyer, Diana Matonis, and Dennis Leppin, Gas Technology Institute, Des Plaines, IL
Yinghong Lu, Green Hi-Tek, Oak Ridge, TN

Abstract

A new Natural Gas Liquids (NGL) separation process is under development that will provide energy and cost savings while providing greater flexibility in terms of the amount of ethane that is recovered or rejected. This allows the gas processor to tailor operations as the value of the NGL components change compared to the value of the natural gas. This will also help maintain a supply of feedstock material to the chemical industry in times of high value natural gas. This process integrates dehydration with NGL recovery in such a way that heat recovery is maximized, pressure losses are minimized, and some high-value equipment such as the compressor, turbo-expander and a separate dehydration unit are avoided.

The paper will present the technical and economic advantages of the process based on a techno-economic evaluation of the process compared to the cryogenic turbo-expander process, laboratory and bench-scale results on component solubilities with various hydrocarbon lean oils, and details on the design of a pilot plant unit under construction.

Project Objective

The overall objective of this project is to develop a new low-cost and energy efficient NGL recovery process – through a combination of theoretical, bench-scale and pilot-scale testing – so that it could be offered to the natural gas industry for commercialization. The new process is based on U.S. patent No. 6,553,784, which if commercialized, has the potential of achieving substantial energy savings compared to currently used cryogenic technology. When successfully developed, this technology will benefit the petrochemical industry which uses NGL as feedstocks and also benefit other chemical industries involving gas-liquid separation and distillation under similar operating conditions. Specific goals and objectives of the overall program include: (i) collecting relevant physical property and VLE data for the design and evaluation of the new technology, (ii) solving critical R&D issues including identification of suitable dehydration and NGL absorbing solvents, corrosion, packing structure and materials, (iii) designing, construction and operation of bench and pilot-scale units for field testing and verifying design performance, (iv) computer simulation of the process using commercial simulators such as Aspen-Plus and HySys, and (v) preparation of a commercialization plan and identification of industrial partners that are interested in applying the new technology.

BACKGROUND

Natural Gas Liquids (NGL) is a collective term for C_2+ hydrocarbons present in the natural gas. Historically, the commercial value of the separated NGL components has been greater than the thermal value of these liquids in the gas. The revenue derived from extracting NGLs is crucial to ensuring the overall profitability of the domestic natural gas production industry and therefore of ensuring a secure and reliable supply in the Lower-48. However, rising natural gas prices have dramatically reduced the economic incentive to extract NGLs from domestically produced natural gas. The successful gas processors will be those who will be using the technologies that are less energy intensive, have lower capital and operating costs and offer the flexibility to tailor the plant performance to maximize product revenue as market conditions change, while still maintaining efficient operation.

The cryogenic turbo-expander technology is the dominant NGL recovery process throughout the world today. This process is known to be highly energy intensive, as it requires substantial energy to recompress the processed gas back to pipeline pressure. The purpose of this project is to develop a new NGL separation process that is flexible in terms of ethane rejection and reduces the current energy consumption by 20-30%, particularly at lower ethane recoveries (< 70%). The new process integrates the dehydration of the raw natural gas stream and the removal of NGLs in such a way that heat recovery is maximized and pressure losses are minimized and some high-value equipment such as the compressor, turbo-expander and a separate dehydration unit are altogether avoided. GTI completed the techno-economic evaluation of the new process based on HYSYS simulation model and purchased equipment cost estimates obtained from equipment suppliers and two different commercial software packages called Aspen-Icarus and Preliminary Design and Quoting Service (PDQ$). For a 100 MMscfd gas processing plant, the annualized capital cost for the new technology was found to be about 10% lower than that of conventional technology for C_2 recovery above 70% and about 40% lower than that of conventional technology for C_2 recovery below 50%. It was also found that at around 40-50% C_2 recovery (which is economically justifiable at the current natural gas prices); the energy cost to recover NGL using the new technology is about 50% of that of conventional cryogenic technology.

PROCESS DESCRIPTION

The proposed NGL recovery process is based on the US patent No. 6,553,784 issued in April 2003 to Prof. Yingzhong Lu, President of Green Hi-Tek (GHT) Co., Oak Ridge, TN. This process uses a highly energy efficient scheme to remove water vapor and NGLs from natural gas stream in a packed column absorber operating at -30 to -40°F. In this process, the separation of trace water vapor and the recovery of NGL are both carried out at refrigeration temperature of -30 to -40°F, which is much easier and economical to obtain compared to –150°F required in conventional cryogenic process described above and the pressure drops are almost negligible compared to more than 500 psi lost during gas expansion in the cryogenic process. These two features of the new process result in substantial energy savings and significantly less capital expenditure because in the new process some high value equipment such as compressor, turbo-expander and a separate dehydration system can be altogether avoided.

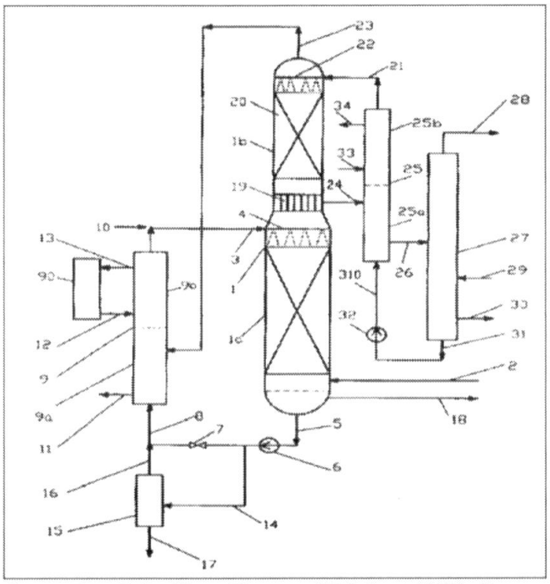

Figure 1 Schematic of the New Energy Efficient NGL Recovery Process

2 – NG Inlet, 1 – Comp. Processor 1a – NG dehydrator; 1b – NGL absorber, 10/27–Regenerators, 9/25–Heat Exchangers, 29 – NGL outlet, 90 – Refrigerator

Figure 1 shows the flow diagram of this novel energy efficient hybrid NGL recovery process. The lower portion (1a) of the main column represents the dehydrator section. The inlet NG stream is brought in contact with the down flowing chilled liquid containing hydrate inhibitor over a high efficiency structured packing. The liquid stream is cooled slightly lower than –30 to -40°F via a propane chiller. Trace water vapors present in the gas stream are condensed at the surface of the liquid film and get carried away with the out flowing liquid. A small fraction of the circulating liquid inhibitor solution enters the regenerator and the wastewater is discharged. The upper portion (1b) of the packed column represents the NGL absorber section. The gaseous NGL components in the cooled and dehydrated natural gas stream are absorbed with the counter-flowing liquid absorbent (e.g., kerosene). The rich absorbent enters the regenerator and the NGL is separated as the product. The gas and liquid flows are so arranged that a good percentage of the energy required to cool the dehydration liquid and NGL absorbent is recovered by gas-liquid heat exchangers 9 and 25.

Because the new process works under constant pressure without any significant pressure loss, no recompression of the residue gas is required resulting in substantial energy and capital cost savings. The energy required for refrigeration of circulating liquid streams is estimated to be less than 30% of the compressor power required for residue gas compression in conventional cryogenic process. This means that if this new technology is successfully commercialized it has an energy saving potential of up to 200 trillion BTU per year by 2020. Moreover, if the schematics of the new and the conventional process shown in Figure 1 and Figure 2 are compared, the new process is seen to be much simpler and involves less costly equipment. An NGL recovery plant built using the new process will be much more compact with a higher turndown ratio – down to a few MMscfd under certain conditions. The smaller units could be installed on skids and operated at the wellhead. It is also attractive for offshore platform applications. Consequently, the economic competitiveness of the new technology is excellent. The new technology is environmentally benign. Compared to conventional cryogenic process and glycol dehydration units, the new hybrid process does not emit benzene, toluene, ethyl-benzene, and xylene (BTEX) pollutants and substantially reduces greenhouse gas emissions due to extremely low energy requirement.

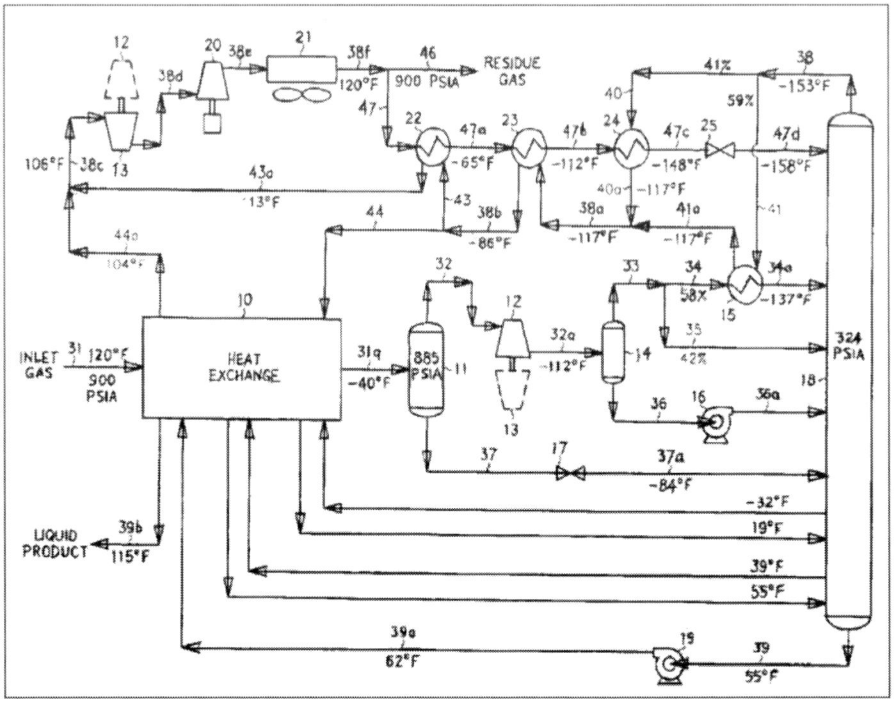

Figure 2 Schematic of the Conventional Cryogenic NGL Recovery Process

31 -- NG Inlet (Solid desiccant dehydrator not shown), 10/15/22-24 Heat Exchangers, 11/14 – gas/liquid separators, 20 – Residue compressor, 12 /13 NG Expander/compressor, 17/ 25 – Expansion valves, 39b – NGL outlet.

SOLVENT TESTING

The original solvent used in our simulation and economic evaluation was decane (C_{10}). With a boiling point of 345°F and molecular weight of 142.28, it met the solvent claims within the process patent. Kerosene was selected as the first alternative solvent for the selective separation of methane from the hydrocarbons. Since kerosene is a blended material, supplies from different refiners may have slightly different constituents at varying concentrations of these constituents. Kerosene is defined as "A refined petroleum solvent (predominantly C_9-C_{16} hydrocarbon, which is typically a mixture of 25% normal paraffins, 11% branched paraffins, 30% monocycloparaffins,12% dicycloparaffins, 1% tricycloparaffins, 16% mononuclear aromatics and 5% dinuclear aromatics" . The effect of these different components and concentrations on the absorption process is unclear.

One means to remove some of the compositional effects of kerosenes is to use a kerosene-like product derived from Fischer-Tropsch (FT) technology. Syntroleum Inc. uses their own Syntroleum Process® to convert stranded natural gas – using air instead of pure oxygen – into ultra-clean liquid fuels. Syntroleum Inc. supplied samples of their kerosene product that contains no aromatics or contaminants. The fuel is paraffinic and isoparaffinic, with no aromatics. This will provide enough information for comparison to the refined kerosene both in simulation and testing. A total of five experiments were conducted to study methane absorption and solubility in both Syntroleum S-1 and Sigma-Aldrich kerosene using the equilibrium and regeneration pressures for the absorption process in our VLE Test Unit as shown in Figure 3.

The vapor liquid equilibrium cell (VLE) was initially evacuated and was filled with exactly 260 ml of the test solvent at room temperature. Then, another vacuum was created, for approximately one minute, to displace air bubbles in the solvent before cooling the system to -40°F. When the VLE temperature was holding stable at -40°F, pure methane at room temperature was introduced into the cell, and the stirrer was turned on at a speed of 175 rpm. After the absorption had reached equilibrium, the VLE was depressurized at -40°F and the solvent was released in to a 1000 ml graduated beaker to record the volume and the mass. The -40°F solvents were tested at both 44 and 400 psia.

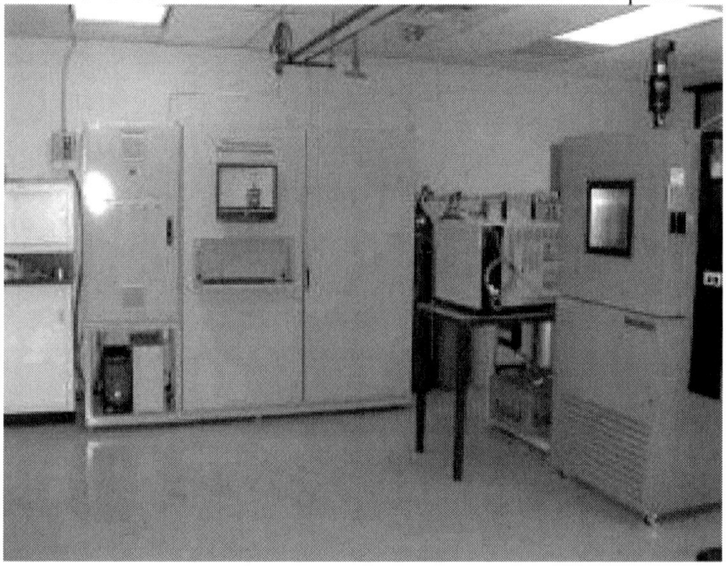

Figure 3 VLE Test Unit

Figure 4 and Figure 5 are representative of the solubilities of methane in both solvents. Tests to investigate and compare solubilities at different temperatures and pressures will be preformed.

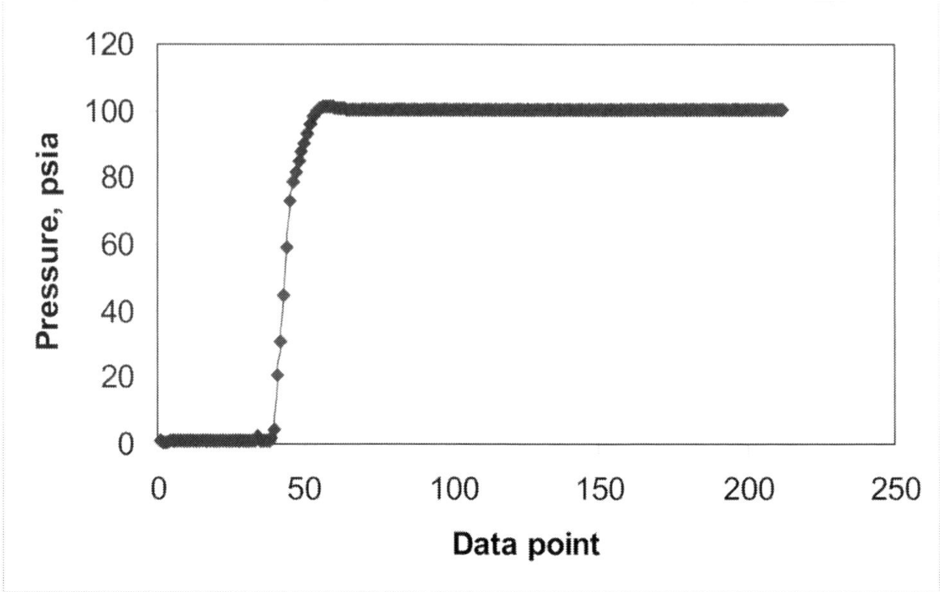

Figure 4 Methane solubility in Sigma kerosene as a function of a dimensionless time unit (time per 2 seconds) at -40°F

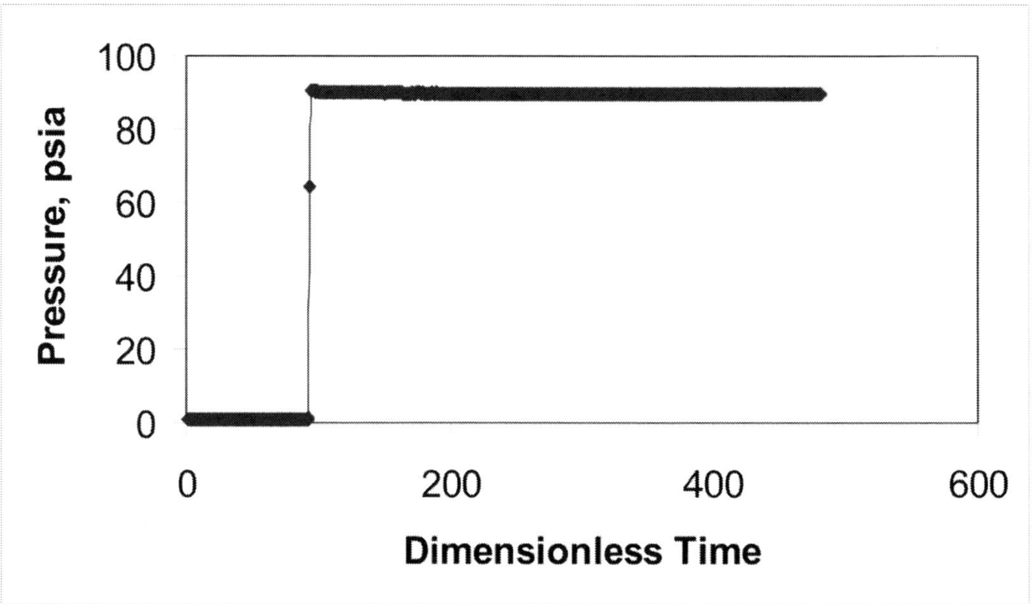

Figure 5 Methane solubility testing in Syntroleum S-1 as a function of a dimensionless time unit (time per 15 seconds) at -40°F

BENCH-SCALE UNIT

A bench-scale unit was designed to evaluate process parameters in a larger system. A bid package was prepared and sent for bid. The process flow diagram (PFD) was completed, reviewed and refined during the course of this quarter as shown in Figure 6 and distributed to several vendors for a budgetary bid. It was used primarily for communications and defining the unit, pressures, flow rates and temperatures. The flow diagram contains only the major components of the core system. The features of the bench-scale unit are given here. Four bids were received and reviewed. The NATCO bid was selected for procurement. The original estimated time for delivery of the bench-scale unit is in April, 2006.

As can be seen from in the figure, the main unit where dehydration and absorption will take place is a 4-inch-outer-diameter by 36-inch-high bench-scale packed bed reactor. It has both co- and counter-current capabilities. The dehydration can be run in counter-current mode, while the cocurrent mode will applied for the absorption section. The flow-rate of simulated feed gas through the packed bed was 20-100 scfh. The operating range for the bench-scale packed bed reactor is from -40 to 300°F and from atmospheric to 1000 psig.

Heat may also be generated when large amounts of solute are absorbed into the liquid phase, due to the heat of solution. The resulting change in temperature along the height of the absorber column may affect performance of this column. This problem can be avoided by adding cooling coils to the column or by using the excess chilled liquid circulator used for the heat exchanger at -40°F. Using this liquid to cool the jacketed small column may be sufficient to control temperature this way. A thermocouple will regulate the circulation, as necessary. The delivery of this system has been delayed significantly since the original quote.

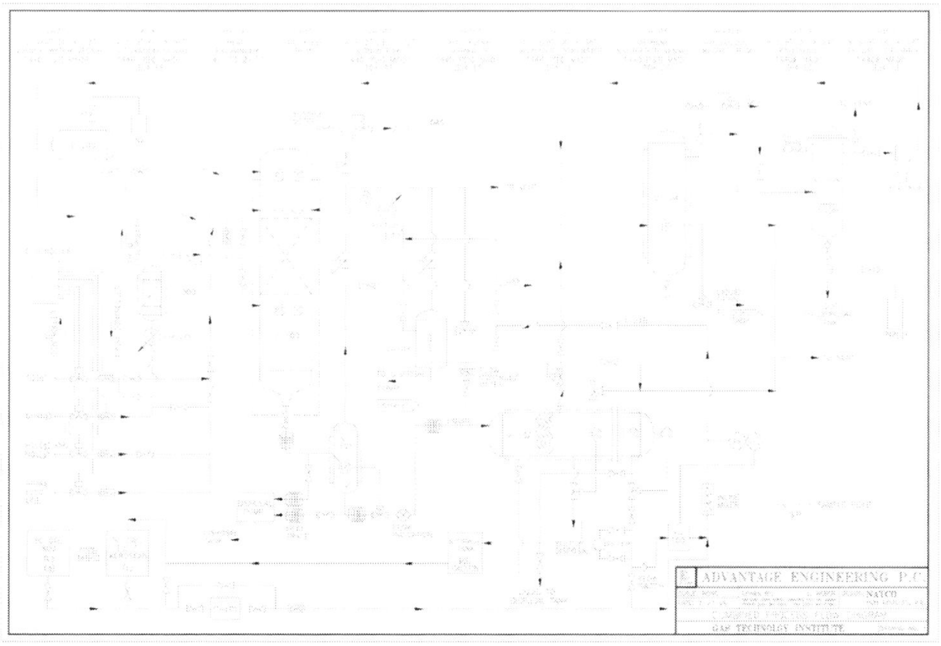

Figure 6 Bench-scale Unit

Pilot Plant Unit

The objectives of the pilot plant may be defined more concretely as: (i) The performance of a direct-contact type dehydrator, operating with two different kinds of hydrate inhibitors – ethylene glycol and calcium chloride; and (ii) The flexibility of variable C_2 recovery rates from 95% to 2% in the absorber using a commercial absorbent operating at equilibrium composition:

Trimeric Corporation has prepared a Process Design Package for the pilot plant. The system will consist of 4 units (i) inlet gas dehydration seen in Figure 7, (ii) lean oil absorption of NGL's seen in Figure 8, (iii) propane refrigeration, and (iv) glycol regeneration. There will be two stages of dehydration; the first stage will be designed to process 1.0 MMSCFD of sweetened natural gas at 100 °F and 900 to 1100 psig. The inlet stream will be saturated with water, ~1500 ppmv, and will have an H_2S content that is less than 4 ppmv. The second stage of dehydration will process 0.036 MMSCFD of gas to an outlet specification of 2 ppmv water. The design rate of the lean oil absorption unit will be set by the capacity of the existing lean oil absorber, V-100, which is 0.036 MMSCFD. Any excess gas will be routed back to the host site. The stream split will be located in the C-111 vapor outlet stream after the NGL knock-out pot. The propane refrigeration unit will be a skid mounted nominal 30 ton package unit. The glycol regeneration unit will consist of a low cost, standard field dehydration reboiler. The requests for bids were submitted to a number of fabrication shops.

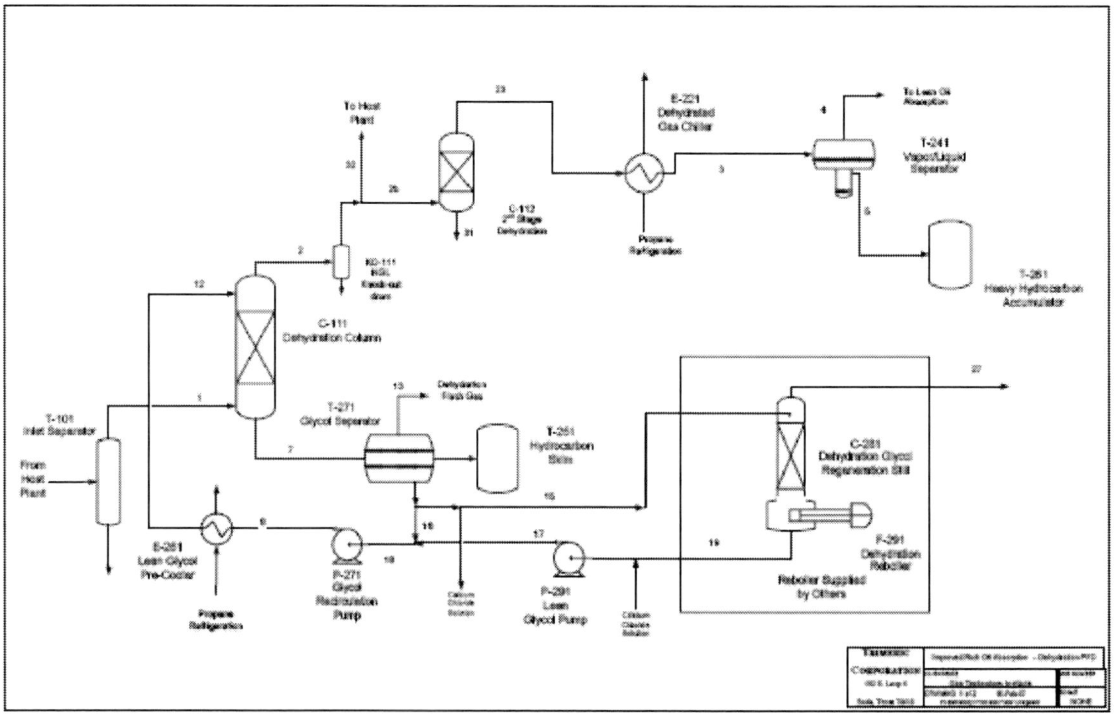

Figure 7 Dehydration Section of Pilot Plant

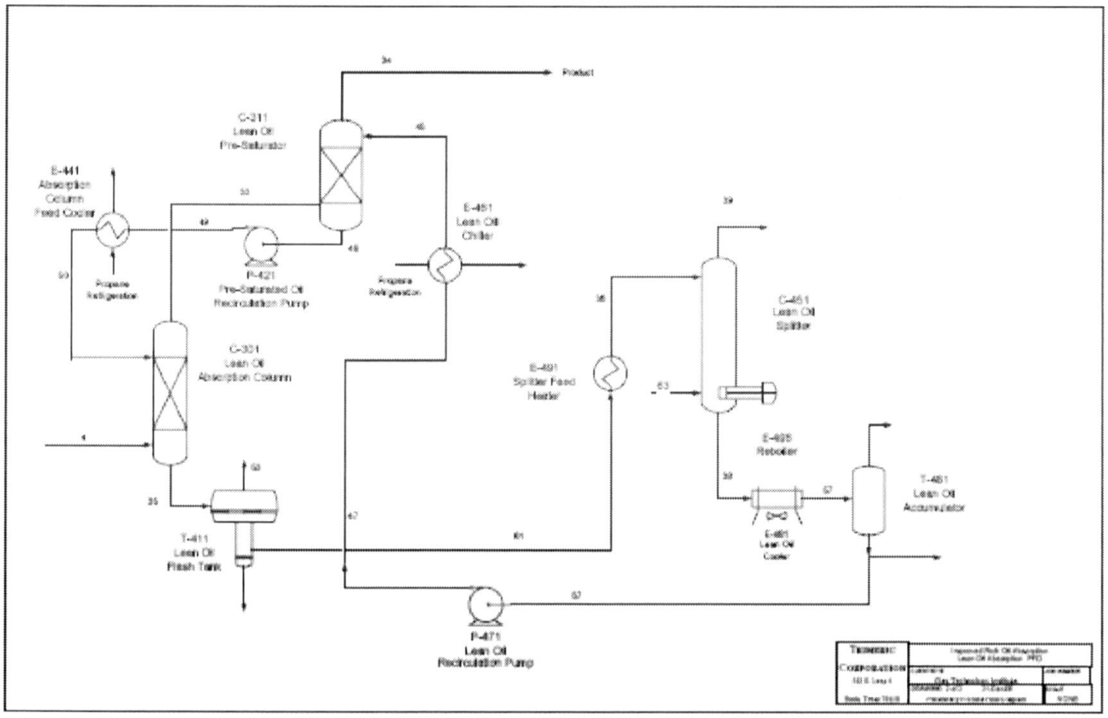

Figure 8 NGL Recovery Section of Pilot Plant

Process Economics

In the first phase of this project, IIT completed a comprehensive computer simulation of the IROA process covering 12 cases, comprising three different NGL contents and each with four operational modes. These were presented by GTI at the AIChE 2005 Spring Meeting and to DOE reviewers.

Although the comprehensive simulation was appropriate for general IROA process evaluation, the flowsheet has not been fully optimized. In particular, the opportunity to better exploit energy use of the system was not studied in depth. There is room for improvements. The following two major areas of potential energy savings have been identified:

(1) Maximizing the residue gas exit temperature could reduce the refrigeration power requirement; and

(2) Separating the NGL condensate from gas chiller to get partial NGL liquid product could reduce the power necessary for liquefying the NGL vapor from ROS.

In Figure 9and Figure 10, the significant energy savings are shown. The updated energy costs are the lowest as compared with the previous simulation energy costs (middle curve) as well as those of the cryogenic process (highest curve). Previously, the IROA energy cost at 95% C_2 was greater than that of the cryogenic process. The updated IROA energy cost at 95% C_2 recovery is now lower. For example in Figure 10, the previous IIT IROA energy cost savings to cryogenic process at 70% C_2 recovery is $13.95/Mgal. The updated IROA energy cost savings to cryogenic process becomes $20.02/Mgal, or a 56.5% increase over the previous IROA energy savings. These updated IROA simulation results predict much better economical performance than previously reported.

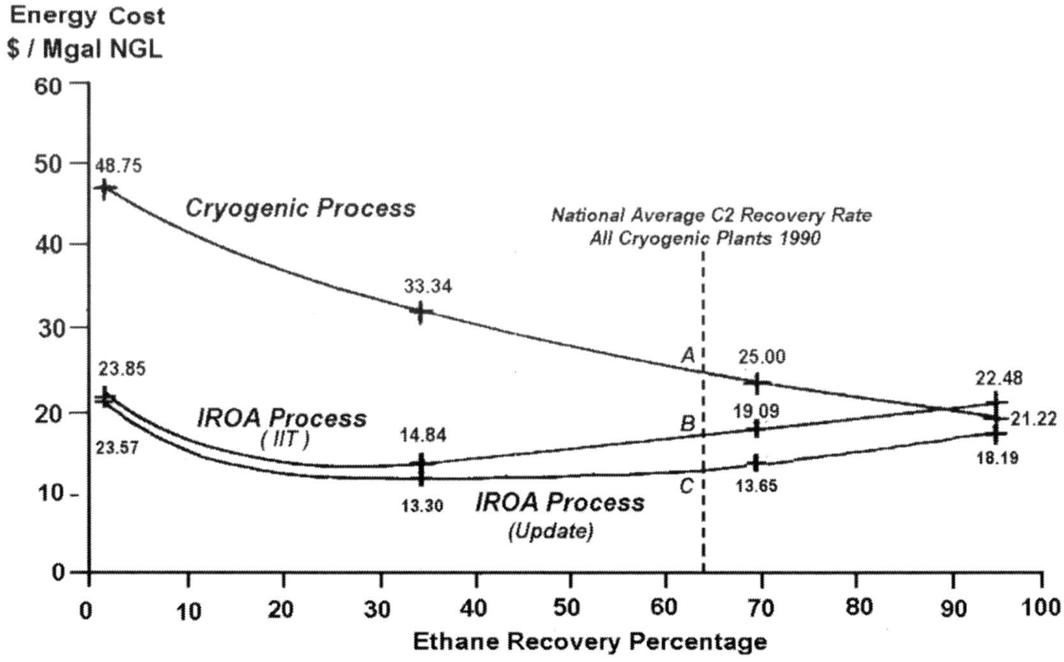

Figure 9 Energy Cost Savings for IROA Process for Rich Gas, 3.8 GPM NGL

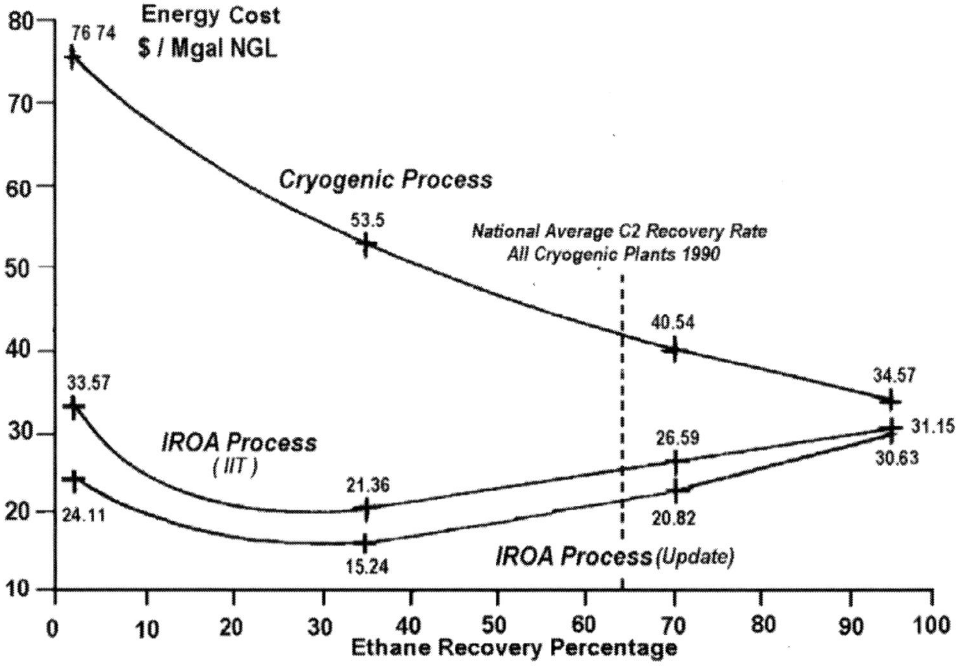

Figure 10 Cost Savings for IROA Process for Average Gas, 2.3 GPM NGL

Oxidative Coupling of Methane to C_{2+} Hydrocarbons

R. G. Anthony, Bedri Bozkurt, Swades Chaudhuri and X. Wu
Kinetics, Catalysis and Reaction Engineering Laboratory
Department of Chemical Engineering, Texas A&M University, College Station, Texas, 77843
Abbas Hassan, Aziz Hassan, and E. Bagherzadeh
Marcus Oil and Chemical Inc., Houston, Texas, 77035.

Introduction

Efficient conversion of methane to C2+ hydrocarbons would be a significant breakthrough in C1 chemistry. The ability to efficiently convert methane to ethane and ethylene is primarily based on an efficient catalyst, but process conditions are also important. In this presentation results from several proprietary catalysts[1] which produce C2+ yields equal to or greater than 25% are presented. Several of the catalyst also results in C2+ yields of 29 and 30%. In addition a very crude kinetic and reactor model is used to estimate the performance of an adiabatic reactor with interstage cooling. The model predicts that C2 yields of 30 to 35% may be feasible.

A major concern in any partial oxidation system is the potential for run away reactions and total combustion. Hence, special care needs to be taken to avoid the flammable zones within the reaction mixture. The state of art in predicting these zones for complex mixtures appears to be limited, but a guide might be the flammability limits of a component in air. Table 1 illustrates the limits for a number of components involved in the OCM reaction.

Table 1: Flammability Limits for Selected Hydrocarbons

Compound	LFL, %	UFL, %	Ign. T, ^{0}C	S.G. rh/rh air	Flame T, ^{0}C
methane	5.3	14	537	0.54	2082
ethane	3.2	12.5	510	1.049	2104
ethylene	2.75	28.6	450	0.973	2343
propane	2.37	9.5	466	1.56	2116
propylene	2	11.1	497	1.49	2254
hydrogen	4.1	74.2	585	0.069	2210

If concentration of hydrogen is less than 4% ignition in air is extremely unlikely. The product carbon monoxide may be produced at a concentration such as ignition might occur since the lower limit for CO ignition in air is 12.5%. However, most of the experiments for evaluation of various catalysts have produced product distributions with concentrations less than LFL for CO and the compounds listed

[1] Patents Pending

above. Never-the-less some conditions has been encountered that have produced run-away reactor conditions.

Stoichiometric Reactions of Interest

Desired Reactions:

$2CH_4 + 0.5\ O_2 \rightarrow C_2H_6 + H_2O$

For each mole of ethane produced only 0.5 moles of oxygen are required, i.e., 2 moles of ethane are produced per mole of oxygen consumed.

$2CH_4 + O_2 \rightarrow C_2H_4 + 2H_2O$

$C_2H_6 + 0.5\ O_2 \rightarrow C_2H_4 + H_2O$

$C_2H_6 \rightarrow C_2H_4 + H_2$

For every mole of ethylene produced one mole of oxygen is required.

Undesired Reactions:

$CH_4 + O_2 \rightarrow CO + H_2 + H_2O$

$CH_4 + 1.5\ O_2 \rightarrow CO + 2H_2O$

$CO + 0.5\ O2 \rightarrow CO2$

$CH_4 + 2O_2 \rightarrow CO_2 + 2H_2O$

These reactions illustrate the desired and undesired reactions, but they do not represent detail mechanisms which are occurring during the OCM (Oxidative Coupling of Methane) reactions. References (1-12) suggest that the C-H bond is activated by oxygen contained in the surface of the catalyst and that ethane is formed from methyl radicals, and then ethane is converted to ethylene by oxidative dehydrogenation or by thermal dehydrogenation. Most of our work appears to indicate oxidative dehydrogenation as the primary path for the formation of ethylene from ethane. However, the details are not addressed in this manuscript.

Experimental Procedures

A tubular reactor packed with 10 to 100 cc of catalyst is used to evaluate the catalyst. Operation was non isothermal because the catalyst was not diluted, which eliminated the option of isothermal operation. As illustrated in Figure 1, a furnace with three heating zones independently controlled was used for evaluation of the catalyst. A schematic of the system is illustrated in Figure 1.

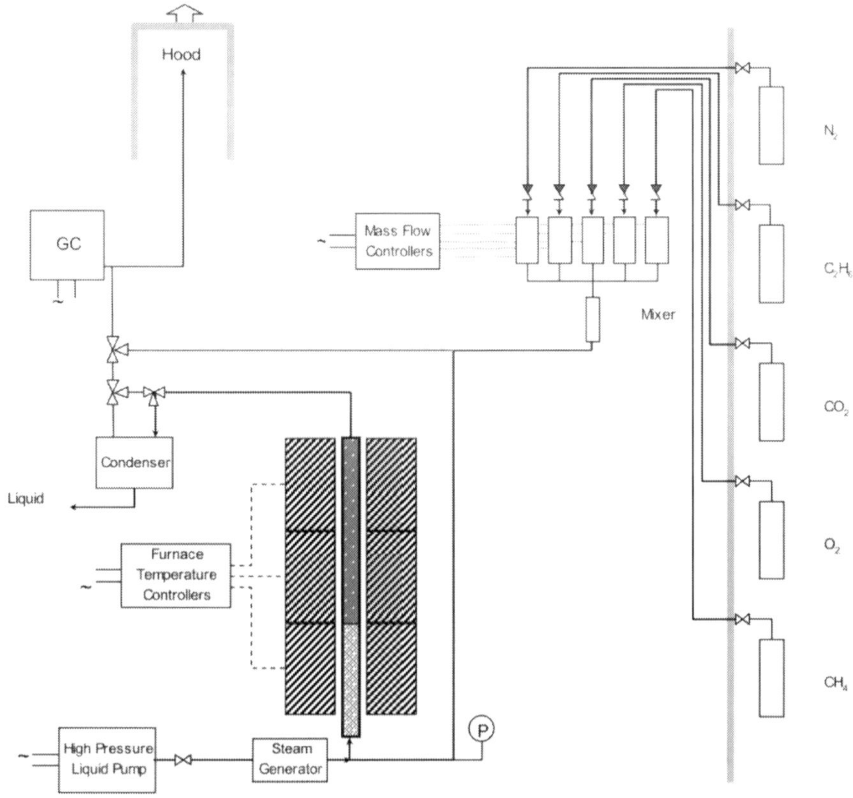

Figure 1: Schematic of OCM Reactor System

Results and Discussion

Data for five proprietary catalysts (patents pending), are illustrated in Table 2. In addition to these five proprietary catalysts, we have at least two dozen different formulations that produce C_{2+} yields of 27% or greater, which will not be presented at this meeting. Even though the C_{2+} yields may be greater than 27%, the different catalysts produced different selectivities of ethane, ethylene and ratios of carbon monoxide to carbon dioxide produced for the various conditions. Column 3 for PPC #1 illustrates the need to have a diluent for the purpose of reducing oxygen concentrations, and increasing C_{2+} yields. Even though no data are presented, steam instead of carbon dioxide is also an effective diluent for the OCM reaction.

TABLE 2: Effect of Operating Conditions and Catalyst on Product Distributions

Catalyst		PPC #1	PPC #1	PPC #1	PPC #1	PPC #2	PPC #2	PPC #3	PPC #4	PPC #4	PPC #4	PPC #5
Feed %,	CH_4	25.00	25.00	58.82	25.00	25.11	33.33	25.00	24.10	25.00	24.39	28.57
	O_2	12.50	12.50	29.41	12.50	15.13	11.11	12.50	15.66	12.50	14.63	15.87
	N_2	12.50	12.50	11.76	12.50	11.32	11.11	12.50	12.05	12.50	12.20	11.11
	CO_2	50.00	50.00		50.00	46.14	44.44	50.00	48.19	50.00	48.78	44.44
	C_2H_6					2.30						
GHSV (hr^{-1}) †		2750	2750	2754	2742	2955	2717	2754	2725	2003	2698	3032
T_{Aver} (C^o)		794	787	770	790	794	794	802	770	767	785	793
T_{Max} (C^o)		819	820	802	805	840	824	833	834	815	841	823
Conv.%,	CH_4	55	52	36	51	44	39	54	52	51	54	48
	O_2	95	86	61	84	97	97	100	82	99	98	93
Sel.%,	C_2H_4	46	44	43	46	84	52	44	48	51	46	48
	C_2	50	48	51	52	50	59	47	52	55	50	54
	C_{2+}	56	52	56	57	56	66	52	56	62	57	59
Yield %,	C_2+	31	27	21	29	28	26	28	29	32	30	28
Molar Ratio of CO/CO_2 (produced)		0.94	1.09	1.15	0.91	0.85	1.15	0.84	1.68	1.40	1.29	1.17
Number of use		1st	2nd	2nd	2nd	7th	6th	1st	6th	5th	2nd	2nd

† Volume of catalysts: 10-35 cc. Each use of the catalyst represents 6 to 10 hours of operation. Therefore, six uses of a catalyst indicates operation at reaction conditions of 36 to 60 hours with cooling and heating cycles between each use.

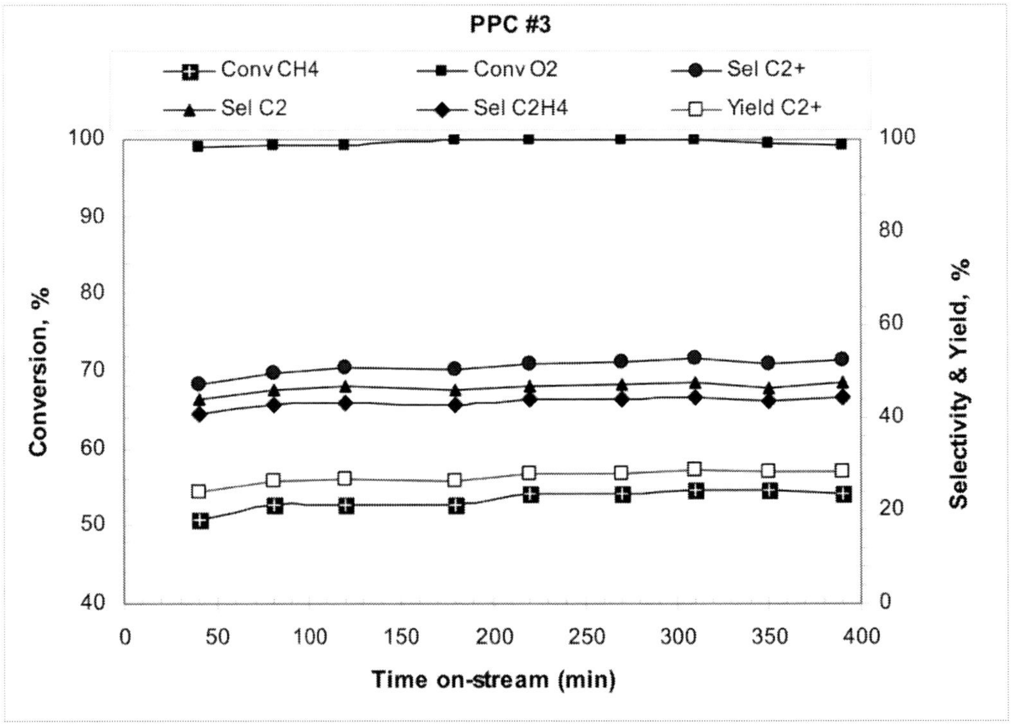

Figure 2: Typical Run for Evaluation of PPC#3. Average Reaction T= 802 ^{0}C

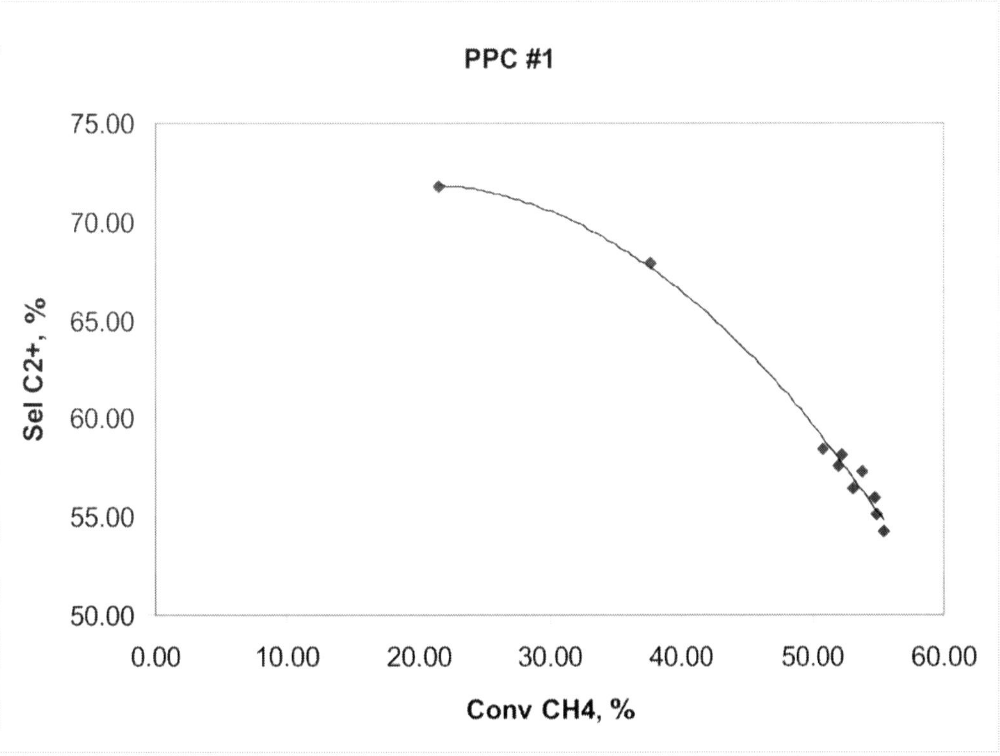

Figure 3: As the temperature is increased, methane conversion increases and selectivities decrease for most catalyst.

Reactor and Kinetic Model

A mechanism based on published literature was used to develop a kinetic model, which will not be presented, and this model was then used with the equations for a plug flow, non isothermal, non adiabatic reactor to simulate some of the experiments. Figure 4 illustrates a simulation of one of the experiments. Obviously, methane and oxygen conversions are under predicted as is the C_{2+} yield. Prediction of the temperature profile also is not particularly satisfying. However, other simulations predicted conversions and yields greater than the experimental values. The reactor model uses the plug flow assumption, which is a gross approximation of the system, since radial temperature and concentrations profiles within the reactor must be occurring. Never-the-less, the model has been used as a first approximation for evaluating potential reaction conditions.

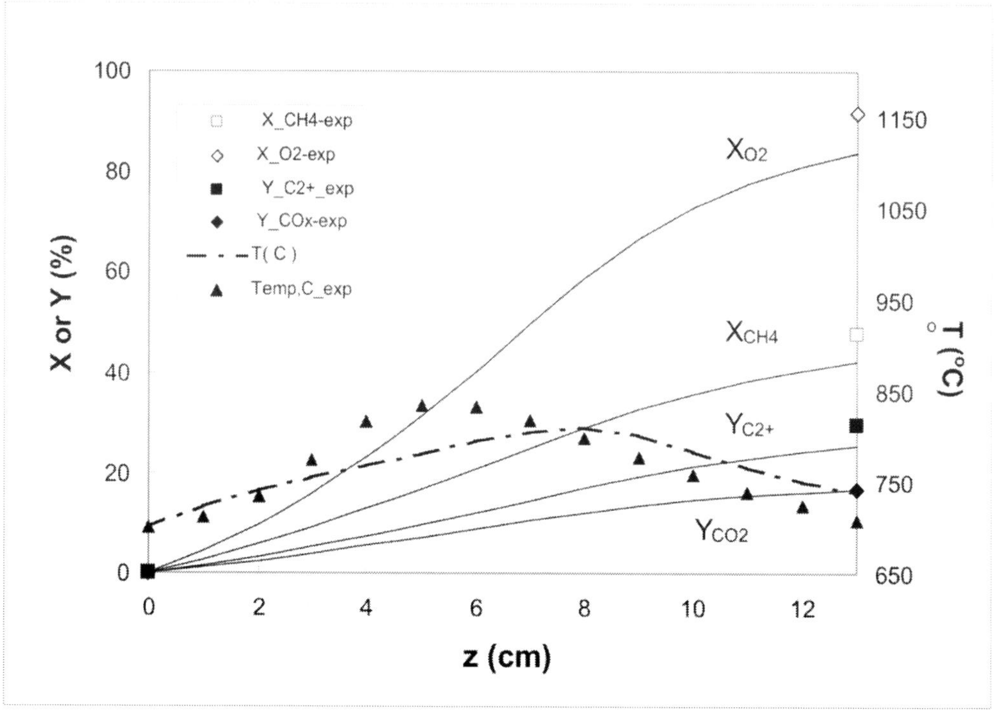

Figure 4: A comparison of experimental data with simulation for PPC#3.

Adiabatic Reactors

Many commercial reactions utilize adiabatic reactors with multiple catalysts beds and interstage cooling. Hence, the potential of operating an adiabatic OCM reactor or a series of OCM adiabatic reactors with interstage cooling was investigated by using the kinetic and reactor model. A stipulation for a potential design was bed depth for the first stage should be at least 0.9 meters or greater. Because of the strong exothermicity of the OCM reactions, the space velocity of 1000 1/h was used. Other feed conditions, such as more dilution of the feed would allow the use of space velocities greater than 1000 1/h. Figure 5 illustrates a simulation of a reactor with four stages with interstage cooling. The four stages were required to achieve a predicted C_{2+} yield of 30% or greater. The simulation illustrates the potential feasibility of operating a commercial scale reactor with interstage cooling.

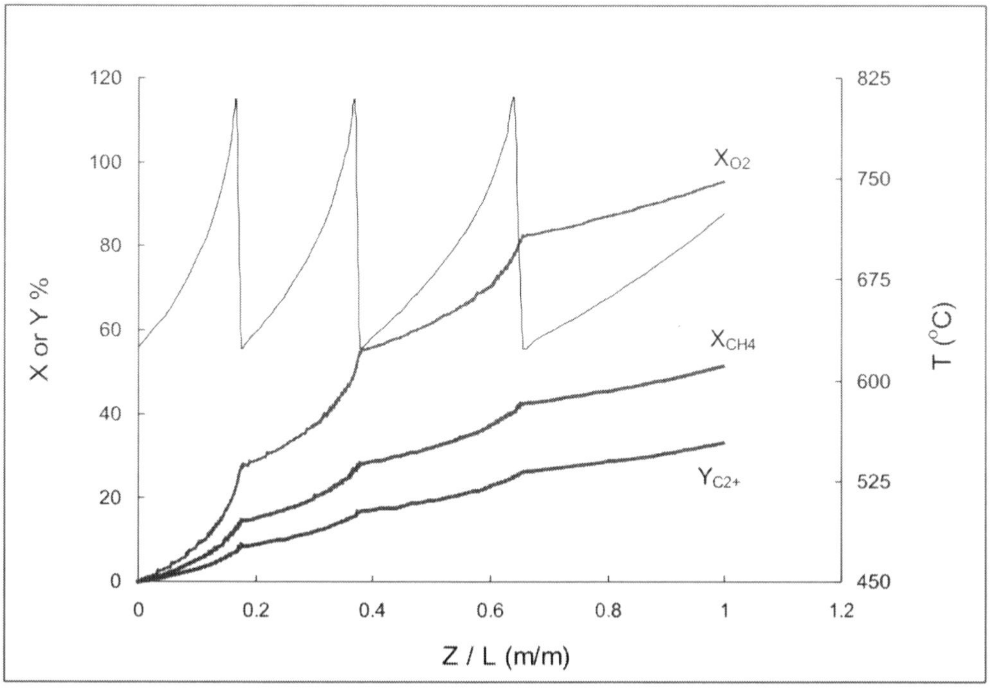

Figure 5: Adiabatic Reactor Simulation for OCM for a GHSV = 1000 1/h.

Key references

1. Sergei Pak, Ping Qiu, and Jack H. Lunsford, *J. Catal.,* **179** (1998) 222.
2. Z. Stansch, L. Mleczko, and M. Baerns, *Ind. Eng. Chem. Res.*, **36** (1997) 2568.
3. Maria Traykova, Nadka Davidova, Jeng-Shiang Tsaih, Alvin H. Weiss, *Applied Catalysis A: General*, **169** (1998) 237.
4. Jay A. Labinger and Kevin C. Ott, *J. Phys. Chem.*, **91** (1987) 2682.
5. R. B. Hall and G. R. Myers, *Methane and Alkane Conversion Chemistry*, M.M. Bhasin and D.N. Slocum, eds., Plenum Press: New York, 1995, page 123-130.
6. Lingjun Chou, Yingchun Cai, Bing Zhang, Jianzhong Niu, Shengfu Ji, Shuben Ji, and Shuben Li, *J. Natural Gas Chemistry*, **11** (2002) 131.
7. Jiaxin Wang, Lingjun Chou, Bing Zhang, Huanling Song, Jun Zhao, Jian Yang, and Shuben Li, *Journal of Molecular Catalysis A: Chemical,* **245** (2006) 272.
8. Longya Xy, Sujuan Xie, Shenglin Liu, Liwu Lin, Zhijian Tian, and Aiming Zhu, Fuel **81** (2002) 1593.
9. Shuben Li, *J. Natural Gas Chemistry*, **12** (2003) 1.
10. Worapon Kiatkittipong, Tomohiko Tagawa, Shugeo Goto, Suttichai Assabumrungat, Kampol Silpasup, and Piyasan Prateserthdam, *Chemical Engineer Journal*, **115** (2005) 63.
11. Travis A. Reine and R. Bruce Eldridge, *Ind. Eng. Chem. Res.*, **44** (2005) 7505.
12. Wolf, E. E., Editor, "Methane Conversion by Oxidative Processes: Fundamental and Engineering Aspects, "(1992) Van Nostrand, Reinhold, NY, NY.

Monetization options for small reserves of low-quality natural gas with high carbon dioxide content

[1]DIPANJAN BHATTACHARYA, [2]RICHARD MALLINSON, [3]ROBERT HUBBARD
[1,2] *School of Chemical Biological and Material Engineering, University of Oklahoma*
[3]*Mewbourne School of Petroleum and Geological Engineering University of Oklahoma, T-301Sarkeys Energy Center, 100 E. Boyd St. Norman, OK 73019-1003 USA.*
dipanjanb@ou.edu mallinson@ou.edu roberth@ou.edu

Abstract:

The feasibility of monetizing small (daily production 10 MMSCFD) stranded natural gas reserves containing carbon dioxide is analyzed. The carbon dioxide being considered is of varying range from 5% to 60%. The price of natural gas is expected to remain on the higher side over the long term. The high prices of oil & gas make the utilization of more remote low quality gas economically attractive. The different options of power generation and chemical production from low quality gas are explored against the base case of producing pipeline quality gas by removing the carbon dioxide. The project economics are evaluated using the approach of netback value. Power generation is considered by evaluating the economic potential of burning low-Btu gas using aero-derivative turbines. Small-scale methanol manufacturing is evaluated by considering the economics of CO_2 reforming to manufacture fuel grade methanol. Reforming with carbon dioxide present has the advantage of generating a stoichiometric ratio of CO/H_2 which favors the overall economics. The base case economics is evaluated by considering amine treating for low carbon dioxide content and hybrid systems comprising amine & membranes for high carbon dioxide content.
Furthermore, different risk scenarios are investigated using statistical techniques so as to rank the different options. Uncertainties like demand, price of end products, natural gas well-head prices, transportation charges and variability is processing costs for different operating conditions are considered for the project risk analysis.

Keywords: Natural gas, remote gas, economics, process analysis& simulation, risk management.

Extended Abstract:

Introduction

The price of natural gas is hovering around $7/MMbtu and is expected to remain on the higher side over the long term. With supply limited, even with increasing imports, interest is generated for monetizing the low quality gas reserves in the U.S. that are stranded by their quality, size or location. These low quality gases are contaminated with carbon dioxide, nitrogen, sulfur compounds etc. The quantity of natural gas reserves contaminated with carbon dioxide is even larger than that for nitrogen at 22% of production and 27% percent of reserves. Essentially the monetary value of these gas reserves is practically negligible and any utilization option which results in a positive netback value for the gas can be a viable option. Our work aims at exploring the different monetizing options for high carbon dioxide content gas using state of the art techniques including risk analysis and price forecasting to determine the utilization options that provide the highest netback value over a range of varying carbon dioxide concentration, resource size, and location options. These probabilistic techniques are applied to study the impact of uncertainties in both well-head price of natural gas and the price & demand of subsequent end products on the overall economics. Efforts are being made to look into the answers with the current technologies that are available and also provide a review of technologies that are on the development stage.

The focus of the study is on small and remotely located gas reserves with the following characteristics:

1) Reserves of 10 Bcf and flow rates of 10 MMscfd.
2) Carbon dioxide content in range of 5% to 60% by molar volume.
3) Minimal H2S or Nitrogen.
4) Absence of NGL's.

Utilization options studied

The primary options looked into are:

1) Treating (separating CO_2 by treating) as a base line for comparison with other options.
2) Power generation (aero-derivative gas turbines)
3) Methanol production by steam/CO_2 reforming

Mode of evaluation

The economic approach taken is the netback value of the gas which is the standard industry method for gas project evaluation. Netback value indicates the value per MMBtu of gas that would accrue to the gas owner /operator after incorporating all the processing/conversion costs (excluding the cost of gas production and exploration).

Deterministic netback value is calculated by using fixed values of production costs, demand, and market prices .But in reality practically every variable has a degree of uncertainty. This uncertainty in the possible value of the inputs is captured as a probability distribution (eg. a bell –curve).

Risk analysis was done by incorporating uncertainty in product price as well as capital and operating costs. For example methanol price and demand can have a huge uncertainty. Tools like Crystal Ball are used to perform Monte Carlo simulation (a technique for simulating real-world situations involving elements of uncertainty) to create thousands of possible alternative outcomes and therefore hedge against risk.

Power Evaluation

Calculations show that LM 6000PC is economic for fields containing up to 45% CO_2 and LM 2500PE is economic beyond that. This is because with increasing CO_2 the fuel heating value comes down and consequently the turbines generate less power and the capacity goes unused in bigger turbines. Figure 1 shows the netback for power generation at different wholesale power prices.

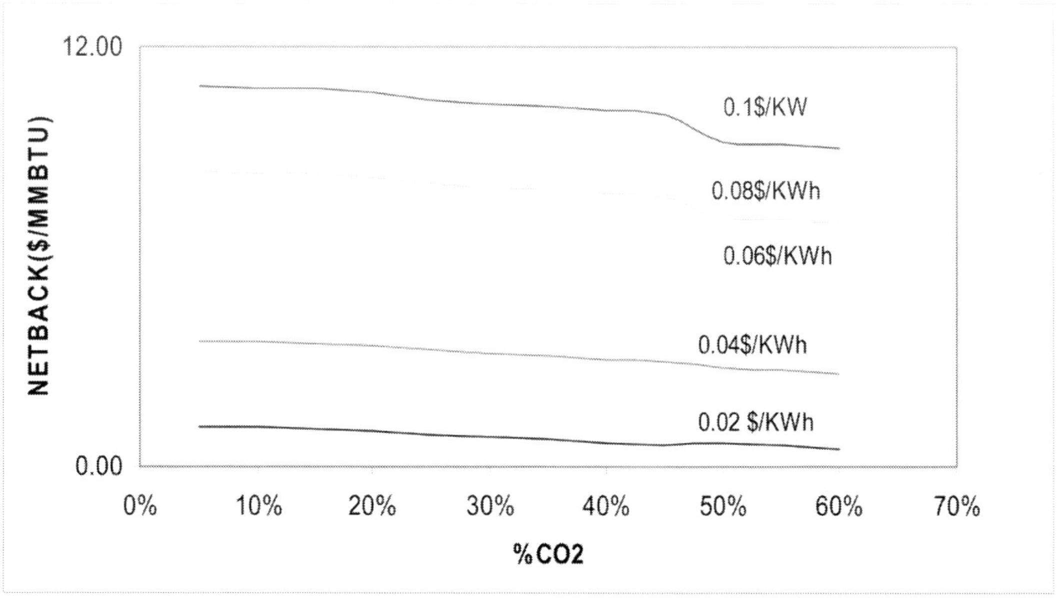

Fig 1: Graphical Representation of Power netback

The figure below shows the capital costs for power plants for different carbon dioxide ranges. LM 6000 turbines are used for around 45% CO_2 and LM 2500 for CO_2 beyond that. Around 45% CO_2, LM 6000 no longer remains economical.

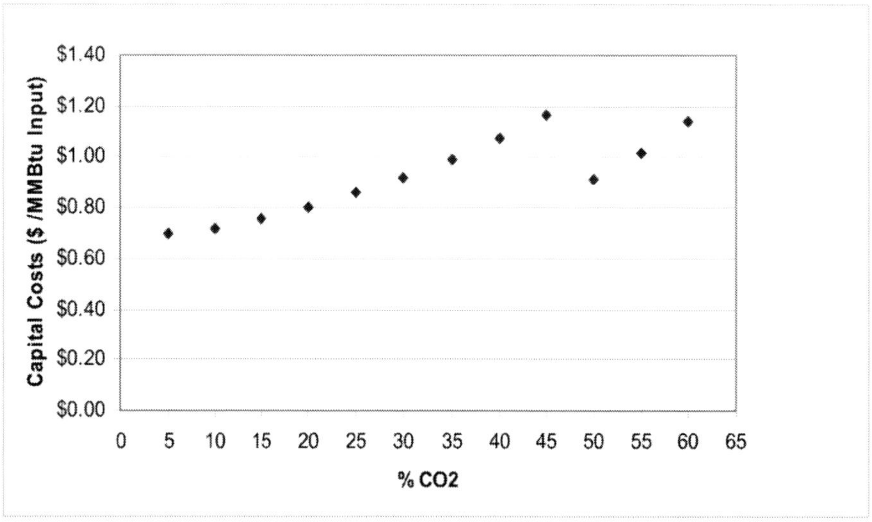

Fig 2: Capital costs for power with varying CO_2

Methanol Evaluation

Traditional methanol production includes the below reactions where there is a stoichiometric imbalance in the methanol synthesis section.

Steam Reforming: $CH_4 + H_2O = CO + 3H_2$ (ΔH = 206.3 KJ/mole)

Methanol Synthesis: $CO + 2H_2 \leftrightarrow CH_3OH$ (ΔH = -91 KJ/mole)

Another option under consideration is dry reforming or CO2 reforming which if practiced can be a potential CO2 utilization option. It is still very much in development stage.

CO_2 reforming: $CH_4 + CO_2 = 2CO + 2H_2$ (ΔH = 247.3 KJ/mole)

Problems with dry reforming are
1) 20% more endothermic than steam reforming
2) High carbon formation

The stoichiometric imbalance can be corrected by having a certain percentage of CO_2 in the reformer feed. This is called Combined Steam and CO_2 reforming.
Combined Steam and CO_2 reforming reaction is as follows:
$3CH_4 + CO_2 + 2H_2O = 8H_2 + 4CO$

It has already been applied in large –scale methanol plant by importing CO_2. This study evaluated the application for natural gas reserves with high CO_2 as a feedstock for methanol plants.

Economic evaluation was done for small –scale methanol plants producing a maximum of 200 T/day. Modular plants can be built for this purpose in the range of 50-200 T/day. Since the optimum carbon dioxide content in the feed is 25%, therefore for CO_2 above 25%, membranes are used to remove CO_2 before feeding the gas to methanol plant.

The reforming section is significantly smaller for the CO_2 reformer plant while the synthesis reactor is slightly larger. With about 60% of the plant investment in the reforming section and only 10-15% in the synthesis section, the overall effect is a substantial cost reduction.
The graph below represents netback for methanol at different methanol wholesale prices.

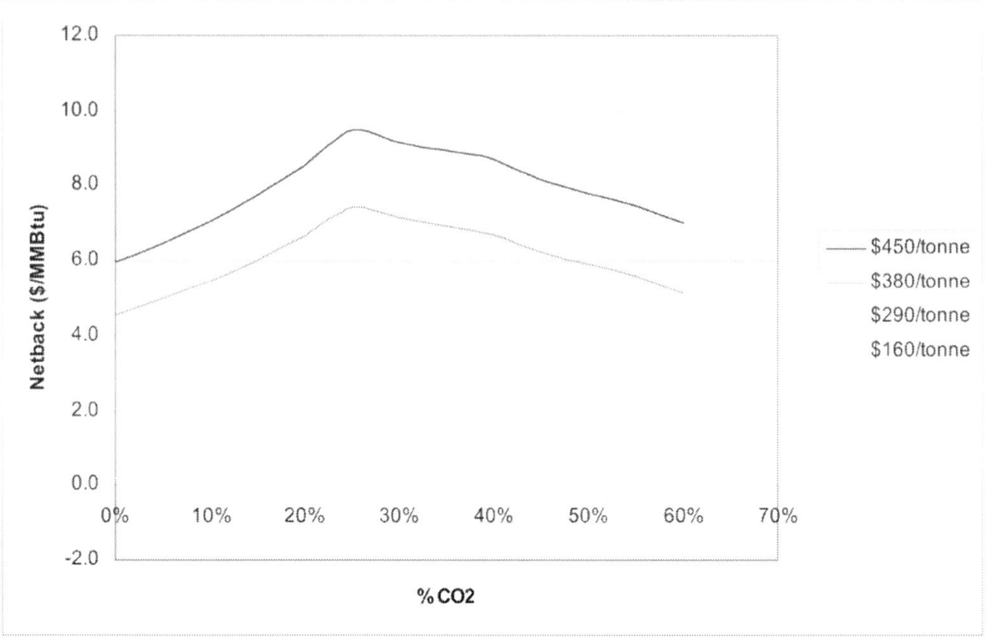

Fig 3: Graphical representation for methanol netback

Gas Processing Evaluation

Amine systems (DEA) is considered for 5%-15% CO_2 .Beyond 15% CO_2, hybrid systems are considered (combination of amine and membrane) for economic analysis. In the hybrid system, the membranes are used to reduce the CO_2 up to 15% and then amines are used to reach the pipeline specification. The figure below shows the total costs for membrane operations to reduce the CO_2 concentration to 15%.

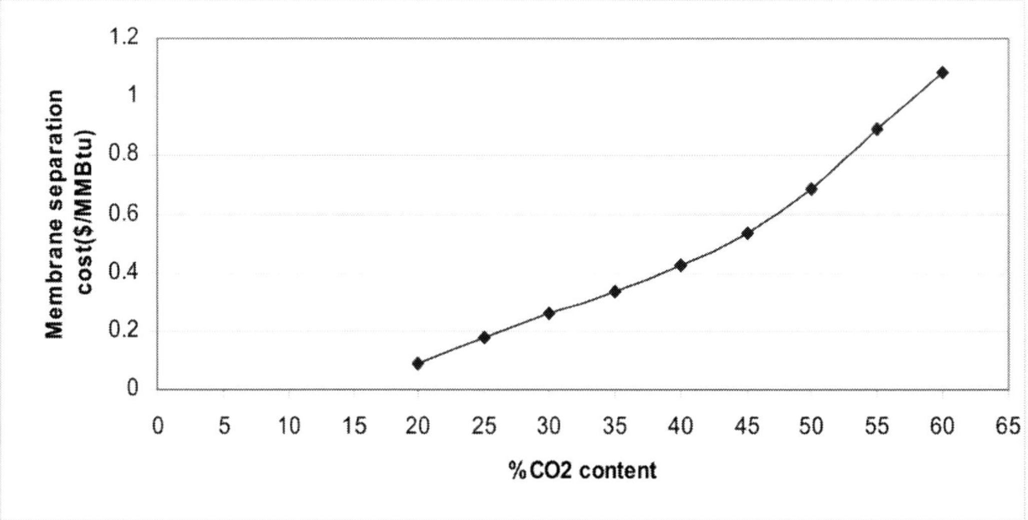

Fig 4: Total costs for membrane systems

The figure below shows the netback value for gas processing at different wholesale prices of natural gas.

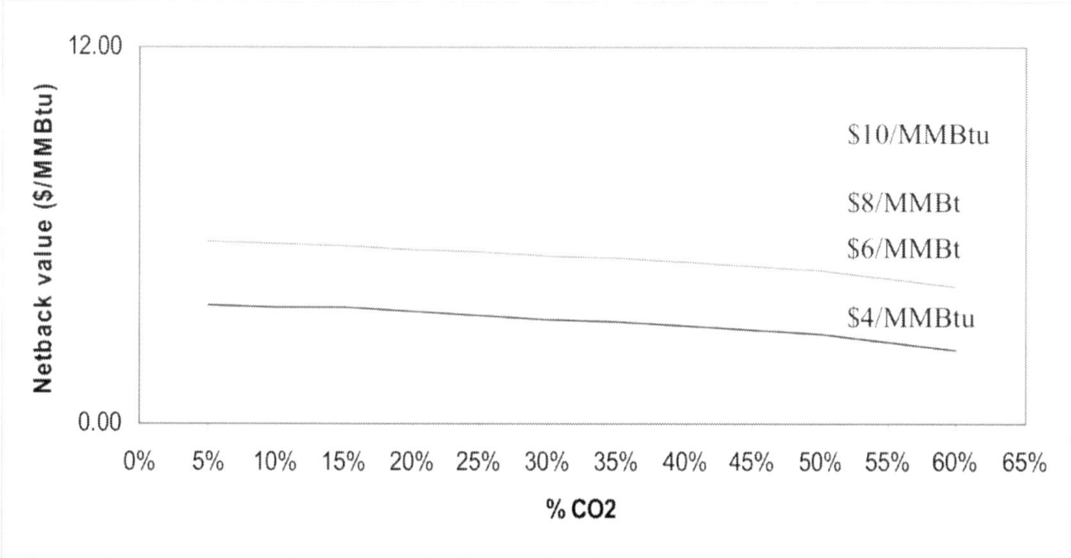

Fig 5: Graphical Representation of Gas processing netback

Uncertainty analysis for all options by Monte Carlo Simulation

Base study conditions for uncertainty analysis:
Mean Methanol Price - $ 250/tonne, Standard deviation - $60/tonne
Mean electricity price- $0.05/kW-hr, Standard deviation – 0.02/kW-hr
Mean natural gas price- $5/MMBtu, Standard deviation – $2.25/MMBtu

The mean prices and standard deviation are derived from historical prices and forecasts. The analysis is conducted by using wholesale prices to simplify the complexities of transportation. Uncertainty in capital and operating costs are incorporated a possible range of variation in capital and operating costs.
The figures below represent the risk curves at different CO_2 content.

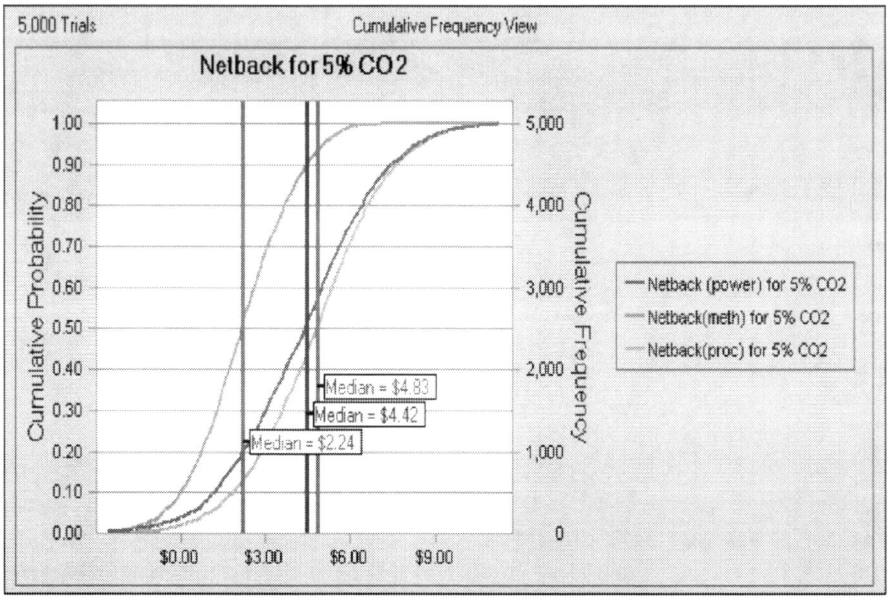

Fig 6: Risk curves for 5% CO2

At 5% CO_2 level it can be seen that methanol has the least netback but risks are relatively low for all options.

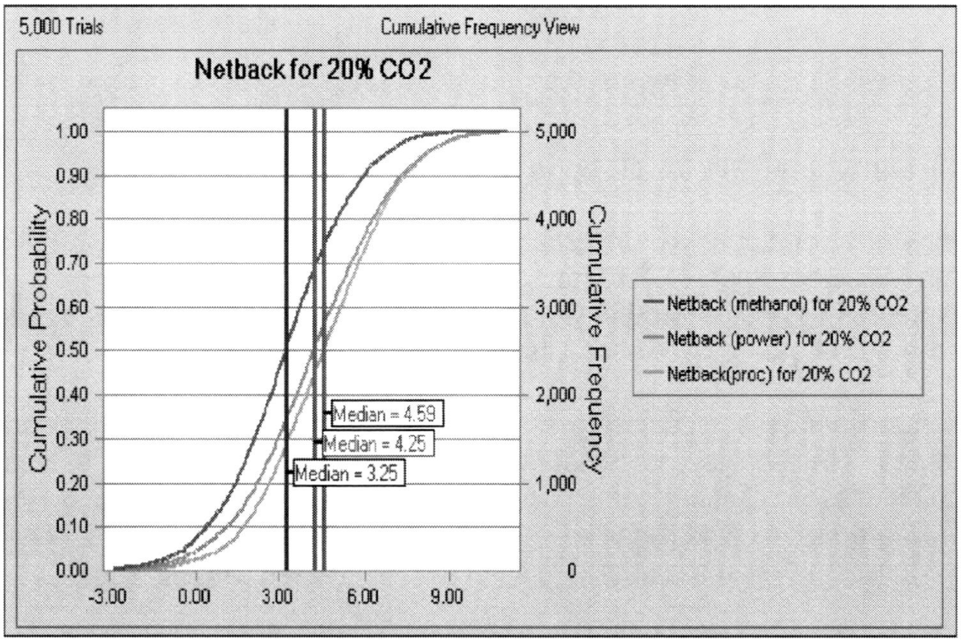

Fig 7: Risk curves for 20% CO_2

At 20 % CO_2 methanol has still lower netback than the than the other two.

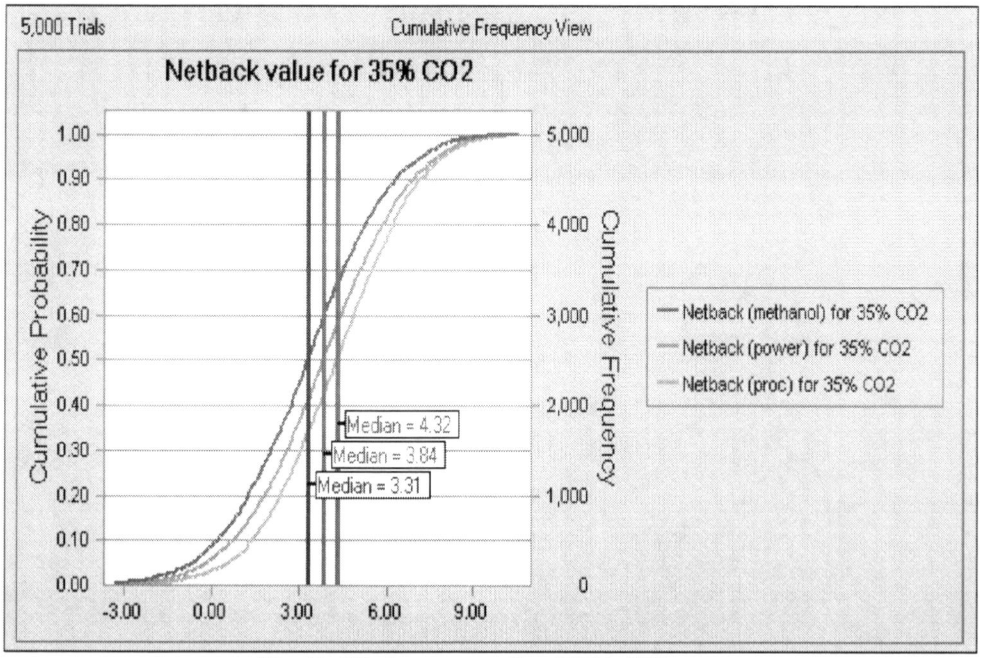

Fig 8: Risk curves for 35% CO_2

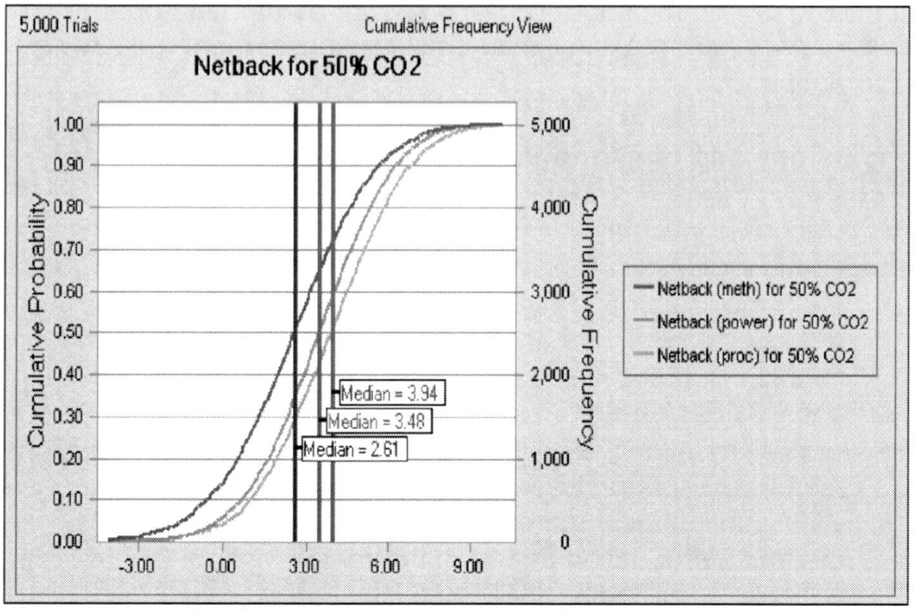

Fig 9: Risk curves for 50% CO_2

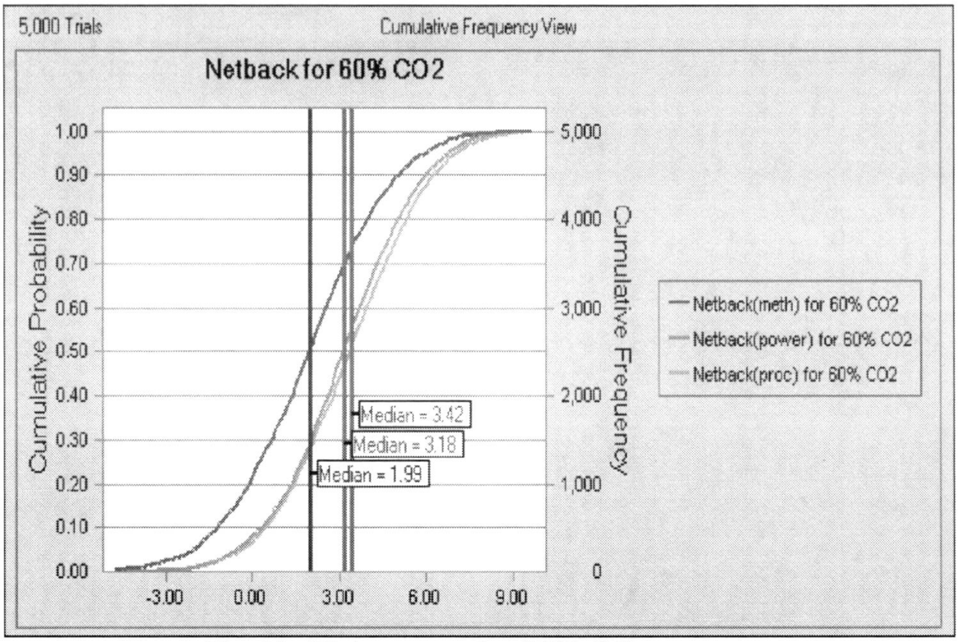

Fig 10: Risk curves for 60% CO_2

It can be seen that at 50%-60% CO_2 content, methanol lags considerably behind the other two processing options.

Depending on the minimum netback sought by the field operator, the risk curves provide decision making capacity by considering a wide spectrum of uncertainties.

Conclusions and Limitations

- Uncertainty analysis renders a much broader view for the project economics.
- Methanol is found to have higher risks compared to power and processing. But, methanol is easier to transport considering the remote location of these fields. Incorporating transportation and logistics into the analysis can render an improved methanol economics.
- Sensitivity study reveals that product prices have the highest impact on the project economics.
- With continuously increasing methanol prices, the economics show that methanol production can be an attractive option. But there is demand side uncertainty for methanol.
- Power generation option with aero-derivatives can be useful for small fields. Aero-derivative being compact can be used over multiple fields over its lifetime.
- Fields deplete at 20%-25% rate but in this study production is assumed constant.

Natural Gas Industry in Iran

Hedayat Omidvar
National Iranian Gas Company
Tehran, Iran
omidvar@nigc.ir

Abstract:

Gas industry in Iran now contributing in supplying energy by 60 percent share in fossil fuel basket has entered the developing stage after 40 years of Natural Gas emerging in energy sector. The growing demand for Natural Gas in residential and industry section and supply pushing for covering this demand, are the reasons for such statement. For responding to this phase of Natural Gas life, Iran needs investment especially in upstream development, technology transfer and facilitating in exports and imports. According to the annual report in long term Middle East will be the second exporter region and Iran which is the first holder of reserves and first producer and consumer in this area is a potential exporter for next decades. The Islamic Republic of Iran is able to produce more than 400 millions cubic meters per day of Natural Gas from giant Pars field. Regarding internal demand a half of this production can be assigned to exports, inter regionally or worldwide.

Flare Systems Design for LNG Facilities

T. J. Bevilacqua
Pressure Relief & Flare Consultant
Berwanger Engineering Services Division
Siemens Oil, Gas, & Marine

AIChE Spring Meeting, April 2007
7th Topical Conference on Natural Gas Utilization
Houston, Texas

Abstract

Liquefied Natural Gas is playing an increasingly important role in satisfying the world's energy needs. The importance of a thorough design of the flare system is no different for this growing segment than for any other part of the hydrocarbon processing industry. In light of the major growth of this sector, however, a careful reconsideration of the key design concepts for flare systems and the application of these concepts to facilities for the production of LNG seem appropriate.

This paper presents a brief review of the considerations that are central to the design of process facility flare systems. It then proceeds to apply these considerations to a generic example of the facility at the heart of the LNG business – the LNG production facility.

Introduction

There is currently a world-wide boom in the design and construction of LNG production facilities. The design, construction, and operation of these facilities are rightfully the focus of our interest and attention. LNG facilities fully capable of meeting well-defined design criteria will play a central role in meeting the world's energy needs in the coming decades.

One part of nearly all hydrocarbon processing facilities is the flare system. This system serves to collect and safely dispose of effluent streams from pressure relief devices, depressuring valves (both for emergency and maintenance purposes), and pressure control valves operating within the facility's basic process control system (BPCS). This system typically comprises several key components – discharge piping leading from individual flow sources (the relief devices, depressuring valves, etc.), header piping into which the individual discharge lines lead, one or more liquid/vapor separators (here typically called "knockout drums"), and means for disposing of the separated phases. These final components are often pumps and associated piping for routing liquids either for reprocessing or into a slops tank for other disposal, and a flare or vent stack for burning or venting the vapors. Concerning these last-mentioned components of a "flare" system, it is important to clarify that essentially all of the discussion in this paper applies equally to effluent handling systems that terminate in either a "flare tip," in which the effluent vapors are combusted, or a "cold vent," in which the vapors are simply sent to the atmosphere.

Flare System Design Considerations

Overview

The design of any process facility "flare" system begins with a thorough examination of all potential effluent streams that will need to be handled – the events that generate these streams, as well as pressures, temperatures, flow rates, chemical compositions, and phase compositions of these streams. This examination, in turn, relies on knowledge of the processes conducted in the facility.

Once the flaring events, source streams, and individual flare systems are defined, the flare systems design effort reduces to an iterative process of sizing each component in the system by initially estimating component dimensions, evaluating the performance of each component against established criteria, and readjusting these dimensions as needed to satisfy each design criterion most economically. As for most aspects of the process industries, these design criteria are established by a combination of government, industry, corporate, and project standards, practices, and specifications. A summary of the typical design criteria for each system component, as well as the sources or bases of these criteria, is given in Table 1.

The following discussion provides a brief overview of each step in the flare system design process – scenario identification, piping sizing, knockout drum sizing, and flare sizing.

Subsystem	Design Criteria	Reference
Piping	Velocity < Mach 0.5 – 0.7	API Std 521 Corporate Stds.
	Developed back pressure does not interfere with performance of source (PRV, PCV, BDV)	API RP 520 ISO 4126
Knockout Drum	Liquid Retention – c. 30 minute event	API Std 521
	Droplet Separation – c. 300 – 600 μm	API Std 521
Flare Stack & Tip	Discharge Velocity – Mach 0.5 – 1	API Std 521
	Thermal Radiation – a function of exposure duration	API Std 521
	Effluent Dispersion Lit Unlit	NIOSH, etc.

Table 1 Typical Flare System Design Criteria

Identification of Flaring Scenarios

In principle, every potential scenario that will generate flow into the flare system should be identified and characterized. In practice, this is both impractical (there are typically a very large number of manually-operated connections into a flare system) and unnecessary (most potential scenarios will not constitute design-case events for any flare system components beyond the "tailpipes" of the individual flow sources involved in those scenarios). It is important, however, that the designer keep in mind that there is more than just a single "design scenario" for all but the most rudimentary flare system. Because the flare system includes several different components that have several independent and sometimes contradictory design criteria, these systems often could be characterized as having at least half a dozen "design cases." An emergency scenario that generates the largest volumetric flow rate of vapor may determine the size of the header piping, but will have no impact on the sizing of the knockout drum. In the same facility, it may be a relatively low flow rate off-design operating scenario that dictates the height of the flare stack, because the acceptable thermal radiation for long-duration flaring can be significantly more restrictive than that for brief flaring episodes. When it comes to evaluating potential design scenarios, the primary requirement is that all possible contingencies be considered, if not fully characterized. The decision of which scenarios to characterize further relies mostly on the need to design the system for all possible *extremes of conditions* – maximum and minimum flow rates, maximum and minimum temperatures, maximum and minimum heating values, the full range of compositions, off-design facility operating conditions, which may involve intermediate flaring rates for extended periods, with particular design criteria (e.g., smokeless combustion) unique to such periods.

The process for identifying credible flaring scenarios is tied to the nature of the flaring source – be it emergency (pressure relief and depressuring devices), operating (pressure control valves), or manual (valves typically used for depressuring/draining in preparation for equipment entry). Manually-operated connections to flare systems rarely constitute design-case sources, because they are individually controlled, handle streams of the same compositions as emergency and operating devices, and flow capacity is typically not part of their design basis. Nonetheless, the flare system designer should always be open to the possibility of a manual connection with large flow capacity or that may be controlling a fluid that is otherwise not tied to the flare system.

In the case of process control valves that discharge to flare systems, consideration should be given to the full range of flaring scenarios and the associated flow rates that are likely to be delivered to the flare. Expected operational cases (start-up, individual component maintenance, etc.) that may involve essentially continuous flaring (for periods of hours to weeks) at rates far below the design maximum rate, or of fluids significantly different from other cases must also be addressed. Consideration of these flaring scenarios may have significant bearing on the selection and design of the flare tip(s), which can be seriously degraded by long-duration operation at "off-design" flow conditions. Such reduced flow operating cases should be discussed with potential flare tip vendors and should likely be part of the tip specification.

It is the emergency flaring cases, of course, that constitute the large-flow "design cases" for flare systems. The process of identifying all credible emergency flaring scenarios involves consideration of single events that may lead to flaring, with due consideration of all direct and indirect consequences of the event. When seeking large flow rate events, this typically means consideration of the consequences of failures of utility systems, because such failures can result in pressure relief from many sources simultaneously. Other initiating events to be considered here include large fires affecting many vessels simultaneously, and single upsets (blockages, control failures) affecting single, large-flow process streams. It is worth noting here that this last consideration can be particularly relevant to large-capacity LNG trains.

As mentioned previously, thoroughness is of prime importance in the identification of design-case flaring scenarios. In the context of utility failures, this manifests as consideration of all utility systems and their users, and of all consequences of their failures – both direct and indirect, resulting in both large and small flaring loads. As we shall see in LNG production facilities, failure of cooling water can generate one or two very large loads (from the main refrigeration systems and from the main gas stream) that dwarf the many other loads (from distillation systems, auxiliary refrigeration systems, etc.). This in no way means that the smaller loads can be ignored as "inconsequential," for together they are likely to be a sizable fraction of the dominant loads.

LNG Production Facilities

Overview

Figure 1 depicts a process block flow diagram of an LNG production facility. It is intended to be a most general example – appropriate for a facility taking as its inlet a full well stream that contains sour, rich natural gas along with hydrocarbon condensate and free water. Thus this facility is in large part a typical gas plant with the added features of very large refrigeration systems, some cryogenic heat exchangers, and systems for handling and storing a product stream that is essentially liquid methane.

Briefly, the process is first to separate the raw well stream into three phases. There is typically a gross split in a Slug Catcher. The separation will probably be refined by some chilling of the gas and flashing, stripping, and distillation of the liquids. The sour water and hydrocarbon condensate are then forwarded to treating units in which the acidic components (CO_2, H_2S, and mercaptans) are removed and forwarded to an Acid Gas Handling unit. The sweetened condensate is stored and shipped as a product, and the sweetened water is disposed of, often by re-injection.

The sour, rich natural gas is first separated into acidic and sweet components by a process appropriate for the composition and flow rate of stream. This is typically via gas/solvent contacting at close to well-stream pressure. The rich amine solvent is then flashed, and regenerated by distillation, with the acid gas stream combined with those from the produced water and condensate units and sent to Acid Gas Handling.

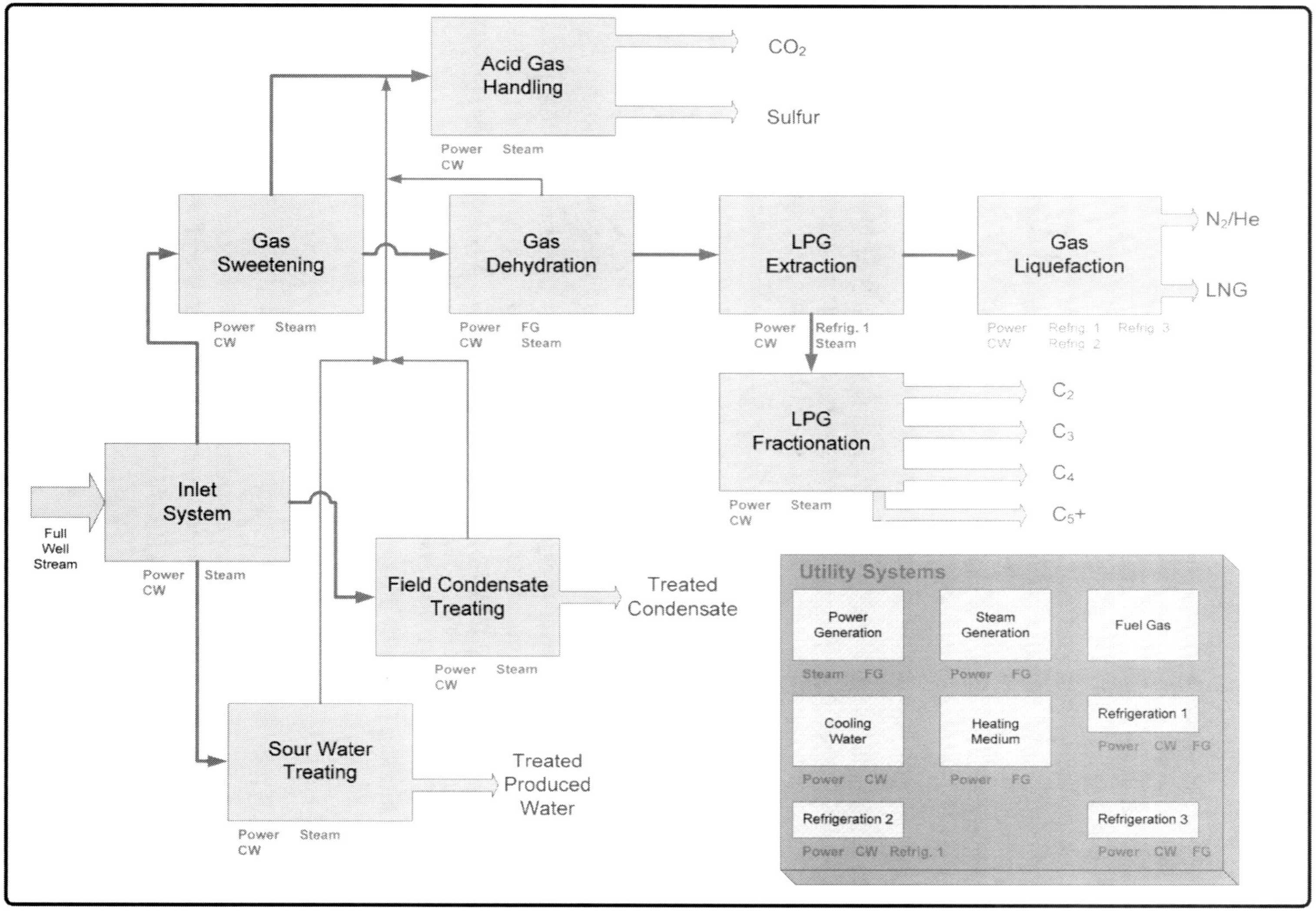

Figure 1 Block Flow Diagram of a Generic LNG Production Facility

The sweet, rich gas passes next to be dehydrated in fixed beds of molecular sieves, which absorb not only the water vapor, but also mercaptans present in the vapor phase. For this reason, the Dehydration unit may also contain a relatively small amine unit to sweeten the wet, sour regeneration gas coming from the beds undergoing regeneration. At the tail end of this unit the sweet, dry gas will pass through a mercury absorption bed to limit the potential for corroding the aluminum heat exchangers downstream.

Now the gas is ready to begin the process of being chilled to liquefaction. If the incoming gas stream contains enough "heavy" alkanes (which here means ethane through perhaps hexane), or the sales contracts call for lean LNG, the initial chilling will be in conjunction with LPG extraction in a demethanizer or de-ethanizer column, with the bottoms LPG product being sent to storage and shipping or to a Fractionation unit. The lean overhead gas is recompressed, and then enters the liquefaction unit, where it is chilled to condensation by a combination of refrigeration systems and turbo- or Joule-Thompson expansions. Nitrogen present in the LNG may then be flashed out of it before the LNG is sent to storage and off-loaded to tanker ships.

The liquefied petroleum gas (LPG) bottoms product of the de-ethanizer or demethanizer column in the LNG process may, if plant economics dictate, be fractionated into separate ethane, propane, butane, etc. products in a series of distillation columns that is standard fair at most larger gas plants.

Acid gas streams, composed primarily of H_2S and CO_2, with traces of hydrocarbons, COS, and mercaptans, are processed into elemental sulfur, typically via a Claus or other process. Sulfur production in this unit may be preceded by an additional gas/solvent contacting operation and the accompanying removal of enriched acid gas from the solvent by distillation.

Finally, the LNG facility will include a set of utilities that support the process operations just summarized. These include, in general, generation (or purchase) and distribution of electric power, steam, and fuel gas. Additional utility systems include cooling water and possibly a heating medium other than steam. And, in general, multiple refrigeration systems that are likely to be inter-dependent. Figure 1 includes a listing of these utility systems that are used in each unit in the LNG facility.

Application of Flare System Design Methods to an LNG Production Facility

The flare system design process begins, as always, with identification of the process consequences of events expected to have impact across the entire facility – because these usually generate the largest loads. The events in this list are failures of utility systems – power, water, instrument air, boiler feed water, etc. It is also important to identify events that may result in smaller total loads but that may discharge loads with "extreme" characteristics – e.g., very large liquid rates, high or low temperatures, corrosive fluids, etc. The full set of characteristics of these streams must be considered in the design of each individual flare system component.

Identification of Flaring Scenarios

Utility Failure – Total Power Failure

Electric power can fail due to a variety of causes – generator failure, single bus bar failure, a cascading failure of multiple buses, etc. When this does occur, all motor-driven equipment and all electric heating stops. Air-cooled exchangers lose most, if not all, of their air-flow, cooling water generally stops flowing, along with column feed, reflux, and products. A unit-by-unit review of the consequences of power failure in the example LNG production facility depicted in Figure 1 is likely to generate the load list shown in Table 2 below.

Unit	Source	Rationale
Inlet System	Condensate Stabilizer	Loss of Offgas Compression
Inlet System	Condensate Degassing Drum	Blocked Liquid Outlet (Overfilling)
Sour Water Treating	Sour Water Degassing Still	Distillation Upset
Gas Sweetening	Absorbent Regeneration	Distillation Upset
Gas Sweetening	Main Gas Flow	Blockage upon Loss of Liquefaction D/S*
Gas Dehydration	Absorbent Regeneration	Distillation Upset
LPG Extraction	De-ethanizer	Blockage upon Loss of Liquefaction D/S*
LPG Extraction	Ethane Column	Distillation Upset
LNG Liquefaction	LNG Flash Drum	Loss of Offgas Compression
LNG Liquefaction	Nitrogen Rejection Column	Loss of Offgas Compression
LPG Fractionation	Debutanizer	Distillation Upset
LPG Fractionation	Depropanizer	Distillation Upset
Acid Gas Handling	Absorbent Regeneration	Distillation Upset
Refrigeration 1	Refrigerant 1 Compressor	Blockage upon Loss of Condensation
Refrigeration 2	Refrigerant 2 Compressor	Blockage upon Loss of Condensation

* The main natural gas stream is expected to be relieved at one or the other of these locations, depending on the design pressures of the equipment and the pressure profile through the process under the upset conditions. Relief into a flare system is *not* expected to occur from both sources.

Table 2 Typical list of pressure relief sources in the event of Total Power Failure

Utility Failure – Cooling Water Failure

Cooling water can fail for a variety of reasons – individual pump failures, local power failures, accidental header rupture, etc. It should come as no surprise that cooling failure should have very significant impact on in an LNG production facility, for which condensation – of both the natural gas feed and of refrigerants – is central to the process. A unit-by-unit review of the consequences of cooling water failure in our example LNG production facility generates a load source list very nearly an identical subset of that in Table 2 above. The cooling water failure list is slightly less extensive because cooling in some individual operations is by aerial coolers, and power failure leads to liquid overfilling of some vessels due to product pump failures.

Characterization of Flaring Streams

A brief review of the sources listed in Table 2 reveals that the flare streams from these sources vary widely in pressure (from a low of 2 or 3 barg in the absorbent regeneration columns in the Acid Gas Handling unit to a high of 100 barg in the Inlet System – or higher, depending on the well pressure) and temperature (less than -50 °C due to auto-refrigerants, up to perhaps 200 °C from some distillation upsets). Also represented in Table 2 are streams rich in water, H_2S, CO_2, light hydrocarbons, and large liquid flow rates (from overfilling either the Condensate Degassing Drum or the Sour Water Still). It is worth recalling at this point in the evaluation that the list in Table 2 contains only those sources expected to flow in a total power failure. There are hundreds of other individual flare streams whose pressures, temperatures, and compositions lie within the range represented in Table 2. However, a review of these individual flaring scenarios finds only one extreme of pressure, temperature, flow, or composition not represented in Table 2: this is an extreme low temperature of roughly -160 °C due to the flow of LNG into the flare system.

Having enumerated the various flaring streams, with their pressures, temperatures, compositions, and flow rates, the first question that must be addressed is, "Can these all be accommodated economically in a single flare system, or does it make sense to install multiple flare systems, each tailored to the streams it serves?" Allowable back pressures on pressure relief valves (PRVs) are proportional to their set pressures, depending on their construction. A single flare system generally would have to be sized to meet the criterion imposed by the PRV with the lowest set pressure. In a facility with set pressures ranging from 2 or 3 barg to perhaps over 100 barg, it is essentially always advantageous to install multiple flare systems. The pressure(s) at which to differentiate the systems are identified on a case-by-case basis. For our LNG facility, there is likely a significant gap in design pressures: solvent regeneration, Claus unit, and perhaps some LPG fractionation equipment in the 2 to 10 barg range, and the rest of the process equipment in the 40 to 100 barg range.

Another reason for installing multiple flare systems is temperature requirements. With low molecular-weight hydrocarbons and relatively high source pressures, most of our flare streams are auto-refrigerating, so much of our flare metallurgy will have to be ready for fluid temperatures below 0 °C, with some of it required to handle temperatures below -46 °C (the limit for low-temperature carbon steel). Thus, we look for opportunities to collect the "warm" streams separately from the cold ones.

The final reason to install multiple flare systems in an LNG facility is to segregate the streams with corrosive components – water, sulfides, CO_2 – from the sweet streams.

Considering all of our flare sources in the light of these reasons for segregated flare systems, we are likely to settle on three systems. The first is for the wet, sour, generally high pressure, potentially cold effluent. These are generally found in the Inlet System unit through the Dehydration unit. This first system is probably best referred to as the "Sour & Cold" flare. The second is for sweet and dry streams only – it takes inlets from essentially everything downstream of the dehydration unit. It can probably be constructed of separate materials in separate sub-headers, each appropriate for a different range of temperatures expected in the different process units. This second system might be called the "Sweet & Dry" flare system. These first two systems will be sized for the very large loads associated with the main natural gas stream and the refrigerant compressor flows. The third system is then tailored to the sour, warm, low pressure streams coming from the acid gas handling systems in the water, condensate, and gas treating units and the sulfur production unit. An appropriate designation for this last system is the "Low Pressure Sour" flare. The loads in this system are significantly smaller than those in the other two, but the back pressure limits are significantly lower, so the line sizes can be of the nearly the same order.

Sizing

Having defined the scenarios, the associated loads, and a first-pass at the flare systems needed to process these loads, the design team can roll up their sleeves and proceed through the sizing calculations for the individual components in each flare system – header, subheader, and discharge lines; knockout drums; and flare stacks. The sizing of these components is done to meet the criteria listed in Table 1 most economically.

Finally, recall that in specifying the performance requirements for the flare tip, it is recommended to provide the full range of streams that the tip will need to burn, along with the design criteria for thermal radiation and noise relevant to each scenario.

Acknowledgement

The author is grateful to the organizers of this session for their invitation to submit this paper and to Siemens Oil, Gas, & Marine for their support of its preparation.

Biography

4615 Southwest Freeway, Houston, TX 77027, USA
Phone: 713-570-2935, Fax: 713-570-1230
E-mail: tom.bevilacqua@siemens.com

Dr. Thomas J. Bevilacqua is a Pressure Relief and Flare Systems Consultant in the Berwanger Engineering Services division of Siemens Oil, Gas, & Marine. Since joining Berwanger, Inc. in 1994 he has designed or conducted design reviews of the pressure relief and flare systems of dozens of facilities for clients in the oil, gas, petrochemical, and chemical industries. He has served several clients as their on-site pressure relief systems consultant; for several of these he has written or revised corporate relief and flare systems design practices and tutorials.

Dr. Bevilacqua is a member of AIChE and ACS. He participates regularly in the DIERS Users Group, and serves on the API Subcommittee on Pressure Relief Systems, on several Working Groups for ISO 4126, and as the Chairman of the US Technical Advisory Group for ISO/Technical Committee 185 (Safety Devices). For the past five years he has coordinated and co-taught the course in Pressure Relief Systems Best Practices for the Mary Kay O'Connor Process Safety Center at Texas A&M University.

Dr. Bevilacqua earned degrees in Chemistry from Villanova University (B.A., 1981), Princeton University (M.A., 1983), and Rice University (Ph.D., 1992), and conducted post-doctoral research as a National Research Council Fellow at the National Oceanic and Atmospheric Administration.

How Big Is the Area required for LNG Receiving Terminal and their Design Considerations

J.C. Kuo, P.E.
Senior Technology Advisor
Chevron Energy Technology Company
Houston, Texas

AIChE Spring Meeting, April 2007
7th Topical Conference on Natural Gas Utilization
Houston, Texas, 22-26, 2007

1.0 Introduction

The purpose of this paper is to outline the general considerations and requirements for developing the layout of the LNG receiving terminal. The paper also presents some equations for determining a quick estimate of the required plot size for the receiving terminal. Equations are based on correlations developed between the receiving terminal capacity and plot space requirements for some of the existing receiving terminals around the world. The information used to develop the correlation is based on the author's personal experience while working on various LNG projects.

2.0 Basis of the receiving terminal layout (6.1,6.2,6.3,6.4,6.5,6.6,6.7,6.8)

The basis of the receiving terminal layout is:

2.1 To ensure the terminal is safe, easy to construct, operate, maintain, and to meet all relevant insurance codes, local laws, and regulations.

All local statutory laws, standards and regulations must be followed such as.

- Australia: AS (Australian Standards)
- Canada: CSA (Canadian Standards)
- China: GB , GBJ and TJ (Chinese Standards)
- European Union : EN (European Norm Standards)
- Great Britain : BS (British Standards)
- India : IS (Bureau of Indian Standards)
- Japan: JGA (Japanese Standards)
- Mexico: NOM (Mexican Standards)

Additionally, codes, guidelines or standards set by individual oil companies or chemical companies should be met. The following organizations provide guidelines for fire protection and safety in chemical plants including receiving terminals. These guidelines directly influence the size of the plot space due to the spacing requirement.

2.1.1 ANSI (American National Standards Institute)
2.1.2 API (American Petroleum Institute)
2.1.3 CFR (Code of Federal Regulations)

2.1.4 EEMUA (Engineering Equipment and Material User Assoc.)
2.1.5 IP (Institute of Petroleum)
2.1.6 IRI (Industrial Risk Insurers)
2.1.7 NFPA (National Fire Protection Association)
2.1.8 OIA (Oil Insurance Association)
2.1.9 OSHA (Occupational Safety and Health Administration)

The documents listed below are commonly referred to in planning for receiving terminal layouts:

- API RP 505-Recommended Practice for Classification of Locations for Electrical Installations at Petroleum Facilities ..."
- API RP 521-Guide for Pressure Relieving and Depressuring Systems.
- API RP 752-Management of Hazards Associated with Location of Process Plant Buildings.
- API RP 2510-Design and Construction of LPG Installations.
- API RP 2610- Design, construction, operation, maintenance and inspection of terminal and tank facilities.
- 49CFR Part 193-Liquefied Natural Gas Facilities.
- 33 CFR Part 127-Waterfront Facilities Handling Liquefied Natural Gas and Liquefied Hazardous Gas.
- EN1473/1160-Installation and Equipment for Liquefied Natural Gas.
- NFPA 30-Flammable and Combustible Liquids Code.
- NFPA 59A-Standard for the Production, Storage and Handling of Liquefied Natural Gas (LNG).

2.2 To provide adequate separation between major facilities, LNG tanks and process areas.

Location and separation of major areas for unloading, LNG storage, process area, utilities, offsite, and etc. shall consider the following:

- Major areas shall be located based on considerations such as prevailing winds, grading of site, toxic, and flammable vapour gas dispersion such as LNG, LPG, hydrocarbon and Mercaptans.

- Process areas, LNG storage tanks, and other flammable liquid storage tanks (e.g. condensate or LPG) should be located so that prevailing winds will not carry flammable vapours to ignition sources such as furnaces, heaters, welding shops, and flares. Such facilities are recommended to be located at the down wind or parallel to the ignition sources. Please refer to the Illustration on Figure 1.

- Disposal of vapours from the flare should also be considered relative to the prevailing winds.

- Offices, shops, and other similar potential ignition sources relative to the process units and storage areas should be considered.

- The site grading such that a storage or process area spill will not migrate towards ignition sources, process areas, or control areas.

- The prevailing wind direction and site drainage should be considered when determining the location of control buildings. Control buildings that are exposed to potential explosions or fires must be designed to withstand potential overpressures explosion and fire effects.

FIGURE 1

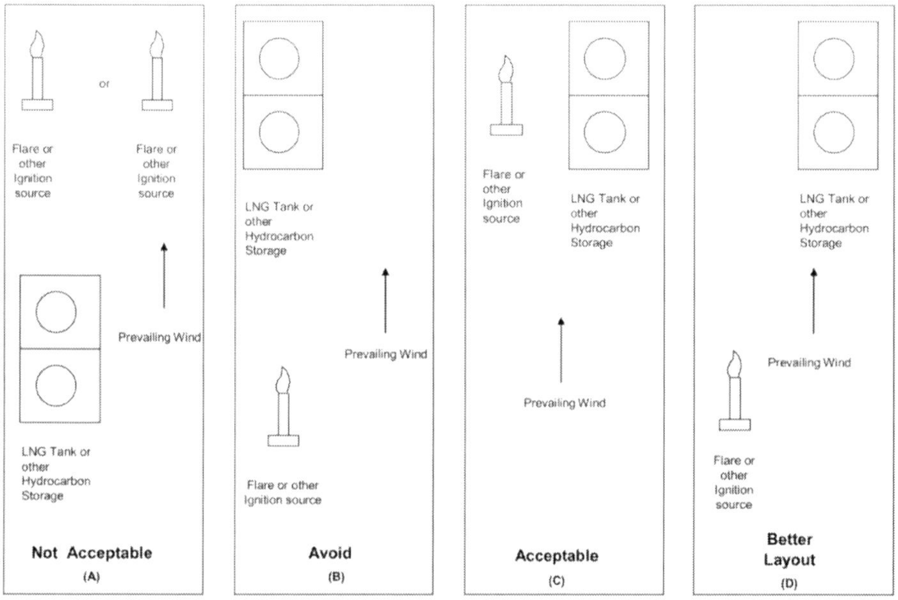

- Electrical power lines, main switchgear, and transformers should be located such that they are not exposed to damage due to the malfunctioning of processing equipment, storage, or loading/unloading facilities.

- Pressurized horizontal product tanks containing natural gas liquids (NGL) or similar volatile liquids should be oriented such that the long axis does not point toward nearby equipment or structures.

Finally, consideration must be given to the subsequent expansion without restricting the operation of the existing facilities. This is because many of the receiving terminals have multiple expansion phases.

2.3 To ensure adequate spill containment

2.3.1 LNG Storage Tanks

LNG storage tanks should be designed with appropriate containment and located remote from process units with common remote impound for a 10 minute design spill.

Separation distance between LNG tanks and other facilities will be set by the thermal radiation limits given in NFPA 59A for a tank top fire on a contained tank. Spill containment should conform to NFPA 59A and all other codes, standards, and recommended practices. Additionally, separation between the tank and the Outer Fence Line may be controlled by vapour cloud dispersion.

2.3.2 Atmospheric Storage Tanks

Atmospheric storage tanks (e.g. diesel) should be located such that spills do not impact on the security of process units, other critical facilities, administration or other occupied buildings and will be enclosed by the dike. Tanks shall be spaced in accordance with related code and standards.

2.3.3 Pressure Storage

Pressure storage of hydrocarbons (e.g. LPG) in spheres and bullets shall be in remote areas, outside the dike enclosure of atmospheric flammable liquid storage.

2.4 To permit safe evacuation of people inside the plant and those living close to the fence-line during a hazardous event.

This is to separate toxic, hazardous materials from high risk facilities and to reduce the potential for event escalation.

- Careful consideration should be given to the location of facilities in relation to property lines and neighbors in order to minimize potential exposures. Exposures from fire, explosions, spills, gas releases, vehicle traffic, etc. should be evaluated in both directions (e.g. exposure to the facility and exposure to neighbors from the facility).
- In addition to the citing criteria noted above, potential escalation of minor events might be limited through the selective inclusion of the flowing into the design:
    - Fire and gas detection initiating alarms
    - Passive fire protection
    - Fire water and other fire suppression systems
    - Installation of Emergency Shutdown and Emergency Depressuring Valves
    - Use of barriers or structures designed for explosion overpressure

2.5 To meet thermal radiation levels and flammability contours from gas or LNG pool fires.

2.6 To limit blast loads in the event of ignition of flammable clouds to acceptable levels.

Additionally,

- To minimize capital and operating costs while satisfying the above safety and maintenance requirements.
- To minimize the engineering and construction bulk material quantities given that all safety, operability and environmental objectives are met.

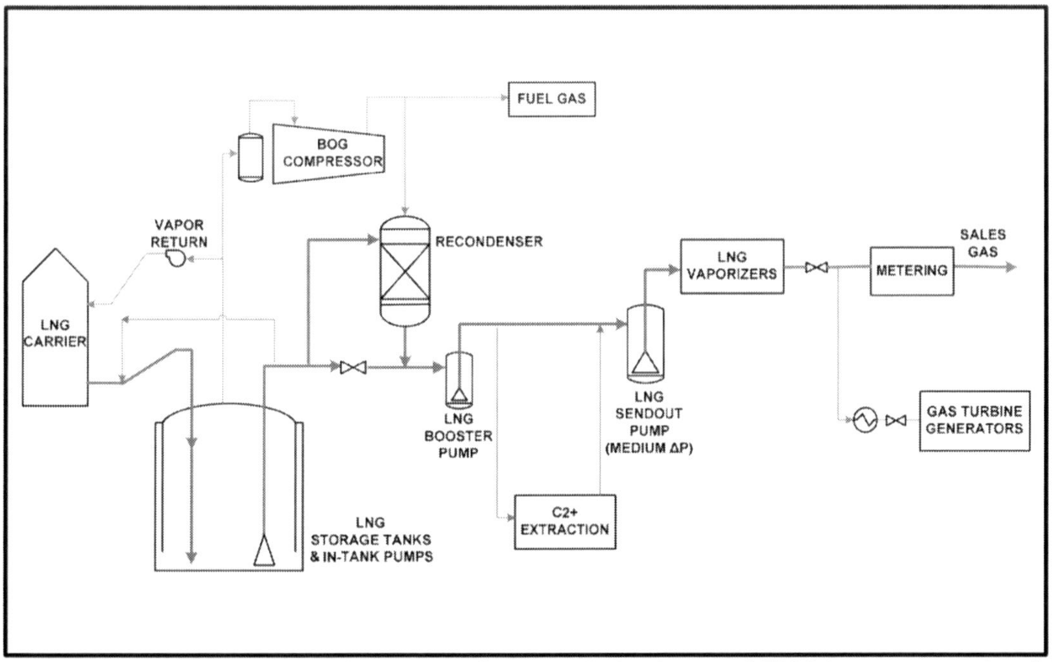

Figure.2 – Typical Terminal Process Schematic Including BTU Management by C2+ Extraction

3.0 General design considerations and requirements for special areas and equipments (6.1,6.2,6.3,6.4,6.5,6.6,6.7,6.8)

Required Studies

First, the plant layout usually needs to accommodate the results of the following studies.
(1) Process Hazard Review,
(2) Vapour and Toxic Gas Dispersion Studies,
(3) Thermal Radiation Exclusion Zones,
(4) Noise Contour Studies, and
(5) Waste Treatment Studies to avoid pollution into the sea and the latest geotechnical and topographical site data.

Process Units

The process and utility units should be preferably divided into discrete areas with their size and grouping of equipment based on:

- commonality of process, operation and maintenance
- acceptability failure consequences in a single fire or explosion
- consideration of size of venting and flaring system
- operational and emergency access

Main Control Room

By providing separation distance between the Main Control Room and the process or storage units, the Main Control Room should be protected from explosion or fire. Alternatively, the building should be built to be blast resistant.

Other Plant Buildings

Substations, according to function, may be located within or near process and utility areas. These buildings, not frequented by plant personnel, will not be designed for blast resistance. The Marine Terminal Building should be designed in such a way as to minimize blast potential. Consideration shall be given to personnel occupancy levels and the criticality of the systems installed within them.

Utilities and Power Generation

Utilities and power generation facilities should preferably be segregated from process units and storage of hydrocarbons.

- Consideration of firewater system sizing. Fire risk area size must be consistent with sizing of flare and firewater system headers.
- As far as possible, areas should be rectangular to permit fire fighting from at least two opposite sides. Consideration should also be given to limit the longer dimension by including additional fire fighting access ways.

Access Roads

Within the perimeter of process units, access roads shall be provided for fire fighting and maintenance. Access roads separating units shall be designed to accommodate trucks, maintenance equipment and mobile cranes.

Fired Equipments

Where practical and economical, fired equipment (e.g. hot oil furnace, gas heater, combustion vaporizer) should be grouped together in an isolated area. Fired equipment shall be separated from other equipment handling flammable products and should be located at the edge of process unit with an access way for maintenance and fire-fighting.

Air Intakes

Air intake for equipment such as furnaces, heaters, ventilation, pressurization, or air conditioning, cooling of large motors, plant air, and air separation plants should be taken from a safe area. In determining a safe area, consideration should be given to normal and emergency exhausts, vents, and possible recirculation patterns around buildings and air coolers. Consideration should also be given to gas detection in the air inlets to such equipment.

Air for all control rooms should be drawn from a safe location. The air intake stack shall be equipped with a combustible gas detector.

Compressor Shed

The compressor shed (covering BOG compressors and their drivers) can be open on all sides or be in enclosed buildings with appropriate detection, alarming and suppression.

Flare Systems

Location, layout, and height of flares should consider factors such as acceptable thermal radiation, noise, and dispersion of flammable and toxic vapours. An agreed-to heat radiation and dispersion model shall be used. Separation distance between the flare riser and the Receiving terminal Fence should follow accordingly with the company or API standards.

A clear zone around flares which is less frequented by plant personnel should be based on the recommended radiation level of API RP 521. There should be enough distance between flares to allow maintenance of one set of flares while the other flares are in operation.

4.0 Plot plan size Requirements [6.1,6.2,6.3,6.4,6.5,6.6,6.7,6.8]

4.1 Parameters affecting the plot plan size.

Some of important parameters affecting the plot plan size are listed below. Section 4.2 has the estimates for the additional plot space required.

4.1.1 LNG tank type: Each client and country specifies their own type. However, for the single containment type tank, it is the single most dominant factor to increase the plot plan size. The vapor dispersion exclusion zone (½ LFL, Lower flammable limit) and the 5.05 Kw/M**2 (1600 Btu/Hr.ft**2) radiation exclusion zone have the major impact. The safety distances from property lines, buildings and public areas will be far greater. Furthermore, the distance required between the centers of the tanks is about 67% more. The tanks space required is about 4 to 8 times larger than the full containment plot size pending on the local code, number of the tank, dike height, tank size and wind velocity. The overall space required is about 2 to 4 times larger than the full containment plot size. Detail calculation is required.

4.1.2 LNG storage capacity philosophy: Each client and country specifies their own requirements. Some countries like to have more LNG tanks to meet the country energy strategy.

4.1.3 LNG BTU management or total NGL recovery: If the BTU management or the total NGL recovery is required for the Incoming LNG then additional plot space is required.

4.1.4 LNG vaporizer type: Five major types of LNG vaporizers are listed as following: (1) Submerged Combustion Vaporizers (SCV) (2) Submerged Combustion Vaporizers with Selective Catalyst Reduction Units (SCVR) (3) Ambient Heating Vaporizers (AHV) (4) Open Rack Vaporizers (ORV) and (5) Indirect Fluid Vaporizers (IFV). For SCV, SCVR and AHV additional plot space is required

4.1.5 There are other parameters which may affect the size of the LNG plot plan. These are climate, flare type (Ground or Elevated type), site conditions including topography and soil conditions, landfill, foundation design, earthquake criteria, elevation, contour, subterranean water, surface drainage, proximities of existing facilities available infrastructure, and limited construction site accommodations.

4.2 Plot Plan Size

The total LNG terminal plot plan was divided into four areas.

(1)The LNG Process area,
(2)The Utilities System area,
(3)The Storage Tank area, and
(4)The Supporting area. This may include construction lay down, construction worker camp, permanent dorm, parking lot, material off loading facilities, and temporary buildings.

Based on the Full Containment Tanks, for a very quick and rough estimate for the four areas mentioned above and/or the individual area, please refer to the followings.

Total four areas (Total Receiving terminal) - (30-54 hectares per BCFD)

(1)The LNG Process Train area,	(6 to 10 hectares per BCFD)
(2)The Utilities System area,	(9 to 16 hectares per BCFD)
(3)The Storage Tank area	(6 to 12 hectares per BCFD)

(4)The Supporting area (9 to 16 hectares per BCFD)

For 1.0 BCFD and up to 5.2 BCFD grass root receiving terminal, 42 hectares per BCFD is a "rule of thumb" for the initial total area estimation.
For Single Containment Tanks, a multiple factor from 4.0 to 2.0 is recommended. Larger terminals will need smaller multiple factor.

BCFD is billion cubic feet per day.
MMCFD is million cubic feet per day.
MTPA is million metric tons per annual
1BCFD=1000 MMSCFD, 1 BCFD is close to about 7.0 MTPA.
One Hectares =100 meters * 100 meters.

For more accurate estimation, it is recommended to use the following correlations.

Based on plot plans of existing receiving terminals, one correlation between the receiving terminal gas send out rate (MMSCFD) and the required receiving terminal plot space (Hectares) was developed for Process area. Utilities System area is based on the ratio of the Process area.

4.2.1 LNG Process area

The LNG process area includes the unit of

- LNG unloading,
- Recondenser,
- BOG compressor,
- LNG send out pump,
- LNG vaporization,
- Gas ordorization,
- Gas metering,
- Substation and/or the local control room,
- Access road around the Process area.

For an estimation of the above Process area required, the following equations can be used for terminal size up to 5.2 BCFD. The LNG process plot space required is given in hectares, and the send out gas rate is given in million standard cubic feet per day (MMSCFD).

Hectares = 0.131*(MMSCFD**0.6)-0.80

Additional space is required for any receiving terminal which has more than those aforementioned units such as the following:

- BTU management by inert injection or C2+ extraction. Add 6%.
- Total NGL recovery by C2+ extraction. Add 12%.
- For Submerged Combustion Vaporizers (SCV) and Submerged Combustion Vaporizers with Selective Catalyst Reduction Units (SCVR). Add 5%.
- For Ambient Heating Vaporizer (AHV). Add 12%

4.2.2 Utilities System area.

The LNG utilities area includes all the services required such as

- Instrument air and Plant air,
- Nitrogen,
- Fuel gas,
- Demineralized water,
- Power generation, Emergency power & UPS
- Cooling water,
- Heat medium,
- Fire water,
- Flare and blow down,
- Drain systems,
- Diesel,
- Waste water/Effluent treatment,
- All sterilized area due to radiation,
- Access road.

For an estimation of the Utilities system area required, it is recommended to use the 150% ratio.

4.2.3 Storage Tank area

4.2.2.1 LNG storage tank (Full Containment Tanks)

Hectares = 0.0093*((Meter**3)**0.6) -8.12

4.2.2.2 Condensate storage tank

Hectares = 0.0082*((Meter**3)**0.6) -2.39

Meter**3 is the volume of the storage tank.

The LNG storage tank is based on the Full Containment Tank. About 15% more plot space is required for the double containment tank due to more space required between each LNG tank. The minimum tank spacing for single containment tank is 1.6 to 1.75 diameters comparing with 1.5 diameters for the full containment tank.

The following table summarizes minimum spacing by tank type:

	Single Containment	Double Containment	Full Containment	In ground Membrane	PC-PC (Pre-stressed Concrete)
Min. Spacing	2.5d	1.6d to 1.75d	1.5d	1.5d	1.5d

Additional area is required for any Receiving terminal which has more than those aforementioned storage tanks such as the export storage for Propane, Butane, LPG or others.

4.2.4 The Supporting area

It is recommended to use the similar plot space as the utility area. However, it should be noted that the area of airport is not included herein.

5.0 Acknowledgements

The author would like to thank his colleagues from various oil companies and engineering contractors for the data and information used in this paper. The author would also like to thank Chevron for providing permission to publish this paper.

6.0 References

6.1 "How Big Is the Area required for LNG Base Load Plant And Their Design Considerations", J.C. Kuo, Presented to the AIChE 2006 National Meeting, LNG section, Apr 23-27, 2006.

6.2 "Tank farm safety design: Updates" J. Swarup, Hydrocarbon Processing, Page 65-77. December 2006.

6.3 "Optimize energy integration for LNG terminal" B. C. Price, Hydrocarbon Processing, Page 43-46. July 2004.

6.4 IRI (Industrial Risk Insurers) General Recommendation for Spacing

6.5 Equipment spacing design guidelines from major oil companies and Engineering Contractors.

6.6 Minimum safety separation distances between units guidelines from major oil companies and Engineering Contractors

6.7 Calculating safety distance in LPG storage and handling installations guidelines from major oil companies and Engineering Contractors

6.8 In house studies and data base from major oil companies and Engineering Contractors

BIOGRAPHY

J. C. Kuo is a Senior Technical Advisor for Chevron Energy Technology Company. He served as the Process Manager/Coordinator for the Gorogn LNG, Delta Caribe LNG, Casotte Landing LNG receiving terminal and Sabine Pass terminal BTU management projects. He also served as the Technical Reviewer/Advisor for Angola, Olokola and Shtokman LNG projects.

He has more than 33 years experience in gas processing, gas treating and LNG. He has BS from Chung Yuan Christian University, Taiwan and MS from the University of Houston all in Chemical Engineering. He also holds an MS in Environmental Engineering from Southern Illinois University, Carbondale. He was the Technology Manager for IPSI/Bechtel and served as the Process Manager for the Pemex Catarell offshore project (1996-2001) and the Egyptian LNG (Idku) train 1 and 2 (2001-2004). He is a register professional engineer (Texas) and member of AIChE.

Understanding LNG Fire Hazards

G. A. Melhem, Ph.D.
H. Ozog
A. S. Kalelkar, Ph.D.

ioMosaic Corporation
93 Stiles Road
Salem, New Hampshire 03079

AIChE Spring Meeting, April 2007
7th Topical Conference on Natural Gas Utilization
Houston, Texas, 22-26, 2007

I. Introduction

Potential hazards resulting from intentional or accidental spilling of large quantities of LNG include thermal radiation from vapor cloud fires (also referred to as flash fires) and pool fires.

There is general agreement among LNG experts regarding the following aspects of potential LNG fire and explosion hazards:

1. Vapors from large, un-ignited spills of LNG cannot travel far into developed areas without finding an ignition source, igniting, and burning back to the source.
2. Once delayed ignition of the vapor cloud occurs, and provided that the cloud is unconfined and rich in methane, the LNG vapors will burn in the form of a vapor cloud fire.
3. A vapor cloud traversing over commercial and/or residential terrain will almost certainly encounter an ignition source early in its downwind drift and the resulting vapor cloud fire will burn back to the source.
4. The vapor cloud fire will burn back to the source and cause a pool fire at the source if the release is a continuous release and the release duration is longer than the time it takes the cloud to find an ignition source.
5. If the vapor cloud is confined and/or the vapors contain large amounts of heavier hydrocarbons[1] (C2+), then the flame can accelerate and result in an explosion. The magnitude of the explosion and explosion damage will depend on several factors including the amount of vapors above the lower flammable limit, the presence of obstacles and degree of confinement, the composition of the vapor cloud, and the strength of the ignition source.
6. If immediate ignition occurs, a pool fire will result. The extent of the pool spreading (diameter) and flame height will depend on several factors including the flow rate of LNG, the spill surface type (water or land), the spill surface geometry, spill surface roughness, release composition, release temperature, ambient wind speed, ambient temperature, and ambient relative humidity.
7. If the liquid pool is unconfined and the inventory of LNG is large, the fire will continue to burn[2] until all the fuel is exhausted by the pool fire. It is not practical or even possible to extinguish

[1] Actual test data with methane-ethane mixtures show that explosion potential increases significantly with fractions of C2+ exceeding 20 percent.

[2] This conclusion is also reached by a recent GAO expert panel study. In addition, the terrorist scenario considered in this paper assumes that the LNG tanker is in port with calm waters (no waves), fully loaded, and ready to unload.

large LNG pool fires resulting from large spills of LNG unless the flow of LNG feeding the pool can be stopped.

The maximum vapor cloud fire hazard area from large LNG spills is typically estimated by calculating a downwind dispersion distance to the lower flammable limit (LFL[3]) and a cross-wind dispersion distance to ½ LFL at low wind speed and stable atmospheric conditions. This maximum fire hazard zone is very unlikely to be experienced in any situation where the cloud drifts over populated areas[4]. As indicated in point 3 above, the cloud will soon encounter an ignition source and burn back to the source well before the maximum hazard area is reached.

Only the outdoor population present within the flammable boundaries of the vapor cloud is assumed to be injured due to (a) short exposure to very high thermal radiation fluxes from the vapor cloud fire, (b) direct flame contact, (c) secondary fires of clothing, and (d) inhalation of hot combustion products. It is assumed that people inside buildings at the time of the flash fire will not be injured[5]. It is also assumed that people inside buildings which are ignited by flash fire or a pool fire will be able to escape from the burning structure without direct thermal impact injuries. This is because the flash fire will ignite buildings from the outside and it will take some time for the fires to penetrate inside.

A. Issues Surrounding Potential Damage from LNG Fires

Published thermal radiation damage criteria often associate a level of damage with a heat flux value or an integrated heat flux value for short duration events such as fireballs. Typical values used and their observed effects are provided by CCPS:

Table 1: Thermal Radiation Flux

Thermal Radiation Flux. kW/m^2	Observed Effect
37.5	Sufficient to cause damage to process equipment
25.0	The minimum energy required to ignite wood at indefinitely long exposure (nonpiloted)
12.5	The minimum energy required for piloted ignition of wood, and melting of plastic tubing. This value is typically used as a fatality number
9.5	Sufficient to cause pain in 8 seconds and 2nd degree burns in 20 seconds.
4.0	Sufficient to cause pain to personnel if unable to reach cover within 20 seconds. However, blistering of skin (second degree burns) is likely; 0% lethality
1.6	Will cause no discomfort for long exposure

There is general disagreement among LNG experts pertaining to the extent of thermal radiation hazard zones resulting from large LNG pool fires due not only to uncertainties regarding flame emissive power but also the limiting thermal radiation impact criteria.

[3] If heavy gas box dispersion models are used, it is recommended that the downwind distance be estimated to ½ LFL to account for peak to mean concentration averaging and pocketing. If well validated CFD models are used, than a downwind distance to LFL provides an adequate representation of the extent of dispersion downwind. The cross-wind distance should be computed to ½ LFL in both cases.

[4] Most experts agree that an LNG vapor cloud will have a near 100 % probability of ignition within one mile of ingress into developed areas.

[5] Although it is possible for people in buildings to be exposed to thermal radiation through windows.

B. Flame Emissive Power

Some experts argue that very large LNG pool fires such as those resulting from a terrorist attack on an LNG tanker will produce sooty flames and the flame emissive power is expected to be much less than 220 kW/m^2. The main argument is that the pool center will be starved from oxygen.

An opposite view, which is more likely to be the correct one, is that the fuel that does not burn at the center of the pool due to oxygen starvation will rise due to thermal buoyancy and burn at a higher elevation as it contacts oxygen there (see references). This is also supported by recent computational fluid dynamics modeling of pool fires[6]. As a result, the flames will be taller and the associated thermal radiation hazard footprints may be higher. In addition, in large scale pool fires field data such as the Montoir field trials, emissive power values of approximately 300 kW/m^2 were reported.

In general, the thermal radiation flux experienced by an observer exposed to a fire decreases in proportion to the square of distance from the fire source.

Figure 1: LNG Pool Fires – Measured Flame Emissive Power

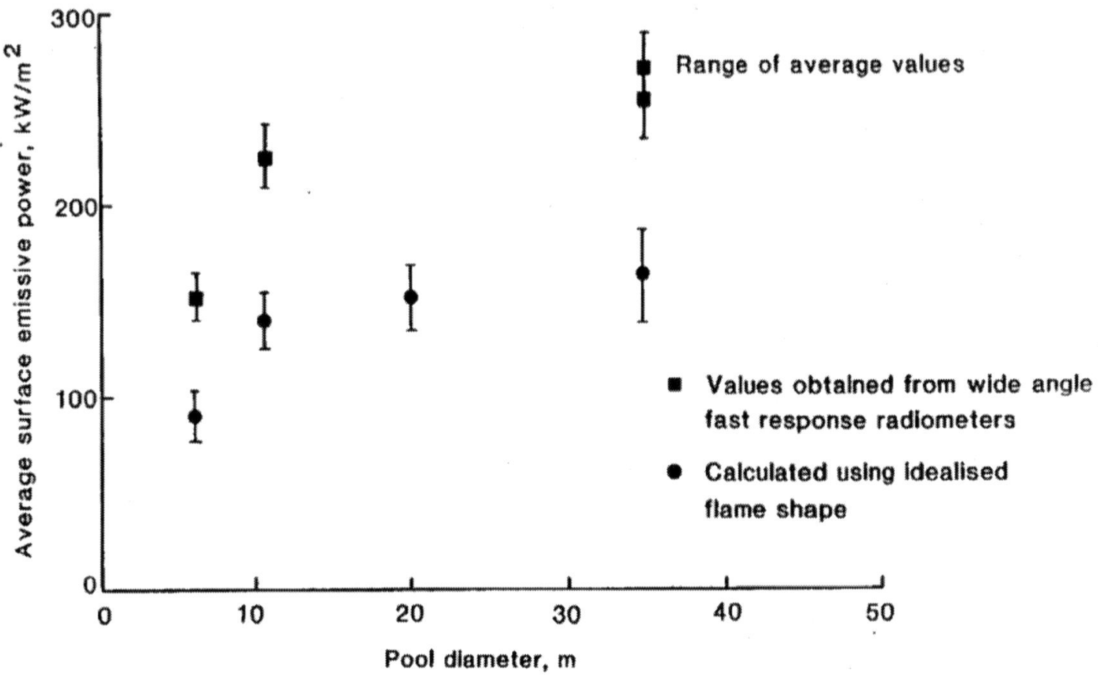

C. Limiting Thermal Radiation Damage Criteria

Some experts argue that a 20-second 5 kW/m^2 limiting thermal radiation exposure criterion is sufficient to establish safe separation distances for the general public. The main argument here is that a typical person will sense pain quickly and can run away fast enough and take shelter. This criterion is adopted for example by NFPA-59 without reference to exposure duration.

An opposite view argues that these criteria cannot be applied to sensitive population or critical areas / infrastructures. Elderly and the very young for example, constitute sensitive populations that may not be able to take cover within 20 seconds when outdoors. "Critical areas" include unshielded areas of critical importance where people without protective clothing can be expected or required at all time including

6 Many fire research scientists consider the computational flame zone to extend to the point where the composition of carbon monoxide is 1000 ppm or less.

during emergencies. "Critical infrastructure" includes buildings or places that are difficult to evacuate on short notice such as sport stadiums, hospitals, schools, play grounds, theaters, etc. As a result a lower criteria is adopted by EN-1473[7] (1.5 kW/m^2), the US Department of Housing and Urban Development (450 BTU/ft2/hr or 1.4 kW/m^2), API-521[8] (1.58 kW/m^2), and the society of fire protection engineers (SFPE Handbook) recommends a level of (2.5 kW/m^2) as a public tolerance limit.

We must recognize that in the specific case of LNG terminals large quantities of LNG will be stored in bulk storage tanks and frequently arriving by ship. Under the right scenario, loss of containment can yield very large pool fires and the extent of the potential hazard zones must be accurately determined in order to establish a prudent estimate of a safe separation distance.

We must also recognize that there are some uncertainties associated with the application of several of the models used to establish safe thermal radiation separation zones. For example, the flame height correlations have not been validated against pool fires that are several hundred meters in diameter.

There are two practical approaches to addressing the issues of thermal radiation damage criteria, assuming we can all agree on what to use as a reasonable value of flame emissive power:

1. Be prudent and conservative. Set the value low enough such that anyone that is continuously exposed will not suffer irreversible injuries.
2. Evaluate the risk accurately. Consider both the exposure duration and the exposure flux (dosage), and consider the demographics of the current and projected population density nearby the proposed facility to be sited, i.e. what fraction of the people will be outdoor, what fraction is sensitive, where the critical locations are, etc. This approach will require a risk tolerability criterion that is acceptable to the community tolerating the risk in lieu of some economic benefit.

The NFPA-59 thermal radiation criteria should not be confused with and/or considered as a risk acceptability criteria. Hazards are just one aspect of risk. Other important aspects of risk management include operational, economic, social, political, and environmental factors as well as the probability of the occurrence of the hazard itself.

<u>The 5 kW/m^2 limiting criterion does not adequately represent the risks presented by an LNG facility to sensitive population and/or critical areas/buildings</u>. Dosage must be considered as mentioned in item 2 above. The most widely recognized and used methods for establishing the impact of thermal radiation on people are those developed by TNO in the Green Book. These methods are referred to as thermal radiation probits or vulnerability models.

D. Thermal Radiation Damage Probits

Probits are used to relate level of injury and exposure duration to a hazardous event of a given intensity. Hazardous events of interest in consequence modeling include dispersion leading to exposure to toxic chemicals, fires leading to exposure to thermal radiation, and explosions leading to exposure to overpressure and flying fragments.

The method of probit analysis was first introduced between 1940 and 1950. A probit (probability unit, Y) is a normally distributed random variable with a mean of 5 and a standard deviation of 1 (see Figure 2).

The mortality response (percent fatality) is expressed as:

[7] Value excludes solar flux.

[8] Value is at any location where personnel with appropriate clothing are continuously exposed. Note that this value includes solar flux.

$$P = \frac{1}{\sqrt{2\pi}} \int_{-\inf}^{Y-5} \exp-\left(\frac{u^2}{2}\right) du = \frac{1}{2} + \frac{1}{2} \operatorname{erf}\left(\frac{Y-5}{\sqrt{2}}\right) \quad (1)$$

For mortality response to a toxic exposure of concentration C and duration t, Y is given by:

$$Y = A + B \ln\left[C^N t\right] \quad (2)$$

If C varies with time, then Y can be expressed as:

$$Y = A + B \ln\left[\int C^N dt\right] \quad (3)$$

Here, the integral containing concentration represents a dose factor. Probit analysis can also be applied to thermal radiation hazards:

$$Y = A + B \ln tI^{4/3} \quad (6)$$

Where, A and B are the probit parameters established from measurements and/or critically evaluated scientific data (see Table 2), I is the radiation intensity in $\mathrm{W/m^2}$, and t is the exposure time in seconds.

Figure 2: Probit Function

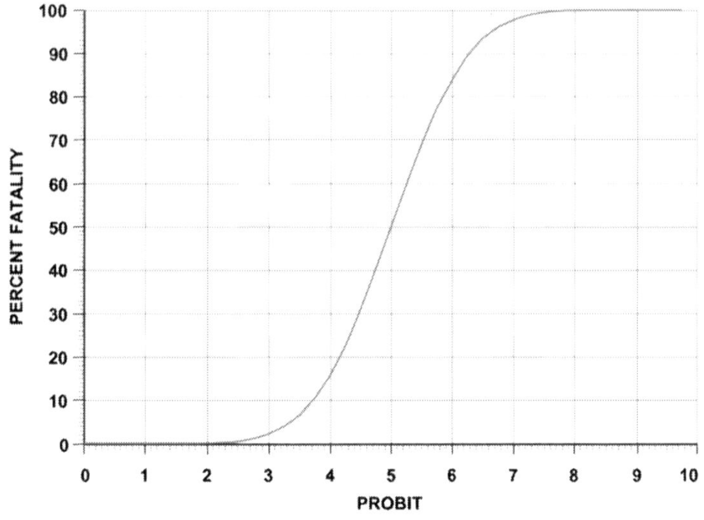

The TNO Green book provides probit functions for first and second degree burns as well as lethality from exposure to heat radiation within the infra-red part of the spectrum (see Table 2). The last probit function reported in Table 2 accounts for clothing protective influence on fatality for humans. It assumes that 20 % of the body area remains unprotected for an average population. As a result, the fatality for protected bodies is about 14 % of the fatality for unprotected bodies.

Table 2: Typical Reaction Times to Thermal Radiation Exposure Levels

Intensity (kW/m^2)	Time to react (s)
22	0.2
18	1.5
11	3.5
8	5.5
5	9.0
2.5	25.0

The influence of running away from a location with high heat radiation to a location where the level of heat radiation is safe (approximately 1 kW/m^2) can also be used for the assessment of injury and fatality from heat radiation. TNO considers 1 kW/m^2 as the maximum heat-flux the skin can absorb during a long time without feeling pain. The probits presented in Table can be modified to take that into account by replacing the exposure time t by an effective exposure time t_e:

$$t_e = t_r + 0.6\frac{x}{u}\left[1-\left(1+\frac{u}{x}t_v\right)^{-5/3}\right] \qquad (7)$$

Where, t_r is the reaction time and is about 5 seconds (see

Table 2), x is the distance to 1 kW/m^2, u is the run velocity in m/s, and t_v is the time to reach 1 kW/m^2. To illustrate the use of the TNO thermal radiation probits, we present in Figure 3 the thermal radiation dosage required to produce a 1 % probability outcome. Note that the probit equations shown in table 2 should not be extrapolated to values less than 1 kW/m^2.

Table 3: Heat Radiation Probit Parameters (taken from the TNO Green Book)

Damage	Probit
First degree burns	$Y = -39.83 + 3.02 \ln tI^{4/3}$
Second degree burns	$Y = -43.14 + 3.02 \ln tI^{4/3}$
Fatality (Unprotected)	$Y = -36.38 + 2.56 \ln tI^{4/3}$
Fatality (Protected)	$Y = -37.23 + 2.56 \ln tI^{4/3}$

Figure 3: Thermal Radiation Flux vs. Exposure Time Leading to 1 % Probability of Injury or Fatality

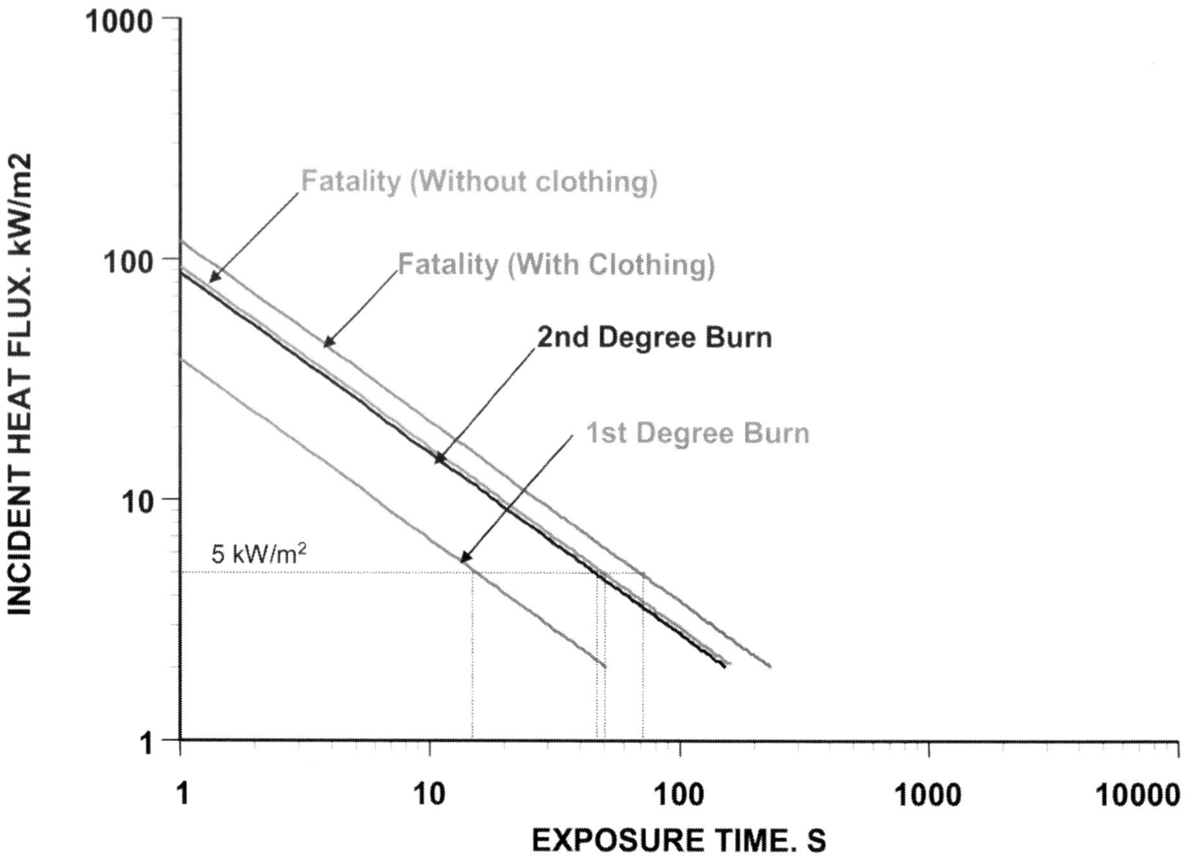

E. Intentional Terrorism Acts against an LNG Tanker

The potential of terrorist attacks against LNG tankers is highly debated by the media and the public. The major concern centers around the large quantity of energy stored on these large ships. New LNG supertankers now under construction will carry up to 265,000 cubic meters of LNG.

In general, and despite the large amount of energy stored on LNG tankers, they are not attractive "mass casualties" targets unless the ship is berthed in ports near highly populated areas when attacked. This is so because the hazard impact zones from a terrorist attack on an LNG tanker will be highly localized near the ship. Many communities opposing the siting of LNG terminals argue that some U.S. ports represent attractive terrorism "economic targets".

We have considered numerous scenarios involving terrorism and acts of sabotage including hijacking, small boat attack, standoff weapon attack, aircraft attack, underwater diver/mine attack, placing solid explosives in the cargo hold areas, blocking the pressure relief valves, starting the pumps and unloading while the ship is berthed, etc. Based on our review we believe there is one maximum potential impact scenario that should be considered as a threat. This scenario involves multiple cascading tank failures[9]. This review will not comment on the likelihood of occurrence of this scenario but will focus on the evaluation of the maximum potential hazard zone that could be realized from this scenario. The scenario involves a fully loaded LNG ship that is berthed before cargo unloading begins.

Before describing this particular scenario in detail we begin by looking at the characteristics and dimensions of different types of existing and planned LNG tankers.

[9] The recent GAO 2007 study considered a terrorist attack on an LNG tanker. The study recommended to DOE that the potential for cascading tank failures should be studied (ranked as the second highest topic requiring more study by GAO's expert panel) in order to better understand the damage to the hull that can be caused by exposure to extreme cold or heat.

F. LNG Tanker Structural and Safety Features

At the beginning of 2005, there were 175 LNG carriers with a combined capacity of 20.68 million m^3. The principal ships in the LNG fleet range from 125,000 m^3 to 150,000 m^3. At the end of 2004, the first orders were placed for LNG carriers of more than 200,000 m^3. Carriers with a capacity of 265,000 m^3 are now under construction. Six of these carriers were ordered recently by Qatar Gas.

LNG carriers are double-hulled vessels specially designed and insulated to prevent leakage and rupture in the event of accidents such as grounding or collision.

The different types of cargo tanks utilized for shipping LNG are illustrated in Figures 4, 5, and 6. The Kvaerner-Moss system employs free standing spherical tanks fitted unto the hull. The second designs incorporate variations of a membrane type tank. In this case, the LNG cargo is contained within thin walled tanks of stainless steel. The tanks are anchored at various points to the inner hull of the double-hulled vessel, and the cargo load is transmitted to the inner hull by the intervening insulation. The world fleet is roughly divided between these two systems.

Boil-off gas during transit is limited by insulation and maximum normal boil-off is about 0.15% of cargo per day. Boil-off gas is used to supplement fuel in the ship's boilers; hence many ship propulsion systems use steam turbines. Diesel propulsion is finding favor, but for those ships onboard boil-off gas recovery systems and gas combustion units are required. Gas turbines are also being considered.

Due to the low density of the cargo, LNG carriers ride with a high freeboard. For this reason, maneuvering in port in windy conditions makes the ships susceptible to being blown to one side of the channel. Therefore, port maneuvering usually requires traffic restrictions and extra tug power for such conditions. In general, this is not a safety concern because LNG ships are normally restricted from entering ports in rough weather.

G. Spherical Tanks

The spherical tanks (see Figure 4) are supported by a cylindrical structure "skirt" attached to the sphere at approximately mid-height. The cylinder sits on the skirt table attached to the side and bottom hull. The sides and bottom are of double hull construction, with longitudinal stiffeners and spaced web frames which act to tie the inner and outer hulls into a box frame arrangement. The webs are open to allow passage of the water ballast (used on the return voyage) to move freely.

The outer hull is typically ¾" plate steel and the inner hull is ½" plate steel. The spacing between hulls is about 187 inches (220cm). The tanks are typically insulated with foam material.

Table 4: Typical Modern LNG Carries Dimensions

	Membrane Carrier		Spherical Carrier	
LNG Carrier Capacity (m^3)	**145,700**	**216,000**	**125,000**	**235,000**
L - Length (m)	277.2	303.0	282.0	328.5
B - Breadth (m)	43.4	50.0	41.6	55.0
D - Depth moulded (m)	26.0	27.0	25.0	32.5
Top of tank above base line (m)	31.0	33.2	37.7	49.0
T - Draft moulded (m)	12.3	12.5	11.5	12.5
Displacement (Tonnes)	116,941	151,599	99,130	178,247
Double bottom height (m)	3.2	3.4	1.4	1.6
Double side width (m)	2.2	2.6	2.4	3.0
Outer side plate thickness (mm)	17-18	16-21	19	18-20
Inner side plate thickness (mm)	14-18	18-19	14-18	14.5-16.5
Transverse frame space (mm)	2,800	4,105	4,180	4,130
Cargo Tank Dimensions				
L - Length (m)	47.6	41.0		
H - Height (m)	27.7	29.8		
B - Breadth (m)	39.0	44.8		
Tank Diameter (m)			35	46
Approx. Volume of Tank (m^3)	43,504	48,174	22,449	50,965

Each tank is contained within water-tight bulkhead compartments. The most common ship design in use has five spherical tanks, each holding 25,000 cubic meters of LNG. Typical dimensions are length between perpendiculars of 936 ft (285m), beam of 143 ft, depth of 82 ft, and draft (constant) of 36 ft.

All of the ship structure comprising the hulls, bottom, bulkheads, and skirts, are fabricated from non-cryogenic steels, with minimum toughness (resistance to brittle fracture) at LNG temperatures. Thus, exposure of these structures directly to LNG is not a design criterion.

A disadvantage of the Moss system is that the spherical tanks do not fit the contour of the ship's hull and consequently a Moss design will result in a ship that is 10% longer for the same cargo capacity. Since shipyards usually specialize in one type or the other, the selection of the design will, to some extent, dictate the shipbuilder. Moss tank diameter is somewhat limited to about 40 meters due to crane lifting limitations.

A disadvantage of the Moss system is that the spherical tanks do not fit the contour of the ship's hull and consequently a Moss design will result in a ship that is 10% longer for the same cargo capacity. Since shipyards usually specialize in one type or the other, the selection of the design will, to some extent, dictate the shipbuilder. Moss tank diameter is somewhat limited to about 40 meters due to crane lifting limitations.

Figure 4: Free Standing Spherical LNG Tank

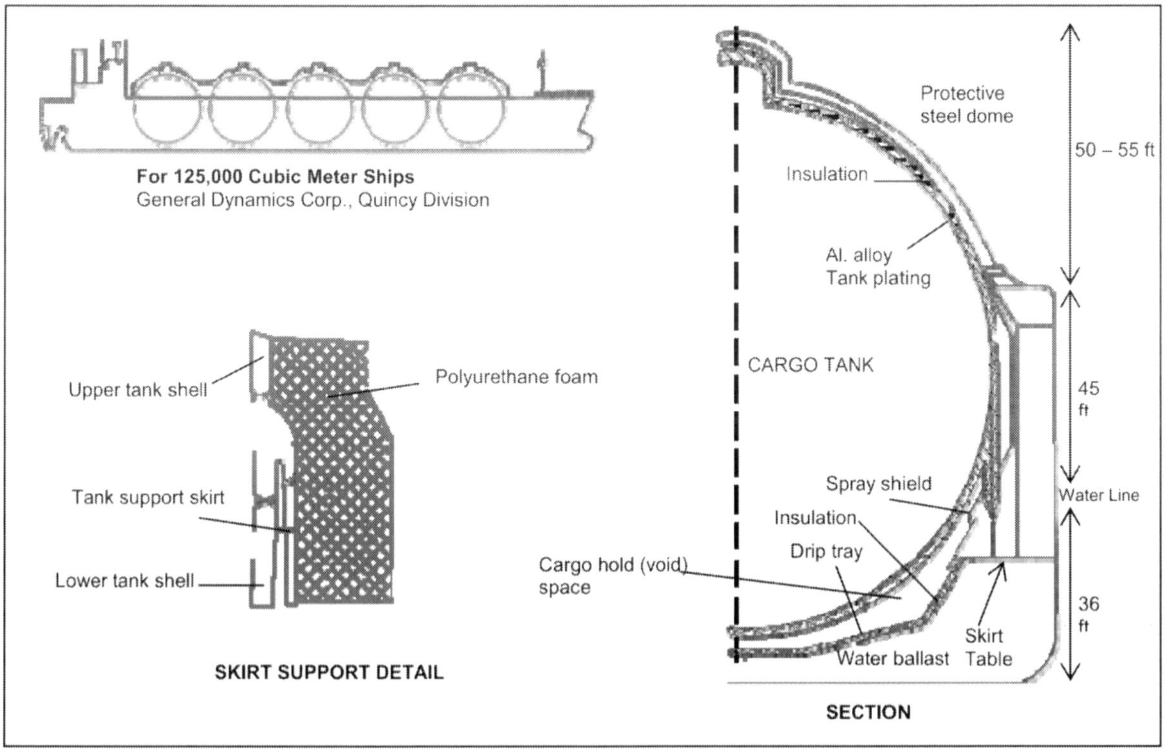

H. Membrane Tanks

The prismatic or membrane tanks are supported by insulation material backed by the inner hull and inner bottom structures (see Figure 4 and Figure 5). The sides and bottom are of double construction, with longitudinal stiffeners and spaced web frames. Watertight bulkheads separate the individual tanks. The double hull and bottom are open between bulkheads to provide space for the water ballast.

A typical design would include 4 to 6 tanks. Typical dimensions would be a length between perpendiculars of 260 meters (850 ft), a beam of 42 meters (138 ft), a depth of 28 meters (92 ft) and a draft of 12 meters (39 ft).

I. Storage Tanks Key Features

Table 5 provides a comparison of key characteristics of the different LNG tank systems. LNG tanks are typically operated at approximately 2 psig independent of carrier type. Membrane tanks are designed for low pressures (3.55 psig max), spherical tanks are designed for 30 psig and prismatic tanks are designed for 10 psig.

Spherical tanks are more difficult to puncture than membrane tanks unless hit at a vulnerable spot or a bigger explosive charge is used. The most vulnerable spot on a spherical tank is the tangent line which is approximately 5 ft below deck or 40 ft above the water line in a typical 125,000 m^3 vessel. In general, spherical tanks are more resilient than membrane tanks as shown in Table 5.

Figure 5: Free Standing Prismatic LNG Tank

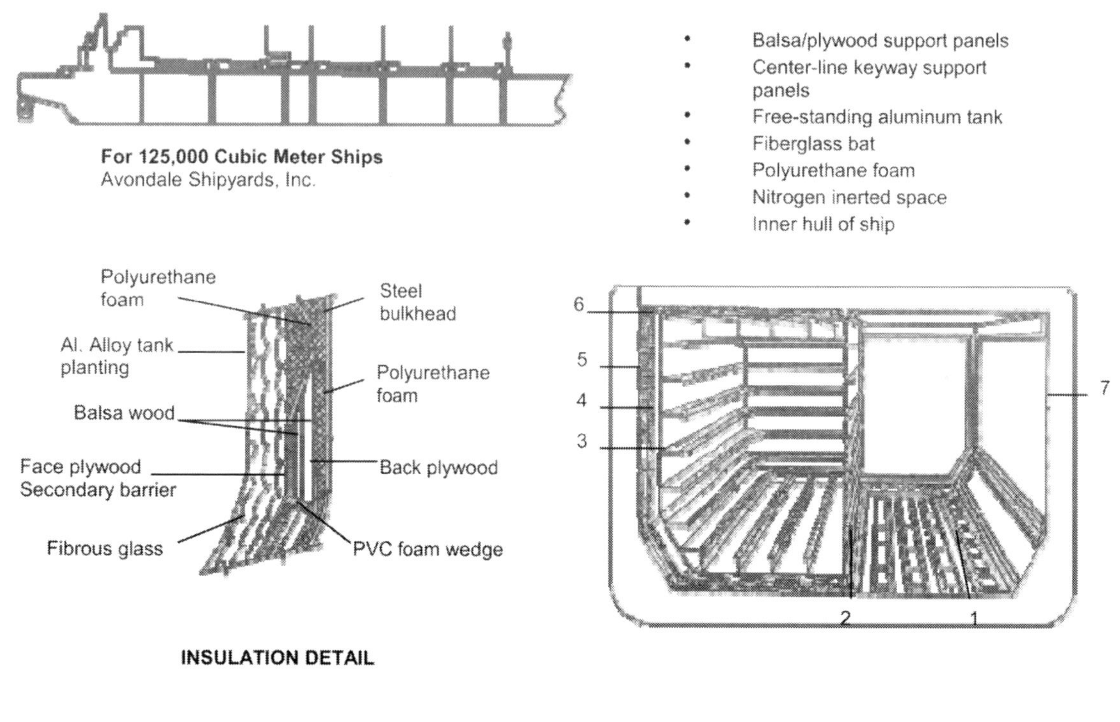

All of the ship structure comprising the hulls, bottom, bulkheads, and skirts, are fabricated from non-cryogenic steels, with minimum toughness (resistance to brittle fracture) at LNG temperatures. Thus, exposure of these structures directly to LNG is not a design criterion.

J. Maximum Impact Scenario Description

For this maximum impact scenario we assume that a terrorist was successful in creating a 1 m diameter hole in the center tank of an LNG tanker at (or slightly above) the water line. The tanker is berthed and fully loaded. We also assume the hole was created using an explosive charge delivered using a small aircraft or other equivalent means and as a result immediate ignition occurs leading to an unconfined pool fire on the water surface. Some experts argue that precise targeting and explosive charge spacing are required in order to cause significant penetration to the inner hull.

Although the maximum impact scenario and the subsequent hazard footprint are developed for a 1 m diameter hole, we also evaluate the impact of larger holes on pool fire diameter and duration of the fire. We also examined the fire behavior for the case of immediate ignition as well as for delayed[10] ignition.

[10] Delayed ignition is assumed to occur immediately after the release ends.

Table 5: Comparative Characteristics of LNG Tank Systems

Characteristics	Free-Standing Tanks		Membrane Tanks
	Prismatic	Spherical	
1. System Designs	Conch Esso McMullen A.G. Weser Hitachi / Esso	Kvaerner-Moss Technigaz Gaz Transport (Cylindrical) Sener	Gaz Transport Gazocean I.H.I (semi-membrane) Bridgestone (semi-membrane)
2. Safety in event of vessel grounding / collision or other emergency	Compared with membrane system less likelihood of hull damage being transmitted to cargo tanks. More efficient use of cubic space.	Safest system in event of grounding or collision – tank structure independent of hull and most void space between vessel hull and cargo tanks. Spherical tanks can be pressurized for emergency discharge in case of cargo pump failure.	Damage to hull of vessel may be more easily transmitted to tank structure than with free-standing tanks. Membrane systems are also more liable to damage or puncture due to causes such as: • surging of cargo in tank • entry of tank for inspection • entry of tank for repair
3. Reliability of Containment System	Most ship years operating experience and most experience without primary barrier failure. Structure can be analyzed and risk of fatigue failures minimized. Tanks can be constructed and 100% inspected prior to installation in vessel.	Tank system easiest to analyze structurally; therefore can be made most reliable.	Structure cannot easily be analyzed and therefore difficult to assure absence of fatigue failures. This could potentially lead to costly off-hire and repair time over the project life.

Figure 6 illustrates the impact of hole size on pool fire diameter from a nominal 25,000 m^3 spherical tank. There are three important observations to be made:

1. Immediate ignition (within 60 seconds) results in smaller pools,
2. As the hole size increases, the maximum pool fire diameter will asymptote,
3. The pool fire resulting from a 1 m diameter hole or larger will be big enough to engulf one entire side of the LNG tanker if the hole is in the center tank[11].
4. The pool fire resulting from a 1 m diameter hole or larger will be big enough to engulf one entire side of the LNG tanker if the hole is in the center tank[12].

This pool fire will be large enough so that it cannot be extinguished. It is prudent to assume that the pool will continue to burn until the LNG supply is exhausted. For a typical 125,000 m^3 Moss spherical vessel the tank inside diameter will be 37.5 m and 9.5 meters will be below the water line. As a result, the release duration from 1 m diameter hole will be approximately 1 hour, and therefore the pool fire will last approximately 1 hour. The associated flame height will be in excess of 200 meters, well above the height of the ship.

[11] Modeling was conducted assuming a wind speed of 2 m/s, F atmospheric stability, and 50 % relative humidity.
[12] Modeling was conducted assuming a wind speed of 2 m/s, F atmospheric stability, and 50 % relative humidity.

If the tanker is a spherical tanker, the cargo hold space will fill with LNG. The hold space is open and the LNG will distribute in the bottom. For either ship design, if the LNG spills into the double sidewall of the hull (ballast[13] areas) and into the water, the tanker might tilt, exposing more of the ship's top to the flames. It is possible for the ship to roll and end up on its side or even sink if enough LNG inventory has been depleted (say 50 %). The initial spill into the hull structure or the cargo hold area would likely produce a brittle fast fracture of the hull steel plates, causing local failure of the hull, the bulkheads, and subsequently the bottom structure.

Concurrently, the fire, impinging on the hull, and radiating onto the top of the ship would be expected to significantly heat the metal structure and cause it to lose its structural integrity. A 20 mm metal plate, the outer hull steel, is calculated to lose 80 % of its tensile strength after approximately five minutes of exposure. Transverse web frames will conduct heat into the inner hull and will create distributed hot spots on the inner hull. This will cause high thermal stress.

It is important to note that for this pool fire engulfment scenario, the relief valves for a typical LNG tank are undersized by at least a factor of 20 if the insulation is not damaged. If the insulation gets damaged by the fire or gets wet then the relief valves would be undersized by a factor of 200. Note that older ships contain non-fire proof insulation. For spherical tankers, the relief valves in the hold space will lift and discharge to the top of the dome/protective cover. These relief valves are only sized for thermal expansion. For either design, the main tank relief valves will also discharge. The discharge will catch on fire causing additional flame radiation contribution being experienced by the top of the tanks.

The bottom of the ship will see cryogenic temperatures while one side of the ship and the top of the ship will see close to flame temperatures.

In this scenario, the failure mechanisms would be expected to propagate along the ship, causing failure of, and spillage from, ALL the adjacent tanks (three or four tanks[14]) near simultaneously. For membrane tankers the center tanks, typically contain 60 % of the cargo. Our expectation is that this complete failure of the ships containment system and of the ship itself would develop quickly, of the order of 5 to 10 minutes due to the initial brittle fracture and to the extreme temperatures from the pool fire.

The insulation on the tanks would not be expected be provide much protection from the intense fire, acting only to prolong the time to failure of the adjacent tankage and of the ship from a few minutes to the expected 5 to 10 minutes stated above.

K. Hazard Footprint Estimates for a Typical 200,000 m^3 LNG Ship

We estimate that the maximum impact hazard footprint for this scenario will result from a pool fire fueled by the simultaneous failure of three or four tanks, releasing approximately 150,000 m^3 of LNG from a 200,000 m^3 tanker. These hazard footprints were estimated assuming a 5 minutes release of 150,000 m^3. The results are shown graphically in Figure 7.

[13] Note that the ballast areas have little water in them when the ship is fully loaded. Water is pumped into the ballast tanks as LNG is discharged to prevent the tanker from rising and stressing the loading arms.
[14] Some new LNG ships with membrane tanks only contain four tanks.

Figure 6: Impact of Hole Size on Pool Fire Diameter from a 25,000 m^3 Tank

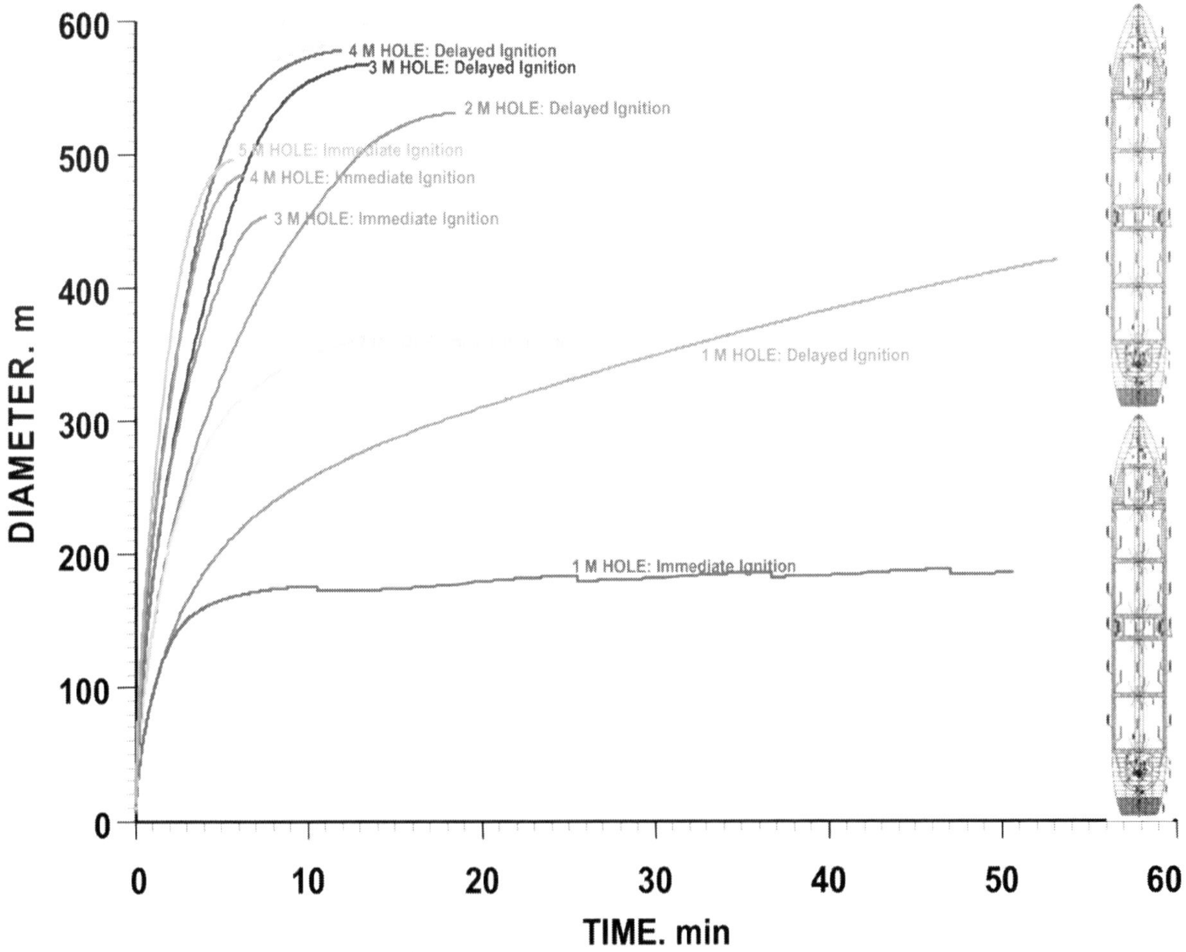

A flame emissive power of 220 kW/m^2 leads to a distance of 3.7 km to the 50 % fatality limit excluding the contribution of solar flux. The % fatality contours shown in Figure 7 should only be used in the context of a detailed risk assessment.

L. Consideration by Regulatory Agencies (NRC, FERC) for Utilizing a Portfolio Approach to Permitting

Most United States regulatory agencies charged with regulating potentially hazardous facilities have a tendency to evaluate each new applicant on a case by case basis. Each new facility that is approved by the agency causes an incremental additional risk to its environment and to the jurisdiction of the regulatory agency. The Nuclear Regulatory Commission was the first regulatory agency to examine risks presented by nuclear power plants and express risk in the form of F-N curves to examine the overall risk for general acceptability. They further had the foresight to anticipate that there may be as many as 100 nuclear plants in the United States and developed design basis criteria and other standards and guidelines so that the risk presented by 100 operating nuclear power plants as measured by F-N curves would be acceptably low. Their pioneering analysis is reported in the Reactor Safety Study (WASH-1400). Today there are 104 operating nuclear operating units at 65 power plant locations in the United States.

Figure 7: Hazard Impact of a Terrorist Attack on a 200,000 m3 LNG Tanker Leading to Pool Fire on Water. Thermal Radiation Hazard Zones expressed using Probits in Percent Fatality.

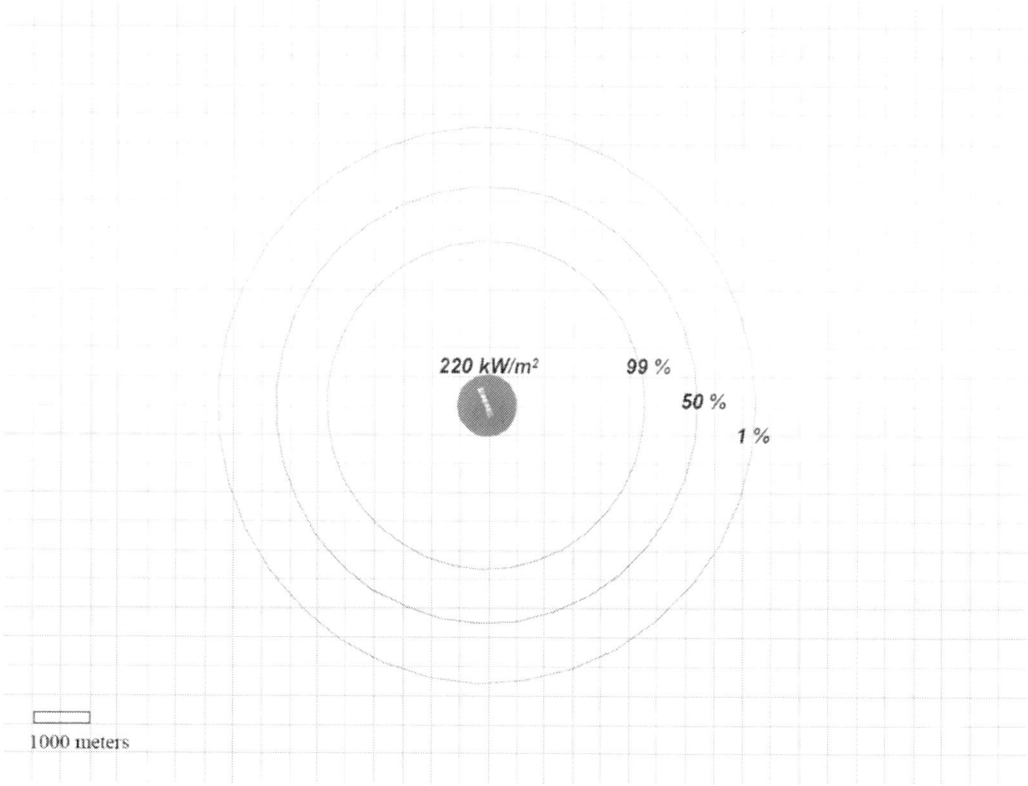

At present FERC does not require a quantitative, F-N style, risk analysis to evaluate applications for onshore LNG facilities, especially those close to highly populated areas. Further, FERC has not, to our knowledge, embarked on an evaluation of what is the total risk presented to the United States from all existing peak shaving and importation LNG facilities and how it changes with each new import terminal that is approved.

A quantitative risk analysis approach to siting approval should be explored. FERC may wish to consider moving from a case by case approval process to considering all anticipated applications for LNG terminals (and peak shaving facilities, if any) and treating them as a portfolio of risk which require management rather than individual projects.

II. Conclusions

Despite the issues raised in this paper regarding the value of flame emissive power, flame height, and pool size for large LNG pool fires, prudent estimates of safe separation distances can be developed for both intentional and accidental release scenarios.

The NFPA-59 thermal radiation criteria should not be confused with and/or considered as a risk acceptability criterion. Hazards are just one aspect of risk. Other important aspects of risk management include operational, economic, social, political, and environmental factors as well as the probability of the occurrence of the hazard itself.

The 5 kW/m^2 limiting criterion does not adequately represent the risks presented by an LNG facility to sensitive population and/or critical areas/buildings. Recognized and peer reviewed vulnerability and/or probit models for establishing the impact dose of thermal radiation such as those developed by TNO represent a better alternative.

References

- "Guidelines for Chemical Process Quantitative Risk Analysis", Second Edition, page 269, American Institute of Chemical Engineers Center for Chemical Process Safety (CCPS), 2000.
- Delichatsios, M. A., "Air Entrainment into Buoyant Jet Diffusion Flames and Pool Fires", Combustion and Flame, Volume 70, Pages 33-46, 1987.
- CPR-16E, "Methods for the determination of possible damage to people ad objects resulting from releases of hazardous materials", First Edition, published by TNO, 1992.
- D. J. Finney. *Probit Analysis*. Cambridge University Press, 1977.
- K. Cassidy and M. F. Pantony, "Major industrial risks - a technical and predictive basis for on and off site emergency planning in the context of UK legislation," Symposium Series No. 110, pages 75-95, Institute of Chemical Engineers, Hemisphere Publishing Corporation, 1988.
- G. Opschoor, R.O.M. Van Loo, and H. J. Pasman. Methods for calculation of damage resulting from physical effects of the accidental release of dangerous materials. In *International conference on hazard identification and risk analysis, human factors and human reliability in process safety*, pages 21–32. Center for Chemical Process Safety, AIChE, 1992.
- G. A. Melhem, A. S. Kalelkar, H. Ozog, and S. Saraf, "Managing LNG Risks: Separating the Facts from the Myths", ioMosaic Corporation Whitepaper, 2006. http://archives1.iomosaic.com/whitepapers/Managing%20LNG%20Risks.pdf
- Congress of the United States Office of Technology Assessment, "Transportation of Liquefied Natural Gas", September 1977.
- United States Government Accountability Office, Report to Congressional Requesters, "Public Safety Consequences of a Terrorist Attack on a Tanker Carrying Liquefied Natural Gas Need Clarification", GAO-07-361, February 2007.
- G. A. Melhem and Henry Ozog, "Managing LNG Risks Training Manual", ioMosaic Corporation, February 2006.
- Nedelka, D., Moorhouse, J., and Tucker, R., "The Montoir 35-m Diameter LNG Pool Fire Experiments", Paper 3 Session III of International Conference on Liquefied Natural Gas: LNG 9 Proceedings, Nice, France, 1989.
- WASH 1400, Reactor Safety Study, Appendix III-Failure Data, WASH-1400 (NUREG-75/014). United States Nuclear Regulatory Commission. October 1975.
- Energy Information Administration, DOE, Nuclear Plants Operating in the United States,http://www.eia.doe.gov/cneaf/nuclear/page/nuc_reactors/reactsum.html. September, 30, 2005.

BIOGRAPHY

Dr. Georges Melhem is president and CEO of ioMosaic Corporation. Prior to ioMosaic Corporation, Dr. Melhem was president of Pyxsys Corporation; a technology subsidiary of Arthur D. Little Inc. Prior to Pyxsys and during his twelve years tenure at Arthur D. Little, Dr. Melhem was a vice president and managing director of Arthur D. Little's Global Safety and Risk Management Practice and its Process Safety and Reaction Engineering Laboratory.

Dr. Melhem is an internationally known pressure relief design, chemical reaction systems, and fire and explosion dynamics expert. In this regard he has provided consulting and design services, expert testimony and incident investigation support and reconstruction for a large number of clients.

Dr. Melhem holds a Ph.D. and an M.S. in Chemical Engineering, as a B.S. in Chemical Engineering with a minor in Industrial Engineering, all from Northeastern University. In addition, he has completed executive training in the areas of Finance and Strategic Sales Management at the Harvard Business School.

Mr. Henry Ozog is a General Partner at ioMosaic Corporation. Prior to joining ioMosaic, Mr. Ozog was a consultant with Arthur D. Little, Inc. for twenty one years, where he managed the process safety consulting business. He also worked for seven years at the DuPont Company as a process and startup engineer.

Mr. Ozog is an expert in process safety and risk management, process hazard analysis (HAZOP, FMEA, FTA), and process safety auditing. He has helped numerous companies and governmental agencies identify process risks and implement cost effective mitigation measures. He teaches courses in each of these areas and is also an instructor for the American Institute of Chemical Engineers' Educational Services.

Mr. Ozog has a B.S. and M.S. in Chemical Engineering from the Massachusetts Institute of Technology. He is a member of the American Institute of Chemical Engineers and serves on various sub-committees for them.

Dr. Ashok S. Kalelkar is a Principal Consultant with ioMosaic. He brings very progressive experience in the area of LNG safety, hazard, and risk management as it applies to LNG Peak Shaving facilities, Import Terminals, Ocean transport and overland shipments by truck.

Prior to joining ioMosaic, Dr. Kalelkar was a senior executive with the firm Arthur D. Little, Inc where he led numerous engagements concerning existing LNG facility safety and operability, proposed new LNG import terminals, state-of-the-art handling of LNG seagoing tankers as well as overland shipments by truck.
The ultimate purpose of most of his work was the sound risk management of the operations being analyzed. He has testified numerous times in public hearings regarding LNG facility siting.

BLEVE Potential due to Marine LNG Vessel Fires

Robin Pitblado
Principal Consultant and Global HSE Lead
DNV (USA) Inc
Houston

AIChE Spring Meeting, April 22-26, 2007
7th Topical Conference on Natural Gas Utilization
Houston, Texas

Introduction

There have been several recent studies assessing hazards associated with marine transportation of LNG [1, 2, 3]. These have all demonstrated large hazards associated with tank leakage or discharge, evaporation, dispersion, ignition, flash fire, and pool fire. These studies did not consider BLEVE events (Boiling Liquid Expanding Vapor Explosions) for marine transport. BLEVE events are well known hazards, often associated with transportation or storage of pressurized flammable liquids. The underlying theory for BLEVE events and several examples are summarized in references [4, 5].

Prof Venart [6], a BLEVE researcher, has questioned the exclusion of BLEVE events from marine hazard studies, specifically quoting the DNV study by Pitblado et al [1]. He noted that on 22 Jun 2002 a truck carrying about 20 tonnes of LNG at -161°C and atmospheric pressure suffered a motor accident at Tivissa, Catalonia, Spain. This accident is analyzed in detail by Planas-Cuchi et al [7]. Before this accident there is no record of an LNG BLEVE onshore, and there has never been a marine accident causing a serious release or thermal consequence of LNG of any type [1, 9].

Prof Venart, in his letter, makes several points:

- He disagrees with industry specialists who claim BLEVE's are not possible with LNG
- LNG vessels may be exposed to long duration high thermal radiation pool fires
- These vessels might not survive long pool fire impingement
- Pressure relief may not be able to limit the pressure rise in a full fire impingement case

He speculates that the tank might BLEVE, involving the entire inventory onboard the vessel, and create a fireball over 2.5km in diameter with a surface emitted power of about 350kW/m2 lasting 54 sec.

The purpose of this paper is to present a response to Prof Venart's published comments and to summarize a more detailed analysis, now in press [8].

Susceptibility of LNG to BLEVE Events

BLEVE events are well known, major hazards. Good summaries of well known BLEVE events and overviews of mechanisms are given in Lees [4] and of the mechanisms in CCPS [5]. A simple definition of BLEVE is given by CCPS.

BLEVE: "any sudden loss of containment of a liquid above its normal boiling point at the moment of its failure. It can be accompanied by vessel fragmentation and, if a flammable liquid is involved, fireball, flash fire, or vapor cloud explosion. "

A fuller BLEVE definition is given in Lees [4] as follows:
"When a vessel containing liquid under pressure is exposed to fire, the liquid heats up and the vapor pressure rises, increasing the pressure in the vessel. When this pressure reaches the set pressure of the relief valve, the valve operates. The liquid level falls as the vapor is released to the atmosphere. The liquid is effective in cooling that part of the vessel wall which is in contact with it, but the vapor is not. The proportion of the vessel wall which has benefit of liquid cooling falls as the liquid vaporizes. After a time, metal which is not cooled by liquid becomes exposed to the fire; the metal becomes hot and weakens and may then rupture. This can happen even if the relief valve is operating correctly. A pressure vessel is designed to withstand the relief valve set pressure, but only at the design temperature conditions. If the metal has its temperature raised, it may lose it strength sufficiently to rupture."

Planas-Cuchi et al [7] describe an LNG road tanker accident that occurred on 22 June 2002 near Tivissa, Catalonia, Spain. They argue that this event exhibited key characteristics of a BLEVE and could thus be the first example of an LNG BLEVE event. Prof Venart quotes this accident to demonstrate that LNG material is potentially subject to BLEVE events. Key facts and tanker design matters are summarized in Table 1 from Ref [7]. There is no prior incident (as demonstrated by the MHIDAS accident database) that has characterized an LNG truck accident as causing a BLEVE.

Table 1. Spanish LNG Tanker Incident

Accident key features	Tanker design matters
• an LNG road tanker rolled over onto its side • the accident dislodged thermal insulation over a space now occupied by vapor • flames appeared immediately between the driver cab and the trailer tank • initially the flames very reported as smokeless • soon after the vehicle tires became involved and both black and white smoke evolved • the tank exploded in two steps 20 min after the initial accident • initially there was a small explosion followed by a hissing sound • subsequently there was a large explosion which created a large white cloud • this white cloud ignited immediately and a fireball ensued • the driver was killed and 2 persons were burned approximately 200m from the truck	• tank dimensions: 13.5m long, 2.33m diameter, volume $56m^3$ • normal fill was 85% liquid (47.6 m^3) and 15% vapor (8.4 m^3) • tank walls – ANSI-304 stainless steel, cylindrical wall thickness 4mm, ends 6mm • insulation – 130mm expanded polyurethane foam, self extinguishing • tank and insulation cover – 2mm aluminum • design pressure 7bar, hydraulic test pressure 9.1bar • LNG cargo conditions (before accident): temperature under -160C, pressure just under 1bar • 5 safety valves – Vapor space: 2 x 1" @ 7bar, 1 x ¾" @9bar, Liquid pipe 2 x ½"@ 10bar • no manhole • heat sources (other than LNG cargo): diesel fuel tank capacity 0.5 m^3, tanker tires, aluminum cover and cab materials

Some debate exists as to the sources of heat for this event. The temperature rise required in the LNG fluid is debatable. Planas-Cuchi suggest the Reid [11] superheat criteria for BLEVE is reached at -117C, a rise of 44C. The LNG fluid vapor pressure is sufficient to lift the relief valves at around -128C a rise of 33C. If the relieved vapors were ignited, this would greatly increase the heat source and temperature, and hence increase the potential for thermal weakening of the steel shell.

A simple scoping heat transfer calculation is presented below. Using the convective heat transfer equation:

$Q = UA\Delta T$

where

U = overall heat transfer coefficient = 570 W/m^2C (Perry's)

A = exposed area (m^2)

ΔT = temperature gradient (C)

Assuming an average temperature seen by the exposed tank surface of 600C (i.e. some hydrocarbon flame at 1100C and some cool sky), the temperature gradient is around 760C. For $1m^2$ exposure this would require 1.6hrs to raise the entire LNG mass to a temperature sufficient to lift the relief valve. An important point is that the entire mass of LNG need not be heated by this amount, heat entering an upper wall surface need heat only that upper layer. The vapor pressure inside the tank will reflect the upper layer properties, not the bulk cooler LNG below. There are 2 unknowns – how much insulation was exposed and how much of the LNG is raised in temperature. As the tank was destroyed in the accident, these cannot be known. However if $2m^2$ of tank were exposed and one quarter of the LNG formed an warmer upper layer then the time required is one eighth = 0.2hr or 12 min. If these values were larger or smaller the time required varies proportionately as per the heat transfer equation.

Once the relief lifts then the pressure in the tank should not rise above 120% of the relief set point and tank failure would be due to thermal weakening of the unwetted steel shell and possibly by the mechanism proposed by Venart. Ignition of the relieved vapor would add to the heat source and thermal input.

Planas-Cuchi et al do not specifically conclude the truck event was a BLEVE. They suggest it meets the criterion for BLEVE as defined by CCPS, but not that of Reid. This may be too formal. The event shared key characteristics of BLEVE events: overpressure, large fragments, fireball and thermal impact, and delay of 20 minutes. These are important factors to be considered by emergency responders, regardless of the whether the event is characterized as a BLEVE or more generally as an ignited explosive rupture. This author agrees with Planas-Cuchi et al and with Prof Venart that this tanker truck event had characteristics of BLEVE involving LNG.

It is worth pointing out that the design of LNG road tankers in the USA is superior in several regards to the Spanish design and makes this event less likely.

Applicability of LNG Marine Vessels

There are two main designs for LNG vessels – membrane and spherical tanks. While these have some differences, they incorporate many additional barriers compared to an LNG truck. A simple count of barriers shows some the comparison.

Table 2. Comparison of Protective Barriers LNG Truck and LNG Marine Vessels

	LNG truck	LNG Membrane Vessel	LNG Spherical Vessel
Threat – external fire	Two threats can cause BLEVE: Pool or jet fire at base or Jet fire from Relief Valve	One threat: Pool fire on water Jet fire from Relief Valve cannot lead to BLEVE on its own	Same as membrane vessel
Barrier 1	Insulation cover – 2mm Al (EU) or steel (USA)	Hull plating – 20+mm steel	Same
Barrier 2	Insulation	Air gap – 2000mm (maybe water if ballast tank)	Same
Barrier 3	Steel wall – 4 to 6 mm	Hull plating – 15-20mm steel, also steel plate above top of tank	Same, but different geometry
Barrier 4	–	Insulation 0.3m – plywood box filled with perlite beads	Air gap due to tank curvature 500mm to 5000mm
Barrier 5	–	Stainless Steel membrane about 1mm	Skirt plating – steel 10+mm (Note 1)
Barrier 6	–	Insulation 0.23m – plywood box filled with perlite beads	Insulation – combustible
Barrier 7	–	Stainless Steel membrane – about 1mm	Tank wall – aluminum 30-65mm
Cargo	LNG	LNG	LNG
Maximum pressure at failure (Note 2)	up to 8barg pressure	0.28-0.30 barg	0.28-0.30 barg

Notes:

1. The situation differs a little for heat entering through the top of the Moss LNG tank as barrier 5 does not apply, but this heat is less (radiant only, not in contact with flames).
2. This assumes operational pressure relief, but relief's are of standard design used in industry where relief's are extremely reliable, and all LNG tanks have multiple relief's

In summary an LPG truck has only 1 barrier against thermal impingement (its steel shell), an LNG tanker truck has 3 barriers, while both LNG marine vessel types have 7 barriers. Every one of these barriers contributes to reducing the risk. A further protection is the limit on internal tank pressure to 0.28-0.30 barg – it will be shown later this is a major limitation on cargo flash and hence BLEVE potential. The initiating events and durations are different and the quality of the barriers are not identical – but they do give an indication that the two situations transport truck vs. marine LNG vessel are quite different.

Table 3. Thermal hazard range for maximum credible events

LNG Marine Event†	250mm hole	750mm hole	1500mm case
Possible cause	Puncture due to striking	Collision / Grounding	Terrorism
Initial discharge rate	226kg/s	2030kg/s	8130kg/s
Duration of event	19hr	2.2hr	0.5hr
Sustainable pool diameter	29m	86m	171m
Thermal hazard range to people	190m	440m	750m

† All punctures above waterline, hole size as diameter, radiation to 5 kW/m^2, wind 3m/s, hazard range measured from center of pool. It is believed that the hazard ranges presented are conservative as the pool flame is likely to be smokier than the luminous assumption used.

Pitblado et al (1) summarized fire threats for typical marine events. These are of a similar order to Sandia (2), but a little smaller in hole dimensions. Duration of exposure is at least as important for thermal escalation as dimension. Large hole events are likely only to be short lasting – as shown by the trend in Table 3. A 5m hole might discharge its contents in only 2-3 minutes. Also the larger fires will most likely be smokey and burn as a mass flame, shorter than traditional pool fire height assumptions. Both of these would significantly reduce thermal escalation potentials.

Can Fire Impact Lead to BLEVE?

Assuming the fire can breach all the ship's structural barriers, and that emergency response is not effective, calculations can demonstrate whether BLEVE is a possible outcome. The hull structure prevents 100% envelopment and the pool fire will only be active on one side of the vessel. The exposed portion of the tank would be well below 10% of the tank area – this is considered in the relief system design as per the IGC Code [10]. The calculation here assumes an instantaneous failure of the LNG tank shell as occurs in a BLEVE event.

Heat from the pool fire, once in contact with the unprotected LNG tank wall, would enter the tank and cause the LNG to boil locally. This analysis assumes all barriers fail somehow – this is unrealistic, but it allows this scoping calculation. Over a short period of time, this will be sufficient to lift the LNG tank pressure relief valves – set for 0.25barg. Allowing a tank maximum pressure of 0.28-0.30 barg it can be calculated that the flash amount on relieving to atmospheric pressure would be 2.4% (i.e. 2.4% of the slightly pressurized LNG liquid would flash to vapor). In practice much less than the whole tank would flash due to the hydraulic head of the LNG liquid. Commercial LNG has a range of densities depending on the quantity of ethane and heavier components, for conservatism a lower range density for nearly pure methane under half that of water is used. For this, each 1m of liquid exerts a static pressure of 0.045barg. Thus at a depth of about 6m the pressure rises to 0.28barg and no flash will occur. Thus the flashing is restricted to the top 6m and averages half the flash at the surface, or 1.2% (i.e. 2.4% at the surface and 0% at 6m depth). In mass terms 1.2% of the top 6m corresponds to 38 tonnes of LNG and might create a sphere of 100% methane vapor about 34m diameter. This is similar in scale to a transport accident and would create a hazard zone of 200m, less than the hazard zone from the pre-existing pool fire. A figure showing the process is given in Figure 1.

This is about the same size as the tank itself and is the same order of size as a propane tanker truck and its consequences might be expected to be similar in terms of thermal radiation from the immediately ignited gas puff. There should be virtually no overpressure as there is no massive tank contents flash expanding at near sonic velocity. There is no confinement at the top of the vessel that could lead to what is known as a partially confined vapor cloud explosion.

Figure 1. LNG Heating Scenario (assume all vessel barriers defeated)

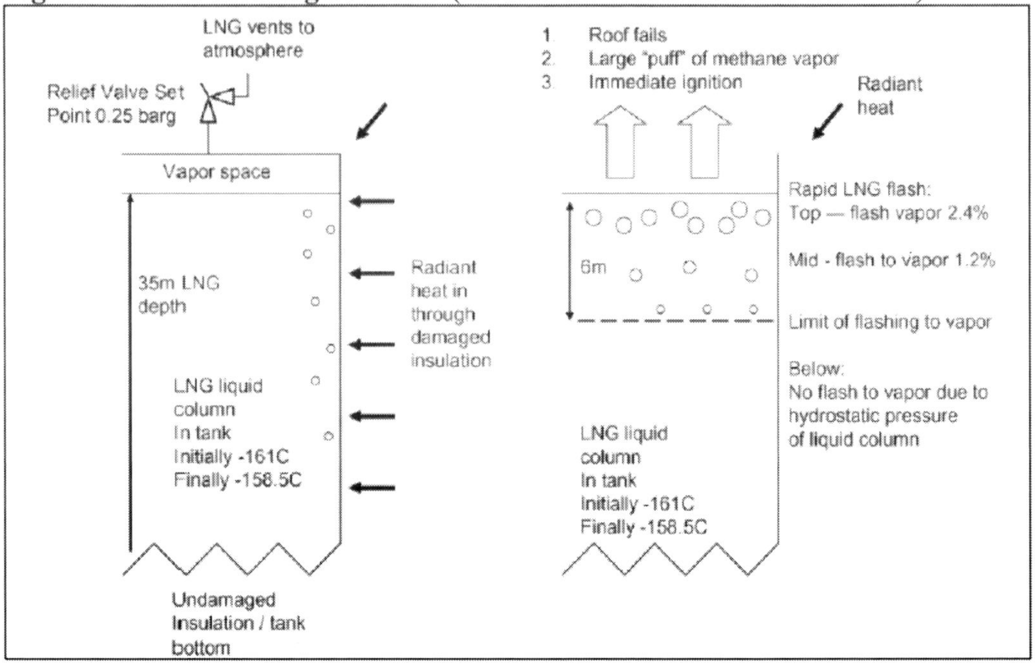

Spherical tanks might in theory be subject to a large BLEVE if their relief system failed as the spherical tank has greater internal strength to be self-supporting, but this would require multiple simultaneous failures of the double or triple relief system, conceivable but virtually unheard of in the process industry. Also tanks are interconnected in the vapor space for boil-off gas collection and this further protects against a relief failure by allowing venting though other tanks.

Conclusion

The Spanish LNG truck incident has been reviewed and this shows that LNG fluid in certain circumstances can give rise to BLEVE-like outcomes.

This paper has reviewed a serious pool fire event affecting an LNG vessel. This would damage the vessel in as yet unproven manner – but a reasonable sequence of events can be hypothesized. This would occur over a significant period of time, of the order of hours, and normal emergency response should ensure no public remains in any significant hazard zone. The pool fire event is serious, but it has been hypothesized that a BLEVE outcome is possible and this is even more serious in consequence zone, up to 2.5km.

This paper has reviewed the potential for a BLEVE of the damaged tank or one of its undamaged neighbors. While tank failure is possible, it would not give rise to a major BLEVE of the type seen in Mexico City. The design of the tank and its relief system limits the flash of slightly pressurized LNG (0.28 - 0.3 barg) to only 1.2% of the upper 6m of tank. Thermodynamic calculations show this could cause only a cloud of 38 tonnes of methane vapor, or a sphere of 34m diameter. This is similar in scale to a transport accident and would create a hazard zone of 200m, less than the hazard zone from the pre-existing pool fire.

Undoubtedly a major LNG accident or terrorism event will give rise to serious dispersion or thermal hazards from the resultant pool fire but these should be localized to the area around the ship and out to several hundred meters. However, a BLEVE event causing a thermal hazard zone in excess of 2.5km is not credible. There is no mechanism for the tank to reach a high enough internal pressure to result in a BLEVE which the design has not eliminated.

As prior studies have noted [1,2], there are many uncertainties associated with accidental events to large marine LNG vessels. Some selected experimental or theoretical calculations would be beneficial for to better establish vessel structural response to external pool fire events.

This paper has demonstrated that the BLEVE potential of large LNG marine vessels, with properly sized relief valves as specified in the applicable IGC code, is greatly reduced in consequence due to design considerations and relief, well below the scale of the initiating pool fire event.

References

1 Pitblado R, Baik J, Hughes G, Ferro C, Shaw S (2005) Consequences of LNG Marine Incidents, AIChE Process Safety Progress 24, 2, pp 108-114.

2 Hightower M, Gritzo L, Luketa-Hanlin A, et al (2004) Guidance on Risk Analysis and Safety Implications of a Large Liquefied Natural Gas (LNG) Spill Over Water, Sandia National Laboratories, SAND2004-6258.

3 ABS / FERC (2004) Consequence Assessment Methods for Incidents Involving Releases from Liquefied Natural Gas Carriers, GEMS 1288209, May 13 2004, and Staff Responses to Comments on Consequence Assessment Methods, Docket AD04-6-000, 39pp, June 18, 2004.

4 Manan S (2005) Lees Loss Prevention in the Process Industries, Third Edition, Elsevier.

5 CCPS (1994) Guidelines for evaluating the characteristics of vapor cloud explosions, flash fires and BLEVEs. Center for Chemical Process Safety, AIChE, New York.

6 Venart J (2005) Letter to the editor, published AIChE Process Safety Progress.

7 Planas-Cuchi E, N Gasulla, A Ventosa, and J Casal (2004) Explosion of a road taker containing liquefied natural gas, J Loss Prev in Process Ind 17, p315-321.

8 Pitblado R, Potential for BLEVE associated with marine LNG vessel fires, Journal of Hazardous Materials 140 (2007) 527–534.

9 Vaudolon, Alain (2000) Liquefied Gases: Marine Transportation and Storage. Witherby & Company Limited, London.

10 IMO (1993) International Code for Construction & Equipment of Ships Carrying Liquefied Gases in Bulk (IGC), Third Edition, Intl Maritime Organization, London.

11 Reid R C (1976) Superheated liquids. American Scientist 64, 146-56 (also discussed in CCPS 1994)

Acknowledgements
Part of this work has been supported by DNV and by Lloyd's Register. Useful comments were provided by Ernst Andre Meyer of DNV and Gordon Milne of Lloyd's Register.

Biography – Dr Robin Pitblado

DNV (USA) Inc, 16340 Park Ten Place, Suite 100, Houston TX 77084, USA
Tel: 1-281-721-6600 E-mail: robin.pitblado@dnv.com

Dr. Pitblado is a Principal Consultant with DNV, based in Houston, and with global responsibility for DNV in the HSE Service Area. He has been actively involved in risk assessment activities for 25 years. He was the technical lead on DNV's recent Joint Sponsor Project assessing hazards of LNG marine transport.

Dr. Pitblado worked initially with Technica and later DNV on a wide range of risk assessment projects while based in London, Los Angeles and Houston. This included several liquefied gas facilities including LNG, LPG and ammonia. This included major studies in Hong Kong, Australia, Europe and in the Middle East. He has carried out integrated assessments of safety management systems, mechanical integrity systems, and risk assessments. He has participated in 6 major accident investigations, including the P36 and several other offshore accidents. He developed a risk knowledge management system for the risk discipline in DNV.

He spent a year with Lloyd's Register in Sydney working mainly on barrier approaches to risk management for the local railway company.

Dr. Pitblado graduated from McGill University with a degree in Chemical Engineering and later a PhD in Chemical Engineering from Sydney University. He is a Fellow of the Institution of Chemical Engineers, a registered safety professional, and a member of AIChE.

Fully Integrated GIS based LNG Consequence Model for Spills on Water

Nicole Whittier[1], Malcolm L. Spaulding[1,2], J. Craig Swanson[1], and Kathy Jayko[1]
[1] Applied Science Associates, Inc,
70 Dean Knauss Drive, Narragansett, RI 02882
[2] Ocean Engineering, University of Rhode Island
Narragansett, RI 02882

AIChE Spring Meeting, April 2007
7th Topical Conference on Natural Gas Utilization
Houston, Texas, April 22-27, 2007

Abstract

LNGMAP, a fully integrated geographic information system (GIS), has been developed to predict the fate and transport of spills of LNG on water. The model contains separate algorithms that are linked together to represent processes that affect LNG once it is released on water. LNGMAP uses spill geometry, environmental data forcing (currents and wind) and physical-chemical properties to simulate fate processes over time after the release, including: (1) time dependent release rate, (2) spreading, (3) transport on the water surface, (4) evaporation from the water surface, (5) transport and dispersion in the atmosphere, (6) and, if ignited, burning and associated radiated heat fields. LNGMAP is a Lagrangian model, when LNG is released it is broken into many tiny individual masses of LNG called spillets. These fate processes then act on each individual spillet. At each calculation time the spillets are integrated together to report the temporal evolution of surface thickness of LNG, vapor concentrations in the atmosphere and thermal intensity on a map-based display.

Due to the lack of field experiment using significant amounts of LNG, LNGMAP was validated by comparisons to predictions of models developed by ABS Consulting and Sandia for time dependent point releases from a draining tank, with and without burning. To illustrate the model predictive capability for realistic emergency scenarios this paper presents results for a tanker entering Block Island Sound. Three hypothetical cases were studied: the first assumes the vessel continues on course after the spill starts, the second that the vessel stops as soon as practical after the release begins (3 minutes), and the third that the vessel grounds at the closest site practical. This model output can be used by decision makers to help select the best course of action to minimize consequences.

1. Introduction

As more natural gas reserves are found in increasingly remote areas, a problem facing the gas industry is how to transport the gas to its end users. Historically, the most common mode of transportation has been through pipelines. However, with large transit distances involved, transport by ship to

Liquefied natural gas (LNG) terminals is the only cost effective alternative. These ships have an excellent safety record and provide a critical link in transporting LNG from production facilities to customer locations. Systematic safety standards and protocols exist for analysis of risk of LNG spills or releases from storage terminals and facilities on land. There are no equivalent standards or guidelines for the evaluation of safety or the consequence of LNG spills in marine waters (Sandia, 2004). As a first step in addressing this issue, ABS Consulting (2004), under contract to the Federal Energy Regulatory Commission (FERC), and Sandia Laboratory, (2004), under contract to the US Department of Energy (DOE), independently reviewed the consequence assessment methodologies that are available to model releases of LNG spills on water. Both studies found that none of the existing models provided a comprehensive tool to assess the consequences of LNG spills. To address this, ASA has developed a state of the art, integrated model system to predict the release, transport, fate (evaporation, dispersion, and burning) and impact of LNG spills in marine waters. An algorithm-based modeling strategy is used that relies on the best understanding of the key transport and fate processes. This is the strategy recommended by ABS Consulting (2004) and the approach used in the model systems they reviewed.

2. Modeling System

2.1 System Overview

LNGMAP, a fully integrated geographic information system (GIS), has been developed to predict the fate and transport of spills of LNG on water. The model contains separate algorithms that are linked together to represent processes that affect LNG once it is released on water. LNGMAP uses spill geometry, environmental data forcing (currents and wind) and physical-chemical properties to simulate fate processes over time after the release, including: (1) time dependent release rate, (2) spreading, (3) transport on the water surface, (4) evaporation from the water surface, (5) transport and dispersion in the atmosphere, (6) and, if ignited, burning and associated radiated heat fields. LNGMAP is a Lagrangian model, when LNG is released it is broken into many tiny individual masses of LNG called spillets. These fate processes then act on each individual spillet. At each calculation time the spillets are integrated together to report the temporal evolution of surface thickness of LNG, vapor concentrations in the atmosphere and thermal intensity on a map-based display.

To validate the particle-based implementation of the individual process algorithms, simulations were performed for case examples presented in other simulation studies (ABS Consulting, 2004; Sandia, 2004). Ideally model validation would use experimental or actual spill data but no data sets exist for large spill events. This validation exercise showed that the present implementation of a particle based strategy gives results that are the same or consistent with more conventional semi-analytic techniques.

More details on the model development strategy, algorithms for the linked process submodels, and validation are found in Spaulding et al.

2.2 Model input

Model inputs are kept to a minimum to simplify the specification of a scenario. The physical and chemical characteristics of LNG and associated rate parameters required for the model are stored in a database to facilitate data organization and updating. Only a few parameters are required to define the LNG release and environmental conditions. These include:

- The release is specified as one of the following:
  - A constant release rate (user specifies rate and duration of release)
  - A time dependent release rate controlled by the tank characteristics, initial depth of fluid in tank, size of tank rupture, and discharge coefficient. The rupture is assumed to be at the water surface.
- The spill location. The location may be stationary or moving. If moving, the vessel's path and speed are required.
- Surface currents, wind speed/direction, sea surface/air temperature, and relative humidity. (The impact of waves on surface transport is approximated through a wind/wave induced drift algorithm that is based on wind speed and direction).

2.2 Model output

The results of a simulation can be displayed in several ways, including:

- Temporal evolution (animations) of spatial distributions (2D) of LNG on the sea surface and in the atmosphere over the simulation period. Given the discrete particle nature of the model, contour maps for liquid gas on the sea surface or gas dispersed into the atmosphere are generated by overlaying a grid system covering the extent of the particle dispersion and interpolating the individual particle masses onto the grid system.
- Time dependent mass balance of LNG for both liquid and gaseous states in various environmental compartments (source, sea surface, atmosphere/burned).
- Time dependent contours of LNG vapor concentration values in the atmosphere with time, highlighting the lower flammability limit (LFL) (5% concentration), upper flammability limit (UFL) (15% concentration), and human effects thresholds.
- Time dependent contours of radiated heat flux in the event of fire, highlighting contours of 38, 25, 12, and 5 kW/m^2 (spanning the range of damage to equipment, ignition of wood structures, melting of plastic and fatalities, and causing skin burns).
- Thermal radiation time history at user-selected locations. Impact on individuals, structures, materials and facilities is best assessed in terms of thermal dose received (thermal radiation level integrated over time).

3. Application of Model System

3.1 Application Overview

A major feature of LNGMAP is its ability to account for the movement of the vessel and the currents onto which the LNG was spilled. This ability allows more realistic pool shapes, vapor clouds and radiated heat zones. To illustrate some of the key features of LNGMAP related to its ability to handle a time varying release from both a stationary and moving source and to incorporate environmental forcing, simulations of three hypothetical scenarios for a vessel entering Block Island Sound, in the vicinity of Montauk Point, and headed toward Long Island Sound (Figure 1) were performed. These hypothetical scenarios are unrelated to any proposed LNG terminals in southern New England.

Three hypothetical cases were studied: (1) the vessel continues on course traveling at 7.5 m/sec after the spill starts (Case 1), (2) the vessel stops as soon as practical after the release begins (3 minutes) (Case 2), and (3) the vessel grounds at the closest site practical (Case 3). This model output can be used by decision makers to help select the best course of action to minimize consequences. In all cases the hole size is 1 m in diameter. The simulation time step was 0.1 sec. The winds are assumed constant from the southwest at 5 m/sec and the Pasquill stability category was moderately unstable (Class B), typical of summer conditions. Tidal currents are provided by a hydrodynamic model of the study area (Ward and Spaulding, 2002) and vary both spatially and temporally. The total volume spilled in each case is 12,500 m3, with an initial tank head of 13 m above the release point. The release is complete over a period of 33.3 minutes. Two series of simulations were performed assuming: (1) the LNG on the sea surface ignited immediately and burned completely and (2) the LNG evaporated and dispersed into the atmosphere (without burning).

Presented below are graphics showing the sea surface area covered by the LNG pool, the radiation contours if LNG is assumed to burn, and the concentrations of LNG in the atmosphere (10 m elevation) if the vapor disperses from the spill. The footprint of the LNG pool, the dispersed cloud and thermal radiation, and the incident energy flux are shown versus time for each case.

3.2 Area of LNG on sea surface

Figure 2 shows the model predicted area of LNG on the water surface over the course of the simulation for the three cases. The graphics were created by overlaying multiple model results, at 4 minute time intervals, during the simulations. They hence constitute snapshots as the release unfolds. The upper panel shows the on-course Case 1 with an insert is provided to allow visualization of the thickness contours. The lower panel provides the results of all cases overlaid to facilitate comparisons. For both the moving cases (Case 1 and 3), the LNG forms an elongated plume behind the vessel. The thickness of the LNG is greatest in the immediate vicinity of the moving source and decreases with distance behind the vessel. The plume width is controlled by the tank release and evaporation rates and the vessel speed. Both cases show a similarly sized area given the fact the vessel and wind speeds and the tank release rate are the same. If the vessel stops however the area is circular in shape and covers a much large area in the immediate vicinity of the stopping location.

3.3 Area of LNG thermal radiation field

Figure 3 shows the model predicted thermal radiation contours highlighting 5, 12, 25, and 38 kW/m2, spanning the range from skin burns, melting of plastic and fatalities, ignition of wood structures, and damage to equipment, respectively, over the course of the simulation for the three cases. The upper panel shows the on-course Case 1 while the lower panel provides the results of all cases overlaid to facilitate comparisons. The results are shown at 3-minute intervals over the time period of the release. For Case 1 the impacted area for a given thermal radiation contour decreases in size with time until eventually disappearing 35 minutes after the release begins. This behavior is a direct result of the declining release rate with time from the vessel. If the vessel stops (Case 2) the radiation field is restricted to the immediate vicinity of the vessel and of course impacts a larger area than for the moving cases. If the vessel is grounded (Case 3), the radiation field along the vessel track shows similar behavior to the continuing on course example (Case 1) while underway and reverts to the stopped case (Case 2) when the vessel eventually grounds. In Case 3, the tanker is still releasing LNG when it grounds and hence a circular radiation field is seen at the end of the vessel track. A close inspection of the figures shows that if the vessel is in motion the radiation contours become elliptical, with the major axis oriented in the direction of vessel motion.

3.4 Area of LNG dispersed

Figure 4 shows the model predicted vapor cloud area and associated concentration contours highlighting the lower flammability limit (LFL) and upper flammability limit (UFL) concentrations, 5 and 15%, respectively, over the course of the simulation for the three cases. The upper panel shows the on-course Case 1 while the lower panel provides the results of all cases overlaid to facilitate comparisons. The results are shown at 3-minute time intervals over the period of the release. For all cases the vapor cloud is transported and dispersed down wind from the release location. The highest concentrations are in the core of the plume with lower concentrations at the margins, and particularly at the downwind end of the plume. The length of the plume is controlled by the release strength, the vertical dispersion coefficient, and degree of stability in the atmosphere. The vessel's forward speed and direction and the wind speed and direction establish the orientation of the vapor cloud to the vessel track. This is nicely illustrated in the on-course case (Case 1) shown in the upper panel of the figure by the change in the orientation of the vapor cloud when the vessel changes from a north-northwesterly to west-northwesterly direction. It is interesting to note that the vapor cloud retains a memory of the change in vessel course (compare orientation of the vapor plume near the vessel track and at the downwind end of the plume after the change of course has occurred). The last two vapor cloud snapshots are detached from the vessel track since the release has ceased and the cloud is being transported away from the vessel by the wind. It is also noted that the last release shown has lower concentrations in the core than its predecessors. This is due to the substantially reduced release rate from the tank near the end of the discharge. The swept area for the stopping case (Case 2) is essentially downwind given the vessel's initial northerly motion and then its fixed location at the stopping point. The plume length increases once the vessel stops. The grounding case (Case 3) has characteristics similar to the on-course case (Case 1) when the vessel is underway and of the stopping case (Case 2) once the vessel grounds. Comparing the three cases, if the vessel continues to move either on course or to a grounding location, the LNG vapor concentrations are lower but distributed over a much wider area than if the vessel stops.

4. Conclusions

A model to predict the transport and fate of marine spills of LNG has been developed within a geographic information system framework. The model allows for releases from both stationary and moving point sources. The latter can be specified by a user defined vessel track. To illustrate the model predictive capability for realistic emergency response scenarios, simulations were performed for a tanker entering Block Island Sound.

The model results have important implications in vessel response during a spill event where the relative benefits of keeping the ship moving must be balanced against the larger swept area of the vapor cloud and / or radiation field. Contingency planning along the ship track may result in different response strategies depending on location. The addition of real time or forecasted meteorological and oceanographic information provided directly to the ship's command center may be important in determining the appropriate response to minimize the hazard.

LNGMAP is designed to be a planning tool that provides more realistic spill consequences than the simple, static approaches found in the ABS (2004) but yet is a much more efficient PC-based calculation than a full CFD approach. Given LNGMAP's modular, algorithm based designed it is ideally suited to incorporate new algorithms as they became available and can readily be used to perform sensitivity analyses. The global re-locatability of the system and ability to link to world wide oceanographic and meteorological models facilitate application to any location.

5. References

M. L. Spaulding, J. C. Swanson, K. Jayko. and N. Whittier, "An LNG release, transport, and fate model system for marine spills," *J. of Hazardous Materials*, in press.

ABS Consulting Inc., "Consequence assessment methods for incidents involving releases from liquefied natural gas carriers," Report prepared for Federal Energy Regulatory Commission, contract FERC04C40196, May 2004.

Sandia National Laboratories, "Guidance on risk analysis and safety implications of a large liquefied natural gas (LNG) spill over water," Sandia Report SAND2004-6258, December 2004.

M. Ward and M. Spaulding, 2002, "A nowcast/forecast system of circulation dynamics in Narragansett Bay," *Proc. of 7th Int'l Conf. on Estuarine and Coastal Modeling*, sponsored by the University of Rhode Island Conference Office, St Pete, Florida, pp.1002 – 1022, November 5-7, 2001.

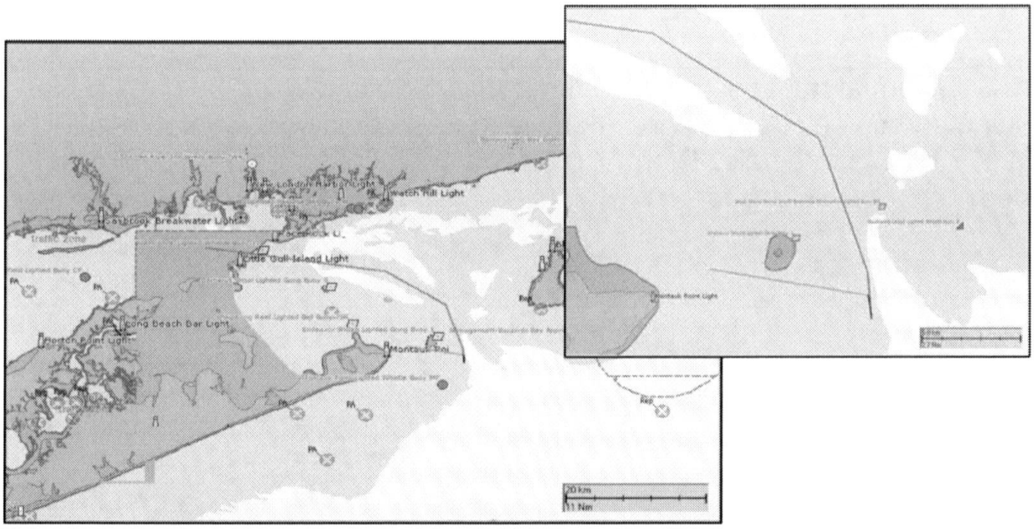

Figure 1. Block Island Sound study area showing the path/location of the LNG tanker if the vessel continues on course (Case 1), if it stops (Case 2), and if diverts its path to intentionally ground (Case 3).

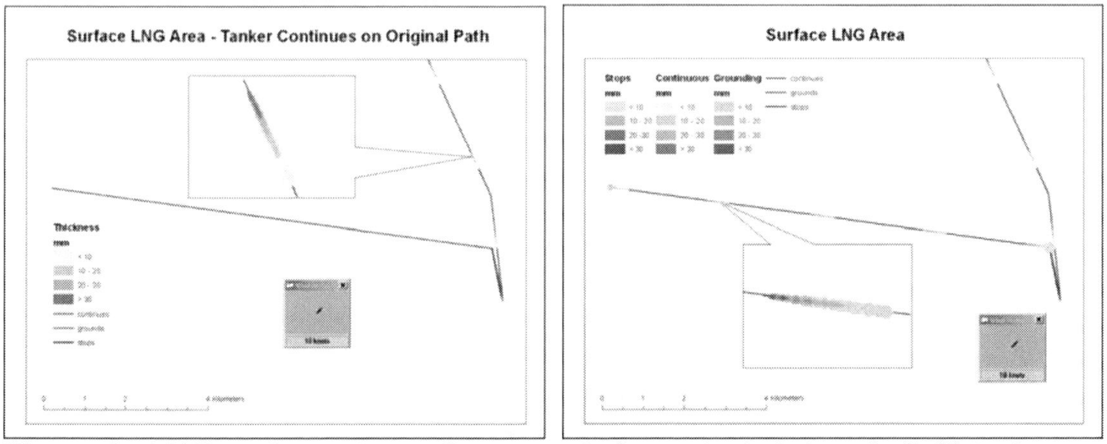

Figure 2. LNG on the water surface area in four minute intervals (left) tanker continuing on course (Case 1), and (right) all cases combined. Thickness

contours for each case are provided. The insert shows a close-up view at one time increment.

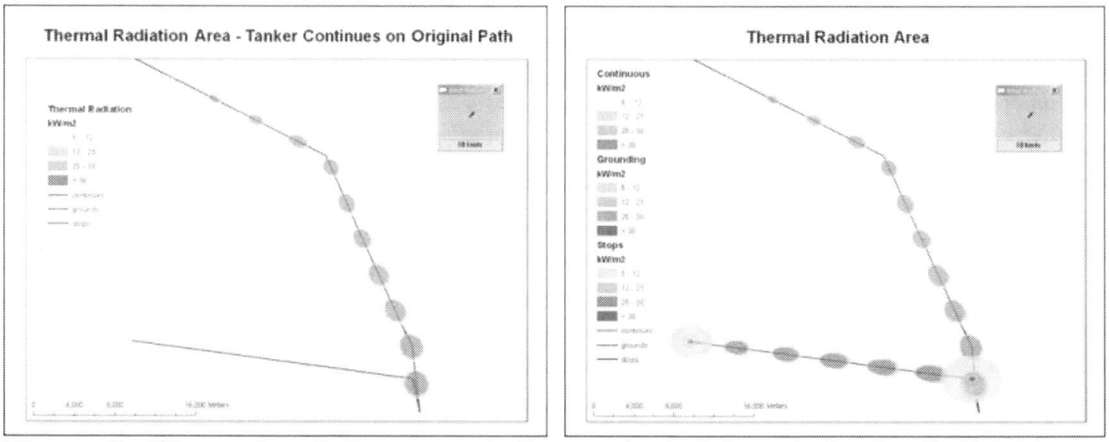

Figure 3. Thermal radiation areas in three minute intervals for the (left) tanker continuing on course (Case 1) and (right) all cases combined. Contours are shown at 3 minute intervals. Thermal radiation contours for each case are provided.

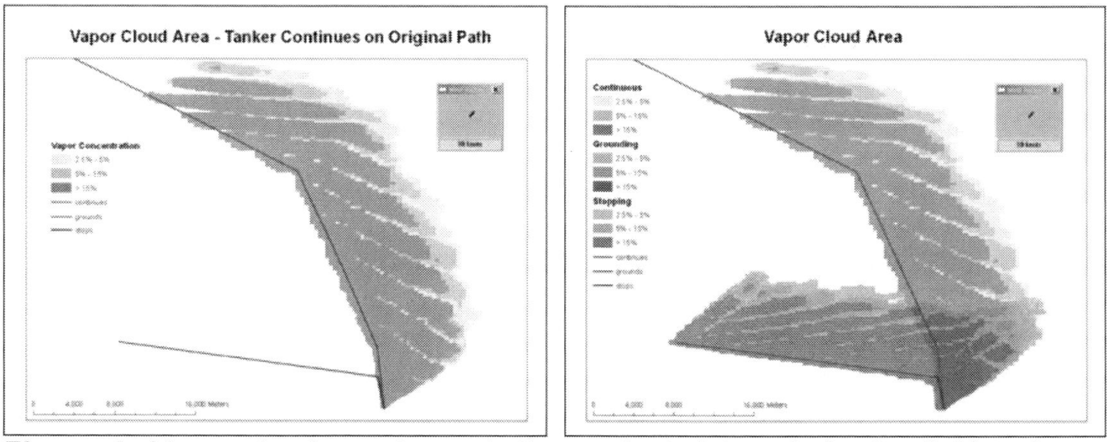

Figure 4. Vapor cloud swept area in three minute intervals for the (left) tanker continuing on course (Case 1), and (right) all cases combined. Contours are shown at 3 minute intervals. Vapor contours for each case are provided.

Biography

Ms. Nicole Whittier, a chemical engineer at ASA, is an expert in modeling the trajectory, fate, and impacts of chemical substances and oil. Her primary responsibilities are evaluating chemical processes and the fates of chemical pollutants to continue with the ongoing development of ASA's suite of water and atmospheric dispersion models. She also focuses on emergency response, data management, and GIS analysis.

Ms Whittier received her Bachelor of Science in Chemical Engineering from the University of Rhode Island and will receive her Masters in Science in Environmental Engineering in December 2007. Currently, she is President of Rhode Island's American Institute of Chemical Engineers chapter and the Rhode Island Society of Environmental Professionals.

Ms Whittier's experience in environmental effects of chemical pollutants and engineering design gives her an exceptional insight in solving realistic environmental problems.

Her e-mail address is: nwhittier@appsci.com

Simulation of Pt/Ni Stratified Bed Catalysts for the Production of Hydrogen by Catalytic Partial Oxidation of Methane

Christa J. Bell
McGill University
Montreal, QC H3A 2B2
christa.bell@mail.mcgill.ca

Corey A. Leclerc
McGill University
Montreal, QC H3A 2B2
corey.leclerc@mcgill.ca

Abstract:

A catalyst consisting of a platinum combustion bed followed by a nickel reforming bed achieves high hydrogen yields with little degradation at a cost that is fractions of the cost of the conventional rhodium catalyst. It has been shown as a proof-of-concept where the platinum and nickel beds were of equal size. This work investigates the effect of relative bed lengths of the platinum and nickel catalysts. Since the oxygen is consumed in millimeters, a platinum catalyst that is much smaller than the nickel catalyst is sufficient to carry out the partial oxidation reaction. Due to the number of process variables, an infinite amount of experiments could be carried out in order to optimize the relative bed lengths. Instead, simulations could be carried out to reduce the financial and time constraints of doing so many experiments. As a first step towards simulation, an elementary step surface reaction mechanism has been developed for reactions of methane on nickel. A screening mechanism was generated from thermodynamic constraints, refined, and validated against experimental results to achieve a working mechanism.

Near-Term Approaches to Nuclear Hydrogen Production

Charles O. Bolthrunis, Shaw Energy & Chemicals, One Main Street, Suite 900, Cambridge MA 02142 aqnd Reiner Kuhr, Stone & Webster Management Consultants, Inc., One Main Street, Suite 900 , Cambridge MA 02142

With the commercialization of the Pebble Bed Modular Reactor (a high-temperature gas-cooled reactor) in South Africa within the next 6 to 7 years and the prospects for demonstration of hydrogen production with water-splitting technology at the NGNP (Next Generation Nuclear Plant) at the Idaho National Laboratory soon thereafter, the path is opening for the use of advanced nuclear reactorsas a source of high-temperature process heat. Steammethane reforming is both an important user of high temperature process heat and currently the dominant commercial method for producing hydrogen. Steam methane reforming using heat from an advanced nuclear reactor is under development and may become an important commercial technology for hydrogen production late next decade. This hyudrogen production technology can serve as an interim strategy for displacing natural gas consumption with nuclear heat and provide an option for large-scale nuclear hydrogen production prior to commercialization of water-splitting technologies.

This paper examines the characteristics of the nuclear heat source, the reforming application, and their integration into a practical system. It also describes the implementation of these innovations in a nuclear hydrogen plant design, identifies plans for key technology development and demonstration steps and discusses the projected markets and economics of this technology.

The advantages of using nuclear energy as a heat source for steam methane reforming lie chiefly in the elimination of natural gas fuel and of the associated carbon dioxide emissions. By investing in a nuclear heat source, the user pre-pays for forty years' supply of natural gas and the associated carbvon dioxide emissions. At $7.00/MMBtu gas in 2007 (Figure 1), the present value of 30-years supply of the gas alone for process heat is $1.55Billion. At $30/ Ton for CO_2 emissions avoidance (Figure 2), another $300 million is added to the present value. The incentive for using a nuclear heat source is large; but there are challenges as well.

In the production of hydrogen and most syngas, the endotherm for the steam methane reforming reaction is supplied in the radiant section of a gas-fired furnace. The flame temperature in the combustion chamber is much higher than the required process temperature and the main design an operating problems revolve around limiting the reformer tube wall temperature and controlling the heat flux profile along the tubes. Control is accomplished by altering the firing rate.

The characteristics of a high-temperature gas-cooled nuclear reactor (HTGR) heat source are very different from a gas-fired furnace and its use in steam methane reforming poses certain design challenges. As implied by its name, an HTGR delivers thermal energy to the user by means of circulating a gas through the reactor core and an intermediate heat exchanger in a closed loop. A second heat transfer loop is used to deliver the heat to the ultimate user. This arrangement avoids the liklihood of contaminating either the nuclear

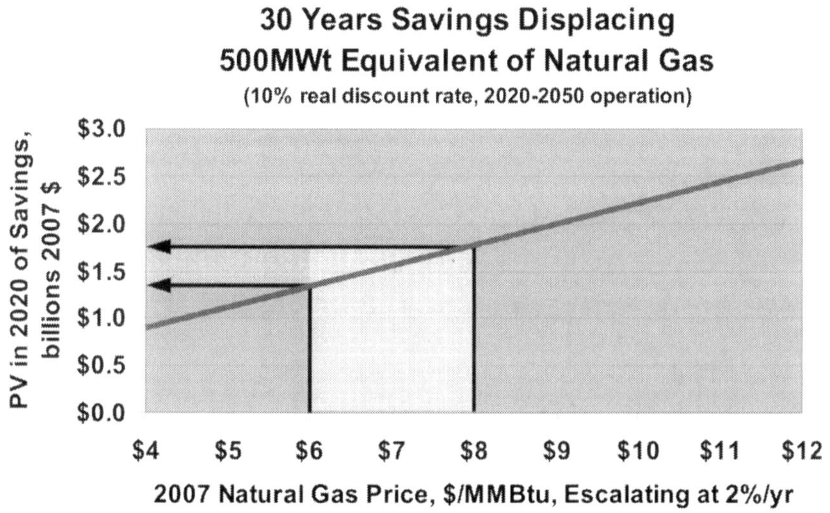

Figure 1 -- PV of Natural Gas Fuel Savings

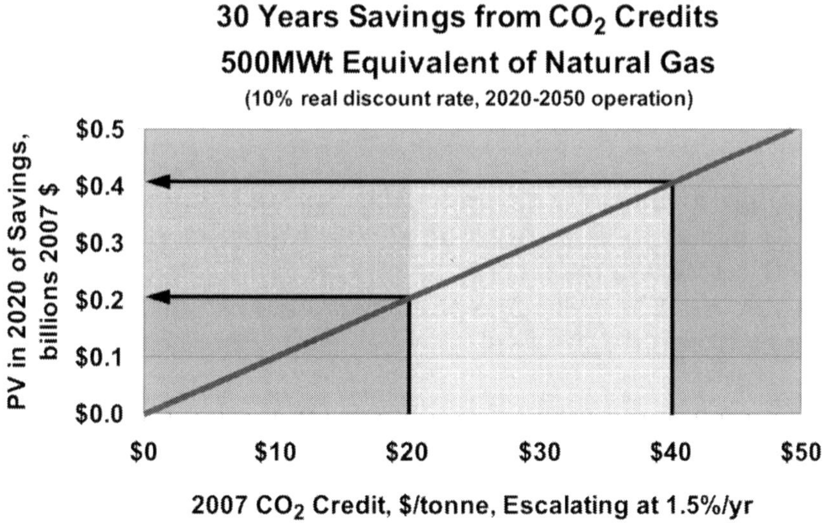

Figure 2 -- PV of CO_2 Credits

reactor with extraneous materials or allowing radioactive nuclides such as tritium to contaminate the heated process. This means that heat is delivered exclusively by forced convection rather than by radiation and the maximum temperature attainable in the process is limited by the maximum temperature reached by the heat transport medium at the exit of the nuclear reactor. In addition, two temperature approaches must be considered, one for the intermediate heat exchanger and one for the convective reformer. The maximum outlet temperature now being considered in PBMRs slated for commercial use is about 900 °C. Since the intermedialte loop will also probably be a gas loop, the maximum practical process temperature will be about 850°C. This means that in order to keep the methane slip at a reasonable level, the steam-to-carbon ratio used in the nuclear steam methane reformer will be somewhat higher than in a conventional unit. The flow scheme for a nuclear powered reformer is shown in Figure 3.

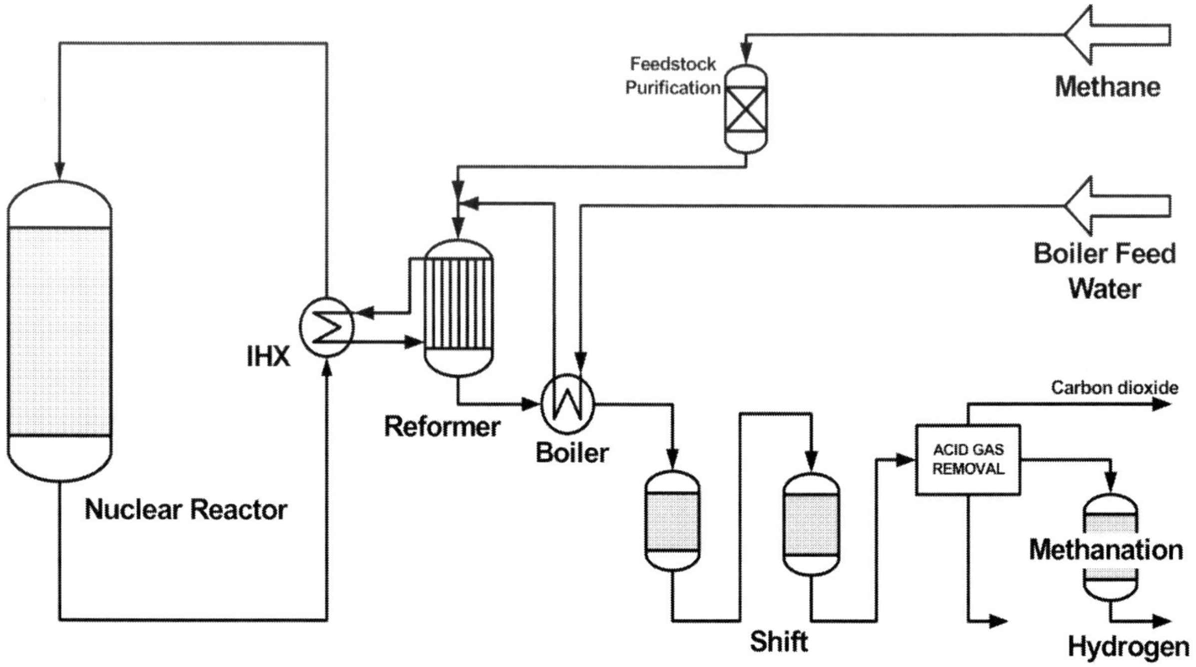

Figure 3 --Nuclear Reformer Flow Sheet

In these reactors, the coolant of choice is helium and the operating pressure is approximately 90 bar. The helium must be returned at a reasonably cool temperature, about 350°C. This means that in addition to recovering the heat from the reformed gas, the secondary coolant must be cooled as well. One proposed heat integration scheme that accomplishes this is shown in Figure 4.

The high temperature and pressure raises challenges in the mechanical design of the reformer. Current reformer tubes that can withstand the operating temperatures are not subjected to 90 bar pressure. In the case of a convective reformer, this pressure would probably be on the outside of the tubes to facilitate catalyst replacement. Moreover, tubes in current use are made of cast alloys and are not available in drawn or rolled forms. A convective reformer operating under these conditions could probably not be fabricated from such cast materials. Yet the materials used would have to have similar creep strength and other properties. These materials have not yet been identified.

Althougth there are several technical challenges, there are many exciting potential applications for this technology. The need for hydrogen made with reduced carbon dioxide emissions will continue to grow. Very large nuclear powered reformers could serve a network of hydrogen pipelines as is contemplated for the hydrogen grid. This could allow the development of the grid before commercial water-splitting technology was available.

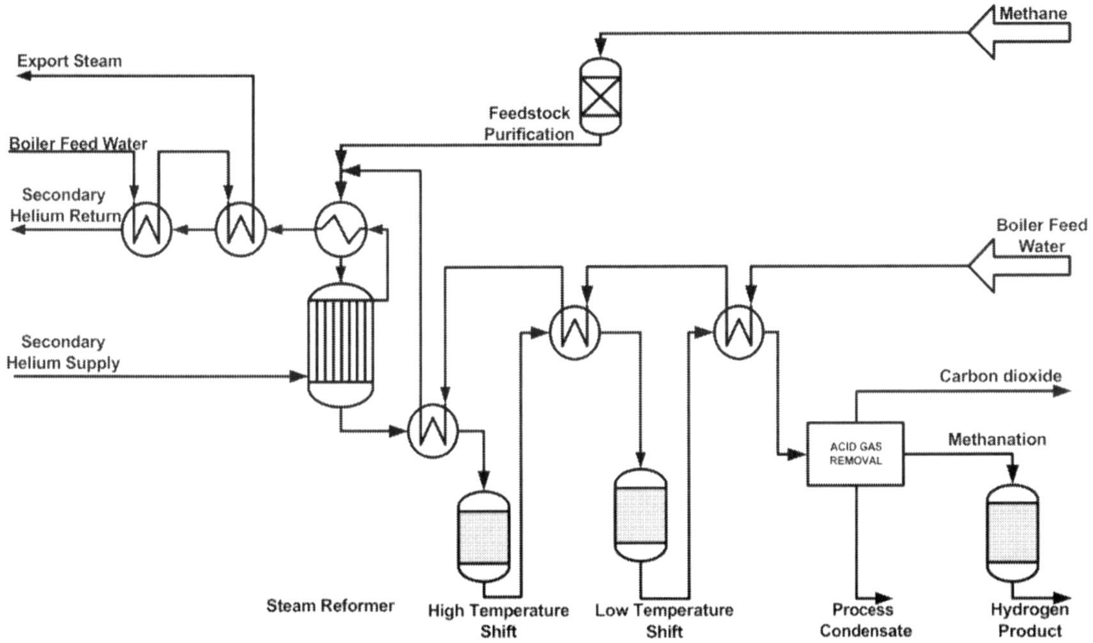

Figure 4 -- Heat integration of a Nuclear Steam Methane Reformer

The Canadian oil sands, and other heavy petroleum deposits require huge amounts of hydrogen for upgrading and the demand will continue to grow as lighter crudes are depleted. Development of these deposits might bve limited by the threat of carbon dioxide emissions increase. Nuclear powered reformers offer a solution to this roadblock again, until water-splitting becomes available. Coal-to-Liquids plants based on Fischer-Tropsch technology produce large amounts of methane. Using nuclear heated steam methane reforming can both reduce the CO_2 emissions from these plants and extend the life of the coal resource. Nuclear powered steam methane reforming may not be the end of the road in buillding the hydrogen economy, but it can be an important step along the way.

Modeling Sulfidation Kinetics and Understanding Sulfidation Equilibrium in Packed Bed Desulfurizers

Jian Sun
School of chemical and environmental engineering, Shanghai University
149 Yanchang Road, Shanghai 200072, China

Summary: After external mass transfer and pore diffusion, sulfidation of porous zinc oxide sorbent with hydrogen sulfide can be described in five steps. They are surface adsorption of hydrogen sulfide gas on zinc oxide sorbent and dissociation of the gas molecule on the sorbent surface, followed by reversible surface reactions, sulfide ion migration under the surface, and sulfidation penetration into the solid crystallite. Based on this understanding of chemistry, an empirical rate law for the intrinsic kinetics of sulfidation process was derived in this study. Kinetics modeling results using this reversible, adsorption, and ion migration (RAIM) model were applied to select experiments of single particle sulfidation and compared with several well-defined sulfidation models from the literature. In modeling the sulfidation with zinc oxide pellets, the saturated sulfur capacity and effective pore diffusivity of gaseous hydrogen sulfide in pellets were calculated using the gPROMS (general PROcess Modelling System) modeling. In modeling the sulfidation in full-scale packed desulfurizers, the gPROMS distributed reactor model was used. Case studies were presented for hydrogen sulfide removal from natural gas in a simulated fuel processing train. Finally, a new understanding on the sulfidation equilibrium in packed bed was proposed.

Introduction

Porous ZnO is a commodity sorbent material widely used in the refinery for removal of gaseous H_2S because it has well predictable reaction kinetics and adsorption (or absorption) capacity. The production cost of ZnO is low compared with other sorbent materials like molecular sieves or zinc-titanium oxide. Sorbent regeneration is usually not economical for ZnO in the present chemical industry. In addition, spent ZnO sorbent is considered as a non-hazardous material for solid waste disposal. ZnO has another important feature: Unlike other metal oxide sorbents from copper or iron, ZnO is relatively more stable and not readily reduced in the presence of H_2. Therefore, it's widely used for desulfurization in fuel processing for the fuel cell industry.

There are basically four well-defined kinetic models in the literature to describe sulfidation of ZnO or ZnO type sorbents with H_2S. They are the generalized random pore (RPM) model (1), unreacted shrinking-core (USC) model (2), grain model (3) and overlapping grain (OGM) model (4). The OGM model has been frequently studied in recent publications (5-8). In this study, let's look into the sulfidation chemistry from another perspective. An empirical rate law for the intrinsic kinetics was developed as a result. Modeling results were compared with those obtained from the sulfidation models form the literature. Using these comparison results as reference or initial values for parameter estimate, the intrinsic sulfidation kinetics and effective diffusivity for a porous ZnO sorbent G-72E were measured in a microflow packed column and calculated with the RAIM rate law. Modeling results were also applied to the design calculation of a full-scale desulfurizer.

Reaction mechanism

According to the specifications by the ZnO manufacturers (Sud-Chemie or SCI), sulfidation of porous ZnO usually applies at temperatures ranging from 150 to 450°C. With increasing temperature, sulfidation kinetics generally increase and the capacity increases to a maximal value followed by a reduction as temperature goes above 450°C due to the sulfidation equilibrium. In a process free of external mass transfer and pore diffusion, sulfidation of solid ZnO by gaseous H_2S on the sorbent surface can be described by the following steps as shown in Figure 1:

Step 1, a H_2S molecule adsorbs and dissociates at two active sites on the solid surface. The two sites can be ZnO or ZnS depending on the surface structure and degree of sulfidation. This reaction can be reversible especially when ZnS is the dominant species on the solid surface, i.e., when sulfidation reaches a high conversion stage.
$H_2S + 2 * \Leftrightarrow H^+ * + HS^- *$

Step 2, adsorbed HS^- reacts with ZnO, which is on the surrounding site if the adsorption site is ZnS, or can be the adsorption site ZnO itself.
$HS^- * + ZnO = ZnS + OH^- *$
$HS^- * (ZnO) = OH^- * (ZnS)$

Step 3, water is formed from yielded OH^- and the neighboring adsorbed H^+, and active sites are reclaimed for the next cycle through water desorption.
$OH^- * + H^+ * = H_2O + 2 *$

What happens if all ZnO on the grain surface is converted to ZnS? Literature suggests that gaseous H_2S diffuses through the solid ZnS layer followed by reaction with ZnO underneath (4-8). The diffusion of gas in solid is modeled in the literature as a very slow process compared with the initial surface reaction. This assumption can be deduced from the commonly used TGA graphs (i.e., sorbent weight gain or ZnO conversion over time) for ZnO sulfidation, which can be elaborated with two slopes of kinetic behavior, the first being the fast surface reaction and the second being gas diffusion in solid.

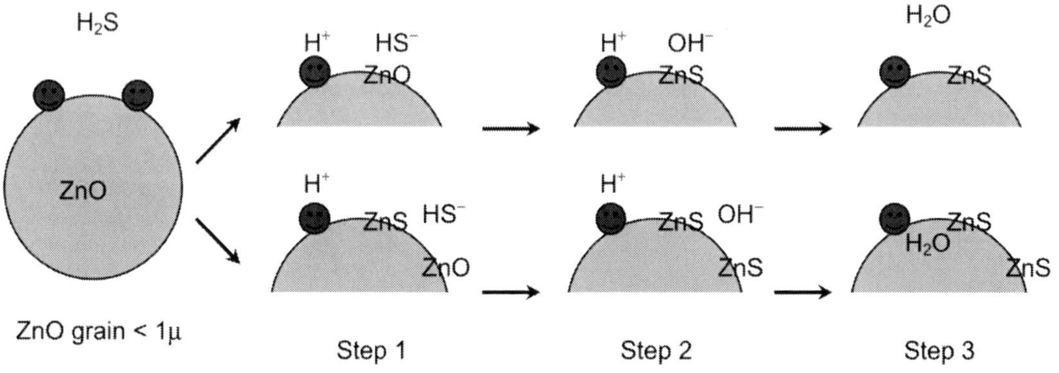

Figure 1. Sulfidation chemistry: surface reaction.

Let's not scrutinize the validity of this assumption that gaseous molecules such as H_2S can diffuse through a dense crystal structure such as ZnS. On the other hand, ion migration or ion exchange theory has been recognized in other fields such as corrosion engineering for decades (9). Ion migration appears to be a valid alternative approach to interpret the sulfidation mechanism, as it can be illustrated by the following steps, i.e., steps of sulfidation of ZnO under ZnS skin layer as shown in Figure 2.

Step 1, H_2S adsorbs on two active sites of ZnS on a completed sulfided grain surface.
$H_2S + 2* \Leftrightarrow H^+ * + HS^- *$

Step 2 (a), H^+ dissociates the adsorption site that is ZnS.
$H^+ * (ZnS) = Zn^{++} * + HS^-$
Note this HS^- is now in (or under) skin and ready to react with ZnO underneath.
$HS^- + ZnO = ZnS + OH^-$
Followed by outward ion migration of OH^- and combining with the adsorbed proton formed in step 2 (b).

Step 2 (b), adsorbed HS^- on surface is readily available to combine with Zn^{++} dissociated by the proton.
$HS^- * + Zn^{++} * = ZnS + H^+ * + *$
This initiates the sulfidation of ZnO second layer.

Step 3, water is formed and active sites are liberated through water desorption. This can be achieved after either OH^- diffusing out followed by water formation
$OH^- + * = OH^- *;$
$OH^- * + H^+ * = H_2O + 2 *$
or proton diffusing in, forming water in the solid followed by water diffusing out thereafter.
$H^+ * = * + H^+; OH^- + H^+ = H_2O$

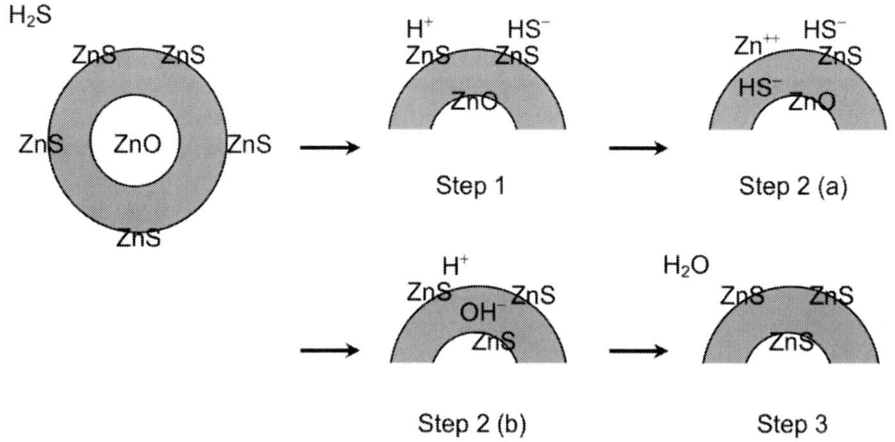

Figure 2. Sulfidation chemistry: ion migration and sulfidation penetration.

Step 2 (a) and (b) presumably take place simultaneously. Ion migration to the third layer and below can be interpreted based on the similar mechanism starting with the inward diffuse of proton to the second sulfided layer.

The rate law

Let's assume that adsorption and dissociation of H_2S on active sites is the rate-limiting step. This is reasonable because all other steps described above are basically near irreversible acid-base reactions. It is more so when the ZnO grain reaches a high conversion. In other words, H_2S adsorption on the solid surface becomes less favorable due to the increasing disappearance of ZnO. Hence, the sulfidation rate law can be simply represented as

$$r = k_1 C N^2 \quad (1)$$

where r is the rate, k_1 is the intrinsic rate constant, C is the concentration of gaseous H_2S and N is the number of active sites on surface. N can be a function of the amount of ZnO in the grain since H_2S has less affinity to ZnS. It also depends on the temperature because higher temperature increases the ion migration (exchange) activity so that the sulfidation of ZnO under skin is accelerated. Ion migration can very well be the other rate-limiting step after majority of surface ZnO is consumed. Since it has a direct impact on N, the effect of ion migration can be lumped into N. Hence, N is a function of the amount of surface ZnO accessible to H_2S, which depends on both the total ZnO amount and rate of ion migration. The rate law can be rewritten as

$$r = k_1 C (k_2 k_3 \rho)^2 \quad (2)$$

where ρ is the bulk (or can be particle) density, representing the changing amount of ZnO in solid grain. k_2 is a constant to linearly correlate N with ρ, k_3 is to correlate N with ion migration rate. The temperature dependence in k_3 can be presented generally as an exponential term as in the Arrhenius law. Combining k's and using mole per unit volume in bulk density.

$$r = kC\rho^2 \quad (3)$$

where k is the apparent rate constant (equivalent to a third order reaction) that includes the kinetic factors discussed above. For the sake of convenience, this rate law is named as the reversible, adsorption, ion migration (RAIM) rate law for ZnO sulfidation.

Sulfidation equilibrium needs to be included when performing the reactor design calculation. Based on the reverse analog, the reverse reaction of sulfidation (i.e., steam oxidation of ZnS; in fact it's used to regenerate spent zinc titanate sorbent with addition of O_2) can be written as

$$r' = k' C\rho^2 = kC\rho^2 / K_{eq} \quad (4)$$

where Keq is the sulfidation equilibrium constant. To model ZnO conversion X in TGA experiments for single particle sulfidation, Equation (3) is integrated to

$$X = \frac{k\rho_0 Ct}{1 + k\rho_0 Ct} \quad (5)$$

where ρ_0 is the initial ZnO bulk density and t is the sulfidation time. The reverse rate Equation (4) is not considered during TGA graph modeling as its runtime is limited.

Model comparison

<u>The intrinsic kinetics</u>

To evaluate intrinsic kinetics for model comparison, select sulfidation experiments of ZnO or ZnO type sorbent and their modeling data were reproduced from Yu and Sotirchos (1). These experiments usually were run on TGA at a controlled temperature and fixed H_2S partial pressure. External mass transfer (film diffusion) resistance is minimal for sulfidation of ZnO powder as concluded from the literature. Figure 3 plots the experimental data and model calculations for sulfidation of G-72D (a predecessor of G-72E), a ZnO sorbent made by United Catalyst (now SCI). The RPM calculation is included in Figure 3 for comparison and it appears to over-predict the kinetics after ~60% conversion. This can be attributed to the RPM presumably not considering the layer diffusion of H_2S (kinetics slower than surface reaction as suggested in the literature) through initially formed ZnS skin on the sorbent surface.

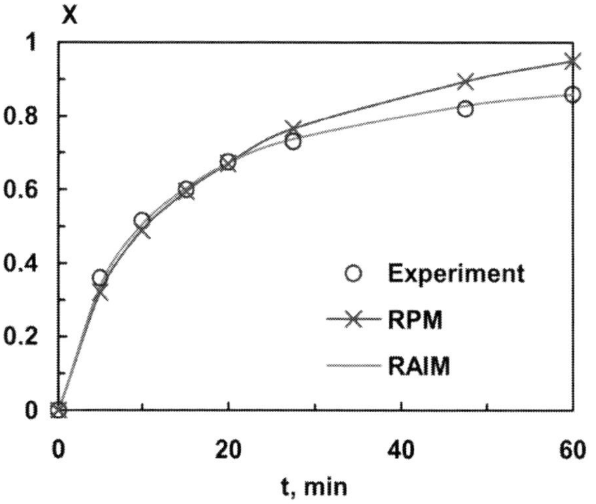

Figure 3. TGA experiment of ZnO conversion (X) over time and model comparison of RPM and RAIM. Experimental data and RPM modeling results were reproduced from Yu and Sotirchos (1). Sulfidation conditions: 0.5% H_2S in N_2 on powdered G-72D (50 mesh or 300 μm particle size) at 400°C. The intrinsic rate of G-72D is calculated as 1 L^2/mol^2 s for the RAIM model. Note that mass transfer resistance is minimal or film diffusion would result in a linear conversion over time.

USC is a conventional gas-solid reaction model and well documented in the literature (10). Many modified USC models for sulfidation have been developed (2, 11). Nevertheless, USC model does not take the internal surface (as indicated from the BET surface area measured for porous materials) into account, presumably resulting in an under-prediction at the beginning and over-prediction towards the end of sulfidation, a result similar to RPM.

In comparison with USC, OGM realistically describes the surface reaction simultaneously over the external solid and internal pore surfaces. It also includes the reduction of surface area due to the molar volume increase of solid (from ZnO to ZnS) during sulfidation. Its sulfidation mechanism, however, assumes that the reaction surface boundary retreats from the skin surface at a same rate for all grain sizes. This assumption can result in a rate contradiction at the intersections of grains of different sizes. Using sophisticated numerical modeling process, OGM model is the latest sulfidation model and frequently studied in the literature.

In another case, Figure 4 plots the comparison of OGM and RAIM on the TGA graph for sulfidation with a zinc-titanium oxide sorbent at 400 and 500°C. The RAIM model gives a good prediction for sulfidation at 400°C, while its prediction at 500°C is not as satisfactory. This can be related to two factors. The existence of titanium in sorbent can change the ZnO sulfidation mechanism, and the applicable temperature range for ZnO sulfidation is below 450°C as recommended by the manufacturers. In sulfidation of zinc-titanium oxide experiment at 500°C, the reaction order for ZnO density in Equation (3) may change with decreasing surface adsorption (with increasing temperature). It seems that the RAIM rate law applies best for ZnO sulfidation at temperatures below 400°C, which is near the simulation temperature for sulfidation reactor design in our case. Note that the rate constant is the only changing variable in Equation (5) for curve fitting in Figures 3 and 4, an advantage for the sulfidation intrinsic kinetics study.

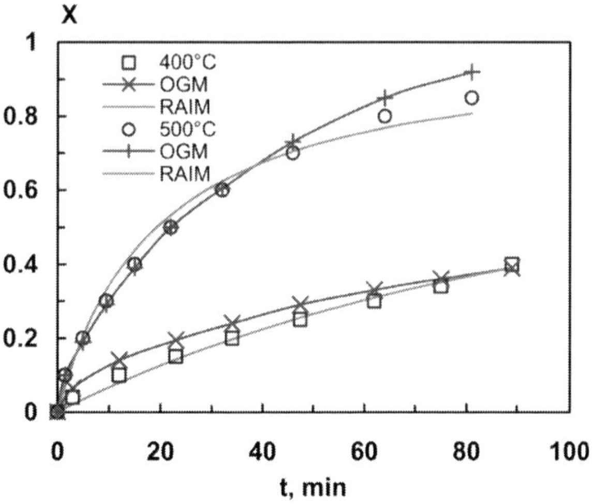

Figure 4. Sulfidation experiment and model comparison of OGM and RAIM. Experimental and OGM data were reproduced from Lew et al. (1992). Symbols square and circle stand for the experimental data points at 400 and 500°C, respectively. Reaction conditions: a powdered zinc-titanium oxide sorbent at 400 and 500°C in 2% H_2S and 1% H_2 in N_2.

Pore diffusion

Pore diffusion is the intermediate step for modeling continuation from kinetics rate law to the packed bed simulation. Sulfidation modeling of porous ZnO pellet using the RAIM rate law was carried out with pore diffusion on the gPROMS platform (pseenterprise.com). Modeling formulations for pore diffusion are omitted in this communication. Using the apparent sulfidation rate constant obtained from the TGA experiments, effective pore diffusivity of H_2S is a modeling parameter in pellet sulfidation.

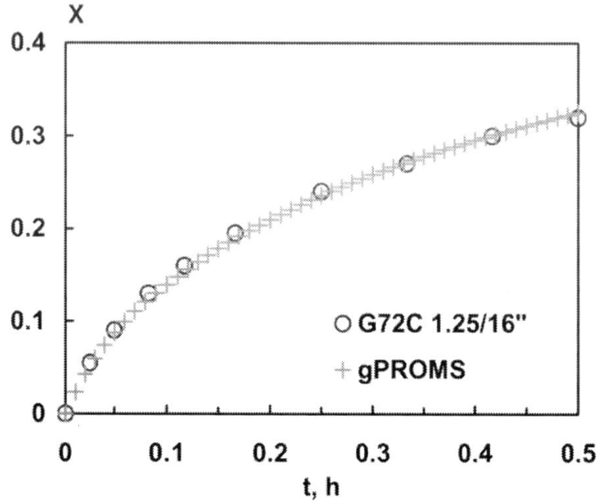

Figure 5. TGA results for G-72C ZnO (1.25/16" pellets) conversion with 1% H_2S in N_2 at 375°C fitted by the gPROMS model. Experimental data were from Gibson and Harrison (3).

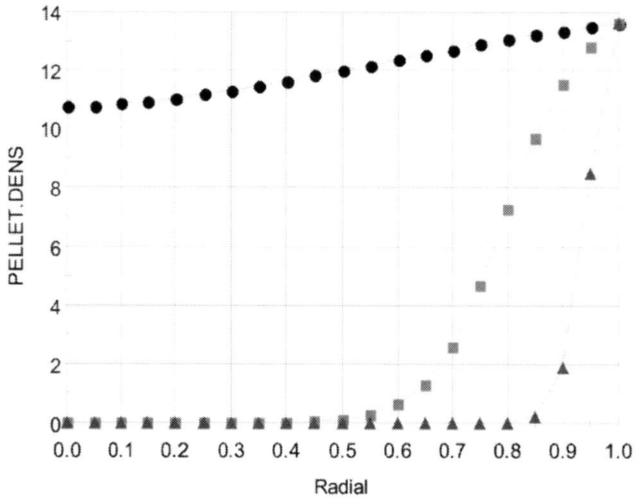

Figure 6. Simulated radial ZnS density (mole/L) profile in a spent G-72C pellet after 15 min sulfidation. With an estimated fixed saturated capacity, the effective diffusivity changes at 1×10^{-5} (circle), 1×10^{-6} (square), and 1×10^{-7} (triangle) m^2/sec.

TGA results were also available in the literature for ZnO G-72C (another predecessor of G-72E) pellets at 375°C (3). Figure 5 reproduces the TGA results of G-72C pellet conversion in 1% H_2S in N_2 at 375°C. It also plots the corresponding gPROMS modeling results. The gPROMS model combines the RAIM rate law for intrinsic kinetics and pellet diffusion with two fitting parameters, effective diffusivity of gaseous H_2S in pores and saturated sulfur absorption capacity of G-72C pellet. In theory, the saturated capacity of ZnO grain of 100% purity is the stoichiometric capacity for ZnO at 100% conversion or 39.5% by weight. For pellets, its saturated sulfidation capacity or maximal conversion is always lower than the stoichiometric value because its sulfidation service time is limited. As a modeling parameter, the saturated capacity for G-72C pellet at the specified condition is calculated as about 24% (or 60% in ZnO conversion). The diffusivity was around 1.67×10^{-6} m^2/sec.

The simulated radial ZnS concentration profile for the sulfided pellet obtained in this study with a changing effective diffusivity 1×10^{-5}, 1×10^{-6}, and 1×10^{-7} m^2/sec is illustrated in Figure 6. The calculated diffusivity 1.67×10^{-6} m^2/sec was found consistent with the observed sulfur profile on the micrograph (3).

Apparatus and sulfur analyzer

A 3/8" microflow packed reactor column (Figure 7) was constructed for sulfidation experiments with porous G-72E ZnO sorbent from SCI. Two cylindrical pellets of 1/16" and 1/8" sizes were tested at 350°C. For sulfidation runs with powder, sample weight was controlled under 0.2 g to minimize the pressure drop and run time. A 50 ppm H_2S feed gas was supplied with a bottled 100 ppm H_2S in methane diluted with an equal molar flow of house N_2.

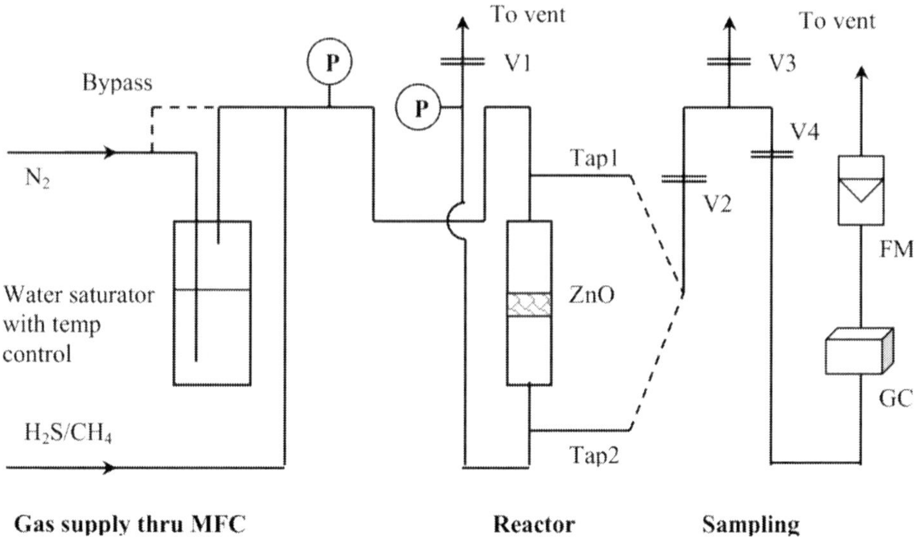

Figure 7. Schematic of the bench scale sulfidation reactor for powdered and pellet G-72E. 'P' stands for pressure gauge, 'V' for valves, 'FM' for flow meters, and 'MFC' for mass flow controller. Water saturator controls the humidity of gas stream to simulate the field operating conditions.

An S-PRO 3200 Series GC System equipped with cryogenic cooling system by OI Analytical was the sulfur analysis instrument used in this study. Integrated with an Agilent 6890N GC, the S-PRO has the capability of sulfur speciation, with a sulfur detection limit of 10 ppbv for individual sulfur compound.

Sample is introduced to the column through the OI volatile interface. Individual sulfur components are chromatographically separated on a Varian CP 8690 column. The entire internal sample pathway is Sulfinert™ treated to minimize sulfur adsorption in the system. The sample is then introduced to the pulsed flame photometric detector (PFPD), which gives a signal proportional to the concentration of sulfur in the sample.

Results and discussion

Intrinsic kinetics with powdered G-72E (dry)

Figure 8 (a) presents one of the measured breakthrough curves for sulfidation of G-72E powder with 50 ppm H_2S at 350°C. A significant amount of COS was observed throughout the test. Since ZnO sorbents can potentially be a reforming and water-gas-shift catalyst (12), it's reasonable to see steam (a sulfidation product) reforming of hydrocarbons (from standard H_2S/methane bottle) on G-72E producing CO thus COS. The COS spike (at 44 h) at the end of run was presumably due to the air exposure (leads to COS formation) of the sample when the inlet H_2S concentration was measured.

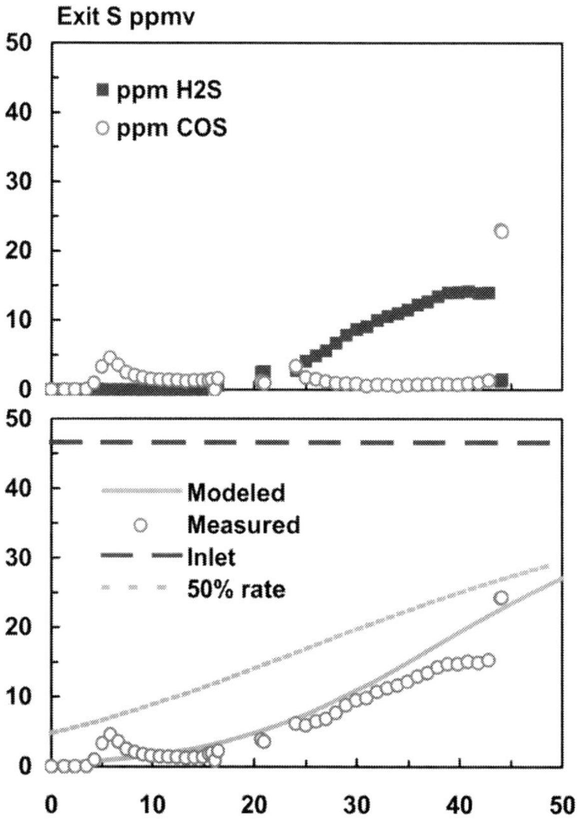

Figure 8. (a, top) Breakthrough curve of powdered G-72E with no steam addition at 350°C. Reaction conditions: 0.165 g (0.11 cc) powder (-200 mesh or < 75 µm) with 50 ppm H_2S in CH_4 and N_2 at 300 cc/min, 3 psig and 350°C. (b, bottom) The corresponding intrinsic kinetics modeling. RAIM rate = 3 L^2/mol^2 s.

The formation of COS was not expected and consideration of its kinetics could multiply the tasks in this sulfidation study. On the other hand, the observation of COS formation was minimal in the subsequent sulfidation tests with pellets. For simplicity, COS concentration was therefore added to H_2S, resulting in a combined breakthrough curve that was modeled as the total sulfur (H_2S + COS) in Figure 8 (b). The RAIM rate for G-72E powder at 350°C was calculated as 3 L^2/mol^2 s using a finite difference calculation. The finite difference method gave good sensitivity for kinetics modeling in this study. For instance, in Figure 8 (b) also plots the modeled curve when the rate was cut in half. With this reduced rate, the saturated ZnO conversion of G-72E would have to reach 130% in order to fit the experimental curve. The finite difference is a common modeling approach and not elaborated in this communication.

Kinetics with powdered G-72E (1.5% steam)

Our simulated reactor design requires sulfidation in humid conditions (e.g., 1.5% steam). In general, addition of steam does not change the chemistry of G-72E sulfidation since sulfidation produces steam. Here is the list of major chemical reactions potentially occurring in the bench scale sulfidation test.

1. $ZnO + H_2S \rightleftharpoons ZnS + H_2O$
2. $ZnO + COS \rightleftharpoons ZnS + CO_2$
3. $HCs + H_2O \rightarrow H_2 + CO$ (HCs: hydrocarbons; steam reforming)
4. $H_2O + CO \rightleftharpoons CO_2 + H_2$ (water gas shift)
5. $H_2S + CO \rightleftharpoons COS + H_2$
6. $COS + H_2O \rightleftharpoons H_2S + CO_2$

Reactions 3 and 4 can be catalyzed by the sorbent. Hydrocarbons (contained in natural gas) come from standard methane in the H_2S/methane bottle.

In the sulfidation of ZnO with no steam addition, zero COS was observed initially followed by a COS spike around 5 h (Figure 8). This can be interpreted in a process with steps as follows; initial formation of steam by sulfidation, CO formation by steam reforming of HCs, and delayed formation of COS by reaction of H_2S with CO. While in the case of sulfidation of ZnO with steam addition (Figure 9), the COS peak came much earlier and stronger because of the presence of steam from the gas feed at time zero and a much higher concentration (1.5%).

Figure 9 (b) plots individual breakthrough curves for H_2S and COS with 1.5% steam addition and modeling of the total sulfur breakthrough curve, respectively. The intrinsic kinetics of sulfidation with H_2S alone obtained in Figure 9 seems approximately the same as in the dry condition (Figure 8). However, roughly 10% of the inlet H_2S was converted to COS at the exit. The presence of COS remains nearly stable throughout the process. Due to this COS influence, the overall kinetics (for both H_2S and COS) with 1.5% steam addition was reduced to 2 L^2/mol^2 s. The influence of COS was again disregarded in the modeling since small COS formation was observed in pellet tests with steam formation. A more complicated rate law would need to be developed to model the breakthrough behavior in Figure 9, which includes the kinetics of catalytic reforming of hydrocarbon and COS formation as well as ZnO sulfidation with COS.

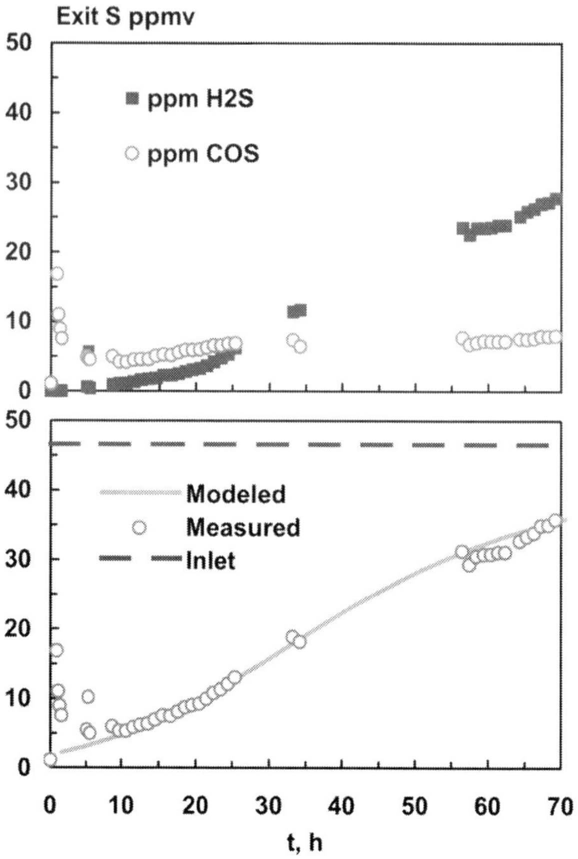

Figure 9. (a, top) Breakthrough curve of powdered G-72E with 1.5% steam addition at 350°C. Reaction conditions: 0.165 g (0.11 cc) powder with 50 ppm H_2S and 1.5% steam in CH_4 and N_2 at 300 cc/min, 3 psig and 350°C. (b, bottom) The corresponding intrinsic kinetics modeling. Rate = 2 L^2/mol^2 s.

Kinetics with pellet G-72E (1.5% steam)

Figure 10 (a) plots the collected GC data for select sulfidation of 3/16" G-72E pellets. Unlike the powder tests, less COS was formed by pellets at initial reaction stage probably because of the reduced availability of 'catalytic' sites. Figure 10 (b) plots the total sulfur (H_2S + COS) measurement and corresponding gPROMS calculation. In general, the diffusivity determines the slope and the saturated sulfur capacity sets the position of modeled breakthrough curve in modeling. The diffusivity and saturated sulfur capacity for the 3/16" pellets are calculated as 2.78×10^{-6} cm^2/sec and 15.4% (41% ZnO conversion), respectively.

To compare the effect of pellet size on reactor design, a 1/16" G-72E pellet sorbent was also tested and modeled. Compared with the 3/16" pellets on the same sorbent volume basis, less total sulfur (COS and H_2S) was observed to slip through the column packed with the 1/16" pellets. Apparently, the smaller pellets were superior in both overall rate and saturated sulfur capacity, assuming similar porous structures. The diffusivity and saturated sulfur capacity were calculated as 6.94×10^{-6} cm^2/sec and 25.7% (Figure 11). In other words, ZnO conversion after saturation was calculated as 68%.

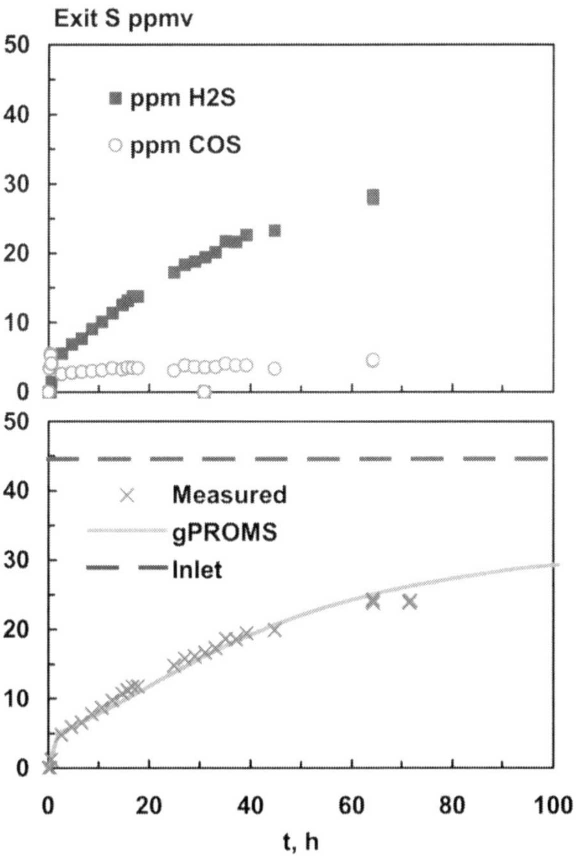

Figure 10. (a, top) Breakthrough curve (COS and H_2S) of G-72E pellets with 1.5% steam addition at 350°C. Reaction conditions: 3 – 6 g 3/16" pellet with 50 ppm H_2S and 1.5% steam in CH_4 and N_2 at 750 cc/min, 2 psig and 350°C. (b, bottom) Breakthrough curve (COS and H_2S) from (a). The gPROMS predicted saturated ZnO conversion is 41%.

<u>Reactor design calculation</u>

In modern fuel processing for H_2 production for fuel cell application, a serial reactor train is desirable. The reactor train is usually equipped with precious metal catalysts in the reformer, water gas shift reactors, or preferential oxidizer stages. Precious metal catalysts are much more active than the conventional catalysts but prone to sulfur poisoning. Hence, desulfurization is critical in fuel processing. Design conditions for our desulfurization simulation are summarized in Table 1.

For a special case, our simulated natural gas contained only one sulfur compound, H_2S (desulfurization of organic sulfur in fuel will be discussed in the future). A sulfur excursion was considered in the model to simulate an extreme upset condition in field operation. Specifically, H_2S level at inlet was simulated to increase from 6 to 30 ppm at 39900 h and decrease back to 6 ppm at 40000 h (known as so-called 100 h sulfur upset). Putting the sulfur upset right before the end of service was the designing worst-case scenario.

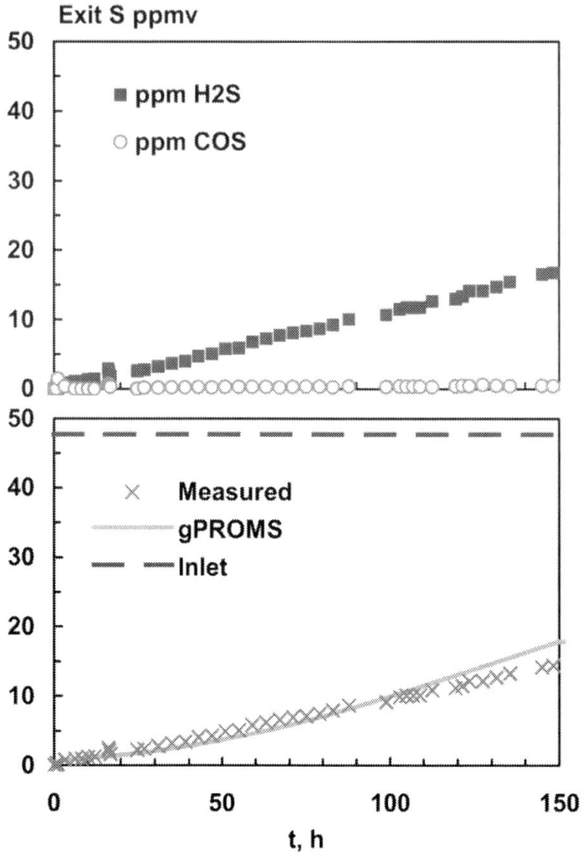

Figure 11. (a, top) Breakthrough curve (COS and H_2S) of G72E pellets with 1.5% steam addition at 350°C. Conditions of a typical sulfidation run are 1.5 – 3.5 g 1/16" pellet with 50 ppm H_2S and 1.5% steam in CH_4 and N_2 at 750 cc/min, 2 psig and 350°C. Virtually zero COS was observed throughout the test. (b, bottom) Modeling breakthrough curve from (a) with COS and H_2S combined.

Table 1. Key design parameters for the simulated desulfurizer

Reaction temperature, °C	350
Reactor pressure, atm	1.23
Desired sulfidation time, h	40000
Fuel molar flow, mole/s	0.79
Maximal sulfur slip, ppb	25
H_2S level in fuel feed, ppm	6
H_2O molar fraction	0.015
Bed voidage	0.5
G-72E 1/16" bulk density, kg/L	0.90
G-72E 3/16" bulk density, kg/L	1.47

Figure 12 presents the predicted breakthrough curve and a close-up near the breakthrough points to illustrate the 100 h sulfur upsets for the full-scale desulfurizers. The H_2S spikes are the simulated sulfur upset, 30 ppm for 100 hour at end-of-life (EOL). As expected, the breakthrough slope is sharper (greater) for the bed packed with the smaller pellets.

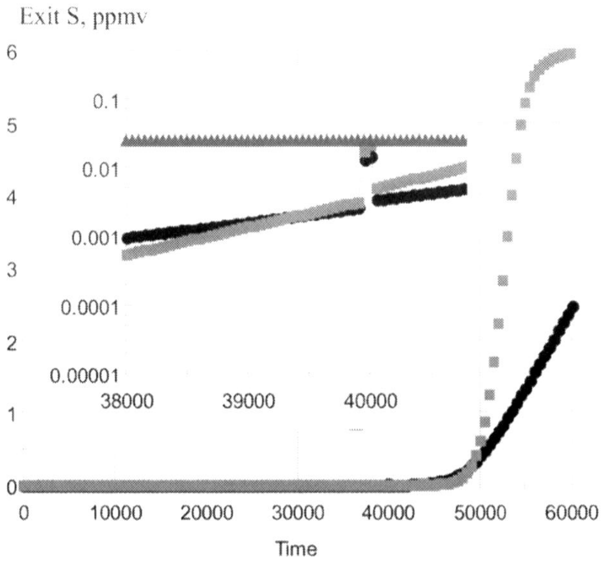

Figure 12. (a, main) The ppm sulfur breakthrough curve for the 3/16" (circle) and 1/16" (square) G-72E pellets in simulated desulfurizer at 350°C. (b, insert) The ppm sulfur is drawn on the log-scale and cropped for a close-up view near the breakthrough point in (a). The horizontal line (triangle) in (b) is the target maximal sulfur slip at exit, 25 ppb. Time on (a) and (b) is in hour.

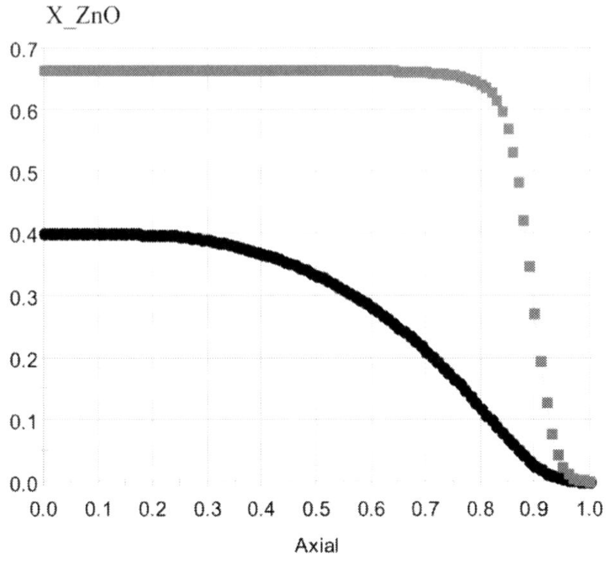

Figure 13. Sulfur loading profile represented as ZnO conversion in simulated desulfurizers packed with the G-72E 3/16" (circle) and 1/16" (square) pellets at EOL. Axial stands for a normalized reactor length.

Figure 13 is the sulfur loading profile in simulated desulfurizers at EOL. The 1/16" ZnO pellets produce a sharp slope or short 'tail' (corresponding to a short sulfidation reaction zone) near the exit. A shorter reaction zone indicates fast sulfidation kinetics with the 1/16" pellet. Although the 3/16" pellets have a higher packing density (1.47 g/cc) than the 1/16" pellets (0.904 g/cc), a larger bed size is predicted for the 3/16" pellets for the same sulfur removal service task (Table 1) because of its slower overall kinetics. Specifically, the reactor volumetric ratio for the desulfurizer packed with G-72E 3/16" pellets to that of G-72E 1/16" pellets is approximately 1.5.

Calculation results for the bed pressure drops were within the design range with a typical reactor (e.g., length/diameter = 2.5) for both pellets. They are not presented here. Note that the 66% ZnO saturated conversion does not reach 68% (Figure 13) as obtained from the parameter estimations. That is due to the sulfidation equilibrium with 1.5% steam in the simulated desulfurizers. In general, the ZnO saturated conversion decreases with increasing steam concentration in the modeled reactor.

On the thermodynamics of ZnO sulfidation

As indicated by the simulation results (Figure 12), the exit H_2S concentration is near zero at the beginning of reactor service for sulfur removal. It then progressively increases to reach a level of around 0.002 ppm at 39900 h. These simulated concentrations are below the equilibrium value, 0.007 ppm, which is calculated using the sulfidation equilibrium constant in 1.5% steam at 350°C (13). The disagreement between modeling and sulfidation thermodynamics is enhanced when running the simulation at higher steam concentrations. Although it has been confirmed through experiments in our unpublished results and the literature (14), this modeling outcome seems to be against the conventional wisdom of sulfidation equilibrium.

Thermodynamic equilibrium of a chemical reaction can be viewed as the state when forward and reverse reaction rates come to an equal point. For gas-solid reaction like sulfidation, its rate is proportional to the reactant concentrations in gas phase and the number of active sites on solid surface. In a typical batch reactor for sulfidation, the principle of thermodynamics is understood. All gas (H_2S and H_2O) molecules can theoretically come in contact with solids (ZnO and ZnS) at any given space and time. In a packed bed like porous ZnO desulfurizer, however, gas molecules come in contact with solids at a specific position in a single pass through the reactor for only one time.

Imagine an upstream position A and a downstream position B in an isothermal ZnO bed (Figure 14). At a certain point of reaction time, sulfidation reaches an equilibrium and the H_2S concentration in gas stream is reduced to 25 ppb at A. This means that the forward reaction rate (proportional to concentrations of H_2S and ZnO active sites) is equal to the reverse reaction rate (proportional to H_2O and ZnS) at A. Gases continue flowing to B at downstream where the numbers of ZnO active sites are greater and ZnS sites are smaller than those at A. Since less ZnO is consumed downstream than upstream, the forward reaction rate is greater than the reverse at B. H_2O concentration of % level can be considered as a constant from A to B. As a result, sulfidation equibrium from A is shifted at B and the H_2S concentration can be reduced further below 25 ppb. In other words, sulfidation reaches a new equibrium state at B.

In theory, zero sulfur concentration can be achieved by maximizing the mass ratio of ZnO/H_2S (minimizing the space velocity) in packed bed. The sulfidation equilibrium at A will eventually reach B when more ZnO is converted to ZnS. In a sense, sulfidation equilibrium can be viewed as a dynamic state in the ZnO packed bed, where sulfidation kinetics play a dominant role in the process.

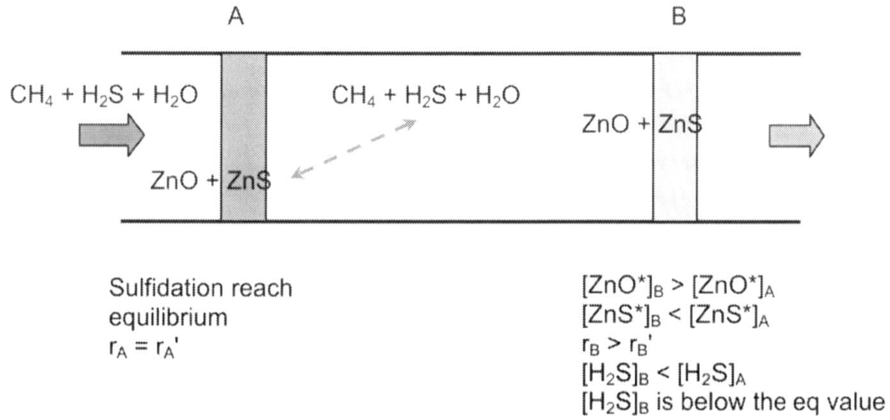

Figure 14. Interpretation of the 'dynamic' equilibrium of sulfidation in packed bed.

Conclusions

We have derived a ZnO sulfidation kinetics equation, integrated it into a distributed gPROMS reactor model, and performed design calculations for desulfurizers in natural gas fuel processing. Although the RAIM modeling results were acceptable for an engineering design point of view, many issues remain untackled. For instance, the validity of the diffusion of protons and water in a solid crystallite presented in mechanistic chemistry needs to be discussed further. In addition, it's uncertain if the RAIM rate law can also include the probability of ion exchange between ZnO and ZnS in a solid. It's well known among many experienced engineers that a sulfided or saturated ZnO bed can regain its H_2S capturing capacity after a hot N_2 treatment. Therefore, the ion migration mechanism derived in this study only covers a part of the complex chemical process of ZnO sulfidation. It's our wish to inspire more studies on this topic. Modeling prediction in simulated desulfurizers should also be subjected to the effect of long service time when ZnO/ZnS ion exchange can have a profound impact in altering the saturated adsorption capacity as well as sulfidation kinetics under heated environment.

Another issue is the effect of COS as observed in our experiments. Clearly, the kinetics study of ZnO sulfidation with COS needs to be investigated, as literature studies of the ZnO sulfidation studies are dominated by sulfidation with H_2S.

Lastly, sulfidation kinetics of G-72E was incomplete in this study (due to limited time budget) as the temperatures being limited to only 350°C. In addition, model validation for parameter justification will be always necessary for the completion of reactor design calculations.

References

1. Yu, H. C.; Sotirchos, S. V. A Generalized pore model for gas-solid reactions exhibiting pore closure behavior *AIChE J.* **1987,** *33*, 382

2. Zevenhoven, R.; Yrjas, P.; Hupa, M. Hydrogen sulphide capture by limestone and dolomite at elevated pressure-II. Particle conversion modelling. *Ind. & Eng. Chem. Res.* **1996,** *35*, 943

3. Gibson, J. B.; Harrison, D. P. *Ind. Eng. Chem. Process Des. Dev.* **1980,** *19*, 231

4. Sotirchos, S. V.; Yu, H. C. Overlapping grain models for gas-solid reactions with solid product. *I & EC Res.* **1988,** *27*, 836

5. Lew, S.; Sarofim, A. F.; Flytzani-Stephanopoulos, M. Modeling of the sulfidation of zinc-titanium oxide sorbents with hydrogen sulfide. *AIChE J.* **1992,** *38*, 1161

6. Konttinen, J.; Zevenhoven, R.; Hupa, M. Hot gas desulfurisation with zinc titanate sorbents in a fluidised bed 1. Determination of sorbent particle conversion rate model parameters. *Ind. & Eng. Chem. Res.* **1997,** *36*, 2332-2339

7. Konttinen, J.; Zevenhoven, R.; Hupa, M. Hot gas desulfurisation with zinc titanate sorbents in a fluidised bed 2. Reactor model. *Ind. & Eng. Chem. Res.* **1997,** *36*, 2340-2345

8. Konttinen, J.; Zevenhoven, R.; Yrjas, P.; Hupa, M. Modelling of sulfided zinc titanate regeneration in a fluidised bed reactor 1. Determination of the solid conversion rate model parameters" *Ind. & Eng. Chem. Res.* **1997,** *36*, 5432-5438

9. Fontana, M. G. *Corrosion Engineering*. McGraw-Hill 3rd edition, New York, November 1985

10. Levenspiel, O. *Chemical Reaction Engineering* 2nd edition, John Wiley & Sons, Inc. New York, 1972

11. Fan, H.; Li, C.; Guo, H.; Xie, K. Microkinetics of H_2S removal by zinc oxide in the presence of moist gas atmosphere. *J. Natural Gas Chemistry* **2003,** *12*, 43-48

12. Raimondi, F.; Schnyder, B.; Kotz, R.; Schelldorfer, R.; Jung, T.; Wambach, J.; Wokaum, A. Structural chenges of model Cu/ZnO catalysts during exposure to methanol reforming conditions. *Surface Science* **2003,** 532-535, 383-389

13. Twigg, M. V. *Catalyst Handbook Second Edition* Manson Publishing, 1996

14. Novochinskii, I.; Ma, X.; Song, C.; Lampert, J.; Shore, L.; Farrauto, R. A ZnO-based sulfur trap for H_2S removal from reformate of hydrocarbons for fuel cell applications. *Proceedings of Topical Conference on Fuel Cell Technology*, AIChE Spring National Meeting, New Orlean, March 10-14 pp. 98-105 2002

Effect of Dopants in the Hydrogen Kinetics and Storage Capacities of Metal Hydrides

Mohammed M. Ghouri, Phani G. K. Dathara, and Daniela S. Mainardi. Louisiana Tech University, 911 Hergot St., Ruston, LA 71272

Among the major problems in making the hydrogen-powered fuel cell vehicles feasible in practice is the on-board storage system for hydrogen. The United States Department of Energy has set target system energy densities at value of 6.5 wt% (usable H2 weight > 6.5% of the storage system weight) and 62-kg H2/m3 by 2010. Ideally, storage and release of hydrogen should take place at temperatures between 0°C and 100°C and pressures of 1–10 bar and on time scales suitable for transportation applications. Although each current storage method possesses desirable attributes, no approach satisfies all of the efficiency, size, weight, cost and safety requirements for personal transportation vehicles. Research needs for improved compressed gas storage include the development of novel materials that are strong, reliable, and low in cost. Solid-state hydrogen storage (storage in metal hydrides, chemicals, and nanostructured materials) offers perhaps the best opportunities for meeting the requirements for onboard storage. While many metal hydrides are reported to be capable of meeting the gravimetric and the volumetric storage densities, favorable kinetics and thermodynamics (dehydrogenation of these metal hydrides at room temperature) is one factor that needs to be resolved in order to make these materials efficiently available for hydrogen storage. There is evidence of the benefits of adding dopants, such as carbon and titanium, to metal hydrides to improve their hydrogen kinetics and storage capacities. For instance, a remarkable improvement in hydrogen adsorption/desorption kinetics by MgH2 metal hydride is observed when it is ball milled with graphite/carbon [1]. Carbon was also shown to improve the hydrogen storage kinetics of Lithium Beryllium Hydrides [2], and Ti-doped sodium aluminum hydride when graphite was used as a co-dopant. In the later case, the graphite addition helped decrease the dehydrogenation temperature by as much as 15° C [3]. Hence, the goal of this work is to understand the effect of dopants, such as carbon and titanium, in the hydrogen storage kinetics of hydrides such as alanates and borates at an atomic level using theoretical/computational methods. First Principles Density Functional Theory and molecular simulations are used to characterize the role of dopants in the desorption kinetics of these hydrides. The effect of dopants on the metal-hydrogen bonding nature is studied using the total and the partial electronic density of states and the electronic charge distributions before and after doping.

1- S. Dal Toè , S. Lo Russo , A. Maddalena , G. Principi, A. Saber, S. Sartori, T. Spataru "Hydrogen desorption from magnesium hydride–graphite nanocomposites produced by ball milling" Materials Science and Engineering B 108 (2004) 24–27. 2- Ghouri M.M. and Mainardi D.S., "The Role of Carbon in the Hydrogen Storage Kinetics of Lithium Metal Hydrides" In: Proceedings AIChE 2006 Annual Meeting, San Francisco, CA, Nov 12-17, 2006. 3- J Wang, A. D. Ebner, T. Prozorov, R. Zidan, J. A. Ritter "Effect of graphite as a co-dopant on the dehydrogenation and hydrogenation kinetics of Ti-doped sodium aluminum hydride" Journal of Alloys and Compounds 307 (2000) 157–166.

New Synthesis Route for Complex Hydride Hydrogen Storage Materials

Jun Wang
University of South Carolina
Columbia, SC 29208
wang233@engr.sc.edu

Armin D. Ebner
University of South Carolina
Columbia, SC 29208
ebnera@engr.sc.edu

James A. Ritter
University of South Carolina
Columbia, SC 29208
ritter@engr.sc.edu

Abstract:

A new synthesis route for complex hydride reversible hydrogen storage materials has been devised. This new route has been demonstrated with the complex hydrides, sodium aluminum hydride and lithium aluminum hydride. For example, it was used to synthesis sodium aluminum hydride and lithium aluminum hydride from NaH and Al powder and LiH and Al powder, respectively. The unique feature of this route is that it enables the synthesis of these complex hydride materials at room temperature and at relatively low hydrogen pressures. The outcome achieved using this new synthesis process is the lowest temperature, lowest pressure route for making these two complex hydrides known to date. This procedure is amenable to fostering the synthesis of other higher hydrogen capacity complex hydrides, like other alanates and boronates. The use of hydrogen for transportation and stationary applications may become more feasible because this synthesis method makes the a complex hydride much more economical, as it uses inexpensive reagents and milder synthesis conditions

AUTHOR INDEX

AUTHOR INDEX

American Institute of Chemical Engineers
3 Park Avenue
New York, NY 10016-5991

ISBN: 978-1-60423-354-4

Regenerative Engineering Society

Held at the 2021 AIChE Annual Meeting

Boston, Massachusetts, USA and Online
7 - 11 November and 15 - 19 November 2021

AIChE
The Global Home of Chemical Engineers